LOEB CLASSICAL LIBRARY
FOUNDED BY JAMES LOEB 1911

EDITED BY
JEFFREY HENDERSON

GALEN

ON TEMPERAMENTS

ON NON-UNIFORM DISTEMPERMENT

THE SOUL'S TRAITS DEPEND ON
BODILY TEMPERAMENT

LCL 546

GALEN

ON TEMPERAMENTS

ON NON-UNIFORM DISTEMPERMENT

THE SOUL'S TRAITS DEPEND ON
BODILY TEMPERAMENT

EDITED AND TRANSLATED BY
IAN JOHNSTON

HARVARD UNIVERSITY PRESS
CAMBRIDGE, MASSACHUSETTS
LONDON, ENGLAND
2020

Copyright © 2020 by the President and Fellows
of Harvard College
All rights reserved

First published 2020

LOEB CLASSICAL LIBRARY® is a registered trademark
of the President and Fellows of Harvard College

Library of Congress Control Number 2020938405
CIP data available from the Library of Congress

ISBN 978-0-674-99738-7

*Composed in ZephGreek and ZephText by
Technologies 'N Typography, Merrimac, Massachusetts.
Printed on acid-free paper and bound by
Maple Press, York, Pennsylvania*

CONTENTS

PREFACE	vii
GENERAL INTRODUCTION	xi
ABBREVIATIONS	cvii
GENERAL BIBLIOGRAPHY	cxi

ON TEMPERAMENTS

INTRODUCTION	2
BOOK I	14
BOOK II	104
BOOK III	210

ON NON-UNIFORM DISTEMPERMENT

INTRODUCTION	282
TEXT AND TRANSLATION	288

THE SOUL'S TRAITS DEPEND ON BODILY TEMPERAMENT

INTRODUCTION	326
TEXT AND TRANSLATION	332

CONTENTS

APPENDIX: TWO SHORT TREATISES

INTRODUCTION 431

1. ON THE BEST CONSTITUTION OF OUR BODY 434

2. ON GOOD BODILY STATE 452

INDEXES 463

PREFACE

Galen's concept of the combination (*krasis*) of the four elemental qualities (hot, cold, wet, and dry) is fundamental to his account of the structure and function of the human body (and of animal and plant bodies generally), which is in turn fundamental to his theory of medical practice. Disturbances of *krasis* (*dyskrasias*) constitute one of his three major categories of disease, as well as being regarded as the cause of deviations from perfect health that fall short of disease and are the province of hygiene. Thus his work ΠΕΡΙ ΚΡΑΣΕΩΝ (*On Krasias*) is of the greatest importance to his theory and practice of medicine. This is the main work included in the present volume. Two other related works are included, dealing with specific aspects of *dyskrasia*, thus amplifying the concept: *On Non-Uniform Distemperment* and *The Soul's Traits Depend on Bodily Temperament*. The former clarifies Galen's distinction between what might be called focal and systemic *dyskrasias*. The latter makes the argument for *dyskrasias* of the brain and its membranes being a major cause of what would now come under the heading of nervous and mental diseases. Also of interest is that this work, probably written some two decades later than the principal work on *krasis*, includes the notion of substance as that in which the ele-

PREFACE

mental qualities inhere, a concept not found in the earlier works.

In Galen's late work, *On the Order of My Own Books*, he refers to two other short works, written at the behest of friends and published by them, which supplement the main treatise on *krasias* and should be read after it. These are *On the Best Constitution of Our Body* and *On Good Bodily State*. The realization that these two short tracts would be a valuable addition to the other three came to me only after I had completed the work initially planned. I thank Jeffrey Henderson for agreeing to their late inclusion as an appendix, and in general for his commitment to giving Galen a greater presence in the Loeb Classical Library over the last decade or so. For almost a century the admirable translation of *De naturalibus facultatibus* by A. J. Brock (1916) had stood as the lone representative of Galen's enormous body of work in the LCL.

On another matter, I would like to take this opportunity to register my disapproval of the recent trend toward using the term "mixtures" to render the Greek *krasias* and the Latin *temperamenta*. "Mixtures" is, in my view, exceptionable in this context. "Temperaments," a term which has a significant history in English, both medically and generally, is more appropriate, although perhaps less than ideal. I am very grateful to two notable scholars of ancient medicine, Vivian Nutton and Alain Touwaide, who were kind enough to offer their thoughts on this issue. The former, accepting my reluctance to use "mixtures," supplied the titles that have been used in the present volume. The latter, who shares my views on "mixtures," proposed a more radical change to "composition." The issue is addressed in section 7 (Terminology) in the General Intro-

PREFACE

duction. In short, my own preference is to use the transliterated Greek term *krasis* in the body of the work and "temperament," sanctioned by long-established use, in the titles, fearing that *krasis* used alone may make the titles opaque to a reader not already familiar with the term.

In translating and commenting on these works, I have approached them more from a medical than a philosophical standpoint. As a doctor, I am fascinated by the prolonged dominance of what might be called the "Galenic paradigm" in medical theory and practice. A somewhat simplistic conclusion could be that it must have had significant success to have withstood the inevitable challenges over many centuries. In the General Introduction I have tried to give a comprehensive account of the system of *krasis/eukrasia/dyskrasia* and the role it played in several aspects of medical theory and practice by quoting from Galen's key works on specific aspects of practice—the classification and causes of diseases and symptoms, diagnosis, treatment and hygiene—and to offer some half-formed thoughts of my own on why it finally yielded its central position.

Once again, I am greatly indebted to my partner, Susie Collis, who read through the translations at various stages of development and offered numerous insightful comments and suggestions, as she has done with previous volumes. I am also very grateful to Elaine Hawkins, a former medical secretary who lives nearby, for typing the manuscript. Our intermittent meetings to exchange tapes followed a walk along a bush track with our two dogs (one, Badhi, now sadly departed and the dedicatee of this volume) and constituted a very pleasant social component of the whole process.

For Badhi

GENERAL INTRODUCTION

Krasis (κρᾶσις: temperament, mixture) and the forms of this, *eukrasia* (εὐκρασία) and *dyskrasia* (δυσκρασία),[1] are foundational concepts in Galen's comprehensive system of medicine. Along with the triad of terms, "capacity" (δύναμις), "function" (ἐνέργεια), and "action" (ἔργον), they are the substance of his physiology, which in turn is coupled with his anatomy, both notional and actual, to provide the basis of what he sees as rational medical practice. Bringing together these concepts of structure and function is an apparatus of definitions and a system of causation.

In Galen's system, *krasis* is the blending of the four elemental qualities—hot, cold, dry, and wet—which, for him, are the fundamental components of all matter, including the identifiable anatomical structures of the body and mind.[2] He allows no mind-body dichotomy, no corporeal/incorporeal division. The range of each of these qualities is infinitely variable, as are their combinations, and

---

[1] The issue of the translation of these terms is addressed in section 7 of this General Introduction, on terminology.

[2] In the first two of the translated treatises, Galen takes these elemental qualities as the foundation of structure. In the third treatise, written some twenty years later, he speaks of a qualityless substance in which these attributes inhere.

GENERAL INTRODUCTION

they are influenced by a variety of factors, both internal and external. *Eukrasia* of the whole body, he claims, is the ideal state, but how often it can be reached and sustained in the individual body is a moot point. *Dyskrasia*, most simply defined as any departure from *eukrasia*, is a common state and, while theoretically abnormal, only becomes a disease state when of such magnitude as to interfere with bodily function. In this form, it represents a major cause of disease, being one of the three components of Galen's tripartite division of the causes of diseases: (a) *dyskrasias*; (b) dysfunction of compound structures (organs): (c) dissolution of continuity.[3] All three are more or less susceptible to correction aimed at the cure of disease and the restoration of health, as defined by Galen.

The three works in the present volume comprise the major statement of Galen's theory of *krasis*: *On Krasias* (*Temperaments, Mixtures*) and two ancillary works. The first of these is *On Non-Uniform Dyskrasia* (*Distemperment*), a short work, a mere twenty Kühn pages, dealing briefly with different *dyskrasias* existing simultaneously, either in the whole body or in a part or parts thereof. The second, *The Soul's Traits Depend on Bodily Krasis* (*Temperament*), puts the case for the application of the concept of *krasis* and its variants to the rational soul, accepting the latter as, in effect, a corporeal entity situated in the brain. Two other short works have been added as an appendix.

---

[3] For details of this classification, see particularly Galen's treatises *De morborum differentiis*, VI.836–80K and *De causis morborum*, VII.1–41K; English translation by Johnston, *Galen: On Diseases and Symptoms*.

GENERAL INTRODUCTION

These are *On the Best Constitution of Our Body* and *On Good Bodily State*, both recommended by Galen as supplementary reading to his main treatise on the subject of *krasias*.

There were practical and theoretical questions pertaining to Galen's concept of *krasis* as a highly relevant medical formulation even at the time of its articulation. Not all of these were satisfactorily addressed, either at the time or subsequently. It was contentious then and is outmoded now. Nonetheless, it proved a firm and clear component of the theory underlying medical practice and, along with the whole body of Galen's medical teaching and practice, held sway not only in Greece and Rome but also in the Near East and later in Western Europe until very recent times. Indeed, the idea of temperaments is still alive today.[4] This makes this triad of books of particular interest not only in relation to Galen specifically but also to the history of medicine generally.

This introduction begins with a brief survey of the antecedents of Galen's definitive concept of bodily structure and the place of the four elements/four qualities therein. Then follows a summary of the three translated treatises and four other works by Galen that are especially germane to the subject. Finally, enumeration of the chief points of views on *krasis* and its variants will be attempted. The four additional works are: (a) the group of four treatises on the classification and causation of diseases and symptoms; (b)

---

[4] See, for example, the recent work by Randy Rolfe, *The Four Temperaments: A Rediscovery of the Ancient Way of Understanding Health and Character* (New York, 2002).

GENERAL INTRODUCTION

the *Art of Medicine*, which deals with diagnosis particularly; (c) the *Method of Medicine*, which covers both diagnosis and treatment; (d) his *Hygiene*, which gives attention to what might be called the "healthy *dyskrasias*" as well as the more commonly discussed morbid *dyskrasias*. The fundamental theoretical work, *De elementis secundum Hippocratem*, and its accompaniment, his commentary on Hippocrates' *Nature of Man*,[5] are not included because the translated works are quintessentially medical works and are obviously directed at doctors, either in practice or in training, although they do have important philosophical implications. There will then be a short account of some contentious issues and points of uncertainty, both practical and philosophical, in his theory, and an outline of the fate of his ideas on *krasis* and its ramifications following his death. Finally, there are four short sections: on terminology, on the people referred to in the translated works, a glossary of diseases and symptoms, and another of medications and foods.

1. GALEN'S ANTECEDENTS

Galen divided all diseases into three classes: (a) *dyskrasias*, (b) diseases of organic parts—subdivided into disorders of conformation (διάπλασις), number (ἀριθμός), magnitude (μέγεθος) and arrangement (συγκείμενος)—

[5] *De elementis secundum Hippocratem* (*On the Elements according to Hippocrates*), I.413–508K; English translation by de Lacy, *Galen on the Elements*, and *In Hippocratis de natura hominis*, XV.1–173K.

GENERAL INTRODUCTION

and (c) dissolution of continuity (λύσις συνεχείας). In the case of the theory of *dyskrasia*, the subject of the treatises translated here, there are several key aspects that embody or are influenced by thoughts of earlier thinkers and writers, some medical and some not. The contributions of the most important of these men are summarized below.

Alcmaeon (5th century BC) is regarded, on the evidence we have, as the first to advance a theory of health and disease based on the concept of balance of qualities. The relevant fragment reads as follows:

> [According to] Alcmaeon, the essential requirement for health is the balance (*isonomia*) of the capacities:[6] wetness, dryness, coldness, hotness, bitterness, sweetness and the remainder, whereas what brings about disease is a preponderance (*monarchia*) in them, for a preponderance of each is destructive. And disease occurs by an excess of hot or cold, and from this through excess or lack of nutriment, and in these [things]—blood, marrow, or brain. Sometimes disease occurs in them from external causes, from the kinds of waters, from place, from fatigue, from necessity, or from things near them. Health, on the other hand, is a balanced mixture of qualities.[7]

Empedocles (ca. 492–432BC) may be regarded as a critical figure in the development of a continuum (four elements/four qualities) theory of matter, a theory taken up

[6] "Capacities" (δυνάμεις) are understood as "qualities."
[7] Fr. B4 DK (1.215–16).

and advanced by two of the three men most influential on Galen's own thinking—Hippocrates and Aristotle. Guthrie, after quoting a fragment from Empedocles, writes:

> How by the mixture of water, earth, air, and sun [fire] there came into being the shapes and colors of all mortal things that are now in being, put together by Aphrodite . . . (fr. 71) The notion of elements has now for the first time acquired a definite meaning as forms of matter which are (a) ungenerated and indestructible, (b) qualitatively unalterable, (c) homogeneous throughout (fr. 17.35).

With Empedocles on one side and his two near contemporaries, Democritus (b. ca. 460 BC) and Leucippus (fl. second half of the 5th century) on the other, there arose the conflict of views between those espousing a continuum theory of matter and those espousing an "atomic" or corpuscular (i.e., particle/void) theory. This conflict extended inevitably into medicine, as evidenced by Galen's *De elementis secundum Hippocratem* and his treatises on the *differentiae* and causes of diseases,[8] in which both theories are considered, with Galen unequivocally stating his adherence to the continuum theory.

Hippocrates (?440–370 BC), as identified by Galen,

---

[8] Galen, *De elementis secundum Hippocratem*, I.413–508K (English translation by de Lacy, *Galen on the Elements*); *De morborum differentiis* (*On the Differentiae of Diseases*), VI.836–80K and *De causis morborum* (*On the Causes of Diseases*), VII.1–41K (English translations by Johnston, *Galen: On Diseases and Symptoms*).

GENERAL INTRODUCTION

was the major precursor of the latter's own views generally and is acknowledged as such by him. A significant number of Galen's own writings were devoted to works from the Hippocratic Corpus. Properly understood and interpreted —that is, according to Galen himself—Hippocrates provided the foundation for all that Galen embraced in the theory and practice of medicine. Three principles of primary importance to Galen were taken from Hippocrates:

1. The humoral theory of bodily composition, as expressed in Hippocrates' *Nature of Man*,[9] with its stated opposition to claims of there being a single basic substance;
2. That each individual disease had a causal explanation that should be sought and, if identified, would be of relevance to treatment;
3. The allopathic principle underlying treatment.

Plato (390–347 BC) was a major influence on Galen in both philosophical and medical matters.[10] De Lacy writes: "Plato is repeatedly praised. He is first among philosophers as Hippocrates is the best of all physicians. Like Hippocrates, he is 'divine.'" Concepts of particular relevance to Galen were:

---

[9] There is, however, a critical difference between Hippocrates' humoral theory as articulated in his *Nature of Man* and Galen's own theory of bodily structure, as is apparent in the quotation from the former in section 7.

[10] See Galen's *De placitis Hippocratis et Platonis* (*On the Doctrines of Hippocrates and Plato*), V.181–805K; English translation by de Lacy, *Galen on the Doctrines*.

GENERAL INTRODUCTION

1. The bodily composition being of the four elemental qualities, hot, cold, wet, and dry, as propounded in the *Timaeus*.
2. The recognition of design in nature, involving the concept of the "Demiurge."
3. The tripartite division of the soul into rational, spirited, and appetitive, located respectively in brain, heart, and liver.
4. Plato's ideas on causation, both in general and in medicine in particular.

The one issue on which Galen disagrees with Plato in relation to the third of the translated treatises is the question of whether the rational soul is corporeal or incorporeal. Clearly, for Galen it is corporeal and subject to the same internal and external influences as the rest of the body, and not incorporeal and immortal as Plato would have us believe. Galen makes his statement of opposition to Plato's position on this matter with, for him, uncommon deference, in *The Soul's Traits Depend on Bodily Temperament*.

Aristotle (384–322 BC), although not always accorded the same reverence as Galen expresses toward Hippocrates and Plato, is nevertheless very influential, especially in the three translated works. This applies particularly in three areas. The first is bodily structure, most notably the concept of *homoiomeres* (uniform parts) and organic parts. For example, in *History of Animals*, we read:

> The parts which are found in animals are of two kinds: (a) those which are incomposite, *viz.*, those which divide up into uniform portions (ὁμοιομερῆ), for example, flesh divides up into flesh; (b) those

> which are composite, *viz.*, those which divide up into non-uniform portions (ἀνομοιομερῆ), for example, the hand does not divide up into hands, nor the face into faces ... Now all the non-uniform parts are composed out of the uniform ones, for example, the hand is composed of flesh, sinews and bones.[11]

The second is his development of the continuum theory of the structure of matter. Lloyd writes:

> At first sight his theory of the ultimate constituents of matter appears to be disappointingly retrograde. After the quantitative, mathematical theories of the atomists and Plato, Aristotle reverted to a qualitative doctrine. All other substances (i.e., apart from heavenly bodies) are thought of as compounds of the four simple bodies, earth, water, air, and fire, and each of these in turn is treated as a combinations of two of the primary opposites: earth is cold and dry, water cold and wet, air hot and wet, and fire hot and dry ...
>
> Aristotle conceives of the problem as being to account for the sensible qualities of physical objects: his very way of stating the question commits him to a qualitative theory. To suggest that these qualities are in turn derived from more fundamental quantitative differentiae would, in his view, be to give the wrong sort of answer to the problem, indeed to mistake the nature of the problem itself.

---

[11] Aristotle, *History of Animals* 1.486a1–15, LCL 437 (A. L. Peck), 2–3 (translation after Peck; see also his introduction, *Parts*, "uniform" and "non-uniform," lxii–lxiv).

GENERAL INTRODUCTION

Moreover, granted that atomism was to prove more fruitful than any qualitative theory of matter, in the short term the doctrine that Aristotle proposed may well have seemed more promising. Certainly the accounts he gave of the ultimate constituents of matter and of the changes affecting the simple bodies stayed close to what could be actually observed. It is obviously true that any physical object may be said to be either hot or cold and either dry or wet, whereas to associate the properties of substances with geometrical shapes must have appeared much more arbitrary. Again, Aristotle could and did offer plausible-seeming interpretation of the changes affecting earth, water, air, and fire.[12]

The third is the distinction between potentiality and actuality, which Aristotle discusses at length in his *Metaphysics*, Book 9, 5–9, and to which Galen refers in a number of places in the translated treatises, but particularly in Book 2 of *On Temperaments*.

Posidonius (ca. 135–51 BC) was a Stoic philosopher of wide-ranging interests. His writings, apparently substantial but largely lost, exist now as fragments collected most recently and comprehensively by Edelstein and Kidd.[13] He is fulsomely praised by Galen in the third of the translated works (*The Soul's Traits Depend on Bodily Temperament*), being described as "the most scientifically knowledgeable of all the Stoics." His relevance to the present matter is twofold. First, he was an adherent of the four

---

[12] G. E. R. Lloyd, *Greek Science* (London, 2012), 98–99.

[13] See Kidd's entry in the *OCD*, where the relevant references to the collections of fragments are also given (1231–33).

GENERAL INTRODUCTION

elements/four qualities theory with an additional important place given to *pneuma* (common among the Stoics) and of *dyskrasia* as a cause of disease; second, after the aforecited favorable remark, Galen states his acceptance, at least in part, of Posidonius' views on the workings of the rational soul.

Athenaeus (?AD 30–70 fl.) was, perhaps, the founder of the Pneumaticist school of medicine. He too accepted the theory of the four elemental qualities but under Stoic influence gave particular prominence to *pneuma* as a fifth elemental component of the body. He defined health as an equilibrium between *pneuma* and the four elemental qualities, and disease as a disequilibrium, or imbalance, of the five components (i.e., a *dyskrasia*) rather than of the four only. His theoretical formulations were based on Stoic concepts, and he was linked by Galen with Posidonius. Whether he was, in fact, one of Posidonius' students, as some suggest, is uncertain, even unlikely.[14]

Kupreeva, after a short translation from the pseudo-Galenic work *Introductio seu medicus*, writes:

> According to Athenaeus, the elements of man are not the four first bodies, fire, air, water, and earth, but their qualities, the hot, the cold, the dry, and the moist, two of which he considers to be productive causes (ποιητικά αἴτια), *viz.* the hot and the cold, and two material, *viz.* the dry and the moist, and he additionally introduces as the fifth, breath (πνεῦμα), which according to the Stoics, pervades everything, by which all things are contained and governed.

[14] Posidonius' dates are uncertain; see Nutton, *Ancient Medicine*, 202–4.

Both Galen and Athenaeus are humoralists. Both seem to think of the four humors—blood, phlegm, yellow and black bile—as the organic equivalents of the four elements. Galen occasionally follows the Pneumaticist usage calling the humors "proximate elements." Both take each of the humors to be a combination of two "organic" elemental qualities: yellow bile of heat and dryness (like Aristotle's fire), blood of heat and moist (like air), phlegm of moist and coldness, and black bile of dryness and coldness.[15]

Whether Galen should be classed as a "humoralist" is debatable. Certainly, the four humors occupied an important place in his pathophysiology, themselves composed of the four elemental qualities, but in his system their role in health and disease was more circumscribed.

2. THE THREE TRANSLATED WORKS

De temperamentis (*On Krasias, On Temperaments*)

There are three books, as summarized below. The first is predominantly theoretical and definitional; the second deals with issues of recognition and introduces the concept of non-uniform *dyskrasia*; the third focuses on *krasis* in nutriments and medications and its relevance in the use of these.

[15] Inna Kupreeva, "Galen's Theory of Elements," in *Philosophical Themes in Galen*, ed. P.Adamson, R. Hansberger, and J. Wilberding (London, 2014), 153–96.

GENERAL INTRODUCTION

Book 1

Galen sets out the foundations of his concept of bodily *krasis* and the states of *eukrasia* and *dyskrasia*—what *krasis* is, what constitutes *eukrasia*, what kinds of *dyskrasia* there are, how these are recognized, and the natural variations of *krasis* in different parts of the human body. First, the *krasis* of all things in general, of animal and plant bodies, and of the human body in particular, is a mixture or blending of the four elemental qualities—hot, cold, dry, and wet. This is not an ordinary mixture such as a person might create, but a thoroughgoing blending that requires a god or Nature for its creation. These four qualities may also be considered as two antitheses (hot-cold, wet-dry), the antithetical qualities in each pair being unable to coexist in a stable way. *Eukrasia* is a state in which there is an appropriate balance of the four elemental qualities in a particular entity—that is, neither of the two qualities in each antithesis prevails over the other. Apart from *eukrasia*, there are eight possible *dyskrasias*, four of which are simple, mono-*dyskrasias* (preponderance of hot, cold, wet, or dry—i.e., of one component of one antithesis with *eukrasia* in the other antithesis) and four compound, bi-*dyskrasias*, in which one quality from each antithesis is simultaneously in excess.

Galen addresses certain terminological issues, identifying three significations of the primary terms, "hot," "cold," "wet," and "dry." The first is absolute and applies only to the elements themselves—fire, air, water, and earth. The second and third are relative; one in relation to the median of the class or kind and one in relation to an individual entity of any class or kind. There is also the

GENERAL INTRODUCTION

relation between the terms *eusarkos* (well-fleshed) and *eukrasia*, which Galen correlates in Chapter 9. The former, however, is a visual and tactile judgment only, while the latter may contain an inferential (rational) component based on function. Galen deals with two possible errors. The first is that of claiming there are only two possible compound *dyskrasias* on the grounds that the wet cannot coexist with the hot, nor the dry with the cold. The second is the attempt to equate each of the compound *dyskrasias* one-to-one with one of the four seasons. Galen himself makes two claims of uncertain validity. One is that the human is the most *eukratic* of all animals and also of all things; the other is that the palmar surface of the hand is the most *eukratic* of all the parts of the human body and is the arbiter of all perceptions, unless damaged in some way, as by undue manual labor.

Book 2

After summarizing the key points from Book 1, Galen adds to the recognition of *eukrasia* by referring not only to the median of hotness and coldness but also to that of hardness and softness, as well as the functions of the soul. *Eukrasia*, he says, is characterized by optimum function and also a balanced state in visible and palpable external features. He gives some consideration to the differences relating to the stages of life. The basic pattern is a beginning, even intra-uterine, of a preponderance of hotness and wetness, passing through the optimum *krasis* (*eukrasis*) in midlife to the preponderance of dryness and coldness of old age. Clinical evaluation of bodily *krasis* is through a number of components: palpation and inspection, the state of the blood vessels, the growth and nature

of the hair, and the various functions. He makes the important point that the differentiation between a healthy and a morbid *dyskrasia* is the finding of impaired function in the latter. This distinction is developed in the final two books of his *Hygiene*.

In Chapter 6 he turns for the first time to the non-uniform *dyskrasias*. He warns several times against drawing conclusions about the *krasis* of the whole body from that of any one part. He gives examples involving the eyes and ears. In evaluating a person, a number of other things must be taken into account: nature, customs, regimen, habitation, and physiognomy. A picture emerges of the combination of history and physical examination in the search for any disorder of bodily *krasis*, whether systemic or focal.

Book 3

In this final book, Galen focuses on nutriments and medications after some prefatory remarks about the Aristotelian terms ἐνέργεια, δύναμις, and ὕλη (actuality, potentiality, and matter), and the four bodily capacities—absorptive, retentive, transformative, and separative, dealt with in detail in his work, *On the Natural Faculties*.[16] Nutriments need modification (working up) and assimilation to be effective, while medications are basically of two kinds—those that overcome and change the body and those that are overcome and changed by the body, then act to putrefy and destroy. There are also those that heat but do no harm and those that both act and are acted upon. With both nutriments and medications there

[16] *De naturalibus facultatibus*, II.1–204K; English translation by Brock, *Galen on the Natural Faculties*.

is likely to be a difference in their effects when administered orally and when applied to the skin. Also, the time of administration is important. Other important distinctions mentioned more than once are that between "in actuality" and "in potentiality," and that between "of themselves" and "contingently." These apply to agents given that are hot, cold, wet, or dry. The action of a substance may be direct or through an intermediary. This third book is particularly about the principles of use of nutriments and medications in the management of *dyskrasias* and is identified by Galen as an important precursor to his *Method of Medicine* and his works on medications.[17]

De inaequali intemperie (On Non-Uniform Distemperment/Dyskrasia)

A non-uniform *dyskrasia* is essentially the existence of different *krasias* (*dyskrasias*) in the same body at the same time. The types of *dyskrasia* are the same eight as previously identified. Galen makes a distinction between non-uniform *dyskrasias* that involve the whole body, giving the examples of generalized edema and most forms of fever, and those that involve one or more specific parts, giving the examples of inflammation and erysipelas, among others. He identifies two forms of causation: (a) primary change in the affected structure due to either internal or

[17] *De methodo medendi*, X.1–1021K (English translation by Johnston and Horsley, *Galen's Method of Medicine*); *De simplicium medicamentorum temperamentis ac facultatibus*, XI.369–802 and XII.1–377K; *De compositione medicamentorum secundum locos*, XII.378–1003 and XIII.1–361K; *De compositione medicamentorum per genera*, XIII.362–1058K.

GENERAL INTRODUCTION

external causes; (b) the inflow of a flux from elsewhere in the body that disturbs the balance. These fluxes are presumably the four humors (blood, yellow and black bile, and phlegm), although inflowing *pneuma* may also be involved. He gives a brief account of non-uniform *dyskrasias* due to heating and cooling, and makes three particular points: (a) Pain in a non-uniform *dyskrasia* is present in a part being affected only while the balance of qualities is changing. Once stability/uniformity is reached, pain stops. (b) Whether a non-uniform *dyskrasia* in a particular part affects adjacent structures depends on the degree of the disproportion. (c) All fevers apart from the hectic are non-uniform *dyskrasias*. He makes the somewhat problematic comment that the non-uniform *dyskrasia* in ague (*epialos*) fevers involves hot and cold—that is, two components from one antithesis. It is not clear here whether he considers the existence of the two components to be simultaneous or sequential, but elsewhere he does state it can be the former.

Quod animi mores corporis temperamenta sequantur (*The Soul's Traits Depend on Bodily Temperament*)

This short work basically makes a single claim, stated at the outset, which is then defended largely by reference to other authorities, as listed below. The claim is that the physical state and functioning of the soul is dependent on the *krasis* of the body as a whole, and that the creation of *eukrasia*, by whatever means, is a prerequisite for excellence of the soul. Two fundamental assumptions are that the soul, in particular, the rational soul, has a substance on which its functions depend, and a range of capacities

that may differ in different individuals or in the same individual at different times. In Galen's view, the rational soul is corporeal, composed of the same types of substances that compose the rest of the body. At the end of Chapter 3, he writes:

> It will also be necessary, then, for those who postulate the soul to be a specific substance, to concede that it is itself a slave to the *krasias* of the body. If in fact they have a power to separate, they compel it to be deranged, take away memory and understanding, and make it more distressed, less courageous and more spiritless, as is seen in the melancholias, while the one drinking wine in moderation possesses the opposite effects to these.

The remainder of the work, Chapters 4 to 9, is taken up with other opinions, alluded to above and listed below, which in sum support his view. These are from the authorities he most respects.

- Plato: Galen agrees with Plato's tripartite division of the soul into rational, spirited, and appetitive, but not with his view that the rational soul is incorporeal and immortal. He offers quotations from Plato's *Laws* and *Timaeus* to establish that he too has accepted that *kakochymia* and wine can affect the soul, as can pursuits and regimen generally.
- Aristotle: Quotations are given from two works, *Parts of Animals* and *History of Animals*, to support two suppositions: (a) that the capacities of the soul at birth follow the *krasias* of the maternal blood; and (b) that the capacities of the soul follow the *krasis* (nature) generally of the body.

GENERAL INTRODUCTION

Hippocrates: Galen offers quotations from *Airs, Waters, Places* and *Epidemics II* to show that Hippocrates subscribed to the four elements/four qualities theory of the basic structure of matter and that both bodily states and mental dispositions are clearly influenced by physical (including climatic) conditions through their effects on the *krasis* of the body.

The Stoics: Galen gives brief consideration to Stoic views, particularly those of Chrysippus and Posidonius, on the origins of good and evil in people. On the issue of nature versus nurture, Galen's view is that there is a basic disposition of the rational soul determined by the *krasis* at the initial genesis, but that this can be changed or modified for better or for worse by nurture, training, and bodily *krasis*.

3. FOUR OTHER RELEVANT WORKS

Four Treatises on the Classification and Causation of Diseases and Symptoms

De morborum differentiis
(On the Differentiae of Diseases)

In this work Galen begins by identifying health as a balance (*summetros*) and disease as an imbalance (*ametros*), following the concept that appears to have originated with Alcmaeon. He then considers the two competing theories of structure, the particle/void (atomic) theory and the four elements/four qualities (continuum) theory. He describes the latter, which he espouses, at some length. Having also listed his three macroscopic structural levels (*homoiomeres*, organs, whole body), he identifies the four primary

diseases of *homoiomeres* as a disproportion of each one of the elemental qualities, hot, cold, dry, and wet—that is, four simple mono-*dyskrasias*. He makes an important distinction in the mono-*dyskrasias*, which he also stresses in the work *On Krasias*: changes in balance of the qualities in the *homoiomerous* bodies may arise primarily due to either internal or external causes, or may arise secondarily due to inflowing material (fluxes), in effect the four humors, blood, phlegm, yellow and black bile, which themselves are composed of the elemental qualities and have a preponderance of two of these (in effect a *dyakrasia*) in each case—respectively, hot and wet in blood, cold and wet in phlegm, hot and dry in yellow bile, and cold and wet in black bile. He writes:

> In the case of the second hypothesis (i.e., the continuum theory), the differentiation of diseases happens to be twofold in that sometimes the *homoiomerous* bodies are changed in their qualities alone, whereas sometimes a certain substance flows into them which has the qualities spoken of. Certainly the second form, bringing about a swelling around the bodies, is obvious to all doctors. For *erysipelata*, inflammatory swellings, oedemas, tumors, glandular swellings, scrofulous swellings, *elephantiases*, *psorai*, *leprai*, *alphoi* and indurations are of this class, and can escape no one. The diseases arising in a *dyskrasia* of the qualities themselves alone are harder to detect, unless at this time a major turning aside toward what is contrary to nature occurs. Under these circumstances it will be readily known by everyone that when heat prevails in the whole body

it is termed a fever, although sometimes it is also clearly manifest in the parts. (VI.848–49K; J, 141)

In the remainder of Chapter 5, he outlines the diseases associated with the individual mono-*dyskrasias*, preserving the differentiation of primary change and change due to inflowing material. At the end of the chapter, he considers what might be called "subclinical *dyskrasias*" as possible causes of dysfunction when no other cause is apparent—he terms this *atonia* (debility) in the case of the stomach.

In Chapter 12, under the heading combined diseases, he lists the four possible combinations of elemental qualities that can be simultaneously disproportionate—hot and dry, hot and wet, cold and dry, cold and wet—stating that these, like the mono-*dyskrasias*, can be either primary changes or due to inflowing material. Also in this chapter, he mentions the possible combination of dissolution of continuity (his third major class of diseases)[18] and *dyskrasia*:

> For it is not impossible for a part to be ulcerated and, at the same time, also more dry than accords with nature, or more wet, or more cold, or more hot; nor is it impossible for a part to be ulcerated and, at the same time, more wet but not to be more hot at all. Therefore, parts that are simultaneously ulcerated and inflamed depart from what accords

---

[18] On this third class of diseases, see *De differentiis morborum* (ch. 12), *De morborum causis* (ch. 11), and *Method of Medicine* (Johnston and Horsley, *Galen's Method of Medicine*, lxxii–lxxiii).

GENERAL INTRODUCTION

with nature in three ways: owing to the ulceration there is destruction of the unity of the specific parts, whilst because they are inflamed they are made hotter and wetter than is natural.

De causis morborum (On the Causes of Diseases)

In this work Galen gives considerable attention to the causes of all four simple (mono-) *dyskrasias*, outlining the causes of each in Chapters 2 to 5. Those he mentions are listed in the following table:

Dyskrasia	Cause
Hot	Increased movement
	Putrefaction
	Proximity to a hotter body
	Constriction
	Foods with a hot capacity
Cold	Idleness
	Disproportionate movement
	Proximity to a colder body
	Constriction and rarefaction
	Foods with a cold capacity
Dry	Increased activity + sweating
	Low food intake
	Dry external conditions
	Thinking too much
	Staying awake too long
	Drying food and fluids

Dyskrasia	Cause
Wet	Excess of fluids
	Excessive bathing
	Luxurious way of life
	Gladness of heart
	Wet foods

In Chapter 6, the four standard compound (bi-) *dyskrasias* are dealt with somewhat peremptorily, as follows:

> It is clear that the causes of combined diseases are undoubtedly combined. For if, on occasion, hotness and dryness conjoin at the same time as a cause, the disease will necessarily be hot and dry, and if hotness and wetness, hot and wet. It is the same in the case of the remaining two conjunctions, wetness and coldness, and dryness and coldness. (VII.20K; J, 168–69)

He then passes on to what he calls non-uniform (anomalous) *dyskrasia*, on which he says:

> Let us now say what it is necessary to add to and define in the discussion. This is that the body is often changed by all the causes that are of the same kind as each other, or often by those that are opposite in their capacities. Of these, sometimes the greater number prevails, or the longer lasting, while sometimes the stronger prevails, or sometimes the body receives damage from both alike. And even if it seems impossible for one and the same body to be made simultaneously more hot and more cold than

is natural, or again more wet and more dry, nevertheless this does also occur. This is properly termed a "non-uniform *dyskrasia*," and those things that are appropriate for us to say about this have been said before separately in another work. (VII.20–21K; J, 169)

On the matter of the fluxes, the second of the two generating causes of *dyskrasias*, and on the qualities of the humors, he says:

It is appropriate to recall again here what was said in the work on the *differentiae* of diseases, to the effect that sometimes change from what accords with nature arises in the four qualities themselves alone, no other substance flowing into them from without, whereas sometimes they are filled with a fluxion which is undoubtedly wet in nature but not wet in capacity. There has been discussion by earlier doctors and philosophers about the capacity of such fluids. We have documented this also in certain other treatises and in those on medications. What from these that is pertinent to the present discussion will be spoken of now. For instance, yellow bile is hot and dry in capacity, black bile is dry and cold, blood is wet and hot, and phlegm is cold and wet. And sometimes each of these humors flows unmixed, but sometimes mixed with others, and the conditions of swollen, indurated and inflamed parts, in consequence, vary still more. (VII.21–22K; J, 169)

GENERAL INTRODUCTION

De symptomatum differentiis (*On the Differentiae of Symptoms*)

There is mention of *dyskrasia* in this work, but not much is said. There is, however, this interesting statement. Speaking of all the *differentiae* of solid bodies pertaining to color, odor, taste, and touch, Galen says:

> So that, whatever among these is contrary to nature is in every case an "offspring" of *dyskrasia*, just as what accords with nature is an "offspring" of *eukrasia*. But every *dyskrasia* is a disease, consequently such symptoms are also "offspring" of diseases. (VII.77K; J, 199)

De symptomatum causis (*On the Causes of Symptoms*)

In Book 1 there is mention of *dyskrasia* in relation to several specific symptoms in different structures. Most notably, in considering the eye, he says that the crystalloid is susceptible to all eight *dyskrasias*, but if slight, they produce little damage to function. In Book 2, speaking about cough, Galen has this to say:

> Why then, in the case of non-uniform *dyskrasias* of the respiratory organs, people cough (for I have said that the doctors of the Pneumatic sect have not worked this out very well), I shall attempt to go over in detail, starting from the substance of the matter, as befits those who intend to speak by means of demonstration. Accordingly, a *dyskrasia* sometimes occurs in relation to singular qualities, when the

bodies themselves are made hotter, colder, drier, or wetter, or also when they suffer this in relation to some conjunction. Sometimes what is *dyskratic* in them [occurs] with a dispersal that is dew-like. A uniform *dyskrasia* is, then, altogether painless, as the hectic fevers and many of the conditions relating to cold show, taking hold of the parts evenly. For such a mixing occurs in each of the parts so disposed as if it were some additional nature. No body is distressed by its own nature. As Hippocrates said, pains occur in those things that are being changed or destroyed in their nature, not in those that have already been changed or destroyed. For in being changed, and in departing from their own nature, bodies are distressed . . . Whenever, then, a non-uniform *dyskrasia* has come into being in any one of the bodies capable of perception, the magnitude of the pain is commensurate with that of the *dyskrasia*. Thus in the sharpest and strongest of the fevers, the solid parts of the animal are distressed, being changed and altered by the heat that is contrary to nature. (VII.175–76K; J, 251–52)

In Book 3, *dyskrasia* is mentioned only occasionally and in none of the instances is the concept elaborated on.

Ars medica (*Art of Medicine*)

A large part of this work is devoted to *krasis/dyskrasia*, focusing particularly on diagnosis. Galen prefaces his consideration of the signs pertaining to the different body parts with a short statement on the importance of know-

GENERAL INTRODUCTION

ing what these parts are (section 5)—a statement he also makes in *On Temperaments*. Sections 6 to 18 (I.319–52K) then describe the signs, summarized as follows:

Section 6. Signs of *krasis* of the brain: the size and shape of the head and the four groups of functions—sensory, motor, hegemonic, and physical. There are also signs due to things befalling the head externally. Diagnosis is by inspection and palpation of the parts of the head and observation of the functions of the brain. In short, a *eukratic* brain will show moderation in all functions and be least affected by external factors.

Section 7. Signs of simple *dyskrasias* of the head: the amount and nature of the superfluities excreted through the recognized channels, the growth and nature of the hair, the complexion, the acuity of the sensations, how the head feels to the one palpating, the appearance of the veins around the eyes, and the sleep pattern.

Section 8. Signs of compound *dyskrasias* of the brain:

Hot and dry: lack of superfluities, acute sensations, extreme wakefulness, rapid development of baldness, hair that is black and curly, a head that is hot to touch and red until the time of full growth.

Hot and wet: healthy complexion and warmth, large veins around the eyes, increased superfluities moderately concocted, straight hair that is light brown, not readily becoming bald.

Cold and dry: a cold head devoid of color, no visible veins around the eyes, lack of superfluities, sometimes overcome by catarrhs and coryzas due to minor causes, faultless sensations in youth.

Cold and wet: lethargy and drowsiness, poor sensa-

tions, excrementitious, easily cooled, full in the head, readily susceptible to catarrhs and coryzas; such people do not become bald.

Section 9. Signs of *krasis* of the eyes: Galen takes the eyes as a model for the organs of sensation. The key features are: how the eyes feel to palpation (hot, cold, hard), size, nature of movements and function. The color of the eyes is also taken to be of some diagnostic importance; Galen addresses this in some detail.

Section 10. Signs of simple *dyskrasias* of the heart: these are to be found in the size, shape, and degree of hairiness of the chest, and in the pulse, respiration, and spirit (calm, fiery, easily angered, brave, cowardly, etc.).

Section 11. Signs of compound *dyskrasias* of the heart: the same observable features as for the simple *dyskrasias* are relevant here.

Section 12. Signs of simple and compound *dyskrasias* of the liver: these include the appearance of the hypochondrium (specifically, the degree of hairiness), the visible veins of the region, the temperature and consistency of the blood, the preponderance of certain humors, and also the predisposition to putrefactive and *kakochymous* diseases. The state of the whole body is also important in diagnosing a *dyskrasia* of the liver.

Section 13. Signs of simple and compound *dyskrasias* of the testes: the important features here are hair growth in the genital region, the amount and consistency of the semen, fertility, and the desire for, and effects of, sexual activity.

Section 14. The state of the whole body: this is generally assessed by examining the most obvious parts. Also,

GENERAL INTRODUCTION

the effect of the dwelling place must be taken into consideration. Galen then describes the features of a well-balanced *krasis* of the whole body in terms of complexion, hair color, and the quantity and quality of the flesh. This is the standard against which bodies that are not well-balanced are measured.

Section 15. Simple *dyskrasias* of the whole body: Galen summarizes the signs of the four simple *dyskrasias* as follows:

> Hot: increased hairiness, decreased fat commensurate with the degree of heat, ruddy complexion, black hair.
> Cold: relatively hairless and fat, cold to the touch, complexion and hair reddish-brown, face may be livid.
> Dry: thinner and harder than the *eukratic*.
> Wet: well-fleshed and soft.

Section 16. Compound *dyskrasias* of the whole body:

> Hot and dry: hairy, hot, hard, thin and devoid of fat, black hair, dark complexion.
> Hot and wet: hairy, soft, well-fleshed, hot to the touch, prone to putrefactive diseases.
> Cold and wet: hairless, pale, soft, dense and fatty, hair and complexion tend to be reddish, although the latter may be livid.
> Cold and dry: hairless, thin, hard and cold to the touch.

Section 17. Signs of *dyskrasias* of the stomach: these are thirst, appetite for foods, and the effects of different sorts of foods and drinks. In the stomach, *dyskrasias* due to disease differ from those that are innate by producing a desire for opposites rather than similars. Galen also makes

GENERAL INTRODUCTION

the point that the state of the chest and lungs has a bearing on thirst and the desire for hot or cold drinks.

Section 18. Signs of *dyskrasias* of the lungs: the most important signs are to be found in the superfluities revealed by the sputum, and in the voice. Obviously, the upper airways also have a considerable bearing on signs detected in the voice.

De methodo medendi (Method of Medicine)

This is Galen's major work on medical practice, so *dyskrasia* naturally receives considerable attention. The first six of the fourteen books were written around the same time as his works on temperaments, while the final eight books were not written until the final decade of the second century. Several excerpts follow. The first two are from Book 2, one on the issue with those who do not subscribe to the four elements/four qualities theory, exemplified by Erasistratus, and one listing those of his own works that he regards as important in understanding the matter of *krasis*. The next excerpt, from Book 3, details the importance of devoting attention to bodily *krasis* in the management of ulcers. The general subject of Books 3 to 6 is the disease class of dissolution of continuity. Book 7, the first of those later written, is, after two introductory chapters, devoted to the *dyskrasias*. The content of the remaining chapters (3–13) is summarized, followed by an extended excerpt from Chapter 9 on the management of a hot *dyskrasia*. Finally, a short excerpt from the start of Book 8, which begins a series of books on fevers, themselves hot *dyskrasias*, is given.

GENERAL INTRODUCTION

1. On adherents and opponents of the continuum theory (*MM*, Bk. 2, X.111–13K; J&H, 1.172–75).

That there are many kinds of pathological *dyskrasia* and that the treatment is different for each, they provide as witnesses not only Hippocrates and a great many other doctors, but also Plato, Aristotle, Theophrastus, Zeno, and Chrysippus, as well as all the noted philosophers. That without the nature of the body being discovered precisely, it is impossible either to find out about the *differentiae* of diseases, or to find suitable means of remedies, they will again provide all these philosophers and doctors now spoken of by me—men who do not, by Zeus, command like these people in the manner of tyrants, but offer demonstrations.

In these matters, then, those who claim that hot and cold are the terms of bath-house attendants and not of doctors will clearly be indulging in ribaldry and making mockery, describing them as fools, or Phrygians, or Scholastics. For they realize that they themselves not only know nothing about demonstration, but they have no understanding at all of what it is. From these jokes there will inevitably arise a source of difference. So let me mention the first of all the things they do not hesitate to say when they talk nonsense without [providing] demonstration, and consider how, inevitably, conflict follows this. Whenever others bring forward a far greater number of more highly regarded witnesses against the undemonstrated statements of these men and

their plausible witnesses, it is necessary for them either to concede that they have been defeated and are worsted in both respects—that they do not attend to demonstration, and that they have been overcome by the quantity of witnesses—or else to declare quite shamelessly that Erasistratus is more credible than all those witnesses. It is necessary here to say that the one disputant was not more credible than the doctors who follow Hippocrates and Mnesitheus, or the philosophers in the circle of Plato and Aristotle, and for the other to argue against this. When this sort of argument has gone back and forth in this ignorant and contentious way, the latter is led to say, "it will not go well for you if you dishonor Erasistratus." You see this sort of thing happening every day, Hiero, in the debates among doctors.

2. On the works necessary for an understanding of Galen's views on *krasis* (*Method of Medicine*, Bk. 2, X.121–22K; J&H,1.188–91).

What else is it apart from *dyskrasia*? For its conformation is obviously not undergoing change, nor is its hollowness, ligaments, or orifice. Indeed, no other part is being destroyed at all—only the natural *krasis* is changing. Thus, all these diseases that occur in *homoiomeres*, as I said, are completely left aside by him. For the damage to a certain function when the blood is blocked at the ends of the arteries is the affection of the arteries as organs, whereas [the affections of arteries] as *homoiomerous* bodies,

the *dyskrasias* in terms of bare qualities, are eight in number and those with fluxions are [also] eight in number.[19]

Anyone who wishes to grasp the demonstrations of these things through scientific knowledge ought to start from the work *On the Elements according to Hippocrates*, and then read through each of the others in turn, as was said earlier. Following that, there is the treatise *On Krasias* (*On Temperaments*), and after these the one *On Non-Uniform Dyskrasia* (*Distemperment*) and those treatises in which each of the affections of the *psyche* is spoken of specifically and in order.[20] Following these, there is [the work] *On the Uses of the Parts*, and following these again, the works *On the Differentiae of Diseases* and *On the Differentiae of Symptoms*.[21]

3. On the importance of bodily *krasis* in the treatment of ulcers/wounds (*MM*, Bk. 3, X.214–17K; J&H, 1.326–31).

What I have shown clearly, I think, is that the person who is going to cure an ulcer (wound) properly

---

[19] The fluxions are yellow and black bile, blood, and phlegm. See *De morborum causis*, VII.22–23K.

[20] *Quod animi mores corporis temperamenta sequantur*, IV.767–822K; *De propriorum animi cuiuslibet affectuum dignotione et curatione*, V.1–57K; and *De animi cuiuslibet peccatorum dignotione et curatione*, V.58–103K. All three are translated by Singer, *Galen: Psychological Writings*.

[21] *De usu partium*, III.1–939K and IV.1–366K; translated by May, *Galen on the Usefulness*.

GENERAL INTRODUCTION

ought to pay close attention to the *krasis* of the body, the seasons of the year and the nature of the parts, and that the primary indicator of the cure is taken from the condition alone. However, it is not yet possible to discover the remedies here before proceeding to the elements of the body and considering the *krasis* of the patient, both of the whole body and of the affected part, and to jointly consider with this right away the *krasis* of the surroundings, which extends to local conditions and regions as well. I shall state at greater length in what follows that contrary indications frequently arise in relation to a single treatment, and how we must handle such [indications], although it would not be out of place to go over them briefly now also. For it is, I think, not surprising for the *krasis* of the patient to be more wet whilst the affected part itself is more dry, or this to be more wet and the whole *krasis* more dry. In like manner, too, in relation to hot and cold, there may be an opposition in *krasis* between the part and the whole. Thus, just as when the whole body is of moderate *krasis*, which we showed was best, we would change nothing in the medications for the sake of the nature of the patient, so, whenever the body is more dry, or more wet, or more cold, or more hot than it should be, it is necessary to increase the potency of the medications to the degree that the body has been taken over by a natural *dyskrasia*. Nor, certainly, will we lose sight of what is a natural *dyskrasia* and what is an unnatural *dyskrasia*. I spoke about these *dyskrasias* in other

[works] and particularly in the treatise *On Non-Uniform Dyskrasia (Distemperment)*.[22]

Suppose, therefore, the whole *krasis* of the patient's body is more wet and, because of this, needs less of the drying medications, whereas the affected part itself is among those things that are more dry in nature, of the sort which I said were less fleshy. Examples are found in the fingers, joints, ears, nose, eyes, and teeth, and, in summary, in a place where there is much cartilage, membrane, ligament, and bones and nerves but very little fat and flesh, or none at all. In these places, the indication from the affected part is different to that from the nature of the patient. As a result, if the part is drier than normal to the same extent that the *krasis* of the patient is wetter in nature than is normal, I undertake neither to add nor take away anything by way of the medication, whereas I would undertake to apply such a medication in the case of a body balanced in *krasis*, when a wound has occurred in a part that is balanced in *krasis*. If the part is drier than normal by as much as the *krasis* of the body is wetter, it is necessary to increase the medication to the extent that the dryness of the part exceeds the whole *krasis*. If, for example, the wounded part is four magnitudes more dry than normal, and the nature of the diseased person three magnitudes wetter than the *eukratic*, it is clear that the part

---

[22] *De inaequali intemperie*, VII.733–52K. English translations by Grant, *Galen on Food and Diet*; Garcia Novo, *Galen: On the Anomalous Dyskrasia*; and in the present work.

which is now wounded will be in need of medication drier by one magnitude than a balanced part. It is, however, patently obvious that all these things are arrived at by guesswork, and that someone practised in calculations about these matters is best able to carry out the estimation.

4. A summary of Chapters 3 to 13 of Book 7 devoted to *dyskrasias* (*MM*, Bk. 7):

Chapter 3. The claim is made that an understanding of *dyskrasia* requires a theory of the underlying structure of the body, and of matter in general. There is reference to the theory of elements described in Galen's *On the Elements according to Hippocrates*. The fundamental aim in preserving the health of the functioning parts of the body is to maintain their *krasis* in a proper balance of the four elemental qualities. Galen provides a general statement of the method of cure of all diseases occurring in *homoiomeres*.

Chapter 4. Galen argues that, from a clear understanding of the underlying general principles, it is possible to move theoretically to the treatment of individual cases. Specific details of the treatment of weakness (*atonia*) of the stomach are given, *atonia* being a result of *dyskrasia*. There is a list of relevant medications and a case report in which Galen again saves the day.

Chapter 5. Some variations in the treatment of the different *dyskrasias* relating to duration and risk are identified.

Chapter 6. Varieties of abnormal dryness (dry *dyskrasias*) are described. The dangers of astringent medica-

tions, foods, and drinks are discussed. There is a further case report of a patient with a dry *dyskrasia* involving the stomach. A detailed consideration of the use of milk, both from asses and women is offered, the key factor being that it must retain its heat when given. The use of honey is described. There is a digression on the optimal care of the animals providing the milk. The case report continues. Aspects of a restorative diet including wine are listed. Some general aspects of a restorative regimen are discussed.

Chapter 7. Cooling is seen as a sequel to chronic dry *dyskrasias*. Once cooling is involved, treatment becomes more complicated. The treatment of a combined *dyskrasia* comprising dryness and coldness is discussed. Details of the various medications and preparations are given, and the use of pitch is described.

Chapter 8. A combined *dyskrasia* of dryness and hotness is considered. The nature of the wine and honey to be used is described. Other useful foods are also considered. Galen provides a detailed case report with a bad outcome and acknowledges his early failings. A second case report follows. Galen, now much more experienced, achieves a better outcome.

Chapter 9. Treatment of a hot *dyskrasia* combined, in the first instance, with a wet component and, in the second instance, with a dry component is discussed. The association with fever is identified. The use of cooling agents and their dangers are considered. A wet *dyskrasia* is seen as the easiest of the *dyskrasias* to cure. Cooling and astringent foods and drinks are the staples of treatment, and these include both water and wine.

GENERAL INTRODUCTION

Chapter 10. Galen provides a summarizing statement on the various *dyskrasias*, single and combined, involving the stomach.

Chapter 11. *Dyskrasias* associated with excess moisture from an external source are described. The main treatment is purging by vomiting. There is further general consideration of *dyskrasias* due to inflowing material. The treatment options in general for a pathological flux are to stop the flux at its source and to prevent possible target structures from receiving the flux. Purging of the whole body is important. The treatment of *kakochymia* is discussed. Medications of value are listed.

Chapter 12. Mixed conditions of the stomach involving both wall and lumen are described. The methodical approach to the sequencing of treatment is outlined. The importance of preserving the patient's capacity (strength) is emphasized. The management of *dyskrasias* of the stomach is taken as the model on which to base the management of *dyskrasias* in other structures.

Chapter 13. The sequence of indications in treating *dyskrasias* is given. Five kinds of indication are listed. Some general observations on the treatment of *dyskrasias* are offered. Specific features of the functions of various organs that bear on the treatment of *dyskrasias* are identified. The importance of preserving the capacity is again stressed. Variations in indications according to the affected place and its sensory capabilities are considered.

5. On the management of a hot *dyskrasia* (*MM*, Bk. 7, X.508–10K; J&H, 2.312–17).

> Now let us assume, in turn, that a hot *dyskrasia* prevails, but that wetness is mixed with it in one

instance, and dryness in another, each in moderation. First, in the case of wetness, we shall treat such a *dyskrasia* more confidently with cold water because no harm is done by this to the parts that are adjacent to it (i.e., to the stomach), which are in a balanced state. This is because, in dry conditions, it is inevitable that not only those parts that are near but also the whole body become more emaciated, whereas when the stomach itself has not as yet been thoroughly dried up, as is now supposed, it is impossible for the whole body to become thin, so there is no harm from the cold drink. If, in this way, there is at some time a severe hot *dyskrasia* in the stomach such as to reach as far as the heart, the person is, of necessity, febrile. As a result, the danger likely to follow such *dyskrasias* is more acute. But the cure is the same in terms of class, and I shall speak of it again in my treatments of fevers. The hot *dyskrasia* conjoined with dryness is subject to these same remedies in terms of class but cooling agents cannot be used confidently in the same way for the reasons I spoke of. However, the wet *dyskrasia*, either existing alone or intermingled with hot or cold, is the most easily curable of all. And because these three *dyskrasias* continually befall the stomach and are the most easily cured, and because we always have the treatment of these in mind, all those who treat the diseases without method transfer the treatment to the other *dyskrasias*, as I said, unaware that there are more. Of course, the remedies of the wet *dyskrasias* existing alone are foods that are drying without heating and cooling strongly, and in addi-

tion to these, there is need of the customary drinks. For when the *dyskrasia* is conjoined with heat, there is the use of astringent foods and drinks. However, these must be astringent without being heating. Furthermore, cold water is suitable for them. The best cures of the wet *dyskrasia* conjoined with cold are all those things that are acrid. Also mix the astringents with them, obviously without what is cooling. And the best cure for these is a very small drink of any of the strongly heating wines. It is clear, too, that such [a wine] should not be new. All the other things applied externally must be analogous to the foods and drinks.

Since enough has been said about these matters, let me return again to the beginning of the discussion and summarize its chief points. These are that you must cool a hot *dyskrasia* but you must heat a cold *dyskrasia*. In like manner, you must dry a wet *dyskrasia* and wet a dry *dyskrasia*. If the *dyskrasia* has occurred by way of some conjunction, mix both indicators, drying and cooling the wet and hot, but drying and heating the wet and the cold. In the same way, too, you must wet and cool the dry and hot but wet and heat the dry and cold. You must be aware that the dry *dyskrasia* is the worst of all the simple *dyskrasias*, and the dry and simultaneously cold *dyskrasia* the worst of the compound *dyskrasias*. With this, let me end the matters concerning *dyskrasia* of the stomach, occurring apart from some excess moisture from without.

6. On fevers as hot *dyskrasias* (*MM*, Bk. 8, X.530–33K; J&H, 2.344–47).

GENERAL INTRODUCTION

Now would be an appropriate time to state that someone might also cure fevers by method in a similar way to other diseases that arise in us due to *dyskrasia*, which I went over in the preceding book. Thus, Eugenianus, that the nature of the simple and primary parts, which Aristotle calls *homoiomeres*,[23] consists in the mixing of hot, cold, wet, and dry with each other, was shown all through one treatise in which I gave them consideration: *On the Elements according to Hippocrates*.[24] That the differentiation of the simple and primary parts from each other lies in their being hotter, or colder, or wetter, or drier, or their having been affected by some conjunction of these [qualities] was shown in the treatise *On Krasias (On Temperaments)*. In the book *On the Differentiae of Diseases* it was shown that there are in all two classes of diseases of the simple parts of the animal. One, which is common to simple and compound, secondary and organic parts, I call "dissolution of union (continuity)." The other, which is selective and specific, consists of a *dyskrasia* of hot and cold, and wet and dry. However, among the *dyskrasias*, there are those that are simple, involving a single opposition, and those that are com-

---

[23] See Aristotle, *Parts of Animals* 2, 648a6–655b27, and *Meteorologica* 4, 10–13, 388a10–390b20.

[24] The works referred to in this opening statement are several of those which form the theoretical foundation of Galen's method of treatment. These are, in order, *De elementis secundum Hippocratem* (I.413–508K), *De temperamentis* (I.509–694K), *De morborum differentiis* (VI.831–80K), and *De causis morborum* (VII.1–41K).

li

pound, involving a double opposition. Thus, in the opposition of hot and cold, two simple *dyskrasias* exist: one, when something becomes hotter than it was to the extent that it is now damaging to its function, and a second, when cold prevails in a similar manner. In the opposition relating to dryness and wetness, there are, in turn, two *dyskrasias*; namely, moist and dry. When the simples combine with each other, four other *dyskrasias* co-exist: wet and cold, wet and hot, dry and hot, and dry and cold. And the causes which generate the *dyskrasias* were spoken of by way of another treatise in which I set out the causes of diseases.

Fever is also one of the diseases involving *dyskrasia* when the heat is raised to such a disproportionate degree that it distresses the person and harms function. I demonstrated that someone was not yet febrile if he does not yet have either of these conditions, even if he happens to have become much hotter than he was. However, since each of the abnormal things which exist in us, whether they obtain a specific or generic name, do not exist outside the body, but the three things, inflammation, pleurisy, and disease also occur simultaneously in one body—that of Dion should this be the case—each [body] demonstrates a specific indication of these and the *differentiae* are nothing less than what is most specific to all of them, it seemed better, on account of this, to deal separately with the *differentiae* of fevers so that, for all the things that are sure to occur, my discussion may be brought to completion here, apart from the refutation of those who have written badly on them.

GENERAL INTRODUCTION

De sanitate tuenda (*Hygiene*)

Considerable attention is given to *krasis* and *dyskrasia/eukrasia* in this work, particularly in the final two books (5 and 6). Chapters 1 and 2 of Book 5 are devoted to general considerations, whereas the remaining ten chapters are focused on the topic of *krasis* and *dyskrasia/eukrasia*. Chapters 3 to 10 are about hygiene for the aged, which, as Galen has pointed out in many places, is essentially a cold and dry *dyskrasia*. He recognizes the difficulty of maintaining health in the face of such a *krasis* and ponders the question of whether old age, as a state of cold and dry *dyskrasia*, should be regarded as a disease, a morbid state between health and disease, or an unstable state of health. Regardless of the answer to this question, it is a state or time of life when people readily become frankly diseased. Nonetheless, amelioration is possible to a degree, and a large part of these sections is devoted to possible measures, including foods and drinks, regimen generally, activities, customs and exercise, baths, and certain medications. In general, correction of the cold and dry *krasis*, as far as it is possible, is through heating and wetting agents of whatever sort. In Chapter 4 he details two interesting cases—Antiochus the doctor and Telephus the grammarian—who were to some extent able to keep the debilitating effects of the cold and dry *dyskrasia* of age at bay.

The final two chapters of Book 5 (11 and 12) are concerned with other *dyskrasias* compatible with health. Chapter 11 begins with a general statement that is particularly apposite:

> The regimen that corrects the *dyskrasia* in many instances extends further. This is because we are

seized by many similar diseases, the person hotter in nature easily becoming ill with hot diseases, the person colder in nature with cold diseases and similarly with the others. So too does each of the *dyskratic* bodies return more quickly to its own particular nature than to the best *krasis*. That which is most midway of all the *dyskrasias*, if it is seized by a disease of the same kind in nature, deviates more, whereas if it is not seized by one of the same kind, it deviates less. Therefore, one must not change the customs, even if they are bad, while bodies are still bad. The exception is those who are more perfectly healthy, but this must only be done when the person takes time away from civil matters, with the intention of being changed.

These things, then, are the common precepts of all the *krasias* that are changed individually to the extent that each *dyskrasia* is wanting in regard to the highest *eukrasia*—and it is no wonder that the *dyskratic* natures, situated in the middle between perfect health and perceptible disease, are corrected by the method of regimen. Nor is it any wonder that some people are obviously benefited and some harmed by the same things. For if all people had a similar constitution to each other, then it would be surprising for some to be benefitted by opposite things and some harmed by these things. However, since the constitutions of the bodies of many people are opposite, it is reasonable that benefit should occur to them from opposites. On which account also, someone might wonder at all doctors who attempt to write treatises on health

without differentiating the natures in the discussion. For just as it is impossible for cobblers to use one last for all people, so too is it for doctors to use a single beneficial kind of life. Because of this, they say it is most healthy for some to exercise excessively every day, whereas for others, there is nothing to prevent them passing their lives wholly in idleness. Also, for some it seems to be most healthy to bathe, whereas for others it does not. And to drink water and wine, and about the other things similarly, not only for hygienic regimes but also for the cures prescribed for diseases. And they write things completely opposite to one another, so it is rare to find one point of agreement among them all. In fact, experience shows that some people are harmed and some are benefited by the same things and similarly with opposites. Anyway, I know of some who immediately become sick, if they remain three days without exercise, and others who continue indefinitely without exercise and yet are healthy; and of some who never bathe, and others who, if they don't bathe, immediately become febrile, like Premigenes the Mytilenean.[25] (VI.362–65K; JHyg., 86–89)

Chapter 12 is a general review of the appropriate measures for preserving health in the different natures/*krasias* and considers also the kinds of disorder that are likely to affect the particular natures/*krasias*.

[25] Apart from the fact that he was a Peripatetic philosopher, I have been unable to find anything of substance about Premigenes in the usual sources. He is mentioned here in the *Hygiene*, but there is no other listing for him in Ackermann's index.

GENERAL INTRODUCTION

Whereas in the first five books there was the assumption of the best constitution, in the sixth, and final, book attention turns to constitutions that fall short of the best and situations in which a person is unable, for one reason or another, to devote sufficient time to the maintenance of health. In Chapter 2, Galen writes:

> The bad (abnormal) constitutions of bodies are twofold in class. Some have the elemental and primary parts of the body, which Aristotle calls *homoiomeres*,[26] mixed regularly while some have them mixed irregularly. I say "regularly" when all the parts of the body are similarly deviated to some *dyskrasia*, being colder, hotter, drier, or wetter than is appropriate or there is a conjunction of colder and wetter, or hotter and drier, or hotter and wetter, or colder and drier. I say "irregularly" when they are not all deviated together, but some are hotter, some colder, some drier, or some wetter; and further, when there is a conjunction, with some being wetter and colder, some drier and hotter, some wetter and hotter and some colder and drier. Furthermore, in respect of the composition of the organic parts, some are compounded regularly and some irregularly.
>
> What I shall now state first is what kinds of constitutions of bodies are the most morbid, just as I showed before with regard to the healthy consti-

[26] Aristotle used this term in relation to both inanimate (*Meteorologica* 388a10–390b20) and animate (*Parts of Animals* 648a6–655b7) things. For a summary of the use of the term up to Galen, see Johnston, *Galen: On Diseases and Symptoms*, 13 and 45.

tution.[27] But the healthy constitution is one single thing, for the best in every class of thing is one, whereas the bad are obviously very many. The generic difference in them is twofold, as I said just now, some having all the parts in a similar *dyskrasia*, and some in a different *dyskrasia*. Of those having the parts in a similar *dyskrasia*, it is clear that the worst are those having severe *dyskrasias*, and particularly the simultaneously cold and dry. It is, however, not easy to enumerate the varieties of the irregular constitutions of bodies, although one might resolve those varieties into two kinds or classes, or whatever one might wish to call them. Thus, the most morbid of these are the ones in which the most important parts are involved in opposing *krasias*, whereas the more moderate are those in which the non-important parts are so disposed. I have already seen some with a cold abdomen but a hot head, just as I have seen others in turn who have a hot abdomen in addition to a cold head. And I saw a stomach which was not hot in nature incidentally and not primarily, nor by the specific reason of the *krasis*, continuously charged with bile, just as I saw another cooled, although not cold by nature. In the same way too, I have sometimes seen a head, liver, and spleen, and some other part, having the benefit of a different *dyskrasia*, although not itself being

---

[27] See Galen's *De optima corporis nostri constitutione*, IV.737–49K. English translations by R. J. Penella and T. S. Hall, *Bulletin of the History of Medicine* 47 (1973); Singer, *Galen: Selected Works*; and also Appendix 1 of the present work.

harmed in nature in respect of the prevailing *dyskrasia*, or being in a state opposed to this. (VI.384–86K; JHyg., 116–21)

He begins his detailed discussion of hygienic measures in the uniform *dyskrasias*, running through the various possible forms, both simple and compound (Chapters 3–8). Then, in Chapter 9, he begins as follows:

Now it is time to pass on to the irregular (non-uniform, anomalous) constitutions of bodies—those which, in fact, are also altogether morbid. Irregular constitutions exist in three forms because the composition of our body is also threefold.[28] First, it is from the primary elements, from which what Aristotle called *homoiomeres* have arisen. Second, it is from these same *homoiomeres*, which are in turn the actual perceptible elements of the *anhomoiomeric* parts, from which the compounding of the organs arises. Third, in addition to these, there is the compounding of the whole body, which is from the organs. What I have termed the third is easier to understand, in terms of being both diagnosed and cared for. The second is more difficult and the first very difficult. It is better, then, to begin from the third, which is easier in respect of diagnosis and care. For example, directly in the case of the head (it is no bad thing to start here), when it becomes *dyskratic* in nature, generates many superfluities

---

[28] For an outline of Galen's tripartite division of bodily structure as a basis for the classification of diseases and symptoms, see Johnston, *Galen: On Diseases and Symptoms*, 69–72.

from which harms occur to whichever of all the organs lying below [the head] some of the superfluity is diverted. The easiest passage is to the mouth and nose, but there is also passage to the eyes, and in some cases to the ears. The esophagus and the rough artery (people also call this "bronchus") receive below the passage of the superfluities to the mouth. The upper end of the rough artery (which is called the "larynx") that is joined to the mouth is the organ of voice, as I have shown in my work, *On the Voice*.[29] This then, the larynx itself, when it is moistened by the things flowing down to it from the head, at first makes the voice hoarse, and as time progresses, small. If the badness comes down to it even more, it is destroyed altogether because the rough artery (trachea and main bronchi) is wet through completely, along with the larynx. (VI.419–21K; JHyg., 170–73)

The topic of fluxes is then continued for the next three chapters (10–13) before the whole work concludes with two chapters on matters unrelated to *dyskrasia*. What can be said about these final two books of Galen's *Hygiene* is that they deal with the preservation of health in people who do not have the best constitution of the body or who cannot follow the ideal regimen. Galen considers in succession the various *dyskrasias* when they fall short of be-

[29] The work *De voce* (Περὶ φωνῆς) was in four books and dedicated to Boethus. The original has been lost, although some fragments and an Arabic summary remain. See V. Boudon, *Galien: Exhortation á l'étude de la medicine, Art medical* (Paris, 2002), 419n3.

ing diseases—that is, when they still permit normal function. Both uniform and non-uniform *dyskrasias* are covered. Two important causes are *kakochymia* and *plethora*. Basic general measures in management include diet, bathing, massage, moderate exercise, and moderation in sexual activity. More specific measures are downward purging and phlebotomy.

4. A SUMMARY OF GALEN'S VIEWS

1. All things, animate and inanimate, at the "microscopic" level are composed of the four elemental qualities—hot and cold, wet and dry—considered as two antitheses.

2. Human and animal bodies, at the "macroscopic" level, are composed of *homoiomeres* and organic structures, the latter being themselves compounded from *homoiomerous* structures.

3. The body as a whole, representing a third level of organization, contains in addition to these *homoiomeres* and organic structures, liquid material in the form of the four humors (blood, phlegm, yellow bile, and black bile) and gaseous material in the form of *pneuma*.

4. The ideal state that allows all bodily functions to be performed normally is a balanced blending of the four elemental qualities in all *homoiomerous* structures (*eukrasia*), organic structures normal in terms of conformation, magnitude, number, and arrangement, and the right state of the four humors (i.e., *eukrasia, eusarkia*, and *euchymia*).

5. Balance of the four elemental qualities in *homoiomerous* structures (*eukrasia*) can be altered adversely in

GENERAL INTRODUCTION

two ways: (a) change involving each of the individual elemental qualities occurring in themselves from either internal or external causes; (b) change caused by an excessive or abnormal inflow of material (a flux), that is, specifically, one or more of the four humors, or *pneuma*, disturbing the balance.

6. A disproportion in the four elemental qualities can involve a single quality or two qualities in combination, one from each antithesis. Such a disproportion results in a *dyskrasia*, of which there are eight kinds—four simple mono-*dyskrasias* (hot, cold, wet, and dry) and four compound bi-*dyskrasias* (hot and wet, hot and dry, cold and wet, cold and dry).

7. *Dyskrasias* are one of three classes of disease, the other two being abnormality of the organic structures in one of the four ways listed in point 4, and dissolution of continuity.

8. *Dyskrasias* occur in *homoiomeres* either alone, as part of an organic structure, or as involved in dissolution of continuity.

9. Recognition of a *dyskrasia* may be through touch, vision, or rational inference.

10. A *dyskrasia* may be "healthy" or "morbid" depending on whether function is impaired or not. In the triad, *eukrasia*, healthy *dyskrasia*, morbid *dyskrasia*, there is a range in each component and indistinct boundaries between them. Old age is an example. This is associated with a cold and dry *dyskrasia* that may or may not be attended by normal function. Galen asks the question—should old age be regarded as a disease?

11. The aim in treating a *dyskrasia* is restoration of a

normal balance (*eukrasia*). Broadly speaking, this is effected by foods and medications employed according to allopathic principles, elimination or redirection of abnormal fluxes, and adjustment of environmental factors.

12. *Dyskrasias* may be uniform in terms of either the whole body (systemic) or a part thereof (focal). Different *dyskrasias* may coexist in the same body (non-uniform *dyskrasia*).

13. The rational soul, considered as a function of the brain, is considered to be like any other part of the body in the case of *krasis*.

14. The ideal is *eukrasia*, identified by the lack of any signs of a preponderance of one or more of the elemental qualities, and the existence of demonstrably normal function in a well-proportioned body of normal outward appearance appropriate for age. The reality is often *dyskrasia*, which is a greater or a lesser departure from *eukrasia* that may or may not impair function and may or may not require some form of treatment or management.

5. THREE QUESTIONS, THEORETICAL AND PRACTICAL

In Galen's theory of *krasis* and its variants (*eukrasia* and *dyskrasia*) as a foundational part of his theory of bodily structure and medical practice—a theory that dominated Western and Near Eastern/Arabian medicine over many centuries—there are, as I see it, three specific problems.

1. How does the continuum theory, based on the four elemental qualities, hold up as an account of bodily structure?

2. What are, and what are the roles of, the four humors (blood, phlegm, yellow bile, and black bile) and of *pneuma*?
3. If *dyskrasias* do constitute a major class of diseases, how reliable is their recognition, given the considerable limitations of evaluative methods?

On the first issue, let us consider Galen's overall picture of bodily structure. He identified three levels of organization of solid material: *homoiomerous* tissue, organic structures, and the whole body. Variably present in these three levels were liquid material (the four humors) and gaseous material (*pneuma*). The solid material in the *homoiomerous* form is, according to Galen in *De temperamentis* and elsewhere, a unique blending of the four elemental qualities (hot, cold, wet, and dry). The first question is how qualities can account for structure. What is the ontological status of the four qualities in Galen's scheme? Now, issues pertaining to substance/essence and attributes/accidents, and the realist/nominalist debate on universals (which the four qualities could be construed as), have exercised Western philosophers for many centuries without a definitive conclusion being reached. In fact, Galen's two most revered philosophical predecessors, Plato and Aristotle, quoted multiple times in these works, were both realists on the matter of universals, although they differed from each other, the former taking universals to be *ante rem*, and the latter taking them to be *in re*. Galen seems to have been aware of the first issue. In the first two works translated, he writes as if the four elemental qualities themselves can account for structure, but in the third (*The Soul's Traits Depend on Bodily Temperament*) he

invokes a qualityless substance or matter in which the qualities inhere.[30] If we extend his identification of *homoiomerous* tissue to include the parenchyma of all organs as a variant of flesh, a position that he seems to take in Book 11 of his *Method of Medicine*, also written late in his career,[31] we can accept the view that all the solid tissue

[30] In the third of the treatises translated, which is dated perhaps two decades later than the works on *krasias* and non-uniform *dyskrasias*, he signals a change in his position, recognizing a substance in which the elemental (and other) qualities inhere. He writes, "Each of the internal organs has a characteristic substance; what exactly this is, let us not yet inquire into. Rather, let us call to mind the common substance of all bodies as was shown by us to be compounded from two principles—matter and form. Matter is without quality conceptually but has in it a *krasis* of four qualities, which are hotness, coldness, dryness, and wetness. From these also, copper, iron, and gold or flesh, sinews (nerves), cartilage and fat, and in short all the things termed *prōtogona* by Plato and *homoiomeres* by Aristotle have come into being" (*The Soul's Traits Depend on Bodily Temperament*, 773K).

[31] "At any rate, nothing prevents you from calling this, for didactic purposes, a 'natural flesh' or also, by Zeus, the 'flesh of this particular part,' there being one for the stomach, and another for the liver, just as there is also one for an artery and one for a muscle. It is only called 'flesh' in muscles; with very few exceptions no one calls any of the others 'flesh.' But the [fleshes] in the viscera, such as the liver, kidneys, spleen and lung, they call parenchymas. That in the intestines, stomach, esophagus and uterus, they leave without a name. But give no thought to the names. Know, however, that the bulk of the substance of each of the parts is mostly filled up by such a basic substance which allows of destruction and generation, as can be seen clearly in hollow wounds and ulcers" (*Method of Medicine*, 11, X.730–31K).

of the body has essential qualities pertaining to temperature, moisture content, hardness, and others, which can exist over a range having a median point and are capable of continuous variation. In Galen's system, the critical qualities are temperature and moisture content. Deviations of these from the median are what result in *dyskrasias*, which can exist with or without causing disease. The determinant of the latter is the function of the part or parts involved. It would also be possible for these deviations to coexist in different parts simultaneously or sequentially. This, in essence, is his model of diseases that do not fall into one of his other two categories of disease—disorders of organic structure and dissolution of continuity, both of which can also coexist with *dyskrasia* in the involved organ or part. Whether he speaks of elemental qualities *per se* or primary qualities in an underlying substance of unknown composition, *dyskrasias* constitute one large class of diseases. They could be caused primarily by internal or external factors, as listed in the table under section 3 above, or by liquid or gaseous material flowing into them and altering one or both of the ranges of primary qualities. They could be treated by choosing appropriate external conditions, regimen generally, other specific measures such as exercise and massage, and by foods and medications employed predominantly on allopathic principles.

The second and related question pertaining to structure is the nature of the four humors—blood, phlegm, yellow bile, and black bile. Galen certainly appears to differ from Hippocrates and Aristotle, his acknowledged predecessors in continuum theory, in the structural relevance of these fluids. All are taken to be essentially liquid, and the two that remain clearly fundamental to theories of

bodily function today—blood and yellow bile—are indeed liquid in normal circumstances, but have no perceptible structural role. The other two—phlegm and black bile—are more elusive. Everyone is, of course, familiar with phlegm, presumably the same phlegm Galen speaks of. A present-day definition would describe it as a viscous fluid produced in the airways that is to a greater or lesser extent pathological. Charles Richet, the noted French physiologist, in 1910 wrote of Galen's phlegm: "This strange liquid, which is the cause of tumors, of chlorosis, of rheumatism, and *cacochymia*—where is it? Who will ever see it? Who has ever seen it? What can we say of this fanciful classification of humors into four groups, of which two are absolutely imaginary?"[32]

Black bile is even more of a problem. Galen devotes a whole short treatise to black bile[33] without clarifying what it actually is and what role it plays in normal physiology, if any. My own interpretation is that phlegm is indeed a product of a pathological respiratory tract, while the black bile described by the ancients perhaps refers to the dark material (i.e., altered blood) found in melaena stools and coffee-ground vomitus. An internal role for these putative humors is posited on this scanty visible evidence. On black bile, Nutton writes:

> Only with *On the Nature of Man*, the text which Galen and subsequent generations believed was quintessentially Hippocratic, does black bile be-

[32] Charles Richet, "An Address on Ancient Humorism and Modern Humorism," *British Medical Journal* 2 (1910): 921–26.
[33] *De atra bile*, V.104–148K; English translation by Grant, *Galen on Food and Diet*.

come an essential humor. By contrast with beneficent blood, black bile was mainly harmful—it was visible in vomit and excreta and later authors described how it hissed and bubbled on reaching the ground, burning up whatever it touched. Modern scholars disagree on what black bile actually was (perhaps some form of dried blood), but once proposed as a humor it fitted neatly into a rational scheme made even more credible by the ease with which it could be extended to cover a whole range of circumstances.[34]

Galen himself says black bile is "similar to thick dark blood which can be seen when it is secreted during either vomiting or diarrhea, but this fluid does not congeal."[35] Certainly he does see a role for both biles in visible pathology, yellow bile being responsible for erysipelas among other things and black bile for carbuncles and cancer *inter alia*.

The third question is, then, how reliably could disturbances of the *krasis* of the body as a whole and of different parts individually (uniform and non-uniform *dyskrasias*) be recognized? Diagnostic methods were limited. There was vision and there was touch; there was evaluation of the materials emanating from the patient, and there was evaluation of functions—for example, of the pulse for cardiac function and the respiration for pulmonary function. Evaluation of some parts/organs was more problematical than others due to the limited understanding of their func-

---

[34] Vivian Nutton in *The Western Medical Tradition*, 24–25.
[35] Grant, *Galen on Food and Diet*, 20.

tion. The signs of a hot *dyskrasia* of the liver, as listed by Paul of Aegina, were as follows:

> The marks of a hot liver are, largeness of the veins, redundancy of yellow bile, and in manhood, of black; the blood hotter than natural, and by means of it the whole body, unless the parts about the heart antagonise; and thick hairs over the hypochondriac regions and over the stomach.[36]

One of the issues in diagnosis of a *dyskrasia* was a conflation of signs presumed to be from the immediate and current dynamic balance of the qualities and those established due to age, location, past habits, race, and the like. For example, each of the stages of life was associated with a specific *dyskrasia*—youth with hot and wet, old age with cold and dry, and so on.

Working on the information gathered from the patient's history and the rather limited physical examination, a decision had to be made on treatment—whether to treat and how to treat. The latter is based on allopathic principles articulated particularly by Hippocrates. On the first question, Galen himself raises the interesting issue of "perpetual affection" ($ἀειπάθεια$) in *De temperamentis* and elsewhere, although he does not develop it. I take it to mean that perhaps *eukrasia* is an ideal state never or only rarely to be truly realized, given the age-related variations of *krasis*, the presence in the body of the four humors, two of which at least are intrinsically *dyskratic*, and the possible presence of one or more of a variety of factors that can adversely affect *krasis*. Certainly in his *Hygiene*,

---

[36] As translated by F. Adams, *The Seven Books of Paulus Aegineta* (London, 1844), 1.98.

he gives considerable attention to what he calls "healthy *dyskrasias*," as opposed to the morbid variety. In fact, the essence of that work is devising a regimen that keeps *dyskrasia* at bay, or serves to minimize its adverse affects. On the second question, foods, drinks, medications, and various activities (bathing, exercise, massage, etc.) can be used as treatment, according to their perceived or theoretical ability to affect the temperature range of the whole body, or an individual affected part. Empirical evaluation of particular substances or measures could lead to a particular treatment regime in a particular situation. To a significant extent both diagnosis and treatment were art rather than precise science—and this is still the case!

This, then, is the paradigm, formalized and documented particularly by Galen, and especially in works such as those here translated, that applied to the practice of medicine in the West and the Near East for the best part of two millennia. Whatever its theoretical shortcomings, it must have worked to some significant extent. There were, of course, no statistical studies until recent times, and there was substantial imprecision in diagnosis, so there is no way of knowing how effective individual treatments were in particular conditions based on these methods. What can be said is that the paradigm lasted for an awfully long time, withstanding a number of challenges of the sort Thomas Kuhn describes in relation to the overthrow of established scientific paradigms![37]

[37] It is of interest to consider what might be called the "Galenic paradigm" in the light of Kuhn's highly influential study of how and why scientific paradigms change; see T. S. Kuhn, *The Structure of Scientific Revolutions* (Chicago, 1962). The *50th Anniversary Edition* (2012) contains additional material, including aspects of the controversy surrounding the term "paradigm."

GENERAL INTRODUCTION

6. AFTER GALEN

Of Galen's many writings, *De temperamentis* and its "satellite" *De inaequali intemperie*, must be numbered among the most enduringly influential. Well into the modern period, when the rest of his works had become largely of historic interest only, books and articles on the subject of temperaments continued to be written, and indeed are still being written today. However, while the temperaments are still being discussed and used in health-related matters, and their attribution as a concept to Hippocrates and Galen is still recognized, the views currently being expressed perhaps owe more to the former than to the latter. As the following excerpt from Hippocrates' *Nature of Man* attests, it is the bodily humors rather than the elemental qualities that are the focus, being used to characterize phenotypes.

> The human body has in itself blood, phlegm, yellow bile and black bile, and these are the nature of the body for a person; it is through these he feels pain and enjoys health, when he has these in a moderation of *krasis* in relation to each other, and of capacity and amount, and they are most completely mixed. On the other hand he feels pain when one of these is deficient, or in excess, or is isolated in the body and not compounded with all the others. For of necessity, when one of these is isolated and stands by itself, not only does the place it left become diseased, but also it produces pain and distress in the place that is overfilled. And truly, when one of those that is overfilling flows out of the body, the emptying (evacuation) produces pain. Conversely, if the

evacuation is inward, and there is transfer and separation from the others, then inevitably to a greater degree in the person, it produces a twofold pain, according to what has been said. In the one case, there is pain in the place left, and in the other, in the place overfilled.[38]

The essence of the Hippocratic view, as stated here, is that the human body is compounded from a mixture of the four humors. There is not a single element or a single humor that is the foundation. Although in this work there is discussion of the four elements (fire, water, earth, and air) and the four elemental qualities (hot, cold, wet, and dry), these are not considered to be the fundamental constituents of the body except insofar as they may be qualities of the humors, which is certainly what Galen takes them to be—that is, qualities inherent in the nondifferentiated (*homoiomerous*) tissues that constitute the body. There is, then, a significant difference between the views of the two men, although they are frequently mentioned together as originators of the enduring concept of humors and temperaments as applied to medicine.

Be that as it may, it is primarily Galen's writings on temperaments that were preserved in Oribasius' *Synopsis*, and subsequently, through Oribasius, by Aëtius of Amida (sixth century) and Paul of Aegina (seventh century). Galen's key works on the subject were incorporated into the *Summaria Alexandrinorum*, which was established in the

---

[38] Hippocrates, *The Nature of Man* 4, LCL 150 (W. H. S. Jones), 10–13. The work is thought to be later than Hippocrates and compiled by Polybius; see Craik, *The Hippocratic Corpus*, 207ff., on the question of authorship.

GENERAL INTRODUCTION

later part of the first millennium AD as the basis of medical teaching. The opening paragraph of Paul's group of sections on temperaments reads as follows:

> That man is in the best temperament of the body when he is in a medium between all extremes, of leanness and obesity, of softness and hardness, of heat and cold, of moisture and dryness; and in a word, who has all the natural and vital energies in a faultless state. His hair also should be neither thick nor thin, and in colour neither black nor white. When a boy, his locks should be rather tawny than black, but, when an adult, the contrary.[39]

There then follow nine sections on the diagnosis of the temperaments that largely reproduce the material in Galen's *Ars medica* via Oribasius' *Synopsis*. Finally, there are three sections on correction of the four simple (mono-) *dyskrasias* (distemperments). Adams' concluding statement reads as follows:

> The modern ideas respecting the temperaments appear to be founded upon the descriptions given by the Arabians of the symptoms which characterise the prevalence of the four humors, as they were called, namely, blood, yellow bile, black bile and phlegm in the body. In proof of this we subjoin

---

[39] Adams, *The Seven Books of Paulus Aegineta*, 1.84 (translation after Adams). This statement goes beyond Galen's basic claim that *eukrasia* is a truly proportional blend of the four elemental qualities, including as it does what might be termed the visible and palpable consequences of *eukrasia*.

Rhases' brief description of them: 'De complexione autem infirmi scias, quod si fuerit albi coloris admixti rubedini, et si fuerit bonae carnis, subtilis cutis, et quando locum fricaveris, rubescit statim, significatur quod materia est sanguinea. Et si corpus fuerit pingue, et albi coloris non mixti rubedine, et corpus nudum pilis, venis strictis, carne molli, occultarum junctuarum, gracilium ossium, et generaliter talis dispositio qualis est in corporibus mulierum, significatur quod materia est phlegmatica. Et si fuerit macrum, citrini coloris, pilosum, et cum crassis venis, et manifestis juncturis, ostendit quod materia est cholerica. Et si fuerit niger color, durities corporis, pilositas, asperitas cutis, significatur quod materia est melancholica.'

It is to be understood, however, that Galen's system of the temperaments was not based, as has been often erroneously represented, upon any hypothesis respecting the humors.[40]

During the later part of the first millennium AD there were translations of the three treatises included in the present volume into Arabic (Hunayn ibn Ishaq for *De temperamentis* and *De inaequali intemperie*, Hubays for *Quod anima mores*); into Syriac (Sergius of Resaina for *De temperamentis*, Ayyub al-Rahawi for *De inaequali intemperie*); and into Hebrew (Simson ben Salomo for *De temperamentis*). Arabian medicine certainly embraced Galen's theory of temperaments, and there was clear awareness of the four qualities, as the following passage shows:

[40] Adams, *The Seven Books of Paulus Aegineta*, 1.105–6.

GENERAL INTRODUCTION

The general principles which constitute the basis of Arabian medicine are the outcome of these conceptions, and the opening chapters of every great systematic work on the subject deal largely with the "Temperaments" or "Complexions" (*Mizáj*), the Natural Properties (*Tabáyi*), and the Humors (*Akláhi*). *Mizáj*, which is still the common word for health in Arabic, Persian, and Turkish, is derived from the root meaning "to mix" and indicates a state of equilibrium between the four Natural Properties or the four Humors; while if this equilibrium is upset by the preponderance of one of the Natural Properties or the Humors, a disturbance entitled "Deflection of the temperamental equilibrium" is produced. But even the normal healthy *Mizáj* is not practically a constant quality, each region, season, age, individual and organ having its own special and appropriate type. Nine types of Complexion are recognized, namely the equable, which is practically non-existent; the four simple complexions, hot, cold, dry, and moist; and the four compounds, namely the hot and dry, the hot and moist, the cold and dry, and the cold and moist. Excluding the rare case of a perfect equilibrium, every individual will be either of the Bilious Complexion, which is hot and dry, the Atrabilious or Melancholic, which is cold and dry; the Phlegmatic which is cold and moist; or the Sanguine, which is hot and moist.[41]

Initial Latin translations were done by the notable Galen translators of the twelfth to the fourteenth centuries:

[41] Browne, *Arabian Medicine*, 119.

of *De temperamentis* by Gerard of Cremona and Burgundio of Pisa, of *De inaequali intemperie* by Niccoló da Reggio and Pietro d'Abano, and of *Quod animi mores corporis temperamenta sequantur* by Niccoló da Reggio. The late fifteenth century saw the start of a veritable explosion of Galen translations coincident with the start of printing. Durling's invaluable list of Renaissance translations of Galen lists only two for *De temperamentis*; one by Thomas Linacre first published in 1521 and accompanied by his translation of *De inaequali intemperie*, and a partial translation by Leonhart Fuchs. There are, however, seventeen editions of Linacre's work listed between 1521 and 1596. Seven Latin translations of *De inaequali intemperie* alone are listed, and for *Quod animi mores corporis temperamenta sequantur* there are five Latin translations and one French. Of particular interest is a later (1881) edition of Latin translations of the two works with a long introduction by J. F. Payne, in the course of which he writes:

> What his (Galen's) general principles were is shown very clearly in the work now reprinted, which is rather physiological or physical than strictly medical. In it we find developed the theory of humors and temperaments, which formed the physiological basis of Galen's system of medicine, and which, conveyed through many popular medical works to the lay public, entered largely into the current philosophy of the time. Hence Linacre speaks of this work as not less necessary to philosophers than to physicians. Some knowledge of these ideas is indispensible for understanding many allusions and metaphors in English writers of the Elizabethan age. Nay

more, a great part of it has passed into our common language. Such words as humor in its many acceptions, and many compounds, temperament, temper, choler, melancholy and others derived all their original significance from the place which they held in the Galenical system. It is perhaps not too much to suppose that this very version may have been among the sources whence such writers as Elyot (who was a pupil of Linacre) in his *Castel of Helth*, Bright, the predecessor of Burton, in his *Treatise of Melancholie*, and later Walkington in the fantastic book called *The Optick Glass of Humours* obtained the ideas, which popularised by them, became the common property of scholars and literary men.[42]

By the sixteenth century, influenced by outbreaks of widespread diseases, apparently contagious, like plague and syphilis, a certain dissatisfaction arose, at least in some doctors, at the perceived inadequacy of Galen's concept of *dyskrasia*. There was also awareness of the difficulty of diagnosis in evaluating Galenic *dyskrasia*. This is, of course, a whole subject in itself succinctly addressed in Chapter 6 of the *Western Medical Tradition* (WMT).[43] Notable contributors on these matters were Girilamo Fracastoro (1478–1553), Jean Fernel (1497–1558), Giambatista da Monte (1498–1552), Johannes Argentarius (1512–1572), and Sanctorius Sanctorius (1561–1636), the las

---

[42] Payne, Introduction to *Galeni Pergamensis, De Tempera mentis et De Inaequali Intemperie*, 44–45.

[43] See pp. 260–64.

GENERAL INTRODUCTION

developing a thermometer to at least provide quantitative evidence of one of the two Galenic antitheses.

The subsequent three centuries (the seventeeth to the nineteenth) saw the major discoveries that finally sounded the death knell of the Galenic concept of *krasis/dyskrasia* of the four elemental qualities in the causation of disease:

The discovery of the circulation of the blood by Harvey (17th century).

The development of the microscope, which allowed the recognition of bacteria (van Leeuwenhoek, 17th century) and of cellular structure (Hooke, 17th century).

The first steps in the triumph of the theory of atoms and chemical elements (Dalton, early 19th century).

The progressive increase in anatomical and physiological knowledge generally.

Nonetheless, the theory of temperaments held tenaciously to a position of some medical significance, albeit increasingly peripheral. An example is the book by the Edinburgh-trained doctor Alexander Stewart, which went into a second edition in 1892. The author focused on the four humors, the preponderance of any one of which was said to be recognizable in outward and visible signs. He did, however, rename the two problematical humors of Hippocrates and Galen, black bile and phlegm, for which neither the ancients nor their successors could find any role in normal physiology.[44] Stewart listed, then, sanguine, bilious, lymphatic, and nervous. Brief mention may also

---

[44] Stewart, *Our Temperaments*, 1892.

be made of a 1947 paper by J. R. Irwin in which he revisits Galen's theory of temperaments and considers some modern articulations of this.[45] Below is a table titled "The Humoral Doctrine of Galen Modified by Irwin," reproduced from his paper. Finally, as previously mentioned, books relating to health and concentrating on the four temperaments are still being produced, although they now fall within the self-help genre.

Element	Air	Earth	Fire	Water
Humor	Yellow Bile (choler)	Black Bile (melancholy)	Blood (heart)	Phlegm/Mucus (coryza)
Temperament	Choleric	Melancholic	Sanguine	Phlegmatic
Quality	Warm-dry	Cold-dry	Warm-moist	Cold-moist
Intensity	Quick-strong	Slow-strong	Quick-weak	Slow-weak

7. TERMINOLOGY

Δύναμις/ἐνέργεια/ἔργον (capacity/function/action): The first of this triad of terms is the most problematical. In the translated treatises, it is used in two distinct senses corresponding to meanings II and IV in LSJ. That is, in the first instance it is the power, ability, capacity, or faculty to effect something, while in the second instance it forms a pair with ἐνέργεια to represent potentiality in contrast to actuality. The actuality is then the potential function in action. In several places Galen himself attempts to clarify the usage of these terms, as follows:

[45] Irwin, "Galen on the Temperaments," 1947.

That is to say, I shall call the action (ἔργον) what has been already brought about and "filled up" by the function (ἐνέργεια) of these; for example, the blood, flesh, and nerve. I term the active movement the function and the capacity (δύναμις) the cause of this.[46]

The so-called *haematopoetic* capacity in the veins and every other capacity are in the category of relative concepts, for they are primarily a cause of the function, and already of the action *per accidens*. But if the cause is relative to something, being alone the cause of this but of nothing else, it is also clear that the capacity is in the category of relative, and whilever we remain ignorant of the true essence of the cause of the functioning, we shall term it a capacity, saying there is a certain *haematopoetic* capacity in the veins, and similarly a digestive capacity in the stomach, and a pulsatile one in the heart, and in each of the others, a specific capacity of the function in relation to the part.[47]

However, in this, many of the philosophers are immediately confused, having an incorrect concept of "the capacity." For just as if some matter were dwelling in the substances, as we dwell in houses, they seem to me to have a mental picture of the capacities, not realising that for each of the things that come about, there is some effecting cause con-

[46] *De facultatibus naturalibus*, I.2.
[47] *De facultatibus naturalibus*, I.4.

ceived of in relation to something, and of this cause, as of any such matter, there is a specific name pertaining to this itself, while in the state (*schesis*), in relation to what comes about from itself, there is a capacity of what comes about, and because of this, we say the substance has as many capacities as it has functions.[48]

Aristotle, whose influence on Galen is especially noticeable in these three passages, considers "capacity" in detail in *Metaphysics* 5.12. It is defined there in the first of the three meanings as follows:

> Capacity then is the source, in general, of a change or movement in another thing or in the same thing *qua* other, and also the source of a thing's being moved by another thing or by itself *qua* other.[49]

I have chosen to translate ἐνέργεια consistently as "function" in accordance with one of the definitions given in LSJ as "physiological function." The reference is to *De sanitate tuenda*, where Galen writes:

> Certainly one must not, therefore, determine those who are healthy and those who are diseased simply by strength or weakness of functions, but one must attribute "in accordance with nature" (*kata phusin*) to those who are healthy in contrast to "contrary to nature" (*para phusin*) to those who are diseased; that is, for the former to be a healthy condition (*di-*

---

[48] *Quod animi mores corporis temperamenta sequantur*, IV.769K.

[49] *Metaphysics* 1019a18–21; translation after Ross in J. Barnes, ed., *The Complete Works of Aristotle* (Princeton, 1984), 2.1609.

GENERAL INTRODUCTION

athesis) in accordance with nature effecting functions, and for the latter to be a diseased condition contrary to nature, harming functions.[50] (VI.21K)

I have distinguished ἔργον from ἐνέργεια by rendering the former "action" and the latter "function." But Galen himself makes the specific point that the two terms are essentially interchangeable when he writes in *De methodo medendi*, in relation to the eye: "For it is agreed then, in this case, by all men, not only by doctors but also by those they meet, that it is its [the eye's] action (ἔργον) to see. And whether I say 'action' (ἔργον) or 'function' (ἐνέργεια) certainly makes no difference now in this case" (X.43K). A similar indifference is displayed by Linacre in his sixteenth-century Latin translation where, within the space of two sentences, he uses *actio* and *functio* interchangeably for ἐνέργεια. Galen himself does define ἐνέργεια as follows in *De methodo medendi*: "Vision is the function of the eyes, speech of the tongue, walking of the legs. Again, the function is this active movement and the movement of these things is a change of what there was before. The active movement is that which is from the thing itself, whereas the passive at any rate is from something external. For example, flying is the function of what flies and walking of what walks" (X.45–46K).

Thus, ἔργον becomes the "action" carried out by the "function," although Galen's own remarks on the equivalence of ἐνέργεια and ἔργον quoted above should be borne in mind. Nevertheless, my own understanding of the use of the three terms in the first sense mentioned is essentially identical with that of Brock in his introduction

[50] *De sanitate tuenda*, VI.21K.

to *De facultatibus naturalibus*.[51] As regards the second use referred to at the outset (LSJ IV), I have chosen "in potentiality/potentially" for δύναμις, and "in actuality/actually" for ἐνέργεια. Aristotle discusses the distinction particularly in *Metaphysics* 9.5–6.

Εὔσαρκος/πολύσαρκος (well-fleshed/excessively fleshy): εὔσαρκος is defined in LSJ as "well-fleshed, in good condition," while πολύσαρκος is defined as "very fleshy" with reference to Hippocrates, *Aphorisms* 4.7, and Aristotle, *History of Animals* 583a7. Galen, in *De temperamentis*, seems to use the former term as the median point of a range between *polysarkia* and severe emaciation. It becomes, then, a feature of *eukrasia* and is exemplified by Polyclitus' spearbearer. In *De temperamentis*, he writes:

> Therefore, those bodily states that are *eukratic* by nature and labor in due measure, become of necessity well-fleshed; that is to say, are well-balanced in every way. Those where the wetness is sufficient while the hotness is lacking of the peak balance, but not by much, become very fleshy. The very fleshy and the *eukratic* in nature live easily and idly. For certainly this was said best by the ancients—that the

[51] See Brock, *Galen on the Natural Faculties*, xxix–xxxi. There he writes: "Any of the operations of the living part may be looked on in three ways, either [a] as a δύναμις, faculty, potentiality; [b] as an ἐνεργεία, which is the δύναμις in operation; or [c] as an ἔργον, the product or effect of the ἐνεργεία." He then continues his analysis by means of a comparison with some concepts advanced by Bergson in his then very influential work, *L'Évolution Creatrice*; see note 1, pages xxx–xxxi.

customs are acquired natures—and perhaps, having said this now once and for all, it will require no further distinguishing and does not require further distinction in relation to each heading, whether this particular person has become colder due to nature, or due to habit, but to leave this to the readers, while for the sake of brevity, I myself recount the characteristic states of the body in each of the *krasias*. Certainly there are some who are thin and have at the same time small veins, but if you cut any one of these whatsoever, fat falls out which has clearly grown under the skin in relation to the internal membrane. Such a thing is found rarely, then, in men but very often in women, for as such it is a sign of both a colder nature and a more idle life. Fat always arises due to cooling of a bodily state, whereas much flesh (*polysarkia*) is the product of a large amount of blood; *eusarkia*, however, is a sign of a *eukratic* nature. Altogether then, those who are very fleshy straightaway have more fat than those who are well-fleshed. The fat does not always increase together in proportion to the flesh, for it is seen among those who are fat that some have a large amount of flesh while some have a large amount of fat. In those in whom both are increased together similarly, the wet is greater than the *eukratic* nature to the same extent as the cold. However, in those in whom the fat is greater, the cold is more or the wet is, just as in those in whom the flesh is more, the wetness is more than it should be but not also coldness. For whenever the hotness remains within the

proper limits, but some larger share of useful blood is added, of necessity excess flesh will be a consequence.

Εὐχυμία/κακοχυμία/πληθώρα (*euchymia/kakochymia/plethora*): Apart from the occasional use in the general sense of "juice" or "flavor," the term χυμός is used to indicate the four basic humors of the humoral component of the four elements/four qualities theory. *Definitiones medicae* has the following explanation: "χυμός in Hippocrates is invariably applied to the humors in the body of which our structure is—that is, of blood, phlegm, and the two biles, yellow and black. In Plato and Aristotle, the gustatory quality which each of these has in us is also termed humor. These are the qualities of sharpness, dryness, harshness, acridness, saltiness, sweetness and bitterness. So Mnesitheus meant in his pathology" (XIX.457–58K).

The following passage from *De methodo medendi* makes clear Galen's ideas on the causal role of the humors in the creation of *dyskrasia* and provides a definition of the terms *kakochymia* and *plethora*, and by extension, *euchymia* (he is speaking here specifically of inflammation).

The *dyskrasia* sometimes arises from something external, and sometimes from the humors in the body. It is external when caused by one of the so-called venomous animals or by a strongly heating or cooling medication, or sometimes by the surroundings; whereas it is due to the actual body of the sick person when it collects bad humors which are dissimilar in their capacities.

Having considered all these things, the first pri-

GENERAL INTRODUCTION

ority in those inflammations that are still in evolution is to eradicate their causes, whilst in those that are already established, it is to treat the inflammations alone. How you must take care of the whole body when it is in a pathological state is something I have spoken about at length, both throughout what has gone before and also in the work *On Plethora*.[52] Now I shall speak of the chief points of the discussions. Whenever the humors are increased to an equal degree to each other, [doctors] call this "abundance" or "plethora." On the other hand, whenever the body is already full of yellow or black bile, or phlegm, or the serous humors, they call such a condition *kakochymia* and not plethora. Plethora is treated by the letting of blood, or by numerous baths, exercises and rubbings, as well as by discutient medications, and, in addition to all fastings, which I covered comprehensively in the treatises on health. *Kakochymia*, however, is treated by the specific evacuation appropriate for each of the humors in excess. There was also discussion about this in the section on prophylaxis in the work *On the Preservation of Health*. (X.890–92K)

Κρᾶσις/εὐκρασια/δυσκρασία (*krasis/eukrasia/dyskrasia*): The term κρᾶσις/κρῆσις is defined in LSJ as "*mixing* or *blending* of things which form a compound . . ." The other meanings given are "temperature of the air," "temperament of the body or mind," "metaph. combination or union." Under "temperament," Hippocrates, *Nature of Man* IV, is cited; this passage is translated in section 6

---

[52] *De plenitudine*, VII.513–83K.

above. For many centuries "temperament" has been the rendering in medical texts, in particular in the title of Galen's book on the subject here translated. In recent times, however, there has been a trend toward using "mixture" instead of "temperament" and to render *dyskrasia* "bad mixture" instead of "distemperment" Both have merits and drawbacks. In the case of temperament, on the positive side there is the long-standing usage and association with the concept, not only in medical literature but also in other writings and in common parlance. On the negative side, there is the association with the four humors rather than the four qualities, present initially, as exemplified in the following excerpt from the pseudo-Galenic *Definitiones medicae*:

> Health is the *eukrasia* of the primary humors in us according to nature, or contrariwise function of the natural capacities. Or health is a *eukrasia* of the four primary elements from which the body is compounded—hot, cold, wet, and dry. Others define it in this way: A harmonious joining of the hot and cold, wet and dry, constituting the man.[53]

In the case of mixture, on the positive side there is the derivation from the verb κεράννυμι, "to mix or blend" (but is there a distinction between this verb and μείγνυμι, "to mix"?). Note that μιγματοπώλης is an apothecary.[54] Also, there is avoidance of the unwonted associations of tem-

---

[53] *Definitiones medicae*, XIX.382K; this definition reflects the difference between the Hippocratic and Galenic views of *eukrasia/dyskrasia*.

[54] See Galen, *De compositione medicamentum secundum locos*, XIII.68K.

GENERAL INTRODUCTION

perament. On the negative side there is the obvious inappropriateness in a medical text. In such a context "mixture" is more suitable for the apothecary's shop than the doctor's office. It is hard to imagine a doctor leaning solicitously forward across the desk, saying: "I'm afraid you have got/are a bad mixture Mrs. Smith."[55]

Galen himself makes a point of saying the compounding of the four elemental qualities is no ordinary mixture:

> Therefore, a thoroughgoing mixing of all these throughout the whole body—I speak of hot and cold, dry and wet—is not possible for a person. When earth is mixed with water to make a paste so it would seem to someone that in this way there has been mixing of all with all, this sort of thing is a juxtaposition on a small scale and not a thoroughgoing *krasis*; a thoroughgoing mixing of both these

---

[55] Alain Touwaide's suggestion of "composition" instead of "temperament" or "mixture" has, in my view, considerable merit. I include the following, slightly edited, comment from our correspondence on the matter: "This is not a simple problem at all! First there is a heavy scholarly tradition. However, as is often the case, traditional translations hide more than they reveal. I don't believe 'temperaments' or 'mixtures' help. I would even say they are misleading. Most of the time the question is about the different proportions of the physiological components of the body, the different elemental components of matter (whatever this matter is), or the different elemental qualities, all of which may be mixed in varying proportions (with the variation being either co-essential to the entity under consideration or accidental, possibly resulting from an illness. . . . I would think the best translation could be 'composition.'" I agree with him that "mixtures" is particularly misleading as a rendering of *krasis* in the work *De simplicium medicamentorum temperamentis ac facultatibus*.

is an action of God and Nature. Still more is this so if the hot and the cold are mixed with each other completely throughout. (I.562–63K)

In my view, the most appropriate course is to retain all three terms in a transliterated form. The term *dyskrasia* (*dyscrasia*) has been in regular medical use for many centuries and remains so today, just as *distemper* remains in use in veterinary medicine. Not least does this serve to give these terms due weight as central technical terms in Galen's physiological theory.

Ὁμοιομέρεια/ὁμοιομερής (*homoiomere/homoiomerous*): Essentially, the term means "having all parts like each other"—that is, of consistently uniform structure. There is general acceptance of the attribution of the term to Anaxagoras—indeed, Galen refers to this in *De placitis Hippocratis et Platonis*, V.3.18 (de Lacy, *Galen on the Doctrines*, 1.308). Usage is, however, particularly associated with Aristotle, in relation to both inanimate things (*Meteorologica* 10–13, 388a10–390b20) and animate things (*Parts of Animals* 2, 648a6–655b27). Galen clearly defines what he means by *homoiomeres* in several places and lists what they are, although the list does vary slightly. In *De morborum differentiis*, Galen lists arteries, veins, nerves, bones, cartilage, ligaments, membranes, and flesh as *homoiomerous* structures and clearly states that they are formed from the primary elements and are themselves the components of organic bodies (III.1, VI.841K).[56] Galen has a specific

[56] Other lists, less comprehensive, can be found in *De elementis secundum Hippocratem*, I.493K, and in *De placitis Hippocratis et Platonis*, VIII.4.7–15; de Lacy, *Galen on the Doctrines*, 2.500.

work on the subject, *Galeni De partium homoeomerium differentia libelli*, not included in Kühn because it is not surviving in Greek.[57]

8. PEOPLE

This list includes all individuals mentioned in the three translated treatises. An asterisk indicates those considered in section 1 (Galen's Antecedents) above. T = *De temperamentis*; N = *De inaequali intemperie*; S = *Quod animi mores corporis temperamenta sequantur*.

Anaxagoras (?500–428 BC) was the first known philosopher to settle in Athens. He posited a pluralistic world in which everything is mixed with everything, mind being the exception and present in all things. On the issue in question, see Aristotle, *Physics* 1.4, LCL 228 (P. H. Wicksteed and F. M. Cornford), 38–51.

T: on perception, 589

Andronicus the Peripatetic (ca. 100–20 BC) was a Peripatetic philosopher from Rhodes. According to Sharples (*OCD*), he "defined the soul as a power resulting from a mixture of the bodily elements and located emotions in an irrational part of the soul."

S: the substance of the soul like the *krasis* of the body, 782

Archimedes (ca. 287–212 BC) was perhaps the most notable mathematician and scientist of ancient times. He is

---

[57] G. Strohmaier, CMG Suppl. O, III (Berlin, 1970), Arabic and Latin.

credited with a significant role in the defense of Syracuse against the Roman fleet. On the matter in question, see D. L. Simms, "Galen on Archimedes: Burning Mirror or Burning Stick?" *Technology and Culture* 32 (1991): 91–96.

T: use of fire sticks, 657

***Aristotle** (384–322 BC) was the greatest influence on these works philosophically.

T: on compound *dyskrasias*, 523; on the four qualities, 535; on body and soul, 566; on old age, 581; on physiognomy, 625; on acquired and innate heat, 628; on the soul not being constituted by the four elements, 636; on "in itself" and contingent, 666, 672

S: concept of *homoiomeres*, 773; soul as a body, 773–74; nutritive and vegetative parts of the soul, 782–83; on homonymy, 783; on the soul's capacities following bodily *krasis*, 791–98

***Athenaeus of Attaleia** (uncertain, probably 1st century AD) was the probable founder of the Pneumaticist medical sect.

T: on compound *dyskrasias*, 522–23

Chrysippus of Soli (ca. 280–207 BC) succeeded Cleanthes as head of the Stoa in 220 BC. Of the Stoics it is predominantly Chrysippus who is mentioned in Galen's writings, although his two predecessors, Zeno of Citium and Cleanthes, as well as the later Stoic Posidonius are also mentioned, as in the works translated. Galen's attitude to Chrysippus is somewhat ambivalent as may be seen by comparing the references to him in his *Method of*

GENERAL INTRODUCTION

Medicine with those in *On the Opinions of Hippocrates and Plato*.

S: on his intelligence, 784; on affections of the soul, 820

Erasistratus (ca. 315–240 BC) was one of the two great Alexandrian doctors of the third century BC. Galen was at odds with Erasistratus on a number of matters, notably his espousal of a corpuscular theory of matter. For a brief summary of the differences, see Johnston, *Galen: On Diseases and Symptoms*, 97–98.

T: on *parenchyma*, 599

Eudemus (2nd century AD) was a Peripatetic philosopher who taught Galen when he first came to Rome in 162. He subsequently benefitted from his pupil's medical expertise, as described in Galen's *On Prognosis, for Epigenes*. See also *On Anatomical Procedures*, II.218K.

T: as an example of *dyskrasia*, 631–32

Heraclitus (fl. ca. 500 BC), according to Nussbaum (*OCD*), "is the first Greek thinker to have a theory of the *psyche* or soul as it functions in a living body. He connects *psyche* with both *logos* and fire, and appears to think of it as a dynamic connectedness that can be overwhelmed by a watery condition which spells death."

S: dryness as an aid to intelligence, 786

Hippocrates (d. 424 BC) was the nephew of Pericles, who died in the battle of Delium.

S: on the folly of his sons, 784

GENERAL INTRODUCTION

*Hippocrates (440–370 BC) was Galen's revered predecessor and the most cited individual.

> T: author of *Nature of Man*, 509; on seasons, 527; on climatic conditions, 530ff; on innate heat, 554; on perception of humors, 603–5; on change, 640, on nutriments, 660; on phlegm, 673
>
> N: on pain arising due to change, 739; on wounds and ulcers, 745
>
> S: on the effects of climate and place on functions of the soul, 798–808

Homer (8th century BC) is quoted several times as follows:

> T: on the North Wind, 513
> S: as the poet, 771; on the wine-like root, 777–78

Medea is a mythological figure portrayed in Euripides' eponymous play. She prepared a potion to protect Jason when he was seeking the Golden Fleece. Subsequently when forsaken by Jason, she killed his new bride, Creon and her own two sons.

> T: medication of, 658

*Plato (390–347 BC) was one of the two men most revered by Galen and is described as "divine." Aspects of his views on the soul were not, however, accepted by Galen.

> T: as a man, 544
> S: followers of, on the excellence of the soul, 768; capacities and divisions of the soul, 771–73; *protogona* 773: soul as immortal and incorporeal, 775; on the

adverse effect of wetness on the soul, 780–82; on the soul being incorporeal and able to survive without a body, 785–87; effects of *kakochymia* on the soul, 789–91; on *krasis* of place, 805–7; influence of nutriments on, 807–9; effects of wine, 808–12; effects of pursuits and studies, 813–16; on good and bad, 819

*Posidonius (ca. 135–51 BC) is described by Galen as the "most scientifically knowledgeable of the Stoics."

S: views on the soul, 819–20

Praxagoras of Cos (325–275 BC) was probably the teacher of Herophilus. His writings survive as fragments only. He is credited with making the distinction between arteries and veins and is notable for his subdivision of the four humors.

N: on hyaloid humor, 749–51

Pythagoras (6th century BC) is famous for his mathematical and other contributions. He is said to have been the originator of the concept of transmigration of the soul.

S: followers of, on the excellence of the soul, 768

Sappho (7th century BC) was a lyric poetess.

S: as the poetess, 771

Stoics belonged to a substantial school of philosophy beginning with Zeno of Citium in the late fourth century BC and continuing in the Roman world during the early centuries AD. Their doctrines are traditionally divided into three groups—logic, physics, and ethics. Galen had sev-

GENERAL INTRODUCTION

eral significant areas of doctrinal agreement with the Stoics, among which were the concept of the four elemental qualities and the nature of causation.

T: compound *dyskrasias*, 523
S: composition of the soul including *pneuma*, 783–84; on goodness, 819–20

Theognis (6th century BC) was an elegiac poet from Megara.

S: on wine, 778

Theophrastus (ca. 371–288 BC) succeeded Aristotle as head of the Lyceum and continued his work in botany and zoology.

T: on compound *dyskrasia*, 523; on the four qualities, 535; as a man, 544

Thucydides (ca 460–400 BC) was the chronicler of the Peloponnesian War. He himself suffered a bout of plague but recovered.

S: on the plague of Athens (430–426 BC), 788

Zeno of Citium (335–263 BC) was the founder of Stoicism. Lehoux (*EANS*, 847) writes: "He is responsible for the original Stoic division into active and passive principles as the basis for physical explanation as well as the doctrines of the interrelation of the four Aristotelian elements with *pneuma*, and the periodic conflagration of the *kosmos*."

S: benefit of wine and wine-like root, 777

GENERAL INTRODUCTION

9. DISEASES AND SYMPTOMS

Anasarca (ἀνὰ σάρκα): The modern definition is "a generalized infiltration of edema into subcutaneous connective tissue" (S). See G, XV.892K on treatment by venesection.

N: as a systemic non-uniform *dyskrasia*, 733

Anthrax (ἄνθραξ): Listed in LSJ as "carbuncle or malignant pustule": the possibility of smallpox has been raised. Also an infected wound.

T: Hippocrates on, 530, 532
N: as a non-uniform *dyskrasia*, 751

Apoplexy (ἀποπληξία): Sudden collapse with paralysis; presumably close to modern usage as a descriptive term. Ascribed to a cold *dyskrasia* by Galen.

T: as a cold disease, 582; as a cold affection, 661

Cancer (καρκίνος): An eroding sore or ulcer. Used by Galen to describe a superficial abnormality attributed to black bile.

N: as a focal non-uniform *dyskrasia*, 733, 751

Cataract (ὑπόχυμα): Taken to be closely equivalent to modern usage; see Galen, X.119, 990, and 1019K; and VII.89, 95K.

S: as a cause of failure of vision, 788

Catarrh (κάταρρος): A nasal discharge or flux from the head; see Galen, VII.107K, XIV.742K.

GENERAL INTRODUCTION

T: as a cold disease, 582; as a cold and wet *dyskrasia* of the head, 634

Convulsions. *See* **Spasms**.

Coryza (κόρυζα): Like catarrh, a flux from the head. The two terms are frequently found together.

T: as a cold and wet *dyskrasia* of the head, 634

Cough (βήξ): As in modern usage.

T: as a cold and wet *dyskrasia* of the head, 634

Delirium (παραφροσύνη): Probably similar to modern usage; that is, confusion, agitation, and excited behavior.

S: due to excess of yellow bile, 777; in old age, 787

Dropsy (ὕδερος): A general term for edema thought by Galen to be due to maldistribution of nutritive materials.

T: cantharides in, 667
N: a focal non-uniform *dyskrasia*, 733

Edema (οἴδημα): A general term for a type of swelling defined by its characteristics on palpation; it is not necessarily limited to localized fluid collections. Galen lists it as disease of *homoiomerous* bodies due to inflowing material (VI.849K). *See also* **Anasarca**; **Dropsy**.

N: as a non-uniform *dyskrasia*, 751

Elephas (ἐλέφας): Taken to be the same as *elephantiasis* but not the same as this in the modern sense. It may refer to true leprosy. See Gr., 168–76.

N: 733

Erysipelas (ἐρυσίπελας): Gr. writes: "The term *erysipe*-

las in Greek medical parlance designates various diseases that 'redden the skin' and also diffuse, purulent inflammations of internal organs, but in its commonest sense designates a group of skin diseases with hot, painful, reddish swelling, now thought to be streptococcal dermatitis." See also G, VI.849K and X.946K ff.

N: 733, 751

Fever ($\pi\upsilon\rho\epsilon\tau\acute{o}\varsigma$): As in modern usage but without the actual measurement of temperature. The various types of fever are described in *De febrium differentiis*, VII.273–405K, and are the subjects of Books 8 to 12 of *De methodo medendi*.

N: ague ($\dot{\eta}\pi\iota\acute{\alpha}\lambda o\varsigma$), 733, 749, 751; bilious remittent ($\kappa\alpha\hat{\upsilon}\sigma o\varsigma$), 750; hectic ($\dot{\epsilon}\kappa\tau\iota\kappa\acute{o}\varsigma$), 733, 743–44, 746, 751; malignant intermittent ($\lambda\iota\pi\upsilon\rho\acute{\iota}\alpha$), 750; tertian and quartan, 751

Gangrene ($\gamma\acute{\alpha}\gamma\gamma\rho\alpha\iota\nu\alpha$): Taken as approximately equivalent to modern usage, that is, "extensive necrosis from any cause" (S). See G, XI.135K (definition), and VII.22, 75K.

N: 733, 751

Heatstroke ($\ddot{\epsilon}\gamma\kappa\alpha\upsilon\sigma\iota\varsigma$): Presumably the same meaning as now. Regarded by Galen as a hot *dyskrasia*.

N: fever in, 748

Herpes ($\ddot{\epsilon}\rho\pi\eta\varsigma$): A skin affection of uncertain nature. For Galen's definition, see XI.74K. See also the pseudo-Galenic *Def. Med.*, XIX.440K.

N: as a non-uniform *dyskrasia*, 733

xcvii

GENERAL INTRODUCTION

Lethargy (λήθαργος): As in present usage but may include drowsiness and forgetfulness. See G, *Method of Medicine*, X.929–31K.

S: due to excess phlegm, 777

Mania (μανία): A disturbance of intellect and conduct without fever. See the pseudo-Galenic *Def. Med.* XIX.416K.

S: in old age, 787–88; abnormal movements of arteries in, 804

Melancholia (μελαγχολία): A state of sadness attributed to an excess of black bile.

S: due to black bile, 777; an affection of the soul, 788

Numbness (ναρκή): More a symptom than a disease. Defined as a combination of disturbed sensation and disturbed movement involving the whole body or the limbs only.

T: as a cold disease, 582

Paralysis (παράλυσις): Similar to modern usage but includes sensory functions as well as motor. See Siegel, *Galen on the Affected Parts*, 227–28.

T: as a cold disease, 582

Phagedaina (φαγέδαινα): A form of eroding ulcer. See G, *Method of Medicine*, X.83K.

N: as a non-uniform *dyskrasia*, 733, 751

Phlegmonē (φλεγμονή): The standard term for inflammation and inflammatory swellings.

T: treatment of, 690–91
N: as a non-uniform *dyskrasia*, 733, 737ff; associated with fever, 746–48; of colon and testes, 749

Phrenitis (φρενῖτις): Inflammation of the brain. Described by Siegel, *Galen on the Affected Parts*, as "delirium with fever." See his discussion, pages 270–72.

S: as an affection of the soul, 788–89

Sore Throat (βράγχος): Sore throat causing hoarseness.

T: as a cold disease, 582

Spasm (σπάσμα): a complex term, listed by LSJ in one sense as "spasm, convulsion" and in another as "sprain or rupture of muscle." Presumably, the former is indicated here.

T: as a cold disease, 582

Tremor (τρόμος): As in the present usage. It was classified by Galen as a symptom due to a disturbance of the motor component of the *psyche*.

T: as a cold disease, 582

Ulcer/Wound (ἕλκος): This is a problematic term. It was used by Galen in a general way to refer to "dissolution of continuity" involving a surface, whether external or internal. It ranges in application from a fresh wound through a chronic and possible infected wound to what now would be referred to as an ulcer.

T: Cheironian and Telephian in *kakochymia*, 664
N: Hippocrates on the term, 745

GENERAL INTRODUCTION

10. MEDICATIONS AND FOODS

Aloes (ἀλόη): *Aloe vera, A. vulgaris*; bitter aloes—a haemostatic for wound conglutination, a purgative, a component of eye salves and ear lotions; D, III.25; G, XI.822K.

S: its various capacities, 769–70

Asphalt (ἄσφαλτος): Asphaltos, bitumen, pitch; different sources; D, I.99; G, XII.375K.

T: readily flammable, 658; potentially hot, 669

Cantharides (κανθαρίδες): Presumably the blister beetle, *Cantharis vesicatoria*. See Hippocrates, *Nat. Mul.*, 32; D, II.65.

T: use in dropsy, cause of ulceration, 667

Castor/Castoreum (καστόρειον): Material derived from the testes of beavers; used in eye and ear salves; D, II.26; G, XII.337–41K.

T: acts on more than is acted on, 675; nutriment and hot medication, 681

Catmint (καλάμινθος): *Nepeta cataria*; various uses, including as a diuretic; D, III.43; Cu, 201.

T: nutriment and medication, 682.

Copper Ore: *See* **Misu**.

Cyrenian Juice (κυρηναικός ὀπός): For a description (a very thin, heating juice), see G, XIII.567K; for use in inflammation, see G, XI.860K.

T: 666

c

GENERAL INTRODUCTION

Dill (ἄνηθον): *Anethum graveolens*; D, III.67; G, XI.832K and XIII.316K; Cu, 105–6.

T: nutriment and medication, 682

Fennel (νάρθαξ): Giant fennel, *Ferula communis*; D, 3.77; Theophrastus, *Historia plantarum* 1.2.7.

T: readily flammable, 658

Fleawort (ψύλλιον): *Plantago psyllinum*; D, II.151; G, XI.740K; Cu, 128–29.

T: effects of fire on, 674

Hellebore (ἑλλέβορος): *Veratrum album, Helleboras niger, H. officianalis*; white and black hellebore; purgative and emetic; D, IV.150–52; G, XI.874K.

T: nutriment for quails, medication for people, 684

Hemlock (κωνειόν): *Conium maculatum*; D, IV.79.

T: cold medication, 649; administration heated, 673; effect of fire on, 674; nutriment for fish, medication for humans, 684
S: cooling agent, 776; effects of, 779

Honey (μέλι): Widely used alone or in mixtures; D, II.101; G, XI.671 and XII.70K.

T: potentially hot, 669; acted on more than acting on, 675

Kostos (κόστος): A root used as a spice; *Saussurea lappa*; D, 1.16; Theophrastus, *Historia plantarum* 9.7.3.

T: as heating medication, 649

GENERAL INTRODUCTION

Lettuce (θριδακίνη): *Lactuca scariola*; D, II.1666; G, XI.887K; Cu, 170–71.

   T: hypnotic, 585; cold medication, 649; variable effect as a cold food, 679; juice of, 680–81

Mandragora/Mandrake (μανδραγόρας): *Mandragora officinalis*; D, IV.76; G, XI.751K and XII.67K; also listed in LSJ as belladonna (*Atropa belladonna*), T, 6.2.9

   T: hypnotic, 585; as a cold medication, 649; administered heated, 673; effect of fire on, 674

Mēdian Juice (ὀπός Μηδικός): Listed in LSJ as a form of Silphium juice, probably *assafoetida*; D, III.94.

   T: possible harm from, 666

Misu (μίσυ): Misy, copperas; a copper ore found in Cyprus; D, V.117; G, XI.688K and XII.241K.

   T: as a hot medication, 649

Mustard (νᾶπυ): Used in plasters; D, II.142; G, I.682K and XI.870K.

   T: hot potentially, 649; nutriment and medication, 682

Nitron (νίτρον): Native sodium carbonate; D, V.130; G, XIII.268K.

   T: hot medication, 649

Olive oil (ἔλαιον): various uses, D, I.30–31.

   T: as hot potentially, 649, 669; as heating, 660

Opium (ὄπιον): D, IV.65; G, XIII.269K. *See* **Poppy**.

cii

GENERAL INTRODUCTION

Oregano (ὀρίγανον): An acrid herb; several varieties; D, III.32–34; G, XIV.140K; Cu, 182–83.

T: nutriment and medication, 682

Parthian Juice (ὀπόν παρθενικιον): Uncertain, possibly feverfew, *Pyrethrum parthenium*; given as a drink to "drive out phlegm"; D, III.155.

T: as heating, 666

Pellitory (πυρέθρον): *Anacyclas pyrethrum*; G, XIII.110.

T: as a hot medication, 649

Pennyroyal (γλήχων/βλήχων): *Mentha pulegium*; D, III.36; G, XI.882K.

T: nutriment and medication, 682

Pine resin (ῥητίνη): Aristotle, *History of Animals*, 617a19; Theophrastus, *Historia plantarum* 9.2.1; D, I.90.

T: as hot potentially, 649, 669; as heating, 659

Pitch (πίσσα): Used as an emollient and in preparation of plasters; D, I.94 and 95; G, XI.734K and XIII.709K.

T: as hot potentially, 649, 669; as heating, 659

Plantain (θρυαλίς): *Plantago crassifoli*; D, II.153.

T: as heating, 659

Poppy (μήκων): *Papaver somniferum*, *P. rhoeas* (wild poppy), *P. argemone* (prickly poppy); D, IV.64–66; G, XII.74K.

T: as hypnotic, 585; as hot potentially, 649, 669; as heating, 659

GENERAL INTRODUCTION

Ptisane (πτισάνη): A medicinal drink made from barley. See Galen's *De ptisana* VI.816–31K, and for powers and uses, XV.452K.

T: administered heated, 673; acted on more than acting on, 675

Purslane (ἀνδράχνη): *Portulaca oluracea*; D, II.151; G, VI.634K, XI.740 and 751K; Cu, 231–32.

T: as a cold medication, 679

Rock Alum (χαλκῖτις): Copper ore; used for erysipelas and herpes; used in eye medications; D, V.114; G, XI.688K and XII.241K.

T: hot potentially, 649

Rosewater (ῥόδινος): A preparation of Rosaceum; D, I.53.

T: whether hot or cold, 685

Rue (πήγανον): *Ruta graveolens*; D, III.52; G, XI.809K.

T: nutriment and medication, 682

Salamander (σαλαμάνδρα): *S. vulgaris*; a kind of newt; D, II.62.

T: as cold, 649

Savory (θύμβρα): *Satureia thymbra*; D, III.37; Theophrastus, *De causis plantarum* 3.1.4

T: nutriment and medication, 682

GENERAL INTRODUCTION

Soapwort (στρούθειον): *Saponaria officinalis*; D, II.163; Theophrastus, *Historia plantarum* 6.4.3.

T: as hot medication, 649

Spignel (μῆον): *Meum athamanticum*; D, I.3; G, XIV.150K; Pliny, *Naturalis historia* 20.253

T: as a hot medication, 649

Spurge (εὐφόρβιον): *Euphorbia resinifera*; D, III.82; G, XIII.848K.

T: as a hot medication, 649

Thyme (θύμον): *Thymbria capitata*; Cretan thyme; D, III.44; G, XI.887K ff.

T: nutriment and medication, 682

Vinegar (ὄξος): Used alone or in mixtures (e.g., oxymel); D, V.21–25; G, XII.90K (medicinal powers).

T: as preservative, 538; whether hot or cold, 685

Wine (οἶνος): For the variety of wines, see D, V.7–82; G, VI.334–39K.

T: as heating, 538; acted on more than acting on, 675
S: for distress and *dysthymia*, 777; Theognis on, 778; Plato on, 808–12; effects on the soul, 821

Yellow flag (ἄκορον): *Iris pseudacorus*; D, I.2; G, XI.819K.

T: as hot when applied, 649

ABBREVIATIONS

Ce   Celsus. *De Medicina*. Translated by W. G. Spencer. Loeb Classical Library. 3 vols. Cambridge, MA: Harvard University Press, 1935–1938.

CMG   Corpus Medicorum Graecorum.

Cu   Nicholas Culpepper. *The English Physician Enlarged (Culpepper's Herbal)*. London: Folio Society, 2007 [1653].

D   Dioscorides. *The Greek Herbal of Dioscorides*, translated by John Goodyer [1653]. Edited by R. T. Gunther. New York: Hafner, 1968 [1933].

EANS   *The Encyclopedia of Ancient Natural Scientists*. Edited by P. T. Keyser and G. L. Irby-Massie. London: Routledge, 2008.

G   Galen. References to Galenic works are indicated by the Kühn volume and page numbers. His three major pharmacological treatises are *De simplicium medicamentorum temperamentis et facultatibus*, XI.379–892K and XII.1–377K; *De compositione medicamentorum secundum locos*, XII.378–1007K and XIII.1–361K; *De compositione medicamentorum per genera*, XIII.362–1058K.

ABBREVIATIONS

Gr	M. D. Grmek. *Diseases in the Ancient Greek World*. Baltimore, MD: Johns Hopkins University Press, 1991.
J	I. Johnston. *Galen: On Diseases and Symptoms*. Cambridge: Cambridge University Press, 2006.
J&H	I. Johnston and G. H. R. Horsley. *Galen: Method of Medicine*. 3 vols. Loeb Classical Library 516, 517, 518. Cambridge, MA: Harvard University Press, 2011.
JHyg.	I. Johnston. *Galen: Hygiene*. 2 vols. Loeb Classical Library 535, 536. Cambridge, MA: Harvard University Press, 2018.
L&S	C. T. Lewis and C. Short. *A Latin Dictionary*. Oxford: Clarendon, 1993 [1879].
LCL	Loeb Classical Library.
LSJ	H. G. Liddell, R. Scott, and H. Stuart Jones. *A Greek-English Lexicon*. 9th ed. (1940), with revised suppl. by P. G. W. Glare. Oxford: Clarendon, 1996.
M	C. C. Mettler. *The History of Medicine*. Philadelphia: Blakiston, 1947.
OCD	*Oxford Classical Dictionary*. Edited by S. Hornblower and A. Spawforth. 3rd ed. Oxford: Clarendon, 1996.
OED	*Oxford English Dictionary*. 12 vols. Oxford: Oxford University Press, 1978 [1933].
S	*Stedman's Medical Dictionary*. 27th ed. Baltimore: Lippincott, Williams and Wilkins, 2000.
Si	R. E. Siegel. *Galen on the Affected Parts*. Basel: S. Karger, 1976.

ABBREVIATIONS

T   Theophrastus. *Enquiry into Plants*. Translated by A. Hort. Loeb Classical Library. 2 vols. Cambridge, MA: Harvard University Press, 1916, 1926.

GENERAL BIBLIOGRAPHY

Adamson, P. E., R. Hansberger, and J. Wilberding. *Philosophical Themes in Galen*. London: Institute of Classical Studies, 2014.

Bazou, A. D. Γαληνοῦ· Ὅτι ταῖς τοῦ σώματος κράσεσιν αἱ τῆς ψυχῆς δυνάμεις ἕπονται. Athens: ΑΚΑΔΗΜΙΑ ΑΘΗΝΩΝ, 2011.

Boudon-Millot, V. *Galien: Introduction general, Sur l'ordre de ses propres livres, Sur ses propres livres, Que l'excellent médicin est aussi philosophe*. Paris: Les Belles Lettres, 2007.

Boulogne, J. *Méthode de traitment*. Paris: Gallimard, 2009.

Brain, P. *Galen on Bloodletting*. Cambridge: Cambridge University Press, 1986.

Brock, A. J. *Galen: On the Natural Faculties*. Loeb Classical Library 71. Cambridge, MA: Harvard University Press, 1963 [1916].

Browne, E. G. *Arabian Medicine*. Cambridge: Cambridge University Press, 1921.

Conrad, L. I., M. Neve, V. Nutton, R. Porter, and A. Wear. *The Western Medical Tradition (800 BC to AD 1800)*. Cambridge: Cambridge University Press, 1995.

Craik, E. M. *The Hippocratic Corpus: Content and Context*. London: Routledge, 2015.

GENERAL BIBLIOGRAPHY

Daremberg, C. *Oeuvres anatomiques, physiologiques et médicales de Galien*. Paris: J-P Baillière, 1854–1856.

De Lacy, P. H. *Galen on the Doctrines of Hippocrates and Plato*. CMG, V.4.1.2. Berlin: Akademie-Verlag, 1978.

———. *Galen on the Elements According to Hippocrates*. CMG, V.1.2. Berlin: Akademie-Verlag, 1996.

Duckworth, W. H. L., M. C. Lyons, B. Towers. *Galen on Anatomical Procedures. The Later Books: IX.6–XV*. Cambridge: Cambridge University Press, 1962.

Edlow, R. B. *Galen on Language and Ambiguity*. Leiden: E. J. Brill, 1977.

Garcia Novo, E. *Galen: On the Anomalous Dyskrasia*. Berlin: Logos Verlag, 2012.

Grant, M. *Galen on Food and Diet*. London: Routledge, 2000.

Green, R. M. *A Translation of Galen's Hygiene*. Springfield, IL: C. C. Thomas, 1951.

Hankinson, R. J. *Galen on the Therapeutic Method: Books 1 and 2*. Oxford: Oxford University Press, 1991.

———. *Galen on Antecedent Causes*. Cambridge: Cambridge University Press, 1998.

———, ed. *The Cambridge Companion to Galen*. Cambridge: Cambridge University Press, 2008.

Helmreich, G. *Galenus: De Temperamentis Libri III*. Leipzig: B. G. Teubner, 1894.

Irwin, James R. "Galen on the Temperaments." *The Journal of General Psychology* 36 (1947): 45–64.

Johnston, I. *Galen: On Diseases and Symptoms*. Cambridge: Cambridge University Press, 2006.

———. *Galen: On the Constitution of the Art of Medicine; The Art of Medicine; A Method of Medicine to Glaucon*.

GENERAL BIBLIOGRAPHY

Loeb Classical Library 523. Cambridge, MA: Harvard University Press, 2016.

———. *Galen: Hygiene*. 2 vols. Loeb Classical Library 535, 536. Cambridge, MA: Harvard University Press, 2018.

Johnston, I., and G. H. R. Horsley. *Galen: Method of Medicine*. 3 vols. Loeb Classical Library 516, 517, 518. Cambridge, MA: Harvard University Press, 2011.

Kudlien, F., and R. J. Durling, eds. *Galen's Method of Healing*. Leiden: E. J. Brill, 1991.

Kühn, C-G. *Claudii Galeni Opera Omnia*. 20 vols. Hildesheim: Georg Olms Verlag, 1997 [Leipzig: 1821–1833].

Marquardt, J., I. Müller, and G. Helmreich. *Claudii Galeni Pergameni: Scripta Minora*. Vol. 2. Leipzig: B. G. Teubner, 1891.

May, M. T. *Galen on the Usefulness of the Parts of the Body*. Ithaca: Cornell University Press, 1968.

Meyerhof, M., and J. Schacht. *Galen über die medizinische Namen*. Berlin: Abhandlungen der Preussischen Akademie der Wissenschaften, 1931.

Nutton, V. *Galen on Prognosis*. CMG, V.8.1. Berlin: Akademie-Verlag, 1979.

———. *Galen on My Own Opinions*. CMG, V.3.2. Berlin: Akademie-Verlag, 1999.

———, ed. *The Unknown Galen*. London: Bulletin of the Institute of Classical Studies, Supplement 77, 2002.

———. *Ancient Medicine*. London: Routledge, 2004.

Payne, J. F. *Galeni Pergamensis De Temperamentis et De Inaequali Intempere, Libri Tres, Thoma Linacro Anglo Intreprete*. Cambridge: Macmillan and Bowes, 1881.

Siegel, R. E. *Galen on the Affected Parts*. Basle: S. Karger, 1976.

GENERAL BIBLIOGRAPHY

Singer, C. *Galen on Anatomical Procedures.* Oxford: Clarendon Press, 1956.

Singer, P. N. *Galen: Selected Works.* Oxford: Oxford University Press, 1997.

———. *Galen: Psychological Writings.* Cambridge: Cambridge University Press, 2013.

Singer, P. N., and P. J. van der Eijk. *Galen: Works on Human Nature.* Vol. 1, *Mixtures (De Temperamentis).* Cambridge: Cambridge University Press, 2018.

Stewart, A. *Our Temperaments: Their Study and Their Teaching.* London: Crosby Lockwood and Son, 1892.

Walzer, R., and M. Frede. *Three Treatises on the Nature of Science.* Indianapolis: Hackett Publishing Company, 1985.

ΓΑΛΗΝΟΥ ΠΕΡΙ ΚΡΑΣΕΩΝ

ON TEMPERAMENTS

INTRODUCTION

Galen's *De temperamentis* (*On Krasias, Temperaments, Mixtures*) is unquestionably among his most important and influential works, providing as it does an account of one of the foundational concepts in his theory of medical practice. *Krasis*, arguably best left untranslated, is seen as a unique blending of the four elemental qualities that form the structural basis of the bodies of all animals, and indeed of all matter, both animate and inanimate. Disorders of *krasis*—*dyskrasias*[1]—constitute one of the three of Galen's classes of disease. In his two great practical works, *The Method of Medicine* (*De methodo medendi*) and *Hygiene* (*De sanitate tuenda*), the diagnosis and treatment of *dyskrasia* and the maintenance of a good (normal)

---

[1] According to Garcia Novo, *Galen: On the Anomalous Dyskrasia*, the first use of the term was by Theophrastus in relation to diseases in plants (*De causis plantarum* 5.8.2). There he writes, on the origins of diseases in plants: "From without, whenever spells of cold or heat are excessive, or rains or dry spells, or some other *dyskrasia* of the air" (LCL 475 [B. Einarson and G. K. K. Link, 1990], 86–87). The term remains in medical use today and is defined in Stedman as follows: 1. A morbid general state resulting from the presence of abnormal material in the blood, usually applied to diseases affecting blood cells or platelets. 2. An old term indicating disease. The *OED* of 1933 defines it simply as "bad temperament" of body, air, etc.

krasis—*eukrasia*—occupy, respectively, substantial parts. His *De temperamentis* is the theoretical basis of such endeavors. In his *Art of Medicine*, he writes:

> There is one book, *On the Elements According to Hippocrates*. Following this, there are three books in the treatise *On Krasias*—two of these concern *krasias* in animals while the third is about *krasias* in medications. On this account also, the work *On the Nature and Powers of Simple Medications* cannot be understood properly without carefully reading the third book of the treatise *On Krasias*. And there is another small book which follows these first two on *krasias*, and was written on the non-uniform *dyskrasias*. Similar to this there are also two other small books: *On the Best Constitution of Our Body* and *On Good Bodily State*.[2]

A very similar statement is made in Galen's *On the Order of My Own Books*, where he adds: "These three very short works were written for friends at their behest and subsequently published by them. Certainly, since the force of these is covered in the work on hygiene, in which the differences of the constitution of our body . . ."[3]

The *De temperamentis* is essentially a theoretical work. The clinical relevance and applications of the concepts contained within it are to be sought in the practical works referred to. There is, however, a summary of the applications of the concepts of *krasis*, *eukrasia*, and *dyskrasia* to

---

[2] *Ars Medica*, I.407–8K.

[3] *De ordine librorum suorum ad Eugeniam*, XIX.56K. (There is a lacuna following "body.")

disease classification, diagnosis, treatment, and hygiene in the four key practical works on these subjects in the General Introduction (section 3). The teaching in *De temperamentis* was preserved by the medical encyclopedists in the several centuries immediately after Galen, then through Syriac, Arabic, and Hebrew translations, and then in Latin translations, as outlined in section 6 of the General Introduction. As late as 1881, an edition of Thomas Linacre's 1521 Latin translation, along with his translation of *De inaequali intemperie*, was published in Cambridge, prefaced by a detailed introduction by the editor, J. F. Payne, although by then the central concept had become outmoded. Payne offers some interesting insights into the ramifications of the concept of *krasis*, or temperament, beyond the narrow confines of medicine. These ramifications are still discernible today.

TEXTS AND TRANSLATIONS

The Greek text used in the present work is that in G. Helmreich's 1904 edition. This in turn is based on four main Greek manuscripts (Laurentianus 74.5; Marcianus 275; Oxoniensis, i.e., Bodleianus 709 = Laudinus 58; and Truvultianus 685) plus the Greek texts in four editions of collected works: Aldina, Basiliensis, Charteriana, and Kühneiana. In the present work, Helmreich's Greek text is compared with Kühn's Greek text and Linacre's 1521 Latin translation as reproduced in Payne's 1881 edition. Significant differences relevant to the translation are indicated in the footnotes. There are two versions of P. N. Singer's English translation; that in his 1997 *Galen: Selected Works* and a revised version in collaboration with

ON TEMPERAMENTS

P. J. van der Eijk in *Galen: Works on Human Nature*, volume 1, published in 2018. Both have been consulted. Neither Durling's work on Burgundio of Pisa's Latin translation nor Tassinari's 1997 Italian translation were consulted. On the style of translation, the principles outlined in relation to the *Method of Medicine* were followed.[4] As previously noted, what were judged to be key technical terms have been transliterated rather than translated on the grounds of their importance to Galen's theoretical formulations and the perceived unsuitability of certain English renderings.

SYNOPSES OF THE THREE BOOKS

Book 1

1. Galen states his adherence to the four elements/four elemental qualities theory of the fundamental structure of matter in general and of animal bodies in particular. He then states his intention to identify all the *differentiae* of the *krasias*.

2. A summarizing statement is given on the *krasias*, beginning with the identification of two opposing positions on compound *krasias*. (a) There are four such *krasias*: hot and dry, hot and wet, cold and dry, cold and wet, which is Galen's own view. (b) There are only two such *krasias*: hot and dry, and cold and wet, since wet cannot persist with hot nor dry with cold. Galen defends his view and counters the opposing view by citing examples such

[4] See *Galen: Method of Medicine*, vol. 1, LCL 516 (Johnston and Horsley, 2011), cxi–cxii.

as baths, climatic conditions, and medications. He elaborates on his argument by considering other qualities/attributes and concludes by restating his claim that there are the four initially stated possible conjunctions of qualities that can exist, as opposed to two only.

3. Galen clarifies *eukrasia* as a state in which neither quality prevails in either of the two antitheses (hot and cold, wet and dry) and there is perfect balance. He laments the neglect and misunderstanding of this term, the latter being particularly to identify *eukrasia* with a *krasis* in which hot and wet predominate in their respective antithesis, exemplified by the characteristic state in childhood and the situation believed to obtain in spring, that is, spring is hot and wet, summer is hot and dry, autumn is cold and dry, and winter is cold and wet.

4. Galen focuses on the error of equating *eukrasia* with the hot and wet *krasis*, and this in turn with spring. There is no need, he says, to assume the four conjunctions of the *krasias* exist in the seasons. Spring is, in fact, in the middle of all the excesses. He stresses the point that it is not necessary to equate the four conjunctions of the disproportionate *krasias* with the seasons.

5. Aspects of the correct usage of terms and the dangers of sophistical reasoning are given. The relativity of descriptive terms must be taken into account.

6. Galen provides a long section about terminology. He identifies three significations of the terms relevant to his exposition—hot, cold, dry, and wet. The first is absolute and applies only to the elements. The second and third are relative; the first in relation to the mean or median of the class or kind, and the second in relation to another individual entity of any class or kind. At the outset, he makes

reference to a similar terminological exercise in his work *On the Diagnosis of Pulses.*

7. Having signaled his intention to deal with the signification of the qualities comprising the *krasias*, Galen begins with a short further consideration of the terminology, and in particular the distinction between "hot" and "hotness" and the same for the other three qualities.

8. In this section Galen offers a clear definition of the term *eukrasia*. He then summarizes the various possibilities. The perfectly balanced state of the four qualities, which is *eukrasia*; the four mono-*dyskrasias*, in each of which one of the primary qualities is excessive; the four possible compound *dyskrasias* comprising two qualities, one from each antithesis, which are in excess—that is, hot and wet, hot and dry, cold and wet, cold and dry. There are, then, nine possible situations pertaining to the *krasis* of a body or part of a body.

9. In a long final section, Galen deals with the issue of training in the recognition of *krasias*. First, there is the recognition of the median in relation to all existing things. He gives the examples of mixing equal amounts of boiling water and ice for the first antithesis, and earth and water for the second. These, however, are really juxtapositions. He addresses the difference between ordinary mixing (a process of which people are capable) and a thoroughgoing *krasis* (a process of which only a god and Nature are capable). The median in humans is the skin, particularly that on the palmar surfaces of the hands, which is "the arbiter of all perception." Not only is it the median *krasis* of the elemental qualities, it is also the median of hardness and softness. He concludes, without presenting evidence, that the human body is the most *eukratic*, not only of all animal

bodies but also of all things. He considers the relationship of *eusarkos* and *eukratos*. He gives examples of other parts of the body and the humors in relation to the skin as the median.

Book 2

1. The section begins with a summary of the contents of Book 1—identification of nine *differentiae* of *krasis,* that is, the four simple *dyskrasias*, the four compound *dyskrasias*, and *eukrasia*; of the human as the most *eukratic* of all things; and the skin on the palmar surface of the hand as the most *eukratic* part of the human, unless altered by use. The *eukratic* person is recognized as being median in terms of thinness and thickness, hardness and softness, and hotness and coldness, and having a soul median in terms of courage and timidity, hesitancy and rashness, compassion and envy. *Eukrasia* is characterized by optimum functioning.

2. Galen presents differences in *krasias* related to various ages/stages of life. The fetus within the womb is wet and hot, as befits something formed from blood and semen; this continues after birth and applies to all structures and all animals. Hotness and wetness progressively change through life to the coldness and dryness of old age, passing through the optimum stage in midlife. Galen comments on the mistaken impression of wetness in old age due to wetness of the superfluities. Contention arises regarding which are hotter, children or those in their prime. Galen then returns to the subject of actuality and potentiality, giving the matter a more extended treatment. After this

somewhat lengthy digression, he comes back to the original question and concludes that neither side is right in their reasoning. Neither is absolutely hotter than the other; the impact of the heat on the one touching differs in the two groups for other reasons.

3. Touch is the only arbiter of a hot or cold body; with a wet or dry body, reason is also involved. Galen gives examples. He then considers the qualities of the various bodily parts including, in addition to the elemental qualities, softness and hardness. He then turns briefly to the humors and the relative contribution of touch and reason to the perception of their qualities.

4. Galen considers the things that follow the various components of the *krasias*. These include the bodily state and the state of the blood vessels (arteries and veins). In evaluation it is necessary to consider the relative roles of nature, custom, and regimen. He then goes over the characteristic bodily states for each of the *krasias*. He also differentiates a healthy *dyskrasia* from a morbid *dyskrasia*. The distinguishing feature of the latter is impairment of function. A brief comment is made on the humors: blood is the most useful and familiar; black bile is a kind of sediment of the blood and is colder and thicker than blood; yellow bile is hotter; phlegm is the coldest and wettest of all.

5. This section considers the variation in hair growth in the different *krasias*. Comparison is made to the growth of plants in the different seasons. The role of the skin in relation to transpiration is also considered. Apart from the growth of hair, its straightness or otherwise, and its color are also relevant in analyzing *krasias*. Hair is then consid-

GALEN

ered in relation to age, place, and bodily nature. Various national characteristics are described. Finally, baldness, particularly age-related, is considered.

6. This long section begins by raising the issue of non-uniform *dyskrasias*. After citing a number of macroscopic examples, Galen states that evidence about the whole should not be obtained from a part. He then considers physiognomy, citing the pseudo-Aristotelian work on the subject. Again he issues the injunction against drawing conclusions on the *krasias* of the whole from the appearance of the part. The example is a very hairy chest associated with a high-spirited or irascible nature and a drier and hotter *krasis*. All the parts and structures must be considered individually, and when touch and vision cannot be used for assessment, function must suffice and the answer provided by reason. Here habitation may contribute significantly. Examples are given. Galen makes a distinction between innate and acquired heat. In summary, an assessment of the *krasis* of each of the parts may be made by tactile and visual examination where possible and by evaluation of function where the former is not possible. There is digression on species variation in the anatomy of the bile ducts. A third sign, the nature of the evacuations, is then considered. Galen reiterates the need to avoid drawing conclusions about the whole from the adjudged *krasis* of a part, providing several examples, including the eyes and the nose. He concludes this long section by commenting on two sources of error—drawing conclusions from the signs and thinking whatever heats also dries. He states his intention to devote the third book to the potency of medications in relation to the four qualities and to add a separate book on non-uniform *dyskrasia*.

ON TEMPERAMENTS

Book 3

1. Galen examines the four elemental qualities in relation to the Aristotelian terms ἐνέργεια, δύναμις, and ὕλη (actuality, potentiality, and matter), giving examples from foods and medications and other external factors. Changing the physical state of agents to allow them to express their potential, and also massage, are considered. He speaks briefly about the four capacities (faculties, powers) of every body: attractive, retentive, transformative, and separative. These are dealt with *in extenso* in his work *De naturalibus facultatibus*, to which he refers.

2. This section deals with nutriments and medications. Nutriments must be suitable for the body being nourished but require both modification (working-up) and assimilation to a varying degree. Medications are largely of two kinds; those that overcome and change the body and those that are changed by the body and then act to putrefy and destroy. Galen mentions two other kinds: those that heat the body but do no harm and those that both act and are acted upon, the last being both medication and nutriment. Three anecdotes are interpolated describing specific instances: a house burned down due to the spontaneous combustion of pigeon excrement; Archimedes using fire sticks; and the poison devised by Medea. Galen refers to Hippocrates' definition of nutriment and the adverse effects of excessive drinking of wine. His final statement is that all the examples given are in accord, "with the theories concerning elements and the *krasias*."

3. Galen considers the different effects of certain substances, both nutriments and medications, when administered orally or applied to the skin, giving examples to ex-

plain why the differences occur. He considers the origin of *kakochymia* (literally, bad humors), listing the several sequelae that are essentially different forms of ulceration. He also considers the juice of the poppy, snake venom, the saliva of rabid dogs, and the cantharides. In particular, he deals with the heating and cooling effects of these substances.

4. Galen considers the distinction between things that are hot, cold, wet, or dry "of themselves" and those that are so "contingently." He continues the differentiation between "in actuality" and "in potentiality" and what this involves in certain instances, for example, olive oil and wine. Detailed consideration is given to so-called deleterious or noxious drugs/medications, particularly in relation to time and amount of administration. He considers the reciprocal action of administered substances—what acts upon the body is also acted upon by the body, although the "reaction" may be imperceptible. The concept of "perpetual affection" is raised. The actions of lettuce are given particular consideration.

5. This section is basically about the evaluation of medications in the *dyskrasias* and particularly hotness and coldness. First, there is the question of whether the medication itself is hot (or cold or wet or dry), and if so, whether it is so in potentiality or in actuality. In assessing a substance, whether medication or food, it should be free of any acquired hotness or coldness before administration, and it should be tested against an absolute or extreme condition. The question of whether it acts in and of itself or incidentally should be addressed; the key determinants of this are the condition and the time. The use of cold water in tetanus is given as an example. Aspects of the

ON TEMPERAMENTS

mechanism of causing change are considered, including the possible involvement of an intermediary. The example of the use of a cataplasm in inflammation is given. Galen finally refers to his works on medications and to his *Method of Medicine* for further details on these matters.

6. This is a short summary, focusing particularly on two things, using hot/hotness as an example. The first is the different ways in which the term is used, and the second, the distinction between actuality and potentiality. The same comments apply to the other three elemental qualities.

ΒΙΒΛΙΟΝ ΠΡΩΤΟΝ

1. Ὅτι μὲν¹ ἐκ θερμοῦ καὶ ψυχροῦ καὶ ξηροῦ καὶ ὑγροῦ τὰ τῶν ζῴων σώματα κέκραται καὶ ὡς οὐκ ἴση πάντων ἐστὶν ἐν τῇ κράσει μοῖρα, παλαιοῖς ἀνδράσιν ἱκανῶς ἀποδέδεικται φιλοσόφων τε καὶ ἰατρῶν τοῖς ἀρίστοις· εἴρηται δὲ καὶ πρὸς ἡμῶν ὑπὲρ αὐτῶν τὰ εἰκότα δι' ἑτέρου γράμματος, ἐν ᾧ περὶ τῶν καθ' Ἱπποκράτην στοιχείων ἐσκοπούμεθα. νυνὶ δ', ὅπερ ἐστὶν ἐφεξῆς ἐκείνῳ, ἁπάσας ἐξευρεῖν τῶν κράσεων τὰς διαφοράς, ὁπόσαι τ' εἰσὶ καὶ ὁποῖαι κατ' εἴδη τε καὶ γένη διαιρουμένοις, ἐν τῷδε | τῷ γράμματι δίειμι τὴν ἀρχὴν ἀπὸ τῆς τῶν ὀνομάτων ἐξηγήσεως ποιησάμενος.

Ἐπειδὰν μὲν γὰρ ἐκ θερμοῦ καὶ ψυχροῦ καὶ ξηροῦ καὶ ὑγροῦ κεκρᾶσθαι λέγωσι τὰ σώματα, τῶν ἄκρως τοιούτων ἀκούειν φασὶ χρῆναι, τουτέστι τῶν στοιχείων αὐτῶν, ἀέρος καὶ πυρὸς καὶ ὕδατος καὶ γῆς· ἐπειδὰν δὲ ζῷον ἢ φυτὸν ἤτοι θερμὸν ἢ ψυχρὸν ἢ ξηρὸν ἢ ὑγρὸν εἶναι λέγωσιν,² οὐκέθ' ὡσαύτως. οὐδὲ γὰρ δύνασθαι ζῷον οὐδὲν οὔτ' ἄκρως θερμὸν ὑπάρχειν ὡς πῦρ οὔτ' ἄκρως ὑγρὸν ὡς ὕδωρ. ὡσαύτως δ'

---

¹ post μὲν add. οὖν, K ² εἶναι λέγωσιν om. K

---

[1] The reference is a general one to the proponents of the four

BOOK I

1. That the bodies of animals are compounded from hot, cold, dry, and wet, and that all the parts are not equal in their *krasis*, has been adequately shown by the ancients—the best of both philosophers and doctors.[1] The probabilities regarding these were also stated by us in another work, in which we examined the elements according to Hippocrates.[2] What follows now in sequence, which is next in order to that, is to discover all the *differentiae* of the *krasias*. In this work, I shall go through how many there are and of what kinds, divided according to class and kind, making a start from the explanation of the terms.

When they say the bodies are compounded from hot, cold, dry, and wet, they say it is necessary to understand such things completely—that is to say, the elements themselves, which are air, fire, water, and earth. When they say an animal or plant is either hot, cold, dry, or wet, it is no longer similar, for no animal can either be completely hot as fire is, or completely wet, as water is. Likewise, no ani-

elements/qualities theory of the basic structure of matter, notably Empedocles, Hippocrates, and Aristotle, as opposed to the proponents of an "atomic" theory, notably Leucippus, Democritus, and in medicine specifically, Asclepiades. For further details, see Galen's *De elementis* referred to in the following note.

[2] Galen, *De elementis secundum Hippocratem*, I.413–508K; English translation by de Lacy, *Galen on the Elements*.

οὐδὲ ψυχρὸν ἢ ξηρὸν ἐσχάτως, ἀλλ' ἀπὸ τοῦ πλεονεκτοῦντος ἐν τῇ κράσει γίγνεσθαι τὰς προσηγορίας, ὑγρὸν μὲν καλούντων ἡμῶν, ἐν ᾧ πλείων ὑγρότητός ἐστι μοῖρα, ξηρὸν δ', ἐν ᾧ ξηρότητος· οὕτω δὲ καὶ θερμὸν μέν, ἐν ᾧ τὸ θερμὸν τοῦ ψυχροῦ πλεονεκτεῖ, ψυχρὸν δ', ἐν ᾧ τὸ ψυχρὸν τοῦ θερμοῦ. αὕτη μὲν ἡ τῶν ὀνομάτων χρῆσις.

2. Ὥρα δ' ἂν εἴη λέγειν ἤδη καὶ περὶ τῶν κράσεων αὐτῶν. ἡ μὲν δὴ πλείστη δόξα τῶν ἐπιφανεστάτων ἰατρῶν τε καὶ φιλοσόφων, ὑγράν τ' εἶναι καὶ θερμὴν κρᾶσιν, ἑτέραν τῆς ὑγρᾶς τε καὶ ψυχρᾶς, καὶ τρίτην ἐπὶ ταύταις τὴν ξηράν τε καὶ ψυχράν, ἑτέραν τῆς ξηρᾶς θ' ἅμα καὶ θερμῆς. ἔνιοι δ' ἐξ αὐτῶν ὑγρὰν μέν τινα καὶ ψυχρὰν ἅμα κρᾶσιν ὑπάρχειν φασὶ καὶ θερμὴν ἅμα καὶ ξηρὰν ἑτέραν, οὐ μὴν οὔτε τὴν θερμὴν ἅμα καὶ ὑγρὰν οὔτε τὴν ψυχρὰν ἅμα καὶ ξηράν. οὐδὲ γὰρ ἐγχωρεῖν οὔθ' ὑγρότητα πλεονεκτούσῃ θερμότητι συνδραμεῖν οὔτε ξηρότητα ψυχρότητι. δαπανᾶσθαι μὲν γὰρ ὑπὸ τοῦ θερμοῦ κρατοῦντος τὴν ὑγρότητα καὶ οὕτω θερμὸν ἅμα καὶ ξηρὸν γίγνεσθαι τὸ σῶμα, μένειν δ' ἄπεπτόν τε καὶ ἀκατέργαστον, ἐν οἷς ἂν σώμασιν ἀρρωστῇ τὸ θερμόν, ὥστ' ἀναγκαῖον εἶναι θερμότητος μὲν ἐπικρατούσης ἕπεσθαι ξηρότητα, ψυχρότητος δὲ πλεονεκτούσης ἀκολουθεῖν ὑγρότητα. οὗτοι μὲν δὴ κατὰ τάδε πεπείκασι σφᾶς αὐτούς, ὡς δύο εἰσὶν αἱ πᾶσαι διαφοραὶ τῶν κράσεων.

---

[3] This paragraph identifies two conflicting views on the pos-

mal can be entirely cold or dry. Rather, the appellations arise from what prevails in the *krasis*, since we call wet a part in which there is a greater amount of wetness, and dry a part in which there is a greater amount of dryness. In like manner too, a part is hot in which the hot prevails over the cold, and conversely cold in which the cold prevails over the hot. This is the actual use of the terms.

2. Now would also be a time to speak about the *krasias* themselves. Certainly, the majority opinion among the most distinguished doctors and philosophers is that there is a wet and hot *krasis*, another that is wet and cold, a third in addition to these that is dry and cold, and another that is dry and at the same time also hot. Some of them say, however, that there is a wet and cold *krasis*, and another that is hot and at the same time dry, but not one that is hot and wet at the same time, nor one that is cold and dry at the same time, because it is not possible for wetness to exist concurrently with overriding hotness, or dryness with overriding coldness. For the wetness is consumed by the predominant heat, and in this way the body becomes hot and dry at the same time, while in those bodies in which the hot remains unconcocted and imperfect, it would be weak, so that it is inevitable that when hotness prevails, dryness follows, and when coldness prevails, wetness follows. Therefore, these men have certainly persuaded themselves on the basis of these things that there are in all two *differentiae* of the *krasias*.[3]

---

sible combinations of the four qualities. Of the four possible combinations of the six theoretical possibilities, the first viewpoint accepts all four, that is, hot and dry, hot and wet, cold and dry, and cold and wet, which is Galen's position, while the second excludes hot and wet, and cold and dry.

GALEN

Ὅσοι δὲ τέτταρας εἶναι νομίζουσι, διχῶς τούτοις ἀντιλέγουσιν, ἔνιοι μὲν εὐθὺς τὸ πρῶτον ἀξίωμα μὴ συγχωροῦντες, ὡς ἐξικμάζεσθαι τὴν ὑγρότητα πρὸς τοῦ θερμοῦ κρατοῦντος ἀναγκαῖόν ἐστιν, ἔνιοι δὲ τοῦτο μὲν συγχωροῦντες, ἀμφισβητοῦντες δ' ἑτέρως. οἱ μὲν δὴ πρῶτοι τοῦ θερμοῦ μὲν ἔργον εἶναί φασι τὸ θερμαίνειν ὥσπερ τοῦ ψυχροῦ τὸ ψύχειν, τοῦ ξηροῦ δ' αὖ τὸ ξηραίνειν ὥσπερ τοῦ ὑγροῦ τὸ ὑγραίνειν. καὶ διὰ τοῦθ' ὅσα μὲν σώματα θερμὰ τὴν φύσιν ἐστὶν ἅμα καὶ ξηρὰ καθάπερ τὸ πῦρ, ᾗ μὲν θερμά, θερμαίνειν, ᾗ δὲ ξηρά, ξηραίνειν. ὅσα δ' ὑγρὰ καὶ θερμὰ καθάπερ ὕδωρ θερμόν, ὑγραίνειν ταῦτα καὶ θερμαίνειν [πέφυκεν ἀεὶ],³ ἐν ἑκατέρας κἀνταῦθα ποιότητος ἔργον ἐχούσης ἀχώριστον. οὔκουν συγχωροῦσιν, εἴ τι θερμόν⁴ ἐστιν, εὐθὺς τοῦτο καὶ ξηραίνειν, ἀλλ' εἰ μὲν ὑγρότης προσείη τῇ θερμότητι, θερμαίνειν ἅμα καὶ ὑγραίνειν ὥσπερ τὰ λουτρὰ τῶν γλυκέων ὑδάτων. εἰ δ' ὥσπερ θερμὸν οὕτω καὶ ξηρὸν εἴη καθάπερ τὸ πῦρ, οὐ θερμαίνειν μόνον, ἀλλὰ καὶ ξηραίνειν εὐθύς, οὐκ ἐκ τῆς θερμότητος τοῦτο λαβόν, ἀλλ' ἐκ τῆς συνούσης αὐτῷ ξηρότητος. ὑπομιμνῄσκουσι δ' ἐνταῦθα τῶν ἐν ἡλίῳ θερινῷ διατριψάντων ἐπὶ πλέον, εἶθ', ὡς εἰκός, αὐανθέντων ὅλον τε τὸ σῶμα καὶ ξηρὸν καὶ αὐχμηρὸν ἐχόντων καὶ διψώντων οὐκ ἀνεκτῶς. ἴασιν γὰρ αὐτοῖς εἶναί φασιν ἑτοίμην τε καὶ ῥᾴστην, οὐκ εἰ πίοιεν μόνον, ἀλλ' εἰ καὶ λούσαιντο θερμοῖς ὕδασι

ON TEMPERAMENTS, BOOK I

Those who think there are four [*krasias*] gainsay these men in two ways. Some right at the start do not accept the first assumption, that the wetness is necessarily dried out by the predominant hotness. Some, however, do accept this, but disagree on other grounds. The first say an action of the hot is to heat, just as that of the cold is to cool, the dry in turn to dry, just as the wet to wet. And because of this, those bodies that are hot in nature and dry at the same time, like fire, heat by virtue of their hotness and dry by virtue of their dryness. Those bodies that are wet and hot, like hot water, naturally always wet and heat, since here too each quality has one inseparable action. They do not, therefore, agree that, if something is hot, it immediately also dries, but if wetness is present with the hotness, it heats and wets at the same time, like baths of sweet, fresh waters[4] do. If, however, just as it is hot, so too it is also dry, like fire, not only does it heat but also immediately dries, and this is not taken from the hotness but from the dryness accompanying it. Here they call to mind those who, when they spend an excessive amount of time in the summer sun, are, as is to be expected, dried out in the whole body and become dry, parched, and intolerably thirsty. They say the cure for them is ready to hand and very easy, not only if they drink but also if they bathe in

---

[4] For the use of γλυκύς in relation to water, particularly the contrast of sweet and fresh with bitter and salt, see Aristotle, *Meteorology* 355a.

---

[3] *H's note on this reads:* ἀεὶ *post* πέφυκεν *om.* LO; *utrumque vocabulum delendum videtur (note 4, p. 3)*

[4] θερμόν, H; θερμαινόν, K

γλυκέσιν, ὡς τῆς ὑγρότητος, εἴτε μετὰ ψυχρότητος εἴτε μετὰ θερμότητος εἴη, τὸ ἑαυτῆς ἀεὶ δρᾶν δυναμένης, ὑγραίνειν τὰ πλησιάζοντα. κατὰ δὲ τὸν αὐτὸν λόγον φασὶ καὶ τὴν ξηρότητα ξηραίνειν ἀεί. τὸν γοῦν βορρᾶν ξηρὸν καὶ ψυχρὸν ἄνεμον ὑπάρχοντα[5] ξηραίνειν ἅπαντα. καὶ τοῦτ' εἶναι τὸ πρὸς Ὁμήρου λεγόμενον·

ὡς δ' ὅτ' ὀπωρινὸς Βορέης νεοαρδέ' ἀλωὴν
αἶψα ξηραίνει.

κατὰ δὲ τὸν αὐτὸν τρόπον καὶ τὸν τῆς μήκωνος ὀπὸν καὶ ἄλλα μυρία φάρμακα ξηραίνειν ἅμα καὶ ψύχειν. ὥστ' οὐκ ἀναγκαῖον, οὔτ', εἴ τι ψυχρόν, εὐθὺς τοῦτο καὶ ὑγρὸν ὑπάρχειν, οὔτ', εἴ τι θερμόν, εὐθὺς καὶ ξηρόν. οὔκουν οὐδὲ τὴν θερμὴν κρᾶσιν ἐξ ἀνάγκης εἶναι καὶ ξηράν, ἀλλὰ δύνασθαί ποτε τὸ μὲν θερμὸν τοῦ ψυχροῦ πλεονεκτεῖν ἐν τῇ κράσει τοῦ ζώου, τὸ δ' ὑγρὸν τοῦ ξηροῦ. καὶ γὰρ δὴ καὶ τὴν γένεσιν καὶ τὴν ἀλλοίωσιν καὶ τὴν μεταβολὴν ἐκ τῶν ἐναντίων εἰς τὰ ἐναντία γίγνεσθαι. τίς γοῦν εἰπών, ὅτι τὸ λευκὸν ἠλλοιώθη τε καὶ μετέβαλεν, ἐγένετο γὰρ θερμόν, οὐκ ἂν εἴη καταγέλαστος; ἐπιζητεῖ γὰρ ὁ λόγος οὐ τὴν κατὰ τὸ θερμὸν καὶ τὸ ψυχρὸν ἀντίθεσιν, ἀλλὰ τὴν κατὰ τὸ χρῶμα.[6] μεταβάλλει γὰρ τὸ μὲν λευκὸν εἰς τὸ μέλαν, ὥσπερ γε καὶ τὸ μέλαν εἰς τὸ λευκόν, τὸ δὲ θερμὸν εἰς τὸ ψυχρόν, ὥσπερ αὖ καὶ τὸ ψυχρὸν εἰς τὸ θερμόν· οὕτω δὲ καὶ τὸ μὲν ὑγρὸν εἰς τὸ ξηρόν, τὸ δ'

ON TEMPERAMENTS, BOOK I

warm, sweet fresh waters, as the wetness itself always has the power to act, whether in association with coldness or hotness, wetting what it comes into contact with. By the same token, they say the dryness always dries. Anyway, the north wind, which is a dry cold wind, dries [and cools][5] everything. And this is what Homer said:

> And when in harvest time the North Wind quickly parches again a well-watered field.[6]

In the same way too, the juice of the poppy and countless other medications dry and cool at the same time. As a result, it is not inevitable that if something is cold, it is also immediately wet, nor if something is hot, it is also immediately dry. It is not therefore necessary that the hot *krasis* is also dry. Sometimes the hotness in the *krasis* of the animal is able to gain ascendancy over the coldness, or the wetness over the dryness. And certainly also the genesis, alteration, and change occur from opposites to opposites. For who would in fact say that white was altered and changed and became hot—would this not be absurd? What the argument requires is not the antithesis pertaining to cold and hot, but that pertaining to color. For white may change to black, just as also black may change to white, while hot may change to cold, just as cold in turn may change to hot. In the same way too, wet may change

---

[5] Added following Kühn.
[6] Homer, *Iliad* 21.346–47, LCL 171 (A. T. Murray), 432–33 (translation after Murray).

---

[5] ὑπάρχοντα H; τυγχάνοντα K
[6] *add.* μεταβολήν K

αὖ ξηρὸν εἰς τὸ ὑγρόν. εἰ γὰρ δὴ φάσκοι τις ἠλλοιῶσθαι τὸ σῶμα τῷ τέως ὑγρὸν ὑπάρχον εἶναι τανῦν[7] λευκὸν ἢ τῷ τέως ξηρὸν ὂν τανῦν φαίνεσθαι μέλαν, οὐκ ἂν ὑγιαίνειν δόξειεν. εἰ δέ γε τὸ τέως ὑγρὸν νῦν ξηρὸν φαίη γεγονέναι ἢ τὸ πρότερον ὑπάρχον μέλαν νῦν εἶναι λευκὸν ἢ | ἐκ θερμοῦ ψυχρὸν ἢ ἐκ ψυχροῦ θερμὸν γεγονέναι, σωφρονεῖν τ' ἂν δόξειεν ὁ τοιοῦτος καὶ λέγειν τὰ εἰκότα.

Τὸ γὰρ μεταβάλλον, ᾗ μεταβάλλει, ταύτῃ μεταχωρεῖν δεῖ πρὸς τοὐναντίον. ἐγένετο γοῦν ἢ γίγνεται μουσικὸς ὅδε, φαμέν, ἐξ οὐ μουσικοῦ δηλονότι, καὶ γραμματικὸς ἐξ οὐ γραμματικοῦ καὶ ῥητορικὸς ἐξ οὐ ῥητορικοῦ· τὸ δ' ἐκ μουσικοῦ γραμματικὸν ἢ ἐκ γραμματικοῦ μουσικὸν ἢ ἐξ ἄλλου τινὸς τῶν ἑτερογενῶν γίγνεσθαί τι λέγειν ἄτοπον. ἐγχωρεῖ γὰρ τὸν τέως γραμματικὸν νῦν μουσικὸν εἶναι προσκτησάμενον τῇ γραμματικῇ τὴν μουσικήν, οὐκ ἀποβαλόντα τὴν γραμματικήν. καὶ μὴν εἰ προσεκτήσατό τι παραμένοντος τοῦ προτέρου, παντί που δῆλον, ὡς οὐκ ἠλλοιώθη κατὰ τὸ μένον· ὥστ' οὐκ ἐκ γραμματικοῦ μουσικὸς ἐγένετο· καὶ γὰρ καὶ νῦν ἔτι γραμματικός ἐστιν· ἀλλ' ἐξ ἀμούσου μουσικός· οὐ γὰρ ἔτ' ἐγχωρεῖ μένειν αὐτὸν ἄμουσον μουσικὸν ἤδη γεγονότα. πασῶν οὖν τῶν μεταβολῶν ὑπὸ τῶν ἐναντίων τε κἀκ τῶν ἐναντίων εἰς τὰ ἐναντία γιγνομένων δηλονότι καὶ τὸ ὑγρόν, | εἰ μεταβάλλοι ποτέ, καθ' ὅσον ὑγρόν, αὐτὸ

[7] τανῦν H; νῦν K

to dry and conversely dry to wet. Assuredly, if someone were to say the body which was at one time wet is now changed to white, or which was at one time dry now seems black, he would not seem to be in his right mind. If, however, what up to that time has been wet is now said to have become dry, or what was previously black is now said to be white, or to have become cold from hot or hot from cold, such a person would seem to be of sound mind and to be saying things that are likely.

For what is changed, in so far as it changes, needs to go toward the opposite to it. Anyway, when we say a particular person has become or is becoming a musician, clearly this is from not being a musician, and the same applies to becoming a grammarian from not being a grammarian and a rhetorician from not being a rhetorician, whereas to say someone has become a grammarian from being a musician, or a musician from being a grammarian, or from something else of the different classes, is to say something absurd. It is possible for someone who at one time was a grammarian to now be a musician, having gained the attributes of a musician besides those of a grammarian, without casting aside his being a grammarian. And it is somehow clear to everyone that, if he has acquired something else while the previous thing remains, he has not undergone a change in what remains. Consequently, he has not become a musician from being a grammarian. Furthermore, he is now still a grammarian, but has become a musician from not being a musician, for it is not possible for him to remain unmusical, if in fact he has already become a musician. Since, therefore, all changes occur by the opposites and from the opposites to the opposites, it is also clear what is wet, if at some time

τε ξηρανθήσεται καὶ τὸ ξηραῖνον αὐτὸ ξηρὸν λεχθήσεται.

Μὴ τοίνυν λεγέτωσαν,[8] φασίν, ὡς θερμὴν ἅμα καὶ ὑγρὰν κρᾶσιν οὐκ ἐγχωρεῖ γενέσθαι. θερμὴν μὲν γὰρ ἅμα καὶ ψυχρὰν εἶναι τὴν αὐτὴν ἢ ὑγρὰν ἅμα καὶ ξηρὰν οὐκ ἐγχωρεῖ· συνυπάρχειν γὰρ ἀλλήλαις οὐχ οἷόν τε καθ᾽ ἓν καὶ ταὐτὸν σῶμα τὰς ἐναντίας ποιότητας· ἅμα δ᾽ ὑγρόν τι καὶ θερμὸν καὶ ψυχρὸν ἅμα καὶ ξηρὸν εἶναι δυνατόν, ὡς ὅ τε λόγος ἀπέδειξε καὶ τὰ μικρῷ πρόσθεν εἰρημένα παραδείγματα. τοιοῦτος μὲν ὁ τῶν προτέρων λόγος.

Ὁ δὲ τῶν δευτέρων οὐδὲν ἄτοπον εἶναί φησιν, ὑποκειμένου τοῦ θερμοῦ δραστικωτάτου τῶν τεττάρων, ὡς μὴ μόνον εἰς τὸ ψυχρὸν ἀλλὰ καὶ εἰς τὸ ὑγρὸν ἐνεργεῖν, εἶναι κρᾶσιν ὑγρὰν καὶ θερμήν, ὅταν εἰς ταὐτὸν ἅμα συνέλθῃ πλῆθος ὑγρότητός τε καὶ θερμότητος εὐθὺς ἐν τῇ πρώτῃ γενέσει τοῦ ζῴου. ὁ δέ γ᾽ ἐκείνων λόγος οὐχ ὡς οὐκ ἄν ποτε γένοιτο καθ᾽ ἓν καὶ ταὐτὸν σῶμα τὸ μὲν ὑγρὸν τοῦ ξηροῦ πλέον, τὸ δὲ θερμὸν τοῦ ψυχροῦ δείκνυσιν, ἀλλ᾽ ὡς οὐκ ἂν διαμεῖναι τοιοῦτον ἄχρι παντός· ἀεὶ γὰρ ἐξικμαζόμενον ὑπὸ τοῦ θερμοῦ τὸ ὑγρὸν ἐν τῷ χρόνῳ ξηρὸν ἀποδείξει τὸ σῶμα καὶ οὕτως οὐκ ἂν ἔτι θερμὸν καὶ ὑγρόν, ἀλλὰ θερμὸν εἴη καὶ ξηρόν· αὐτὸ δ᾽ αὖ πάλιν τοῦτο τὸ θερμὸν καὶ τὸ ξηρὸν ἐπὶ προήκοντι τῷ χρόνῳ ψυχρὸν ἔσται καὶ ξηρόν. ἐπειδὰν γὰρ ἐκβοσκήσηται

[8] λεγόντων, K

ON TEMPERAMENTS, BOOK I

it changes in relation to how much wetness there is, will be dried, and what dries it will be said to be dry.

Accordingly, they should not state that it is impossible for a *krasis* to be hot and wet at the same time. The same *krasis* cannot be hot and cold at the same time, or wet and dry at the same time, for the opposite qualities cannot exist with each other in one and the same body. However, it is possible for something to be wet and hot at the same time and cold and dry at the same time, as the argument demonstrated and the examples stated a little earlier. Such then is the argument of the first [group].[7]

That of the second group states nothing that is strange, since heat is assumed to be the most active of the four qualities, such that if it not only acts on the cold but also on the wet, there is a wet and hot *krasis* whenever an abundance of wetness and an abundance of hotness come together straightaway in the first genesis of the animal. The argument of those people does not show that there could never be more wet than dry, or more hot than cold in one and the same body, but that it would not remain in such a state continuously, for the wetness is always dried out by the heat, and over time will make the body dry, and in this way it would no longer be hot and wet, but hot and dry, while this hot and dry body will in turn, with the further progression of time, become cold and dry. When the heat

517K

---

[7] This is Galen's own position, the first group being presumably his noted predecessors referred to in section 1. A member of the second group would be Erasistratus, who comes in for severe criticism in a number of places; see, for example, *De naturalibus facultatibus*, I.16.

τὴν ἰκμάδα πᾶσαν αὐτοῦ τὸ θερμόν, ἄρχεσθαι τοὐντεῦθεν ἤδη φασὶ καὶ αὐτὸ μαραίνεσθαι μηκέτ᾽ εὐποροῦν τροφῆς, ὅθεν ἐξήπτετο. θαυμαστὸν οὖν οὐδὲν εἶναι νομίζουσι καὶ κατ᾽ ἀρχὰς εὐθὺς ἐν τῇ πρώτῃ γενέσει τοῦ ζῴου συνδραμεῖν εἰς ταὐτὸν ἢ τὸ ὑγρὸν τοῦ ξηροῦ πλέον ἢ τὸ θερμὸν τοῦ ψυχροῦ. δυνατὸν δὲ κἂν τῷ χρόνῳ προϊόντι γενέσθαι τὴν τέως ὑγρὰν καὶ θερμὴν κρᾶσιν αὖθις ξηρὰν καὶ θερμήν, ὥσπερ αὖ πάλιν τὴν ξηρὰν καὶ θερμὴν ἀποσβεννυμένου τοῦ θερμοῦ ψυχρὰν καὶ ξηρὰν ἀποτελεσθῆναι.

Ὅτι μὲν οὖν ἐγχωρεῖ θερμὴν ἅμα καὶ ὑγρὰν εἶναί τινα καὶ ψυχρὰν | καὶ ξηρὰν ἑτέραν τῆσδε κρᾶσιν, ἐκ τούτων ἐπιδεικνύουσιν. ὅτι δὲ πλείους τῶν τεττάρων διαφορὰς κράσεων ἀδύνατον ὑπάρχειν, ἐκ τῶνδε πειρῶνται διδάσκειν. ὑποκειμένων γάρ, φασί, τεττάρων ποιοτήτων εἰς ἀλλήλας τὸ δρᾶν τε καὶ πάσχειν ἐχουσῶν, θερμότητός τε καὶ ψυχρότητος καὶ ξηρότητος καὶ ὑγρότητος, ἀντιθέσεις γίγνεσθαι δύο, τὴν μὲν ἑτέραν, ἐν ᾗ τὸ θερμὸν ἀντίκειται τῷ ψυχρῷ, τὴν δ᾽ ἑτέραν, ἐν ᾗ τὸ ξηρὸν τῷ ὑγρῷ, καὶ διὰ τοῦτο τέτταρας ἀποτελεῖσθαι τὰς πάσας συζυγίας. ἓξ μὲν γὰρ γίγνεσθαι τῶν τεττάρων ἀλλήλαις ἐπιπλεκομένων τὰς συζεύξεις, ἀλλὰ τὰς δύο τούτων ἀδυνάτους ὑπάρχειν. οὔτε γὰρ ὑγρὸν ἅμα καὶ ξηρὸν οὔτε θερμὸν ἅμα καὶ ψυχρὸν δύναται γενέσθαι σῶμα. λείπεται δὴ τέτταρας εἶναι συζυγίας κράσεων, ὑγρὰς μὲν δύο, ξηρὰς δὲ δύο θερμότητι καὶ ψυχρότητι διῃρημένας.

3. Ἃ μὲν οὖν οἱ χαριέστατοι τῶν πρὸ ἡμῶν ἰατρῶν

ON TEMPERAMENTS, BOOK I

has consumed all the wetness, they say it will from that time then begin to waste away, no longer having the ready store of nourishment on which it depended. There is nothing surprising, then, in their thinking that, immediately at the start, in the first genesis of the animal, the wetness is greater than the dryness and the hotness greater than the coldness, coming together in the same body. It is possible, with the progression of time, for wet and hot *krasis* to exist at one time and to become in turn dry and hot, just as also in turn it is possible, when the hot is quenched, for the dry and hot *krasis* to be rendered cold and dry.

That it is possible, therefore, for one *krasis* to be hot and wet at the same time, and another to be cold and dry, they demonstrated from these things.[8] That it is impossible for there to be more *differentiae* than the four *krasias*, they attempt to teach from the following. For they say, since there are four underlying qualities which are able to act on each other and be acted on—hotness and coldness, and dryness and wetness—two antitheses arise, the one being the hot opposed to the cold and the other the dry opposed to the wet, and because of this they make four conjunctions in all, for six conjunctions arise when the four qualities are combined with each other, but two of these are impossible. For a body cannot become simultaneously wet and dry or hot and cold. What remain, then, are four conjunctions of *krasias*; two are wet while two are dry, differentiated by hot and cold in each case.

3. These then are what the most distinguished of the

[8] Apparently Galen is referring to the first group here, although this is not made clear in the Greek.

GALEN

τε καὶ φιλοσόφων εἰρήκασι, ταῦτ᾽ ἐστίν. ἃ δ᾽ ἐγὼ παραλιπεῖν αὐτοὺς νομίζω, λέγειν ἤδη καιρός.[9] ἓν μὲν δὴ καὶ πρῶτον, ὅτι τὴν εὔκρατον,[10] ὥσπερ οὐχ ἁπασῶν τῶν εἰρημένων ἀρετῇ θ᾽[11] ἅμα καὶ δυνάμει προὔχουσαν, ἐπελάθοντό τε καὶ τελέως παρέλιπον, ὥσπερ οὐδ᾽ ὅλως οὖσαν, καίτοι μηδὲ φθέγξασθαί τι χωρὶς ἐκείνης ὑπὲρ τῶν ἄλλων δυνάμενοι. τὸ γοῦν ἐν τῇ θερμῇ κράσει πλεονεκτεῖν τὸ θερμὸν ἔν τε τῇ ψυχρᾷ τὸ ψυχρὸν οὐδ᾽ ἐπινοῆσαι δυνατὸν ἄνευ τοῦ προτέραν ὑποθέσθαι[12] τὴν εὔκρατον. οὐδὲ γὰρ οὐδὲ τὴν ὑγιεινὴν δίαιταν εἰς ἄλλο τι βλέποντες ἐξευρίσκουσιν ἢ εἰς τὴν εὔκρατον ἐκείνην φύσιν, τὸ μὲν θερμότερον τοῦ δέοντος σῶμα κελεύοντες ἐμψύχειν, τὸ δ᾽ αὖ ψυχρότερον θερμαίνειν, ὡσαύτως τὸ μὲν ὑγρότερον ξηραίνειν, τὸ δὲ ξηρότερον ὑγραίνειν, ἀντεισάγοντες ἀεὶ δηλονότι τῷ πλεονάζοντι τὸ λεῖπον, ὡς εὔκρατόν τινα καὶ μέσην ἐργάσασθαι κατάστασιν. ἣν οὖν ἀεὶ μεταδιώκουσι καὶ πρὸς ἣν ἀποβλέποντες ἐπανορθοῦνται τὰς δυσκράτους, ἐγὼ μὲν ἠξίουν ἁπασῶν πρώτην λέγεσθαι πρὸς αὐτῶν. οἱ δ᾽ ἄρα τοσοῦτον ἀποδέουσι τοῦ μεμνῆσθαι πρώτης, ὥσθ᾽ ὅλως παραλείπουσιν αὐτήν. ἀλλ᾽ οὐ παραλέλειπται, φασί τινες ἐξ αὐτῶν, ἐν γὰρ τῇ θερμῇ καὶ ὑγρᾷ περιέχεται. καὶ πῶς οὐχὶ πέντε λέγετε τὰς πάσας εἶναι κράσεις, ἀλλὰ τέτταρας, εἴπερ τῆς ἀρίστης μέμνησθε; δυοῖν γὰρ θάτερον,

---

[9] *The first two sentences in section 3 here are given as the final sentence of section 2 in* K.

ON TEMPERAMENTS, BOOK I

doctors and philosophers before us have spoken of. Now is an appropriate time for me to state what I think they have passed over. | First and foremost, they forgot about and completely left out the *eukratic* [nature], as if it were not superior in excellence and power to all those mentioned. It is as if it did not exist at all, and indeed, apart from that, they are unable to say anything about the others. At any rate, it is not possible to conceive of the hot being predominant in the hot *krasis* or cold in the cold *krasis*, without the prior postulation of the *eukratic*. Nor do people discover the healthy regimen by looking at anything other than that *eukratic* nature. They order cooling for the body that is hotter than it should be, and conversely, heating for the body that is colder than it should be, just as they do drying for that body which is wetter and wetting for that body which is drier, always introducing what is lacking to what is obviously in excess, so as to create a *eukratic* and median state. Therefore, I think what should be said first of all by these people is what they are always pursuing and look toward restoring the *dyskratic* [states] to. However, they fail to mention this first to such an extent that | they leave it aside completely. Some of them say it has not been left aside, for it is contained in the hot and wet. But how do you not say there are not five *krasias* in all, but four, if you make mention of the best? One or other of [two possibilities must obtain]: either one

---

[10] *add.* φύσιν *post* εὔκρατον, K
[11] ὥσπερ οὐχ ἁπασῶν τῶν εἰρημένων ἀρετῇ θ' H; ὥσπερ οὐκ ἀρετῇ τε K
[12] ὑποθέσθαι H; ἐπιθέσθαι K

ἢ τῶν δυσκράτων ἀνάγκη παραλελεῖφθαι μίαν ἢ τὴν εὔκρατον. ἐγὼ μὲν δὴ σαφῶς οἶδα τὴν εὔκρατον αὐτοὺς παραλιπόντας ἐξ ὧν ἀξιοῦσιν. ἐπειδὰν γὰρ θερμὴν καὶ ξηρὰν ἢ ψυχρὰν καὶ ὑγρὰν ἤ τιν' ἄλλην λέγωσι κρᾶσιν, οὐ τῶν ἄκρων ἡμᾶς ἀκούειν χρῆναι ποιοτήτων, ἀλλὰ κατὰ τὴν πλεονεκτοῦσαν ἀεὶ γίγνεσθαι τὴν προσηγορίαν.

Εἰ δ' οὐ βούλονται τὴν εὔκρατον παραλελεῖφθαι, τῶν ἄλλων τινὰ δειχθήσονται παραλιπόντες. ἔστω γὰρ εὔκρατον εἶναι τὴν ὑγρὰν καὶ θερμήν, ὥσπερ αὐτοὶ βούλονται. παραλελοίπασιν ἄρα[13] τὴν ἀντικειμένην τῇ ψυχρᾷ καὶ ξηρᾷ δυσκρασίᾳ, ἐν ᾗ τὸ ὑγρὸν πλεονεκτεῖ καὶ τὸ θερμόν. ἀλλ' αὕτη, φασίν, ἐστὶν ἥδε. καὶ πῶς ἐνδέχεται καὶ πλεονεκτεῖν ἅμα καὶ μὴ πλεονεκτεῖν τὸ θερμὸν καὶ κρατεῖσθαι καὶ μὴ κρατεῖσθαι τὸ ψυχρόν; εἰ μὲν γὰρ εὔκρατός ἐστιν, οὐδὲν οὐδενὸς ἀμέτρως ἐπικρατεῖ, εἰ δὲ δύσκρατος, ἀνάγκη πλεονεκτεῖν τι τῶν ἐκ τῆς ἀντιθέσεως. ἀλλ' αὐτὸ τοῦτο, φασίν, ἴδιόν ἐστι τῆς εὐκράτου τὸ κρατεῖν ἐν αὐτῇ τὸ μὲν θερμὸν τοῦ ψυχροῦ, τὸ δ' ὑγρὸν τοῦ ξηροῦ. κρατήσαντος γὰρ δὴ τοῦ ψυχροῦ μετρίως μέν, οὐκ ἀγαθὴν εἶναι τὴν κρᾶσιν, ἔτι δὲ μᾶλλον, νόσον ἤδη γίγνεσθαι, καθάπερ, εἰ καὶ σφοδρῶς κρατήσειε, θάνατον. οὕτω δὲ κἀπὶ τοῦ ξηροῦ συμπίπτειν ἐν ἀρχῇ μὲν δυσκρασίαν, ἐπὶ πλέον δὲ νόσον, ἐπὶ πλεῖστον δὲ κρατήσαντος θάνατον, ὥσπερ οὐχὶ κἀπὶ τῆς ὑγρᾶς καὶ θερμῆς ταῦτα συμπίπτοντα. τίς γὰρ οὐκ ἂν ὁμολογήσειεν, ἐπειδὰν μὲν ἐπ' ὀλίγον ᾖ τὸ θερμὸν τοῦ

ON TEMPERAMENTS, BOOK I

of the *dyskrasias* must be left aside, or the *eukrasia*. I know for sure it is the *eukratic* that they leave aside from those they deem worthy of mention, for when they speak of hot and dry, or cold and wet, or some other *krasis*, we don't need to understand the extreme qualities but always that which is termed predominant.

If, however, they do not wish the *eukratic* to be left aside, they will be shown to have omitted one of the others. Let it be the case that the wet and hot *krasis* is *eukratic*, as they themselves wish. Then clearly they will have omitted the *krasis* antithetical to the cold and dry *dyskrasia* in which wetness and hotness prevail. But this, they say, is the same. But how is it possible for the hotness to prevail and not prevail at the same time, and the cold to be overcome and not to be overcome? For if there is *eukrasia*, no quality prevails disproportionately over any other, while if there is *dyskrasia*, one of those from the [two] antitheses necessarily prevails. But this very thing is, they say, characteristic of the *eukratic* and in this the hot prevails over the cold and the wet over the dry. For if the cold prevails moderately, it is not a good *krasis*, while if it prevails still more, it already becomes a disease, just as if it were to prevail strongly, it would be fatal. And the same happens in the case of the dryness prevailing—in the beginning there is *dyskrasia*, while if it is greater, disease, and if to the greatest extent, death, as if these things don't happen in the cases of the wet and hot prevailing. For who would not agree that, when the hot should happen to gain

---

[13] *post* ἄρα *add.* σαφῶς K

ψυχροῦ τύχῃ πλεονεκτῆσαν ἢ τὸ ὑγρὸν τοῦ ξηροῦ, δυσκρασίαν οὕτω γιγνομένην, ἐπειδὰν δ' ἐπὶ πλέον, νόσον, ἐπειδὰν δ' ἐπὶ πλεῖστον, θάνατον; ὁ γὰρ αὐτὸς ἐπ' ἀμφοῖν λόγος. ἢ μηδὲ τὰς ἀμέτρως ὑγρὰς καὶ θερμὰς καταστάσεις αἰτιώμεθα μηδ' ὅσα μεθ' ὑγρότητος ἀμέτρου νοσήματα συνίσταται θερμά, μηδὲ ταῦθ' ὁμολογῶμεν εἶναι νοσήματα.

Πρὸς δὴ τοὺς τοιούτους λόγους ἀπομαχόμενοί τινες τῶν ἀπ' Ἀθηναίου τοῦ Ἀτταλέως ὁμόσε χωροῦσιν οὔτε κατάστασιν ὑγρὰν καὶ θερμὴν μέμφεσθαι λέγοντες οὔθ' εὑρεθῆναί τι νόσημα φάσκοντες ὑγρὸν καὶ θερμόν, ἀλλὰ πάντως ἢ θερμὸν καὶ ξηρὸν ὑπάρχειν ὡς τὸν πυρετόν, ἢ ψυχρὸν καὶ ὑγρὸν ὡς τὸν ὕδερον, ἢ ψυχρὸν καὶ ξηρὸν ὡς τὴν μελαγχολίαν. ἐπιμέμηνται δ' ἐνταῦθα καὶ τῶν ὡρῶν τοῦ ἔτους, ὑγρὸν μὲν καὶ ψυχρὸν εἶναι τὸν χειμῶνα φάσκοντες, ξηρὸν δὲ καὶ θερμὸν τὸ θέρος καὶ ψυχρὸν καὶ ξηρὸν τὸ φθινόπωρον, εὔκρατον δ' ἅμα καὶ θερμὴν καὶ ὑγρὰν ὥραν εἶναί φασι τὸ ἔαρ. οὕτω δὲ καὶ τῶν ἡλικιῶν τὴν παιδικὴν εὔκρατον θ' ἅμα καὶ θερμὴν καὶ ὑγρὰν εἶναί φασιν. δηλοῦσθαι δὲ τὴν εὐκρασίαν αὐτῆς νομίζουσι κἀκ τῶν ἐνεργειῶν τῆς φύσεως ἐρρωμένων τηνικαῦτα μάλιστα. καὶ μὲν δὴ καὶ τὸν θάνατόν φασιν εἰς ξηρότητα καὶ ψῦξιν ἄγειν τὰ τῶν ζῴων σώματα. καλεῖσθαι γοῦν ἀλίβαντας τοὺς νεκροὺς ὡς ἂν οὐκέτι λιβάδα καὶ ὑγρότητα κεκτημένους οὐδεμίαν, ἐξατμισθέντας θ' ἅμα διὰ τὴν ἀποχώρησιν τοῦ θερμοῦ καὶ παγέντας ὑπὸ τῆς ψύξεως. ἀλλ' εἴπερ ὁ θάνατος,

ON TEMPERAMENTS, BOOK I

a slight predominance over the cold, or the wet over the dry, in this way a *dyskrasia* arises, while if still more, a disease, and if to the greatest extent, death? It is the same argument in both cases, for we neither inculpate immoderately wet and hot states, nor those hot diseases which coexist with immoderate wetness, nor do we agree these are diseases.

In regard to such arguments, certain contentious followers of Athenaeus the Attaleian resile on the issue, saying that neither a wet and hot state is deserving of blame, nor is any disease discovered that is wet and hot, but in every case there is either hotness and dryness as in fever, or coldness and wetness as in dropsy, or coldness and dryness as in melancholia. Here, however, they also make mention of the seasons of the year, saying wet and cold relate to winter, dry and hot to summer, cold and dry to autumn, while they say spring is *eukratic* and at the same time a hot and wet season. In this way too, of the ages, they say childhood is *eukratic* and at the same time hot and wet. They think the *eukrasia* of this is shown by the functions of nature being particularly strong at that time. Furthermore, they say death leads bodies of animals to dryness and coldness. At all events, the dead are called *alibantes* (corpses)[9] as they would no longer have any acquired moisture and wetness, being simultaneously turned into vapor through the departure of the heat and being congealed by the cooling. But, they say, if death is such as this,

---

[9] On the term ἀλίβας in this sense, see Plato, *Republic* 387C.

φασί, τοιοῦτος, ἀναγκαῖον ἤδη τὴν ζωήν, ὡς ἂν ἐναντίαν οὖσαν αὐτῷ, θερμήν τ' εἶναι καὶ ὑγράν· καὶ μὴν εἴπερ ἡ ζωή, φασί, θερμόν τι χρῆμα καὶ ὑγρόν,[14] ἀνάγκη πᾶσα καὶ τὴν ὁμοιοτάτην αὐτῇ κρᾶσιν ἀρίστην ὑπάρχειν· εἰ δὲ τοῦτο, παντί που δῆλον, ὡς εὐκρατοτάτην, ὥστ' εἰς ταὐτὸ συμβαίνειν ὑγρὰν καὶ θερμὴν φύσιν εὐκράτῳ καὶ μηδὲν ἄλλ' εἶναι τὴν εὐκρασίαν ἢ τῆς ὑγρότητός τε καὶ θερμότητος ἐπικρατούσης. οἱ μὲν δὴ τῶν ἀμφὶ τὸν Ἀθήναιον λόγοι τοιοίδε. δοκεῖ δέ πως ἡ αὐτὴ δόξα καὶ Ἀριστοτέλους εἶναι τοῦ φιλοσόφου καὶ Θεοφράστου γε μετ' αὐτὸν καὶ τῶν Στωϊκῶν, ὥστε καὶ τῷ πλήθει τῶν μαρτύρων ἡμᾶς δυσωποῦσιν. ἐγὼ δὲ περὶ μὲν Ἀριστοτέλους, ὅπως ἐγίγνωσκεν ὑπὲρ θερμῆς καὶ ὑγρᾶς κράσεως, ἴσως ἄν, εἰ δεηθείην, ἐπὶ προήκοντι τῷ λόγῳ δείξαιμι· δοκοῦσι γάρ μοι παρακούειν αὐτοῦ.

4. Τὸ δέ γε νῦν ἔχον πειράσομαι πρῶτον ἐνδείξασθαι τοῖς λέγουσι ταῦτα, πῇ ποτε σοφίζονται σφᾶς αὐτούς, εἶτ' ἐφεξῆς ἀποδεῖξαι τὸν ἅπαντα λόγον εἰς ἓν ἀθροίσας κεφάλαιον. ὅτι μὲν δὴ τὸ ἔαρ οἴονται θερμὸν εἶναι καὶ ὑγρὸν ἅμα καὶ εὔκρατον, ἐνταῦθα σοφίζονται προφανῶς. οὔτε γὰρ ὑγρὸν ὡς ὁ χειμὼν οὔτε θερμὸν ὡς τὸ θέρος, ὥστ' οὐδέτερον ἀμέτρως. ἀμετρίας δ' ἦν ἕκαστον τῶν τοιούτων ὀνομάτων καὶ κατ' αὐτοὺς ἐκείνους δηλωτικόν. ἐσφάλησαν δὲ διχῶς, πρῶτον μὲν ἐκ τοῦ βούλεσθαι πάντως ἐν ταῖς ὥραις εὑρεῖν τὴν τετάρτην συζυγίαν τῶν κράσεων, ἔπειτα δ' ἐκ τοῦ θερμότερον ἢ κατὰ τὸν χειμῶνα καὶ ὑγρότερον

ON TEMPERAMENTS, BOOK I

of necessity life, being the opposite to this, is hot and wet. And if life, they say, is something hot and wet, it is necessary in every way also that the *krasis* most similar to this is the best. If this is so, it is clear to everyone that it is the most *eukratic*, so that a wet and hot nature corresponds to *eukratic*, and the *eukratic* is nothing else than the predominance of the wet and hot. Such then are the arguments of the followers of Athenaeus. However, it seems in a way that both Aristotle the philosopher and Theophrastus after him are of the same opinion, and also the Stoics. As a consequence, they put us to shame by the number of witnesses. Regarding Aristotle, and how he knew about the hot and wet *krasis,* I shall show perhaps, if the need arises, in the discussion to come, for they seem to me to misunderstand him.

4. Now I shall attempt first to point out to them the things they say, and how at times they deceive themselves, then next to demonstrate the whole argument collected together under one heading. Certainly, in that they think spring is hot and wet, and at the same time *eukratic*, they are here obviously deceiving themselves. For it is neither wet like winter nor hot like summer, so that in neither [quality] is it disproportionate. And in relation to those people themselves, each of such terms is indicative of disproportion. However, they are mistaken on two counts: first, from altogether wishing to find in the seasons the fourth conjunction of the *krasias,* and then to postulate the spring to be both hotter in relation to the winter and

---

[14] *post* ὑγρόν *add.* ἐστιν K

ἢ κατὰ τὸ θέρος[15] ὑπολαμβάνειν ὑπάρχειν τὸ ἔαρ. ἀλλ' οὔτ' ἀναγκαῖον ἐν ταῖς ὥραις ὑποτίθεσθαι τὴν τετάρτην συζυγίαν τῶν κράσεων, εἰ μὴ καὶ φαίνοιτο, καὶ τὸ παραβάλλειν αὐτὸ ταῖς ἑκατέρωθεν ὥραις οὐδὲν μᾶλλον ὑγρὸν καὶ θερμὸν ἢ ξηρὸν ἀποδείξει καὶ ψυχρόν.[16] εἰ μὲν γὰρ ἀμετρίας ἐστὶν ὀνόματα τὸ θερμὸν καὶ τὸ ὑγρόν, οὐκ ἀληθεύεται κατ' αὐτούς· σύμμετρον γὰρ ἐν ἅπασι τὸ ἔαρ. εἰ δ' ὅτι θέρους μέν ἐστιν ὑγρότερον, χειμῶνος δὲ θερμότερον, ὑγρόν ἐστι καὶ θερμόν, οὐδὲν ἧττον αὐτὸ ψυχρὸν καὶ ξηρὸν νομίζεσθαι προσήκει, διότι θέρους μέν ἐστι ψυχρότερον, χειμῶνος δὲ ξηρότερον. ἢ τίς ἀποκλήρωσις, ἓν μὲν τῶν ἐκ τῆς ἀντιθέσεως ἀπὸ τοῦ χειμῶνος, ἓν δ' ἀπὸ τοῦ θέρους λαμβάνειν; ἐν ἑκατέροις γὰρ ἀμφοτέρων διαφέρον οὐκ ἐξ ἡμίσεος ὀφείλει τὴν παραβολὴν ἀλλ' ὁλόκληρον ἴσχειν. καὶ μὴν εἴπερ οὕτω γίγνοιτο, τἀναντία φήσομεν ὑπάρχειν αὐτό.[17] θερμὸν μὲν γὰρ ἔσται καὶ ξηρόν, εἰ τῷ χειμῶνι, ψυχρόν, δ' αὖ καὶ ὑγρόν, εἰ τῷ θέρει παραβάλλοιτο. κατ' οὐδετέραν οὖν τῶν παραβολῶν ὁλοκλήρως γιγνομένην ὑγρὸν ἔσται καὶ θερμόν. εἰ δ' ἔξεστιν ἐκείνοις ἐξ ἑκατέρας αὐτῶν ἥμισυ λαβοῦσιν ὑγρὸν ἀποφαίνειν αὐτὸ καὶ θερμόν, ἐξέσται δήπου καὶ ἡμῖν ἐπὶ θάτερον ἥμισυ μετελθοῦσι ξηρὸν καὶ ψυχρὸν ἀποφῆναι, ξηρὸν μὲν ὡς πρὸς τὸν χειμῶνα, ψυχρὸν δ' ὡς πρὸς τὸ θέρος. ἅπαντ' οὖν οὕτως ἔσται τὸ ἔαρ, ὑγρὸν καὶ ξηρὸν καὶ ψυχρὸν καὶ θερμόν.

Ἀλλ' οὐδὲ κατ' αὐτοὺς ἐκείνους οἷόν τ' ἐστὶν ἐν ἑνὶ

ON TEMPERAMENTS, BOOK I

wetter in relation to the summer. But it is not necessary to assume in the seasons the fourth conjunction of the *krasias*, if it does not appear, and comparing this to the seasons on either side will demonstrate nothing more than it is wet and hot rather than dry and cold. For if the terms hot and wet refer to disproportions, they are not true in themselves, | for spring is moderate in all respects. If, however, because it is wetter than summer and hotter than winter, it is wet and hot, it is no less fitting to think it cold and dry, since it is colder than summer and drier than winter. Isn't it unreasonable to take one of the terms of the antithesis from winter and one from summer? For in each there is a difference of both terms, so it is of benefit for the comparison to be from the whole and not from a part. And if it does occur in this way, we shall be saying it is the opposite to this, for it will be hot and dry if compared to winter, but conversely cold and also wet if compared to summer. In respect of neither of the comparisons, when carried out properly, will it be wet and hot. If, however, it is possible for those people to take half from each of these, and to declare it wet and hot, it will also be possible, I presume, for us that they follow the other half and declare it dry and cold—dry in relation to winter and cold in relation to summer. Therefore, in this way, spring will be all these—wet and dry, cold and hot.

But | it is not possible, even for those people them-

---

15 *post* τὸ θέρος: ὑπολαμβάνειν ὑπάρχειν τὸ ἔαρ. H; ὑπάρχειν αὐτό. K

16 ψυχρόν H; ξηρόν K

17 αὐτό H; αὐτῷ K

καὶ ταὐτῷ πράγματι τὰς τέτταρας ἐπικρατῆσαι ποιότητας. οὔκουν οὔτε θέρει παραβάλλειν οὔτε χειμῶνι τὸ ἔαρ, ἀλλ' αὐτὸ καθ' ἑαυτὸ σκοπεῖσθαι δίκαιον. οὐδὲ γὰρ οὐδὲ τὸν χειμῶνα διὰ τοῦτο λέγομεν ὑγρὸν εἶναι καὶ ψυχρόν, ὅτι τῶν ἄλλων ὡρῶν ἐστιν ὑγρότατός τε καὶ ψυχρότατος, ἀλλὰ τοῦτο μὲν ἄλλως αὐτῷ συμβέβηκεν· ὅτι δὲ πλεονεκτεῖ κατ' αὐτὸν ἡ μὲν ὑγρότης τῆς ξηρότητος, ἡ δὲ ψυχρότης τῆς θερμότητος, διὰ τοῦθ' ὑγρὸς καὶ ψυχρὸς εἶναι λέγεται. κατὰ ταὐτὰ δὲ καὶ τὸ θέρος, ὅτι κἀν τούτῳ τὸ μὲν ὑγρὸν ἀπολείπεται τοῦ ξηροῦ, τὸ δὲ ψυχρὸν τοῦ θερμοῦ, διὰ τοῦτο θερμὸν εἶναι λέγεται καὶ ξηρόν. καὶ γὰρ καὶ δίκαιον ἐκ τῆς οἰκείας φύσεως ἑκάστην τῶν ὡρῶν ἐξεταζομένην, ἀλλὰ μὴ πρὸς ἄλλην τινὰ παραβαλλομένην ἢ θερμὴν ἢ ψυχρὰν ἢ ξηρὰν ἢ ὑγρὰν ὀνομάζεσθαι. καὶ δὴ καὶ σκοπουμένῳ σοι κατὰ τάδε φανεῖται τὸ ἔαρ ἀκριβῶς μέσον ἁπασῶν τῶν ὑπερβολῶν. οὔτε γὰρ ὡς ἐν χειμῶνι πλεονεκτεῖ τὸ ψυχρὸν ἐν αὐτῷ τοῦ θερμοῦ οὔθ' ὡς ἐν θέρει πλεονεκτεῖται. κατὰ ταὐτὰ δὲ καὶ ξηρότητός τε καὶ ὑγρότητος ἰσομοιρία τίς ἐστιν ἐν αὐτῷ μήθ' ὡς ἐν θέρει κρατοῦντος τοῦ ξηροῦ μήθ' ὡς ἐν χειμῶνι τοῦ ὑγροῦ καὶ διὰ τοῦτ' ὀρθῶς εἴρηται πρὸς Ἱπποκράτους· "ἦρ δὲ ὑγιεινότατον καὶ ἥκιστα θανατῶδες."

Ἀλλὰ καὶ τὸ φθινόπωρον ἧττον μὲν ἢ τὸ θέρος θερμόν, ἧττον δ' ἢ ὁ χειμὼν ψυχρόν. ὥστε ταύτῃ μὲν οὔτε θερμὸν ἁπλῶς οὔτε ψυχρόν, ἀμφότερα γάρ ἐστι, καὶ οὐδέτερον ἄκρως. ἕτερον δέ τι πρόσεστιν αὐτῷ

selves, for the four qualities to prevail in one and the same thing. It is not right, therefore, to compare spring to summer or to winter but to consider it in relation to itself. Because of this, we do not say winter is wet and cold because it is wettest and coldest compared to the other seasons. This is otherwise incidental to it. It is because, in relation to itself, wetness prevails over the dryness and coldness over the hotness. Because of this, it is said to be wet and cold. The same applies to the summer because in this the wetness is lacking compared to the dryness and the coldness compared to the hotness, and for this reason it is said to be hot and dry. For it is also right to examine carefully each of the seasons from the characteristic nature but not to compare it to some in order to term it hot or cold or dry or wet. And certainly, to you investigating these things, the spring appears precisely in the middle of all the excesses. For | in the spring, the cold does not prevail over the hot, as it does in winter, nor is it prevailed over as in summer. In the same way too, dryness and wetness are of equal parts in it, since the dryness does not prevail as in summer nor the wetness as in winter, and because of this, it was correctly said by Hippocrates that, "Spring is the most healthy and least deadly."[10]

But the autumn too is less hot than the summer, while it is less cold than the winter. As a consequence of this, it is neither absolutely hot nor absolutely cold, for it is both but neither to the extreme. There is, however, another bad

---

[10] Hippocrates, *Aphorisms* 3.9. The aphorism in full reads: "It is in autumn that diseases are most acute and, in general, most deadly; spring is most healthy and least deadly"; see LCL 150 (W. H. S. Jones), 124–25 (translation after Jones).

κακόν, ὅπερ ἐπεσημήνατο καὶ Ἱπποκράτης ἐν Ἀφορισμοῖς εἰπών· "ὁκόταν τῆς αὐτῆς ἡμέρης ὁτὲ μὲν θάλπος, ὁτὲ δὲ ψῦχος ποιέῃ, φθινοπωρινὰ τὰ νοσήματα προσδέχεσθαι χρή." καὶ τοῦτό γ᾽ ἐστὶ τὸ μάλιστα νοσῶδες ἐργαζόμενον τὸ φθινόπωρον, ἡ ἀνωμαλία τῆς κράσεως. οὐκ ὀρθῶς οὖν εἴρηται ψυχρὸν καὶ ξηρόν, οὐ γάρ ἐστι ψυχρὸν αὐτὸ καθ᾽ αὑτὸ θεωρούμενον, ὥσπερ ὁ χειμών, ἀλλὰ τῷ θέρει παραβαλλόμενον ἐκείνου ψυχρότερον. οὐ μὴν οὐδ᾽ ὁμαλῶς εὔκρατον, ὡς τὸ ἔαρ,[18] ἀλλ᾽ ἐν τούτῳ δὴ καὶ μάλιστα διενήνοχεν ἐκείνης τῆς ὥρας, ὅτι τὴν εὐκρασίαν τε καὶ τὴν ὁμαλότητα διὰ παντὸς ἴσην οὐ κέκτηται. πολὺ γὰρ θερμότερόν ἐστι κατὰ τὴν μεσημβρίαν ἢ κατὰ τὴν ἕω τε καὶ τὴν ἑσπέραν. ὑγρότητος δὲ καὶ[19] ξηρότητος οὐκ ἀκριβῶς μέν ἐστι μέσον, ὡς τὸ ἔαρ, ἀλλ᾽ ἐπὶ τὸ ξηρότερον ῥέπει. λείπεται δὲ κἀν τούτῳ τοῦ θέρους, οὐ μὴν τοσοῦτόν γ᾽ ὅσον θερμότητι. δῆλον οὖν, ὡς οὐδὲ τὸ φθινόπωρον ἁπλῶς οὕτω ῥητέον, ὡς ἐκεῖνοι λέγουσι, ψυχρόν τ᾽ εἶναι καὶ ξηρόν. ἄκρως μὲν γὰρ οὐδέτερόν ἐστιν, ἐπικρατεῖ δ᾽ ἐν αὐτῷ τὸ ξηρὸν τοῦ ὑγροῦ καὶ δικαίως ἂν λεχθείη ταύτῃ μὲν ξηρόν, ἐν δὲ τῇ κατὰ θερμότητα καὶ ψυχρότητα διαφορᾷ μικτὸν ἐξ ἀμφοῖν καὶ ἀνώμαλον.

Ὥστ᾽ εἴπερ τὰς τέτταρας συζυγίας τῶν κράσεων εἰς τὰς τέτταρας ὥρας διανεῖμαι σπουδάζουσιν, ἴστωσαν οὐ μόνον ἦρι κακῶς προσάψαντες ὑγρότητα καὶ

ON TEMPERAMENTS, BOOK I

aspect to autumn, which Hippocrates also remarked on, saying in the *Aphorisms*: "Whenever heat and cold occur during the same day, one must expect the autumnal diseases."[11] And this, the non-uniformity of the *krasis*, is what makes the autumn particularly baneful. Therefore, it is not correctly described as cold and dry, for it is not to be considered cold in itself, like the winter, but when compared to summer | it is colder than that. Nor again is it uniformly *eukratic*, like the spring, but in this differs most from that season because it has not acquired the equal *eukrasia* and evenness throughout. For autumn is much hotter at midday than it is at dawn or dusk, and it is not precisely midway between wetness and dryness as spring is, but tends toward being drier. However, even in this, it is wanting compared to summer, but not in fact as much as it is wanting in hotness. Clearly, then, one must not speak absolutely of autumn as being cold and dry in this way, as those men say. It is neither to the extreme, although in it the dryness prevails over the wetness in this, and it could legitimately be described as dry on this basis, while in the difference pertaining to hotness and coldness, it is a mixture of both and non-uniform.

As a consequence, if they hope to apportion the four conjunctions of the *krasias* to the four seasons, they should know that not only do they wrongly attribute the wet and

---

[11] Hippocrates, *Aphorisms* 3.4, LCL 150 (Jones), 122–23.

---

[18] *post* τὸ ἔαρ,: ἀλλ' ἐν τούτῳ δὴ καὶ μάλιστα H; ἀλλ' ὅτι μάλιστα τούτῳ K

[19] *post* ὑγρότητος δὲ καὶ: ξηρότητος οὐκ ἀκριβῶς μέν ἐστι μέσον, H; ξηρότητι μέσον μέν οὐκ ἐστιν ἀκριβῶς, K

θερμότητα κράσεως, ἀλλὰ καὶ φθινοπώρῳ ψυχρότητα καὶ ξηρότητα. | καίτοι γ᾽, εἰ καὶ τοῦτο συνεχωρεῖτο ξηρὸν εἶναι καὶ ψυχρόν, οὐκ ἦν ἀναγκαῖον εὐθέως ὑγρὸν εἶναι καὶ θερμὸν τὸ ἔαρ. οὐ γὰρ εἰ τέτταρες αἱ πᾶσαι συζυγίαι κράσεών εἰσιν ἀμέτρων, ἤδη καὶ πάσας ἀναγκαῖον εἰς τὰς τέτταρας ὥρας νενεμῆσθαι. ἀλλ᾽ εἴπερ ἄρα τάξις τίς ἐστιν ἐν τῷ κόσμῳ καὶ κατὰ τὸ βέλτιον, οὐ[20] τὸ χεῖρον ἅπαντα κεκόσμηται, πιθανώτερον ἦν, εὐκράτους μὲν τὰς πλείους ὥρας γίγνεσθαι, μίαν δ᾽ ἐξ αὐτῶν, εἴπερ ἄρα, τὴν δύσκρατον· οἱ δὲ τοὐναντίον ἐπιδεῖξαι σπεύδουσιν, ὡς οὐδεμία τῶν ὡρῶν ἐστιν εὔκρατος, ἀλλ᾽ ἐξ ἀνάγκης ἐν αὐταῖς ἐπικρατεῖ νῦν μὲν τὸ ψυχρόν, αὖθις δὲ τὸ θερμόν, καὶ νῦν μέν, εἰ τύχοι, τὸ ξηρόν, αὖθις δὲ τὸ ὑγρόν. ἐγὼ δὲ τοσοῦτον ἀποδέω ἢ θερμὸν καὶ ὑγρὸν ἀποφαίνειν τὸ ἔαρ, ἢ ὅ τί περ ἂν εὔκρατον ᾖ, θερμὸν καὶ ὑγρὸν εἶναι συγχωρεῖν, ὥστε πᾶν τοὐναντίον ἀποφαίνομαι χειρίστην εἶναι κατάστασιν κράσεως τοῦ περιέχοντος ἡμᾶς ἀέρος τὴν θερμὴν καὶ ὑγράν, ἣν ἐν μὲν ταῖς ὥραις οὐκ ἂν εὕροις ὅλως, ἐν δὲ ταῖς νοσώδεσι καὶ | λοιμώδεσι καταστάσεσιν ἐνίοτε συμπίπτει, καθάπερ που καὶ Ἱπποκράτης ἐμνημόνευσε λέγων· "ὗεν ἐν καύμασιν ὕδατι λάβρῳ δι᾽ ὅλου." τοῦτο γὰρ ἴδιόν ἐστιν ὑγρᾶς καὶ θερμῆς καταστάσεως ὕειν συνεχῶς ἐν καύμασιν. εἰ δ᾽ ἤτοι μόνον εἴη καῦμα, καθάπερ ἐπὶ τοῦ κατὰ φύσιν ἔχοντος θέρους, ἢ ὕοι μὲν ἀλλ᾽ ἐν κρύει, καθάπερ ἐν χειμῶνι, θερμὴν καὶ ὑγρὰν οὐχ οἷόν τε τὴν τοιαύτην εἶναι κατάστασιν. ἆρ᾽ οὖν ἀνο-

hot *krasis* to spring, but also the cold and dry to autumn.
And indeed, even if the latter were agreed to be dry and
cold, it would not be immediately necessary for spring to
be wet and hot. For if there are in all four conjunctions of
disproportionate *krasias*, it is not now necessary for all
these to be distributed to the four seasons. But if there is,
then, some order in the universe, and it is for the better,
and all has not been ordered for the worse, it would be
more credible for the greater number of the seasons to be
eukratic, and only one of these, if any, *dyskratic*. They,
however, hope to demonstrate the opposite, which is that
none of the seasons is *eukratic* but of necessity in these,
now the cold prevails, and then in turn the hot, and now,
should it so happen, the dry, and then in turn, the wet. I,
however, fall so far short of calling spring hot and wet, or
agreeing that what is *eukratic* is hot and wet, that I declare
it to be the total opposite, and that the hot and wet *krasis*
of the ambient air is the worst condition which you would
not find in the seasons generally, but sometimes occurs
in the morbid and pestilential conditions, as Hippocrates
also mentioned somewhere, saying: "Rain falling violently
during hot weather continuously."[12] This is characteristic
of wet and hot climatic condition—rain falling continuously in times of burning heat. If there is either burning
heat alone, as is the case of a normal summer, or it rains,
but in cold weather, as in winter, it is not possible for there
to be such a hot and wet climate. Does he say, then, that

[12] This is the second sentence in the quotation given in full below (*Epidemics* 2.1).

---

[20] *post* οὐ: κατὰ (K) *om.*

σον ἐκεῖνο τὸ θέρος, ἐν ᾧ, φησίν, ὗεν ἐν καύμασιν ὕδατι λάβρῳ δι' ὅλου; καὶ μὴν ἄνθρακας ἐν τούτῳ γενέσθαι διηγεῖται, σαπέντων δηλονότι τῶν ἐν τῷ σώματι περιττωμάτων καί τινας ἰχῶρας θερμοὺς καὶ ὑγροὺς ἀμέτρως γεννησάντων. εἴσῃ δ' ἐξ αὐτῆς τῆς ῥήσεως, εἰ πᾶσαν αὐτήν σοι παραγράψαιμι, τόνδε τὸν τρόπον ἔχουσαν·

ἄνθρακες ἐν Κρανῶνι θερινοί. ὗεν ἐν καύμασιν ὕδατι λάβρῳ δι' ὅλου. ἐγίνοντο δὲ μᾶλλον νότῳ καὶ ὑπεγίνοντο μὲν ὑπὸ τὸ δέρμα ἰχῶρες. ἐγκαταλαμβανόμενοι δ' ἐθερμαίνοντο καὶ κνησμὸν ἐνεποίεον· εἶτα φλύκταιναι ὥσπερ πυρίκαυστοι | ἐπανίσταντο καὶ ὑπὸ τὸ δέρμα καίεσθαι ἐδόκεον.

ἀλλ' ἐνταῦθα μὲν ὡς ἂν μιᾶς ὥρας μετακοσμηθείσης ἧττον τὸ κακόν. εἰ δὲ δύο ἢ τρεῖς ὑπαλλαχθεῖεν ἢ καὶ σύμπαν τὸ ἔτος ὑγρὸν καὶ θερμὸν γένοιτο, μέγιστον ἀνάγκη συμπεσεῖν οὕτω λοιμόν, οἷον ἐν τῷ τρίτῳ τῶν Ἐπιδημιῶν διηγεῖται. παραγράψω δὲ πρῶτα μὲν ἃ περὶ τῆς τῶν ὡρῶν ἀκοσμίας εἶπεν, ἐφεξῆς δὲ καὶ τὰ περὶ τῆς ἐπιγενομένης φθορᾶς τῶν ἀνθρώπων. ἐν ἅπασι δ' αὐτοῖς προσέχειν σε τὸν νοῦν ἀκριβῶς ἀξιῶ καὶ σκοπεῖσθαι πρῶτον μέν, ὁποῖόν τι πρᾶγμα θερμὴ καὶ ὑγρὰ κρᾶσίς ἐστιν, ὡς οὐδὲν ἦρι παρόμοιον, εὐκράτῳ χρήματι· δεύτερον δ' ὡς ἀναγκαῖον ἐν αὐτῇ

ON TEMPERAMENTS, BOOK I

a summer is disease free in which there is "continuous violent rain in times of burning heat?" And further, he describes in detail the occurrence of *anthraces* (pustules) in this, when there are obviously putrefying superfluities in the body, when certain hot and wet ichors are disproportionally generated. You will understand this from the statement itself, which I shall add for you in full in the following manner:

> *Anthraces* in Kranon in summer. Fierce rain falling in hot weather continuously. This occurred most with the south wind, and after it occurred, ichors collected under the skin. Being trapped within, they were burning and set up itching. Then blisters, like those caused by burns, arose and seemed to burn under the skin.[13]

531K

But here, as only one season is changed, the trouble is less. If, however, two or three are changed, or the whole year becomes wet and hot, inevitably in this way a very great plague necessarily happens, as described in detail in the third book of *Epidemics*.[14] I shall add first those things he said about the disorder of the seasons, and next also those concerning the subsequently occurring destruction of the people. In all these I think it worthwhile for you to direct your attention precisely and consider first what kind of thing a hot and wet *krasis* is—that it bears no resemblance to spring, which is something *eukratic*, and

---

[13] Hippocrates, *Epidemics* 2.1.1; see *Hippocrates* 7, LCL 477 (W. D. Smith), 18–19. The Greek text is slightly different here.

[14] The following quotations are from Hippocrates, *Epidemics* 3.2–4, LCL 147 (Jones), 238–41 (translations after Jones).

σήπεσθαι πάντα. ἄρχεται μὲν οὖν ὁ Ἱπποκράτης τῆς διηγήσεως ὧδε· "ἔτος νότιον, ἔπομβρον, ἄπνοια διὰ τέλεος." εἶτα τούτοις ἐπιφέρει τὰ κατὰ μέρος ἁπάσης τῆς καταστάσεως, ὄμβρους πολλοὺς ἐν θερμῇ καὶ νοτίᾳ καταστάσει γενέσθαι γράφων, εἶτ' αὖθις ἐπὶ τέλει τὸν σύμπαντα λόγον εἰς ἓν κεφάλαιον ἀγαγὼν οὕτω φησί· "γενομένου δὲ τοῦ ἔτεος ὅλου νοτίου καὶ ὑγροῦ καὶ μαλακοῦ" τάδε καὶ τάδε συνέπεσεν, ἃ σύμπαντα μὲν γράφειν ἐν τῷδε τῷ λόγῳ μακρόν. ἔνεστι δὲ τῷ βουλομένῳ λαβόντι τὸ τρίτον τῶν Ἐπιδημιῶν ἀναγιγνώσκειν τὰ κατὰ μέρος εἰς ἓν ἅπαντα κεφάλαιον ἀναγόμενα, σηπεδόνα μεγίστην, ἧς καὶ αὐτῆς ὀνομαστὶ πολλάκις ὁ Ἱπποκράτης ἐπιμέμνηται, ποτὲ μὲν ὡδίπως λέγων·

ἦν δὲ καὶ τὸ ῥεῦμα τὸ συνιστάμενον οὐ πύῳ ἴκελον, ἀλλὰ σηπεδών τις ἄλλη καὶ ῥεῦμα πολὺ καὶ ποικίλον·

ποτὲ δὲ πάλιν ὡδί·

καὶ ἐν αἰδοίοισιν ἄνθρακες οἱ κατὰ θέρος καὶ ἄλλα, ἃ σῆψις καλέεται,

καὶ ὡς ἐκ τῆς σήψεως ταύτης πολλοῖς μὲν βραχίων καὶ πῆχυς ὅλος ἀπερρύη, πολλοῖς δὲ μηρὸς ἢ τὰ περὶ

---

[15] The Greek term is νότιος, which may mean "wet, rainy" or "southerly," the latter indicating southerly winds bringing rain. I

ON TEMPERAMENTS, BOOK I

secondly, in this everything necessarily putrefies. Hippocrates, then, begins the narrative as follows: "The year was wet[15] and raining with no winds throughout." Then he adds to these the aspects of the whole condition individually, writing that there were many rains in hot and southerly conditions. Then, in turn, bringing the whole discussion to a conclusion under the one heading, he speaks in this way: "Since the whole year | was southerly, wet and mild," such and such things occurred. To write out all these things in this work would be a lengthy matter. It is possible, however, for someone who wishes to do so to take up the third book of the *Epidemics* and read individually all those things brought under the one heading—severe putrefaction—which itself Hippocrates often mentions by name. On one occasion, he speaks somewhat as follows:

> The flux which formed was not like pus but was a different sort of putrefaction with a copious and varied flux.

And again, as follows:

> There were many pustules (*anthraces*) in summer on the genitals, and others which are called putrefaction.[16]

And from this sepsis in many cases an arm or forearm died away completely, while in many the thigh or those struc-

---

have followed Jones in rendering "southerly" here "wet"; the Latin text printed in Kühn (KLat) has *austrinus* (from the south).

[16] Hippocrates, *Epidemics* 3.7, LCL 147 (Jones), 246–47. There are textual variations.

κνήμην ἀπεψιλοῦτο καὶ πούς ὅλος. ἀλλὰ καὶ σαρκῶν καὶ ὀστῶν καὶ νεύρων[21] ἐκπτώσεις ἐγίγνοντο μεγάλαι.

Καὶ ὅλως οὐδέν ἐστιν εὑρεῖν ὧν ἔγραψε παθημάτων, ὃ μὴ σηπεδόνος ἔκγονον ὑπάρχει, δεόντως. οὔτε γὰρ ὑπὸ ξηροῦ τι σήπεσθαι πέφυκεν οὔθ' ὑπὸ ψυχροῦ. μάθοις δ' ἄν, εἰ θεάσαιο τά τε κρέα καὶ τἆλλα σύμπαντα τὰ πρὸς τῶν ἀνθρώπων ταριχευόμενα, τὰ μὲν ἁλσί, τὰ δ' ἅλμῃ, τὰ δ' ὄξει, τὰ δ' ἄλλῳ τινὶ τῶν ξηραινόντων σκευαζόμενα καλῶς. μάθοις δ' ἂν καὶ ὡς ἐν τῷ βορρᾷ, ξηρῷ καὶ ψυχρῷ τὴν φύσιν ὑπάρχοντι, ἄσηπτα μέχρι πλείστου διαμένει πάντα· σήπεται δ' ἑτοίμως ἐν νοτίαις καταστάσεσιν. ἔστι γὰρ οὖν καὶ οὗτος ὁ ἄνεμος ὑγρὸς καὶ θερμός.

Ὥστε πᾶν τοὐναντίον ἡμεῖς ἀποφαινόμεθα τοῖς ὑγρὸν καὶ θερμὸν ὑπολαμβάνουσιν εἶναι τὸ ἔαρ. οὔτε γὰρ τοιοῦτόν ἐστιν οὔθ' ὑγιεινὸν ἂν ἦν, εἴπερ ἦν τοιοῦτον. οἱ δὲ καὶ τοιοῦτον εἶναί φασιν αὐτὸ καὶ διὰ τοῦθ' ὑπάρχειν ὑγιεινόν, ἐν ἀμφοτέροις ἁμαρτάνοντες, ὅσα τε ταῖς αἰσθήσεσιν ἔνεστι διαγνῶναι καὶ ὅσα τῷ λόγῳ διασκέψασθαι. ταῖς μὲν γὰρ αἰσθήσεσιν ἔνεστιν ἐναργῶς μαθεῖν εὔκρατον ἀκριβῶς αὐτό, τῷ λόγῳ δ' ἐξευρεῖν, ὡς διὰ τοῦτ' ἐστὶν ὑγιεινόν, διότι μηδὲν ἐπικρατεῖ τῶν τεττάρων. εἰ δέ γ' ἤτοι τὸ θερμὸν ἐπεκράτει πολλῷ τοῦ ψυχροῦ ἢ τὸ ὑγρὸν τοῦ ξηροῦ, σηπεδόνων τ' ἂν ἦν εὔφορον οὕτω καὶ πασῶν τῶν ὡρῶν νοσωδέστατον. ἀλλ' ἡ τῆς τῶν τεττάρων κρά-

---

[21] καὶ ἄρθρων add. K

tures around the calf, or the whole foot were stripped bare. But also major decay of flesh, bones and sinews occurred.[17]

All in all, one finds none of the affections he wrote about which are not the product of putrefaction—and this is as it should be. For there is nothing that is naturally putrefied by dryness or by coldness. You would learn this, if you were to look at the fleshes and all the other things preserved for people, some of which are properly prepared with salt, some with brine, some with vinegar, and some with another of the drying agents. You would also learn that in the north wind, which is dry and cold in nature, everything remains free of putrefaction for a very long time, whereas in the climatic conditions in south winds, there is readily putrefaction, for this wind is wet and hot.

As a consequence, we offer an opinion completely opposite to those who take the spring to be wet and hot. For it is not such as this, nor would it be healthy if it were. However, those who say it is such as this, and because of this is healthy, err on both counts—those things it is possible to recognize by the senses and those considered by reason. It is clearly possible to learn through the senses that it is perfectly *eukratic*, while it is possible to discover by reason that because of this it is healthy, on which account none of the four [elemental qualities] prevails [in it]. If, in fact, either the hot were to prevail greatly over the cold, or the wet over the dry, it would be productive of putrefactions, and in this too the most baneful of all

[17] Hippocrates, *Epidemics* 3.4, LCL 147 (Jones), 240–43. This is not an exact quotation.

σεως ἰσομοιρία τῆς τ' εὐκρασίας αὐτοῦ καὶ τῆς ὑγιείας αἰτία. πόθεν οὖν ἐπῆλθέ τισιν ἰατροῖς τε καὶ φιλοσόφοις ὑγρὸν καὶ θερμὸν ἀποφήνασθαι τὸ ἔαρ; ἐκ τοῦ βουληθῆναι δηλονότι τὰς τέτταρας συζυγίας τῶν κράσεων εἰς τὰς τέτταρας ὥρας διανεῖμαι. τοῦτο δ' αὐτὸ πάλιν ἐκ τοῦ παραλιπεῖν τὴν πρώτην ἁπασῶν, τὴν εὔκρατον, ἐγένετο. καὶ γὰρ οὖν καὶ διαιτημάτων καὶ φαρμάκων ἁπάντων τε τῶν ὄντων εἰς τὰς τέτταρας ταύτας συζυγίας ἀνάγουσι τὰς διαφοράς.

5. Ὧι καὶ δῆλον, εἰς ὅσον οἱ περὶ φύσεως[22] λογισμοὶ σφαλέντες τῆς ἀληθείας εἰς τὰς ἰάσεις βλάπτουσι καὶ βέλτιόν ἐστι δυοῖν θάτερον, ἢ μηδ' ὅλως ἅπτεσθαι τῶν τοιούτων λόγων, ἀλλ' ἐπιτρέψαι τῇ πείρᾳ τὸ πᾶν, ἢ πρότερον ἐν τῇ λογικῇ θεωρίᾳ γυμνάσασθαι. τὸ δὲ μήτε τῇ πείρᾳ προσέχειν τὸν νοῦν ἐπιχειρεῖν τε θεωρίᾳ φυσικῇ πρὸ τοῦ τὸν λογισμόν, ᾧ μέλλοιμεν εὑρίσκειν αὐτήν, ἀσκῆσαι πρεπόντως εἰς τὰ τοιαῦτ' ἀναγκαῖον ἀπάγειν | σοφίσματα, καὶ περί τε τῶν φαινομένων ὡς ἀναισθήτους ἀναγκάζει διαλέγεσθαι μάρτυρά τε καλεῖν Ἀριστοτέλην παρακούοντας ὧν διδάσκει. πολλαχῶς γὰρ ἐκεῖνος οἶδε καὶ τὸ θερμὸν λεγόμενον καὶ τὸ ψυχρὸν καὶ τὸ ξηρὸν καὶ τὸ ὑγρόν· οἱ δ' οὐκ ἀκούουσιν αὐτοῦ πολλαχῶς ἀλλ' ὡσαύτως ἀεί. καὶ μὲν δὴ καὶ ὡς οὐ ταὐτόν ἐστιν ἢ οἰκείῳ τινὶ καὶ συμφύτῳ θερμῷ θερμὸν ὑπάρχειν ἐπικτήτῳ τε καὶ ἀλλοτρίῳ διῆλθεν Ἀριστοτέλης· οἱ δὲ

[22] post περὶ φύσεως add. ἀνθρώπου K

the seasons. But the equal distribution of the four [components of the] *krasias* is the cause of its *eukrasia* and of health. How then did spring come to be declared wet and hot by certain doctors and philosophers? Obviously from the wish to apportion the four conjunctions of the *krasias* to the four seasons. However, this itself arose in turn from the omission of the first *krasis* of all—the *eukratic*. Therefore, they also refer the differences of regimens, medications, and indeed all things to these four conjunctions.

5. It is also clear from this how much those who reason about Nature, when they fall short of the truth, bring harm to the cures. It is better to do one or other of two things—either to have nothing at all to do with such arguments but to rely completely on experience, or to first become practiced in logical theory. However, not to direct one's attention to experience, and to turn one's hand to the physical theory before properly training the reasoning faculty, with which we intend to discover this, inevitably leads to such captious arguments. And people feel compelled to debate about the phenomena as if devoid of sense perception, calling on Aristotle as a witness, although misunderstanding what he teaches. For that man knew that hot, cold, dry, and wet are spoken of in many ways, but they don't understand his many ways, [taking the terms] to always be the same. Certainly, Aristotle went over in detail that specific and innate heat is not the same as acquired and alien heat; they also misunderstand this.[18] Further, in

[18] Aristotle addresses the issue of the uses of hot/hotter and cold/colder particularly in *Parts of Animals* 2.2; see 648a20–649b8, LCL 323 (A. L. Peck), 120–29.

καὶ τούτου παρακούουσιν. ἔτι δὲ πρὸς τούτοις ὁ μὲν Ἀριστοτέλης, ὡσαύτως δὲ καὶ ὁ Θεόφραστος, εἰς ὅ τι χρὴ βλέποντας ἢ εὔκρατον ἢ δύσκρατον ὑπολαμβάνειν εἶναι τὴν φύσιν, ἀκριβῶς εἰρήκασιν· οἱ δ' οὐδὲ τοῦτο γιγνώσκουσιν, ἀλλ' ὅταν ἀκούσωσί που λεγόντων αὐτῶν ὑγρὸν εἶναι καὶ θερμὸν τὸ ζῷον ἢ τὴν τοῦ παιδὸς κρᾶσιν ὑγρὰν καὶ θερμήν, οὔθ' ὅπως εἴρηται ταῦτα συνιᾶσιν ἐμπλήκτως τε μεταφέρουσι τὸν λόγον ἐπὶ τὰς ὥρας ὥσπερ ταὐτὸν ὂν ἀλλ' οὐ μακρῷ διαφέρον ἢ τὴν οἰκείαν κρᾶσιν ὑγρὰν εἶναι καὶ θερμὴν ἢ τὴν τοῦ περιέχοντος ἡμᾶς ἀέρος. οὔτε γὰρ ταὐτόν ἐστιν οὔθ' ὁμοίως ὑγρὰ καὶ θερμὴ ζῴου κρᾶσις ἀέρος ὑγρᾷ καὶ θερμῇ κράσει λέγεται.

Τί δὴ τὸ τούτων ἁπάντων αἴτιον, ἤδη διηγήσομαι καὶ δείξω σαφῶς τοῖς προσέχειν τὸν νοῦν βουλομένοις, ὡς μικρὰ πταίσματα τῶν ἐν ἀρχῇ τῆς λογικῆς θεωρίας διδασκομένων αἴτια μεγίστων ἁμαρτημάτων γίγνεται, καὶ κινδυνεύει πάντα τὰ κακῶς πραττόμενα κατά τε τὰς τέχνας ἁπάσας καὶ μέντοι καὶ κατὰ τὰς ἐν τῷ βίῳ πράξεις ἕπεσθαι σοφίσμασιν. ἕπεται τοιγαροῦν ἤδη καὶ τάδε τὰ σοφίσματα τῷ μὴ διελέσθαι περὶ τῶν σημαινομένων ὀρθῶς, ἀλλ' οἰηθῆναι τὸ θερμὸν λέγεσθαι διχῶς, τὸ μὲν ὡς ἄκρατον καὶ ἄμικτον καὶ ἁπλοῦν, τὸ δ' ὡς ἐν τῇ πρὸς τοὐναντίον ἐπιμιξίᾳ πλεονεκτοῦν. ὅτι δὲ καὶ παραβάλλοντες ἑτέρᾳ κράσει πολλάκις ἑτέραν ἀποφαινόμεθα τὴν ἑτέραν αὐτῶν εἶ-

ON TEMPERAMENTS, BOOK I

addition to these things, Aristotle and likewise also Theophrastus, in looking at the matter, took it as necessary to assume the nature to be either *eukratic* or *dyskratic*, and they have spoken precisely on this. But these people don't recognize this; rather, whenever they hear these things said, i.e., the animal is wet and hot, or the *krasis* of the child is wet and hot, they don't understand how these things are being said, and transfer the concept stupidly to the seasons as if it is the same and not something very different, to say either our own specific *krasis* is wet and hot, or that of the ambient air is. For *krasis* of an animal is wet and hot is not the same as, nor said similarly to, *krasis* of air is wet and hot.

536K

I shall now set out in detail what the reason is for all such things, and I shall show clearly to those who wish to pay attention that, when there are small false steps in the beginning of the teachings of logical theory, they are causes of very great errors, and there is a danger of everything being done badly in all the arts, and indeed also in relation to the activities in life, and this will follow sophistical reasoning. So, for example, these very sophisms follow not making distinctions correctly regarding the significations, but the "hot" being thought to be said in a twofold manner—as something uncompounded, unmixed and simple, and as something prevailing over its opposite in a combination of elements.[19] However, often when we compare one *krasis* to another, and say one of them is

---

[19] On ἐπιμειξία as a combination of elements, see Galen, VI.587K.

ναι θερμὴν[23] ἐν ἴσῳ τῷ θερμοτέραν, ἐπιλανθάνονται τοῦδε. καὶ μὴν οὕτω τὰ ζῷα θερμὰ καὶ ὑγρὰ λέγεται πρὸς τῶν παλαιῶν, οὐ κατὰ τὴν ἰδίαν κρᾶσιν ἁπλῶς, ἀλλὰ τοῖς τε φυτοῖς καὶ τοῖς τεθνεῶσι παραβαλλόμενα. καὶ γὰρ τῶν τεθνεώτων τὰ ζῷα καὶ τῶν φυτῶν ἐστιν ὑγρότερα καὶ θερμότερα.

Καὶ μὲν δὴ καὶ τῶν ζῴων αὐτῶν ἀλλήλοις κατ' εἴδη παραβαλλομένων, ξηρότερον μὲν κύων, ὑγρότερον δ' ἄνθρωπος. εἰ δὲ μύρμηκι καὶ μελίττῃ παραβάλλοις τὸν κύνα, ξηρότερα μὲν ἐκεῖνα, τὸν κύνα δ' ὑγρότερον εὑρήσεις. ὥστε ταὐτὸν ζῷον ξηρὸν μὲν ὡς πρὸς ἄνθρωπον ὑπάρχει, ὑγρὸν δ' ὡς πρὸς μέλιτταν· οὕτω δὲ καὶ θερμὸν μὲν ὡς πρὸς ἄνθρωπον, ψυχρὸν δ' ὡς πρὸς λέοντα· καὶ θαυμαστὸν οὐδέν, εἰ τὰ πρὸς ἕτερόν τι λεγόμενα τὰς ἐναντίας ἅμα κατηγορίας ἐπιδέχεται. οὐ γὰρ τοῦτ' ἄτοπον, εἰ ταὐτὸν σῶμα θερμὸν ἅμα λέγεται καὶ ψυχρόν, ἀλλ' εἰ καὶ πρὸς ταὐτόν· οὐδὲ γὰρ ὅτι δεξιὸς ἅμα καὶ ἀριστερὸς ὁ αὐτὸς ἄνθρωπος, ἄτοπον,[24] ἀλλ' εἰ πρὸς τὸν αὐτόν· εἰ δὲ πρὸς ἄλλον μὲν δεξιός, πρὸς ἄλλον δ' ἀριστερός, οὐδὲν ἄτοπον. οὕτως οὖν καὶ κύων ὑγρὸς ἅμα καὶ ξηρὸς καὶ ψυχρὸς ἅμα καὶ θερμός, ἀλλ' οὐ πρὸς ταὐτό. πρὸς μὲν γὰρ ἄνθρωπον ξηρός, ὑγρὸς δὲ πρὸς μύρμηκα, καὶ πρὸς μὲν ἄνθρωπον θερμός, ψυχρὸς δὲ πρὸς λέοντα. καὶ γὰρ δὴ καὶ θερμὸς μὲν ὡς ζῶν· εἰ γάρ τι τέθνηκεν, οὐ

---

[23] post θερμὴν: ἐν ἴσῳ τῷ θερμοτέραν, ἐπιλανθάνονται τοῦδε. H; ἐπιλανθάνονται καὶ τοῦδε. K

ON TEMPERAMENTS, BOOK I

"hot" in an equivalent sense to "hotter," they overlook this. And further in this way, the hot and wet animals are said by the ancients not absolutely in relation to the specific *krasis*, but when compared to plants and dead bodies. For truly, living things are wetter and hotter than dead bodies and plants.

Furthermore, when animals themselves are compared with each other according to kinds, the dog is drier and the human wetter. If, however, you were to compare the dog to an ant or a bee, you will find the latter are drier and the dog wetter. As a consequence, the same animal is dry compared to a human, but wet compared to a bee. In the same way too, a dog is hot compared to a human but cold compared to a lion. It is not surprising, if what is said in relation to one thing allows of opposite predicates at the same time. For it is not strange if a particular body is called hot and cold at the same time, provided it is not in relation to the same thing. For it is not strange for the same man to be called dexterous and clumsy[20] at the same time, if in fact he is dexterous relative to one man and clumsy relative to another. In this way, then, a dog is wet and dry at the same time, and cold and hot at the same time, but not in relation to the same thing. On the one hand, it is dry in relation to a human and on the other, wet in relation to an ant, and hot relative to a human but cold relative to a lion. Furthermore, it is hot when alive but not

---

[20] These two terms could be read as "right" and "left"—the point is the same either way.

---

[24] post ἄτοπον,: ἀλλ' εἰ πρὸς τὸν αὐτόν· εἰ δὲ πρὸς ἄλλον μὲν δεξιός, H; εἰ δὲ γε πρὸς ἄλλον μὲν δεξιός, K

θερμόν· οὐ θερμὸς δ' ὡς πρὸς ἕτερον, εἰ οὕτως ἔτυχε, κύνα. ταυτὶ μὲν οὖν ἅπαντα πρὸς ἄλληλα ἐκ παραβολῆς λέγεται, τὰ δ' ὡς ἐν ζῴων γένει καθ' ἕτερον τρόπον, ὥσπερ αὖ καὶ ὅσα κατ' εἶδος ζῴου. κύων γὰρ ὡς πρὸς μὲν μύρμηκα καὶ μέλιτταν ὑγρός, ὡς δ' ἐν ζῴων γένει ξηρός. αὐτῶν δὲ τῶν κυνῶν κατ' εἶδος ὁδὶ μὲν ξηρός, ὁδὶ δ' ὑγρός, ἄλλος δέ τις, ὡς κύων, εὔκρατος.

6. Λέλεκται μὲν οὖν ἐπὶ πλεῖστον ἡμῖν ὑπὲρ ἁπάσης τῆς τοιαύτης χρήσεως τῶν ὀνομάτων ἐν τῷ δευτέρῳ Περὶ διαγνώσεως σφυγμῶν· ἀνάγκη δ', ὡς ἔοικεν, εἰπεῖν τι καὶ νῦν ὑπὲρ αὐτῶν διὰ κεφαλαίων, ὅσον εἰς τὰ παρόντα χρήσιμον. τὸ μὲν ἁπλῶς ξηρόν, ὃ πρὸς μηδὲν ἕτερον λέγεται, μόνοις τοῖς στοιχείοις ὑπάρχει, πυρί τε καὶ γῇ, καὶ μὲν δὴ καὶ τὸ ὑγρὸν ὕδατι καὶ ἀέρι. κατὰ ταῦτα δὲ καὶ περὶ θερμοῦ καὶ ψυχροῦ χρὴ νοεῖν· οὐδὲν γὰρ τῶν ἄλλων σωμάτων ἀκριβῶς οὔτε θερμὸν οὔτε ψυχρόν ἐστιν ἀλλ' ἢ τὰ στοιχεῖα μόνα. ὅ τι δ' ἂν τῶν ἄλλων λάβῃς, ἐκ τούτων κέκραται καὶ διὰ τοῦτο κατὰ τὸ δεύτερον σημαινόμενον οὐκέθ' ἁπλῶς θερμὸν ἢ ψυχρὸν ὡς ἄμικτόν τε καὶ εἰλικρινές, ἀλλ' ὡς ἤτοι πλέονος μὲν τοῦ θερμοῦ, τοῦ ψυχροῦ δ' ἐλάττονος, ἢ τοῦ μὲν ψυχροῦ πλέονος, ἐλάττονος δὲ τοῦ θερμοῦ μετέχον ἕκαστον τῶν ἄλλων ἢ θερμὸν ἢ ψυχρὸν λέγεται. δύο μὲν δὴ ταῦτα σημαινόμενα τοῦ θερμοῦ καὶ ψυχροῦ καὶ ξηροῦ καὶ ὑγροῦ, τὸ μὲν ἁπλῶς λεγόντων ἡμῶν ἄμικτόν τε καὶ εἰλικρινές, ἕτερον δὲ μικτὸν μὲν ἐκ τῶν ἐναντίων, ἀλλὰ τῷ τοῦ πλεονεκτοῦντος ὀνόματι προσαγορευόμενον. οὕτα

ON TEMPERAMENTS, BOOK I

hot if it has died, or not hot with regard to another dog, if this should be the case. All these, then, are said from comparison with one another, just as in another way in a class of animals, and in turn in relation to a kind of animal, for a dog is wet relative to an ant and a bee, whereas in a class of animals it is dry. However, of dogs themselves, in relation to kind, this one may be dry and that one wet, while some other is, as a dog, *eukratic*.

6. I have spoken to the greatest extent about every such use of the terms in the second book of *On the Diagnosis of Pulses*.[21] It would, however, seem necessary to say something about these now under the main points, as far as is useful for present purposes. The absolutely dry, which is said in regard to nothing else, is of the elements alone—fire and earth—and further, the wet of water and air. It is necessary to conceive of hot and cold in the same way, for none of the other bodies is entirely either hot or cold, but only the elements. Any of the other things you might apprehend has been compounded from these [elements], and because of this is no longer absolutely hot or cold in relation to the second signification, as uncompounded and pure, but each is said to be either hot or cold compared to others, if it partakes of more of the hot but less of the cold, or more of the cold but less of the hot. Certainly, there are these two significations of hot and cold, and dry and wet; one when we speak absolutely of something uncompounded and pure, and the other when we speak of something compounded from opposites, applying the name to

---

[21] Galen, *De dignoscendis pulsibus*, Book 2, VIII.823–77K; see particularly VIII.828K.

μὲν οὖν ὑγρὸν αἷμα καὶ φλέγμα καὶ πιμελὴ καὶ οἶνος ἔλαιόν τε καὶ μέλι καὶ τῶν ἄλλων τῶν τοιούτων ἕκαστον λέγεται. ὀστᾶ δὲ καὶ χόνδροι καὶ ὄνυχες καὶ ὁπλαὶ καὶ κέρατα καὶ τρίχες καὶ λίθοι καὶ ξύλα καὶ ψάμμος καὶ κέραμος ἐλάττονα μὲν ὑγροῦ μοῖραν ἔχει, πλείονα δὲ ξηροῦ καὶ διὰ τοῦτο πάλιν ἅπαντα τὰ τοιαῦτα ξηρὰ προσαγορεύεται. μύρμηξ δὲ ξηρὸν καὶ σκώληξ ὑγρὸν ὡς ζῷα καὶ πάλιν ἐν αὐτοῖς τοῖς σκώληξιν ὁδὶ μὲν ξηρότερος, ὁδὶ δ' ὑγρότερος ἢ ἁπλῶς ὡς σκώληξ ἢ τῷδέ τινι παραβαλλόμενος ἑτέρῳ.

Αὐτὸ δὲ δὴ τοῦτο τί ποτ' ἐστίν, ὅταν οὕτω λέγωμεν, ὡς σκώληξ ὑγρός, ὡς ἄνθρωπος θερμός, ὡς κύων ψυχρός, εἰ μή τις ἀκριβῶς ἀκούσειέ τε καὶ νοήσειεν,[25] ἅπαντα συγκεχύσθαι τὸν λόγον ἀνάγκη. τὸ δ' ἀκριβῶς ἐστιν, ὃ κἂν τῷ δευτέρῳ Περὶ διαγνώσεως τῶν σφυγμῶν λέγεται, τὸ κατὰ γένος ἢ εἶδος ὀνομάζεσθαι τηνικαῦθ' ἕκαστον οὐ θερμὸν μόνον ἢ ψυχρὸν ἢ ξηρὸν ἢ ὑγρόν, ἀλλὰ καὶ μέγα καὶ μικρὸν καὶ ταχὺ καὶ βραδὺ καὶ τῶν τοιούτων ἕκαστον, ὅταν ὑπὲρ τὸ σύμμετρόν τε καὶ μέσον ᾖ, οἷον ζῷον θερμόν, ὅταν ὑπὲρ τὸ μέσον ᾖ τῇ κράσει ζῷον, ἢ ἵππος θερμός, ὅταν ὑπὲρ τὸν μέσον ἵππον ᾖ. τὰ μέσα δ' ἐν ἑκάστῳ γένει τε καὶ εἴδει τὰ σύμμετρά ἐστιν· ἴσον γὰρ ἀπέχει τῶν ἄκρων ἐν ἐκείνῳ τῷ γένει τε καὶ εἴδει. γένος μὲν οὖν τὸ ζῷον, ἵππος δὲ καὶ βοῦς καὶ κύων εἴδη. καὶ δὴ καὶ μέσον μέν ἐστι τῇ κράσει καθ' ὅλον τὸ γένος τῶν ζῴων ὁ ἄνθρωπος· ἐν γὰρ τοῖς ἐφεξῆς τοῦτο δειχθή-

ON TEMPERAMENTS, BOOK I

what is predominant. In this way then, blood, phlegm, fat, wine, olive oil, and honey and each of the other such things are called wet, while on the other hand, bone, cartilage, nails, hooves, horns, hair, stones, wood, sand, and clay have a lesser part of wetness, but a greater part of dryness, and because of this, contrariwise all such things are called dry. An ant is dry and a worm wet, as animals, and in turn, among worms themselves, one is drier while one is wetter, either absolutely as a worm or when compared to some other worm.

What this actually is, when we say in this way that a worm is wet, a human hot, or a dog cold, if someone doesn't understand and conceive precisely beforehand, inevitably the whole argument is confused. The "precisely" which is stated in the second book of *On the Diagnosis of the Pulses*, is applied to each term under the circumstances in relation to a class or kind, not only to hot or cold, dry or wet, but also to large and small, rapid and slow, and each of the other such things, whenever it is above the moderate and median. For example, an animal is hot whenever it is above the median animal in the *krasis*, or a horse is hot, whenever it is above the median horse. In each class and kind, there are median and moderate things. They are equally distant from the extremes in that class and kind. Animal, then, is a class; horse, ox, and dog are kinds. Furthermore, the human is median in *krasis* in relation to the whole class of animals. This will be

---

[25] *post* νοήσειεν,: ἅπαντα H; πρότερον, ἀνάγκη K

σεται. μέσος δ' ὡς ἐν ἀνθρώποις κατ' εἶδος ὁ καλούμενος εὔσαρκος· οὗτος δ' ἐστίν, ὃν οὔτε παχὺν οὔτε λεπτὸν ἔχομεν εἰπεῖν οὔτε θερμὸν οὔτε ψυχρὸν οὔτ' ἄλλῳ τινὶ τῶν ἀμετρίαν ἐνδεικνυμένων ὀνομάτων προσαγορεῦσαι. ὅστις δ' ἂν ὑπὲρ τοῦτον ᾖ, πάντως οὗτος ἢ θερμότερός ἐστιν ἢ ψυχρότερος ἢ ξηρότερος ἢ ὑγρότερος. ὀνομάζεται δὲ πῇ μὲν ἁπλῶς ὁ τοιοῦτος, πῇ δ' οὐχ ἁπλῶς· ἁπλῶς μέν, ὅτι θερμὸς ἢ ψυχρὸς ἢ ξηρὸς ἢ ὑγρὸς εἶναι λέγεται, μηκέτι παραβαλλόμενος ἀφωρισμένως ἑνὶ τῷδε. καθ' ἕτερον δὲ τρόπον οὐχ ἁπλῶς, ὅτι τῷ συμμέτρῳ τε καὶ μέσῳ παντὸς τοῦ εἴδους παραβάλλεται. οὕτω δὲ καὶ κύων ζῷον ξηρὸν ἁπλῶς μέν, ὡς ἄν τῳ δόξειε, λέγεται, μηκέτι παραβαλλόμενος, εἰ τύχοι, μύρμηκι, καθ' ἕτερον δὲ τρόπον οὐχ | ἁπλῶς, ὅτι τῷ συμμέτρῳ τε καὶ μέσῳ τῇ κράσει[26] τῶν ζῴων ἁπάντων, ὅ τί ποτ' ἂν ᾖ τοῦτο, παραβάλλεται.

Πρόδηλον οὖν ἤδη γέγονεν, ὡς ἕκαστον τῶν οὕτω λεγομένων ἢ ἐν ἑνὶ παραβάλλοντες ὅτῳ δήποτε θερμὸν ἢ ψυχρὸν ἢ ξηρὸν ἢ ὑγρὸν ὡς πρὸς ἐκεῖνο λέγομεν ἢ τῷ μέσῳ, καθ' ὅπερ ἂν εἶδος ἢ γένος ᾖ τὸ λεγόμενον, οἷον εἶδος μὲν ἵππον καὶ βοῦν καὶ κύνα καὶ πλάτανον καὶ κυπάριττον καὶ συκῆν, γένος δὲ ζῷον ἢ φυτόν. ἐπὶ τούτοις τρίτον ἄλλο σημαινόμενον ἦν τῶν ἁπλῶς λεγομένων, ἃ δὴ καὶ τὰς ἀμίκτους τε καὶ ἄκρας ἔφαμεν ἔχειν ποιότητας ὀνομάζεσθαί τε στοι-

---

[26] post τῇ κράσει: παντὸς τοῦ εἴδους add. K

shown in what follows. Median in humans, as a kind, is called *eusarkos* (well-fleshed).[22] This is a person whom we could say is neither fat nor thin, neither hot nor cold, nor could be called by any other of the terms indicating disproportion. Someone who is beyond this is, at all events, either hotter, colder, dryer, or wetter. Such a person is named on the one hand absolutely and on the other hand not absolutely. Absolutely, because he is said to be hot, cold, dry, or wet, when not yet compared individually to someone else. In the other manner, not absolutely, because there is comparison to the moderate and median of the whole species. In this same way too, a dog is said to be a dry animal absolutely, as it would seem to be, not yet being compared, as may happen, to an ant, while in another manner not absolutely, because it is compared to the moderate and median *krasis* of all animals, whatever this might be.

It has, then, already become clear that each of the things spoken of in this way, either when we compare one thing with any other one thing, we say hot, cold, dry, or wet in relation to that, or to the median, in relation to whatever kind or class the thing spoken of is. For example, horse is a kind (species), as are ox, dog, plane tree, cypress, and fig tree, whereas animal or plant is a class. In addition to these, there is another, third signification of those things said absolutely, which in fact we said are the uncompounded and extreme qualities termed elements. Fur-

---

[22] On Galen's apparently idiosyncratic use of the term *eusarkos*, see the introductory section on terminology.

χεῖα· καὶ μὲν δὴ καὶ τὰς ποιότητας αὐτὰς ὀνομάζομεν ἐνίοτε τοῖς ποιοῖς σώμασιν ὁμωνύμως. ἀλλὰ περὶ μὲν τούτων μετ᾽ ὀλίγον ἐροῦμεν. εἰς δὲ τὰ παρόντα, τῶν ποιῶν σωμάτων τριχῶς λεγομένων, ἐπισκοπεῖσθαι προσήκει, πῶς ἐν ἑκάστῃ ῥήσει κέχρηταί τις τῇ προσηγορίᾳ, πότερον ὡς ἁπλοῦν τι καὶ ἄμικτον δηλῶν ἢ ὡς πρὸς τὸ σύμμετρον ὁμογενὲς ἢ ὁμοειδὲς παραβάλλων | ἢ ὡς πρὸς τὸ τυχὸν ὁτιοῦν· οἷον ὅταν ὀστοῦν εἴπῃ τις ξηρὸν ἢ ψυχρὸν ἁπλῶς οὑτωσὶ μόνον ὀνομάσας ἄνευ τοῦ προσθεῖναι λέοντος ἢ κυνὸς ἢ ἀνθρώπου, δῆλον, ὡς πρὸς τὴν ὅλην φύσιν ἀποβλέπων ἁπάντων τῶν ἐν τῷ κόσμῳ σωμάτων ἐπινοεῖ τι μέσον, ᾧ παραβάλλων αὐτὸ ξηρὸν εἶναί φησιν. ἂν δέ γ᾽ εἴπῃ[27] τὸ τοῦ λέοντος ὀστοῦν [ἢ τοῦ ἀνθρώπου ἢ τοῦ κυνὸς][28] ξηρὸν εἶναι, δῆλον, ὡς ἐν αὐτοῖς πάλιν τοῖς τῶν ζῴων ὀστοῖς τῷ μέσῳ παραβάλλει. καὶ χρὴ κἀνταῦθά τι νοῆσαι, πάντων τῶν ζῴων τῶν μὲν μᾶλλον τῶν δ᾽ ἧττον ἐχόντων ὀστᾶ ξηρά, μέσον εἶναι τὴν κρᾶσιν ὀστοῦν ἔν τινι γένει ζῴων, οἷον ἀνθρώπων, εἰ τύχοι, καὶ τούτῳ τἆλλα παραβαλλόμενα τὰ μὲν ξηρά, τὰ δ᾽ οὐ ξηρὰ προσαγορεύεσθαι. καὶ μὲν δὴ κἀν τοῖς ἀνθρώποις αὐτοῖς πάλιν ὁ μέν τις ξηρόν, ὁ δ᾽ ὑγρὸν ὀστοῦν ἔχειν λεχθήσεται, τῷ μέσῳ παραβαλλόμενος ὡς ἐν ἀνθρώποις.

[27] εἴπῃ om. K
[28] There is variable omission of the latter two terms in this list (see H's note, p. 22). All three are included in the translation.

ON TEMPERAMENTS, BOOK I

thermore, we also sometimes name the qualities themselves homonymously to the kinds of bodies. But we shall say more about these a little later. For our present purposes, since we speak of the kinds of bodies in a threefold way, it is appropriate to consider how someone has used the name in each statement—whether as indicating something absolute and uncompounded, or as comparing it to what is moderate of the same class or kind,[23] or as whatever else there might happen to be. For example, whenever someone speaks of dry or cold in respect of bone, applying the term absolutely only in this way, without adding lion, dog, or human, it is clear that he is looking at the whole nature of all bodies in the world, conceiving of some median, in comparison to which he says this is dry. If, however, he were to say the bone of the lion, the human, or the dog is dry, it is clear that he is in turn making the comparison to the median in the actual bones of the animals. And it is necessary even here to consider that since the bones of all animals are dry, some more so and some less so, and that the bone that is median in *krasis* is in some class of animals, for example human, should this be the case, and in comparison to this, some of the others are called dry and some not dry. And furthermore, in humans themselves in turn, one bone will be said to be dry and another wet compared to the median in humans.

[23] Used in the present context, these are Aristotelian terms; see, for example, *Metaphysics* 1032a24, *Categories* 5b19, *Generation of Animals* 715a22 and 747b30.

Ὅτι δ' ἐν ἅπασι τοῖς οὖσι τὸ μέσον τῶν ἄκρων ἐστὶ τὸ σύμμετρόν τε καὶ κατ' ἐκεῖνο τὸ γένος ἢ εἶδος εὔκρατον, ἀεὶ χρὴ προσυπακούειν ἐν ἅπαντι τῷ λόγῳ, κἂν παρελθόντες ποτὲ τῇ λέξει τύχωμεν αὐτό, καὶ δὴ καὶ τούτων οὕτως ἐχόντων, ὅταν ὑγρὰν εἶναί τις εἴπῃ τήνδε τὴν κρᾶσιν ἢ θερμήν, ἐρωτᾶν, ὅπως εἴρηκεν, ἆρά γε τῷδέ τινι παραβάλλων ἀφωρισμένως ἑνί, καθάπερ, εἰ τύχοι, τῷ Πλάτωνι τὸν Θεόφραστον, ἢ κατὰ γένος ὁτιοῦν ἢ εἶδος· [ἢ γὰρ ὡς ἄνθρωπον ἢ ὡς ζῷον ἢ ὡς οὐσίαν ἁπλῶς].[29] τὸ γὰρ δὴ τρίτον σημαινόμενον ἑκάστου τῶν τοιούτων ὀνομάτων, ὅπερ ἁπλοῦν ἐλέγομεν εἶναι καὶ ἄμικτον, οὐκ ἔστιν ἐν τοῖς κεκραμένοις, ἀλλ' ἐν αὐτοῖς τοῖς πρώτοις, ἃ δὴ καὶ στοιχεῖα προσαγορεύομεν, ὥστε τριχῶς ἑκάστου τῶν ποιῶν σωμάτων λεγομένου τῶν δύο μόνων ἡμᾶς χρῄζειν εἰς τὴν περὶ κράσεων πραγματείαν ἢ πρὸς τὸ τυχὸν ὁτιοῦν παραβάλλοντας ἢ πρὸς τὸ σύμμετρον ὁμογενές.

Ἐπεὶ δὲ πολλὰ τὰ γένη, καθάπερ οὖν καὶ τὰ ἄτομα, δυνήσεται ταὐτὸν σῶμα καὶ θερμὸν καὶ ψυχρὸν καὶ ξηρὸν καὶ ὑγρὸν εἶναι κατὰ πολλοὺς τρόπους. ἀλλ' ὅταν μὲν ἑνὶ τῷ τυχόντι παραβάλληται, πάνυ σαφές ἐστιν, ὡς ἐγχωρεῖ τἀναντία λέγεσθαι ταὐτόν, οἷον Δίωνα Θέωνος μὲν καὶ Μέμνονος ξηρότερον, Ἀρίστωνος δὲ καὶ Γλαύκωνος ὑγρότερον. ὅταν δὲ πρὸς τὸ σύμμετρον ὁμογενὲς ἢ ὁμοειδές, ἐνταῦθ' ἤδη συγχεῖσθαί τε καὶ ταράττεσθαι συμβαίνει τοὺς ἀγυμνά-

ON TEMPERAMENTS, BOOK I

That in all these the median of the extremes is the moderate and *eukratic* in relation to that class or kind (genus or species) must always be understood in the whole argument, even though we might sometimes happen to omit this in the statement. And actually, these things being so, whenever someone says this particular *krasis* is wet or hot, what must be asked is how he has said this—whether he is making a comparison with some one thing separately, for example comparing Theophrastus to Plato, or to any class or kind whatsoever, or as human, or as animal, or as substance absolutely. For certainly the third signification of each of such terms, which we said to be simple and uncompounded, is not in those things that have been compounded, but in the primary [bodies] themselves, which we also call elements. Consequently, it is said in a threefold manner of each of the bodies with qualities, but only two are useful to us for the matter concerning *krasias*, either when making a comparison to anything whatsoever or to the moderate of the same class (genus).

However, since there are many classes, just as there also are many indivisible things, the same body will be able to be hot and cold and dry and wet in many ways. But when it is compared to any one thing, it is very clear that it is possible to state the opposite in respect of this—for example, that Dion is drier than Theon or Memnon, but wetter than Ariston and Glaucon. When it is said with reference to the moderate (well-balanced) of the same class or kind, here already those who are unpracticed

---

[29] H *has: verba* ἢ γὰρ ὡς ἄνθρωπον—ἁπλῶς *ab interprete inserta esse videntur (p. 22).*

στους. ὁ γὰρ αὐτὸς ἄνθρωπος ὑγρὸς ἅμα καὶ θερμὸς εἶναι δύναται καὶ ξηρὸς καὶ ψυχρός, ξηρὸς μὲν καὶ ψυχρὸς ὡς πρὸς τὸν σύμμετρον ἄνθρωπον παραβαλλόμενος, ὑγρὸς δὲ καὶ θερμὸς ὡς πρὸς ἄλλο τι ἢ ζῷον ἢ φυτὸν ἢ οὐσίαν ἡντινοῦν, οἷον ὡς πρὸς μὲν ζῷον, εἰ τύχοι, μέλιτταν τε καὶ μύρμηκα, πρὸς δὲ φυτὸν ἐλαίαν ἢ συκῆν ἢ δάφνην, πρὸς οὐσίαν δέ τιν' ἑτέραν, ἢ μήτε ζῷόν ἐστι μήτε φυτόν, οἷον λίθον ἢ σίδηρον ἢ χαλκόν. ἐν τούτοις δὲ τὸ μὲν πρὸς ἄνθρωπον παραβάλλειν πρὸς ὁμοειδές ἐστι παραβάλλειν,[30] τὸ δὲ πρὸς μέλιτταν ἢ μύρμηκα πρὸς ὁμογενές, ὡσαύτως δὲ καὶ πρὸς φυτὸν ὁτιοῦν. ἔστι γὰρ ἀνωτέρω τοῦ ζῴου τοῦτο τὸ γένος, ὥσπερ οὖν καὶ αὐτοῦ τούτου λίθος καὶ σίδηρος καὶ χαλκὸς ἐκ τῶν ἄνωθεν γενῶν. καλείσθω γοῦν ⟨πρὸς⟩ ὁμογενὲς ἕνεκα συντόμου διδασκαλίας ἡ τοιαύτη σύμπασα παραβολή, τοσόνδε μόνον ἐν αὐτῇ διελομένων ἡμῶν, ὡς, ἐπειδὰν μὲν ἁπλῶς οὐσία τις εὔκρατος λέγηται καὶ ταύτης δέ τις ἑτέρα ξηροτέρα καὶ θερμοτέρα καὶ ψυχροτέρα καὶ ὑγροτέρα, τὴν μὲν εὔκρατον ἐνταῦθα τὴν ἐκ τῶν ἐναντίων ἀκριβῶς ἴσων συνελθόντων ὀνομάζομεν, ὅσον δ' ἀπολείπεται τῆσδε καὶ πλεονεκτεῖ κατά τι, τῷ τοῦ πλεονεκτοῦντος ὀνόματι προσαγορεύομεν· ἐπειδὰν δ' ἤτοι φυτὸν εὔκρατον ἢ ζῷον ὁτιοῦν εἴπωμεν, οὐκέθ' ἁπλῶς ἀλλήλοις ἐν τῇ τοιαύτῃ λέξει[31] τἀναντία παραβάλλομεν, ἀλλὰ πρὸς τὴν τοῦ φυτοῦ φύσιν ἢ τὴν τοῦ ζῴου τὴν ἀναφορὰν ποιούμεθα, συκῆν μὲν εὔκρατον. εἰ τύχοι, λέγοντες, ὅταν, οἵᾳ μάλιστα πρέπει τὴν φύ-

come to be confused and troubled. For it is possible for
the same person to be wet and hot at the same time, and
dry and cold—dry and cold as compared to the moderate
(well-balanced) person, while wet and hot as compared to
some other animal or plant or any existing thing whatsoever. For example, as regards an animal, it may be a bee
or ant, while as regards a plant, it may be an olive, fig, or
laurel tree, and as regards some other substance, which is
neither animal nor plant, like stone, iron, or copper. In
these, the comparison to a human is a comparison to the
same kind (species), while that to a bee or ant is to the
same class, and similarly also to a plant of whatever kind.
For | this class is higher than that of animal just as also that
of stone, iron, and copper is from the classes higher than
this itself. Anyway, for the sake of brevity in teaching, let
every such comparison be called homogenous, while we
make a distinction in this only as much as when some
substance is called *eukratic* in an absolute sense, while
some other substance is drier, hotter, colder, and wetter
than this, we term the *eukratic* here the precisely equal
coming together of the opposites, while as much as it falls
short of this and predominates relatively, we call by the
name of what predominates. However, when we say either
a plant or an animal is *eukratic* in any way whatsoever, in
such a statement we are not comparing opposites with
each other absolutely, but are making the reference to the
nature of the plant or animal, calling a fig tree *eukratic*,
for example, when it is such as to particularly resemble the

---

30 παραβάλλειν H; παραβολὴ K
31 λέξει H; τάξει K

σιν ὑπάρχειν συκῇ, τοιαύτη τις ᾖ, κύνα δ' αὖ καὶ σῦν καὶ ἵππον καὶ ἄνθρωπον, ἐπειδὰν καὶ τούτων ἕκαστον ἄριστα τῆς οἰκείας ἔχῃ φύσεως. αὐτὸ δὲ δὴ τοῦτο τὸ τῆς οἰκείας φύσεως ἔχειν ἄριστα ταῖς ἐνεργείαις κρίνεται. καὶ γὰρ καὶ φυτὸν καὶ ζῷον ὁτιοῦν ἄριστα διακεῖσθαι τηνικαῦτά φαμεν, ὅταν ἐνεργήσῃ κάλλιστα. συκῆς μὲν γὰρ ἀρετὴ βέλτιστά τε καὶ πλεῖστα τελεσφορεῖν σῦκα· κατὰ ταὐτὰ δὲ καὶ τῆς ἀμπέλου τὸ πλείστας τε καὶ καλλίστας ἐκφέρειν σταφυλάς, ἵππου δὲ τὸ θεῖν ὠκύτατα καὶ κυνὸς εἰς μὲν θήρας τε καὶ φυλακὰς ἄκρως εἶναι θυμοειδῆ, πρὸς δὲ τοὺς οἰκείους πρᾳότατον.

Ἅπαντ' οὖν ταῦτα, τά τε ζῷα λέγω καὶ τὰ φυτά, τὴν ἀρίστην τε καὶ μέσην ἐν τῷ σφετέρῳ γένει κρᾶσιν ἔχειν ἐροῦμεν οὐχ ἁπλῶς, ὅταν ἰσότης ἀκριβὴς ᾖ τῶν ἐναντίων, ἀλλ' ὅταν ἡ κατὰ δύναμιν αὐτοῖς ὑπάρχῃ συμμετρία. τοιοῦτον δέ τι καὶ τὴν δικαιοσύνην εἶναί φαμεν, οὐ σταθμῷ καὶ μέτρῳ τὸ ἴσον, ἀλλὰ τῷ προσήκοντί γε καὶ κατ' ἀξίαν ἐξετάζουσαν. ἰσότης οὖν κράσεως ἐν ἅπασι τοῖς εὐκράτοις ζῴοις τε καὶ φυτοῖς ἐστιν, οὐχ ἡ κατὰ τὸν τῶν κερασθέντων στοιχείων ὄγκον, ἀλλ' ἡ τῇ φύσει τοῦ τε ζῴου καὶ τοῦ φυτοῦ πρέπουσα. πρέπει δ' ἔσθ' ὅτε τὸ μὲν ὑγρὸν τοῦ ξηροῦ,[32] τὸ δὲ ψυχρὸν τοῦ θερμοῦ πλέον ὑπάρχειν. οὐ γὰρ ὁμοίαν χρὴ κρᾶσιν ἔχειν ἄνθρωπον καὶ λέοντα καὶ μέλιτταν καὶ κύνα. πρὸς δὴ τὸν ἐρόμενον, ᾗστινός ἐστι κράσεως ἄνθρωπος ἢ ἵππος ἢ βοῦς ἢ κύων ἢ ὁτιοῦν ἄλλο τῶν πάντων, οὐχ ἁπλῶς ἀποκρι-

nature of a fig tree, and the same applies to a dog, pig, horse, and human, when each of these has the best of the characteristic nature. This itself—having the best of the characteristic nature—is judged by the functions. For surely, we say any plant or animal whatsoever is in the best state when it functions best. Thus, excellence of a fig tree is when it brings forth the best and most numerous figs. In the same way too, we say this of a vine that produces the largest number of and the best grapes, and of a horse that runs fastest, and of a dog that is most courageous in hunting and guarding, but is most gentle toward members of the household.

All these things—I speak of animals and plants—we will say have the best and median *krasis* in their own class, not in an absolute sense, when there is an exact equality of opposites, but when there is a balanced state in respect of capacity in them. Justice is such a thing, we also say, evaluating it not by the equality in balance and measure, but its appropriateness and relation to worth. Therefore, there is equality of *krasis* in all *eukratic* animals and plants, not in relation to the mass of the compounding elements, but in the appropriateness to the nature of the animal and plant. Sometimes the wet is more suitable than the dry, and sometimes the cold more than the hot. For it is not necessary for a human, lion, bee, and dog to have a similar *krasis*. Certainly, to someone asking, one must not answer in an absolute sense what kind of *krasis* a human, horse, ox, dog, or any of the other things has. It is not pos-

---

[32] *post* τοῦ ξηροῦ,: τὸ δὲ ψυχρὸν τοῦ θερμοῦ H; τὸ δὲ θερμὸν τοῦ ψυχροῦ K

τέον. οὐ γὰρ ἐγχωρεῖ τὰ πολλαχῶς λεγόμενά τε καὶ κρινόμενα καθ' ἕνα τρόπον ἀποκρινάμενον ἀνέγκλητον εἶναι. χρὴ τοίνυν δυοῖν θάτερον, ἢ πάσας ἐπέρχεσθαι τὰς διαφορὰς ἢ περὶ τίνος ἤρετο πυθόμενον ἐκείνην μόνην εἰπεῖν. εἰ μὲν γὰρ ὡς ἐν ζῴοις πυνθάνοιτο, τίνος εἴη κράσεως, ἐπὶ τὸ μέσον ἁπάντων τῶν ζῴων τῇ κράσει βλέποντα τὴν ἀπόκρισιν ποιητέον· εἰ δ' ἁπλῶς τε καὶ ὡς πρὸς ἅπασαν οὐσίαν, οὕτως ἤδη παραβάλλειν τἀναντία τῶν ἐν αὐτῷ πρὸς ἄλληλα καὶ σκοπεῖσθαι, μηκέτι πρὸς τὰς ἐνεργείας ἀναφέροντα τὴν κρᾶσιν, ἀλλὰ πρὸς τοὺς ὄγκους τῶν στοιχείων. εἰ δ' ἀφωρισμένως τῷδέ τινι παραβάλλων | ἤρετο, πρὸς ἐκεῖνο μόνον αὐτὸ παραβλητέον. ἔτι δὲ μᾶλλον, εἴ τινος τῶν ἀτόμων οὐσιῶν, οἷον, εἰ τύχοι, Δίωνος ἢ τοῦδέ τινος τοῦ κυνὸς ἐρωτηθείημεν ὁποία τις ἡ κρᾶσίς ἐστιν, οὐχ ἁπλῶς ἀποκριτέον. ἀφορμὴ γὰρ οὐ σμικρὰ τοῖς σοφισταῖς ἐντεῦθεν εἰς τὸ συκοφαντεῖν. εἰ γὰρ εἴποις ξηρᾶς καὶ θερμῆς κράσεως εἶναι τὸν Δίωνα, ῥᾷστον αὐτῷ, προχειρισαμένῳ τοῦτο μὲν[33] τῶν θερμοτέρων τε καὶ ξηροτέρων ἐκείνου τὴν κρᾶσιν ἀνθρώπων ὁντινοῦν, ὑγρὸν καὶ ψυχρὸν ὡς πρὸς ἐκεῖνον ἀποφῆναι τὸν Δίωνα, τοῦτο δ' ἄλλο τι φυτὸν ἢ ζῷον, οἷον, εἰ τύχοι, λέοντά τε καὶ κύνα, καὶ τούτων ὑγρότερόν τινα καὶ ψυχρότερον ἀποδεῖξαι τὸν Δίωνα.

Χρὴ τοίνυν, ὅστις μήθ' ἑαυτὸν ἐξαπατῆσαι βούλεται μήθ' ὑπ' ἄλλου σοφισθῆναι, τοῦτον ἀπὸ τῶν ἁπλῶς λεγομένων θερμῶν καὶ ψυχρῶν καὶ ξηρῶν καὶ

ON TEMPERAMENTS, BOOK I

sible for there to be one kind of answer that is beyond reproach in respect of those things spoken about and judged in many ways. Accordingly, it is necessary to do one or other of two things: either to go through all the differences, or to choose the one inquired about and speak only of that. If the inquiry is what kind of *krasis* there is in animals, one must make the answer, looking to the median in the *krasis* of all the animals. If, however, [the question is posed] absolutely and in regard to every substance, in this way now [it is necessary] to consider the comparison of the opposites of these with each other in the thing itself, no longer referring the *krasis* to the functions, but to the masses of the elements. On the other hand, if what is being raised for comparison is one of these separately, one must make the comparison to that alone. Still more, if we are asked what kind of *krasis* there is of one of the indivisible beings, like for example Dion or some particular dog, we must not answer in an absolute sense. Therefore, there is no small pretext for the Sophists to indulge here in their quibbles. For if you were to say Dion is of a hot and dry *krasis*, it would be a very easy matter for him to bring forward some man of those who in *krasis* are hotter and drier than Dion, and declare that Dion is wet and cold in relation to that man. Or he may bring forward some other plant or animal—for example, as might be the case, a lion and a dog—and show Dion to be colder and drier than these.

Accordingly, it is necessary, for someone who does not wish to deceive himself or to be tricked by someone else to begin this from the substances said to be hot, cold, dry,

---

33 τοῦτο μὲν *add.* H (*see his note, p. 25*)

ὑγρῶν οὐσιῶν ἀρξάμενον οὕτως ἐπὶ τὰς ἄλλας μετιέναι καὶ πρῶτον μὲν αὐτὸ δὴ τοῦτ᾽ ἐπ᾽ αὐτῶν διορίσασθαι, τὸ μηδὲ ταύτας, εἰ καὶ ὅτι μάλιστα δοκοῦσιν ἁπλῶς λέγεσθαι, πεφευγέναι τὴν πρὸς τὸ σύμμετρον ὁμογενὲς παραβολήν. ὥσπερ γὰρ κύνα μέσον ἁπάντων κυνῶν τῇ κράσει λέγομεν, ὅταν ἴσον ἀπέχῃ τῶν ἄκρων, οὕτω καὶ οὐσίαν ἐροῦμεν εἶναι μέσην τῇ κράσει τὴν ἴσον ἀπέχουσαν τῶν ἄκρων, ἃ δὴ καὶ πρῶτα πάντων ἐστὶ στοιχεῖα. ἴσον δ᾽ ἀφέξει δηλονότι τῶν ἄκρων, ἐξ ἴσου κερασθέντων ἁπάντων. τὴν οὖν ὑπερβάλλουσαν ἢ ἐλλείπουσαν τῆσδε θερμὴν ἢ ψυχρὰν ἢ ξηρὰν ἢ ὑγρὰν εἶναι φήσομεν, ἅμα μὲν τῇ μέσῃ παραβάλλοντες, ἅμα δὲ καὶ τῶν ἐναντίων στοιχείων ἐξέτασιν ἐπ᾽ αὐτῆς ποιούμενοι. καὶ δὴ καὶ κατὰ τοῦτο μὲν ἁπλῶς ἐροῦμεν αὐτὴν ἢ θερμὴν ἢ ψυχρὰν ἢ ξηρὰν ἢ ὑγρὰν ὑπάρχειν. ἐπειδὰν δὲ τῇ μέσῃ κράσει παραβάλλωμεν, οὐχ ἁπλῶς, ἀλλ᾽ ὅτι πρὸς τὸ σύμμετρον ὁμογενὲς οὕτως ἔχει.

Γένος δ᾽ ἦν αὐτῶν ἡ οὐσία. πάντα γὰρ ὑπὸ ταύτην πέπτωκεν ὡς ἀνωτάτω τι γένος, ἔμψυχά τε καὶ ἄψυχα, καὶ κοινόν ἐστιν ἥδε καὶ ἀνθρώπου καὶ κυνὸς καὶ πλατάνου καὶ συκῆς καὶ λίθου[34] καὶ χαλκοῦ καὶ σιδήρου καὶ τῶν ἄλλων ἁπάντων γένος. ὑπ᾽ αὐτῇ δ᾽ ἐστὶν ἕτερα γένη πολλά, τὸ μὲν ζῷον ὄρνιθός τε καὶ ἰχθύος, τὸ δὲ φυτὸν δένδρου τε καὶ βοτάνης, ἀετοῦ δὲ καὶ κόρακος ὄρνις, καὶ λάβρακος καὶ φυκίδος ἰχθύς. ὡσαύτως δὲ καὶ τὸ μὲν δένδρον ἐλαίας τε καὶ συκῆς γένος, ἡ βοτάνη δ᾽ ἀναγαλλίδος τε καὶ παιωνίας.

ON TEMPERAMENTS, BOOK I

and wet absolutely in this way, then to move on to the others, first distinguishing this itself from these, even if they particularly seem to be said absolutely, avoiding the | comparison to the mean of the same class. For just as we say a dog is median in *krasis* of all dogs, whenever it is equidistant from the extremes, in this way too we shall say a substance equidistant from the extremes is median in *krasis*, these in fact also being the primary elements of all things. Clearly something will be equally distant from the extremes from an equality of all compounding things. We shall speak, then, of the excess or deficiency of this particular thing in respect of hot, cold, dry, and wet, making at the same time a comparison with the median and a close examination of the opposite elements in this case. And indeed, on this basis, we shall say this is hot, or dry, or cold, or wet absolutely. When, however, we make a comparison with the median *krasis*, not absolutely but because it is so in relation to the mean of the same class, it is like this.

The substance, however, is a class of these. For all fall under this as the highest class, both animate and inanimate things, and this is common for human, dog, plane tree, | fig tree, stone, copper, iron, and a class of all the others. Below it, however, there are many other classes. Animal includes bird and fish, while plant includes tree and herb, bird includes eagle and raven, and fish includes bass and wrasse. Similarly also the class tree includes olive and fig, while that of herb includes pimpernel and peony.

---

[34] καὶ λίθου *add.* H *(see his note, p. 26)*

ἔσχατα δὲ γένη ταῦτα καὶ διὰ τοῦτο καὶ εἴδη προσαγορεύεται, κόραξ καὶ φυκὶς καὶ συκῆ καὶ ἀναγαλλίς· οὕτω δὲ καὶ ἄνθρωπος καὶ βοῦς καὶ κύων. ἄνωθεν μὲν κατιόντων ἔσχατα γένη ταῦτα καὶ διὰ τοῦτο καὶ εἴδη προσαγορεύεται, κάτωθεν δ' ἀνιόντων ἀπὸ τῶν ἀτόμων οὐσιῶν πρῶτα. καὶ δέδεικται δι' ἑτέρου γράμματος, ὡς εὐλόγως οἱ παλαιοὶ ταῦτα σύμπαντα τὰ μεταξὺ τῶν ἀτόμων τε καὶ τῶν πρώτων γενῶν εἴδη θ' ἅμα καὶ γένη προσαγορεύουσιν.

7. Ὁπότ' οὖν διῄρηται τὰ σημαινόμενα καὶ ὡς οὐχ ἁπλῶς ἀποφαίνεσθαι χρὴ θερμόν τι καὶ ψυχρὸν καὶ ξηρὸν καὶ ὑγρὸν σῶμα, σαφῶς ἐπιδέδεικται, ζητητέον ἐφεξῆς αὐτῶν τὰ γνωρίσματα. καίτοι κἀνταῦθα χρὴ πρότερον ὑπὲρ τῶν ὀνομάτων διελέσθαι τῶν ἐμπίπτειν μελλόντων ἐξ ἀνάγκης εἰς τὸν ἐφεξῆς λόγον ἐξαπλῶσαί τέ τι πρᾶγμα, δυνάμει μὲν ἤδη προαποδεδειγμένον, οὐ μὴν ἐναργῶς γε πᾶσι[35] τοῖς ἀναγιγνώσκουσι τόδε τὸ γράμμα νοηθῆναι δυνάμενον. ὑπὲρ τῶν ὀνομάτων οὖν πρῶτον εἰπόντες οὕτως ἐπανίωμεν ἐπὶ τὸ πρᾶγμα.

Τὸ θερμὸν καὶ τὸ ψυχρὸν καὶ τὸ ξηρὸν καὶ τὸ ὑγρὸν ὅτι μὲν οὐχ ἕν τι σημαίνει παρὰ τοῖς Ἕλλησιν,[36] ἐπειδὰν ἐπὶ σωμάτων λέγηται, δέδεικται πρόσθεν· ὅτι δὲ καὶ τὰς ἐν τοῖς σώμασι ποιότητας αὐτὰς μόνας

[35] *post* γε πᾶσι: τοῖς ἀναγιγνώσκουσι τόδε τὸ γράμμα H; τοῖς ἐντυγχάνουσι τῷδε τῷ γράμματι K
[36] παρὰ τοῖς Ἕλλησιν *add.* H (*see his note, p. 27*)

ON TEMPERAMENTS, BOOK I

These classes are the ultimate, and because of this are also called kinds (species)—raven, wrasse, fig, and pimpernel. The same also applies to human, ox, and dog. Therefore, proceeding downward from above, these classes are lowest, and because of this are also called kinds (species), whereas proceeding from below upward from the undivided substances, they are primary. And it has been shown through another work that the ancients reasonably called all these between the undivided (individual things) and the primary classes, kinds and at the same time classes.[24]

7. Therefore, when the significations are determined and it has been clearly demonstrated that a body must not be called something hot, cold, dry, or wet in an absolute sense, | the next things we must seek are the signs of these. And indeed, even here, we must first make a distinction regarding the names that are going to come up of necessity in the discussion to follow, and explain some matter which has already been demonstrated beforehand authoritatively, although not so clearly that it can be conceived of by all those reading this particular work. Having spoken about the names first, let us therefore return to the matter.

The hot, the cold, the dry, and the wet, and that they do not signify one thing among the Greeks when said regarding bodies, has been shown before. That they also sometimes name in this way the actual qualities alone in

[24] See Galen, *De differentiis pulsuum* VIII.601 and 630K, and *De methodo medendi* X.135 and 143K.

ἄνευ τῶν δεδεγμένων αὐτὰς οὐσιῶν οὕτως ὀνομάζουσιν ἐνίοτε, τοῦτο μὲν οὔπω εἴρηται πρόσθεν, ἤδη δ' αὐτὸ λέγεσθαι καιρός. ὥσπερ οὖν τὸ λευκὸν ὄνομα κατά τε τοῦ χρώματος ἐπιφέρουσιν, ἐπειδὰν οὕτω λέγωσιν, ἐναντίον ἐστὶ τὸ λευκὸν χρῶμα τῷ μέλανι, κατά τε τοῦ δεδεγμένου τὸ χρῶμα σώματος, ἐπειδὰν τὸ τοῦ κύκνου σῶμα λευκὸν εἶναι φάσκωσιν, οὕτω καὶ τὸ θερμὸν ὄνομα κατά τε τῆς ποιότητος ἐπιφέρουσιν, ὡς εἰ καὶ θερμότητα προσηγόρευον, ἀλλὰ καὶ τοῦ σώματος, ὃ τὴν θερμότητα δέδεκται. τὴν γὰρ δὴ ποιότητα παρὰ τὸ δεδεγμένον αὐτὴν σῶμα χρὴ νομίζειν εἶναί τινα φύσιν ἔχουσαν ἰδίαν, ὡς ἐν τοῖς Περὶ τῶν στοιχείων λόγοις δέδεικται. θερμότης μὲν οὖν ποιότης, ἡ δ' αὐτὴ καὶ θερμὸν ὀνομάζεται, καθάπερ λευκότης καὶ λευκόν. αὐτὸ δὲ τὸ σῶμα τὸ θερμὸν ἓν τοῦτο μόνον ὄνομα κέκτηται τὸ θερμόν, ὥσπερ τὸ λευκόν. οὐ μὴν οὔτε θερμότης οὔτε λευκότης αὐτὸ τὸ σῶμά ποτε προσαγορεύεται. κατὰ δὲ τὸν αὐτὸν τρόπον καὶ ψυχρὸν καὶ ξηρὸν καὶ ὑγρὸν ὀνομάζεται τό τε σῶμα αὐτὸ καὶ ἡ ποιότης· οὐ μὴν καὶ ψυχρότης ἢ ξηρότης ἢ ὑγρότης ἔτι καλεῖται τὸ σῶμα καθάπερ ἡ ἐν αὐτῷ ποιότης.

Τούτων οὖν οὕτως ἐχόντων εὔλογον, ἐπειδὰν μὲν ἤτοι θερμότητά τις ἢ ψυχρότητα διαλεγόμενος εἴπῃ, μηδὲν γίγνεσθαι σόφισμα· μόναι γὰρ ἐκ τῶν τοιούτων ὀνομάτων αἱ ποιότητες δηλοῦνται· θερμὸν δ' εἰπόντος ἢ ψυχρόν, ἐπειδὴ καὶ ἡ ποιότης οὕτω καὶ τὸ σῶμα τὸ δεδεγμένον αὐτὴν ὀνομάζεται, πρόχειρον γίγνεσθαι

ON TEMPERAMENTS, BOOK I

the bodies without the substances that have received these qualities is something not yet said before. Now is the time for this to be said. Therefore, just as they apply the term "white" referring to the color when they say in this way "the white color is opposite to the black," and in relation to the body that has taken up the color, when they say the body of the swan is white, in this way too, | they apply the term "hot" in relation to the quality, as if calling it hotness, but also referring to the body that has received the hotness. For certainly the quality must be thought of as being apart from the body that has received it, and as having some specific nature, as has been shown in the work on the elements.[25] Hotness, then, is a quality which itself is also termed hot, just as with whiteness and white. However, the body which is itself hot has acquired this single name of hot alone, just as with whiteness and white. The body itself is never called "hotness" or "whiteness." And in the same way too, the body itself and the quality are termed cold, dry, and wet. The body itself is not called "coldness," "dryness" and "wetness" just as the quality in it is.

553K

These things being so, it is reasonable, when someone conversing says something is either hotness or coldness, nothing sophistical arises, for the qualities alone are indicated by such terms. However, when someone says hot or cold, since both the quality and the body that has received it in this way | are so named, it becomes easy for someone

554K

[25] Galen, *De elementis secundum Hippocratem*, I.461–65K.

τῷ κακουργεῖν ἐθέλοντι τὸ μὴ δηλούμενον ὑπὸ τοῦ λέγοντος ἀκούειν, ἵν' ἔχῃ σοφίζεσθαι. τοιοῦτον γάρ τι δρῶσι καὶ οἱ πρὸς τὸν ἀφορισμὸν ἀντιλέγοντες, ἐν ᾧ φησιν Ἱπποκράτης· "τὰ αὐξανόμενα πλεῖστον ἔχει τὸ ἔμφυτον θερμόν." οὐ γὰρ σῶμά τι θερμὸν ἔμφυτον τῷ ζῴῳ λέγεσθαι πρὸς Ἱπποκράτους ἀκούσαντες οὐδὲ ζητήσαντες, ὅ τί ποτ' ἐστὶ τοῦτο, κατὰ τῆς ποιότητος μόνης, ἣν δὴ καὶ θερμότητα καλοῦμεν, εἰρῆσθαι τοὔνομα δεξάμενοι τὴν ἀντιλογίαν οὕτω ποιοῦνται. καὶ φαίνεται καίτοι σμικρὸν ὂν τὸ διαστέλλεσθαι τὰς ὁμωνυμίας ἐν τῇ χρείᾳ τῶν πραγμάτων ἱκανῶς ἀξιόλογον ὑπάρχον.

8. Ἀλλ' ἐπειδὴ καὶ τοῦτο σαφῶς ἤδη διώρισται, ἐπὶ τὸ ὑπόλοιπον αὖθις ἐπανιτέον. οὔσης γάρ τινος ἀκράτου καὶ ἀμίκτου ποιότητος, θερμότητός τε καὶ ψυχρότητος καὶ ξηρότητος καὶ ὑγρότητος, ὅσα ταύτας ἐδέξατο σώματα, θερμὰ δηλονότι καὶ ψυχρὰ καὶ ξηρὰ καὶ ὑγρὰ τελέως τε καὶ ἀκριβῶς ἐστι.[37] ταυτὶ μὲν οὖν μοι νόει τὰ τῶν γιγνομένων τε καὶ φθειρομένων ἁπάντων στοιχεῖα, τὰ δ' ἄλλα σώματα τά τε τῶν ζῴων καὶ τὰ τῶν φυτῶν καὶ τὰ τῶν ἀψύχων ἁπάντων οἷον χαλκοῦ καὶ σιδήρου καὶ λίθων καὶ ξύλων, ἐν τῷ μεταξὺ τῶν πρώτων ἐκείνων τετάχθαι. οὐδὲν γὰρ αὐτῶν οὔτ' ἄκρως θερμὸν οὔτ' ἄκρως ψυχρὸν οὔτ' ἄκρως ξηρὸν οὔτ' ἄκρως ὑγρόν ἐστιν, ἀλλ' ἤτοι μέσον ἀκριβῶς ὑπάρχει τῶν ἐναντίων, ὡς μηδὲν μᾶλλον εἶναι θερμὸν ἢ ψυχρὸν ἢ ξηρὸν ἢ ὑγρόν, ἢ θατέρων τῶν ἄκρων προσκεχώρηκεν, ὡς μᾶλλον εἶναι θερμὸν

else wishing to be captious not to understand what is being indicated by the speaker, so there is sophistry. Assuredly, those gainsaying the aphorism in which Hippocrates says, "Things that are growing have the greatest amount of innate heat"[26] are doing such a thing. For they do not understand some hot body innate to the animal as what is being referred to by Hippocrates, nor do they seek after what this is in relation to the quality alone, which in fact we also call hotness, accepting the name stated to effect the refutation in this way. And indeed, it is apparent that the distinguishing of the homonyms, small though it may be, is of considerable importance in the use of the matters.

8. But since this has already been clearly determined, we must go back again to what remains. For since hotness, coldness, dryness and wetness are each some pure and unblended quality, those bodies which receive them will clearly be hot, cold, dry, and wet completely and precisely. Therefore, suppose for me that these are the elements of all the things that come into being and pass away. The other bodies, however—those of animals and plants and of all the inanimate things like those of bronze, iron, stones, and wood—are placed in between those primary [elements]. For none of the [bodies listed] are either completely hot or completely cold, or completely dry or completely wet, but are either precisely midway between the opposites so as to be no more hot, cold, dry, or wet, or to approach one or other of the extremes, so as to be more

---

[26] Hippocrates, *Aphorisms* 1.14.

[37] ἐστι H; ἔσται K

ἢ ψυχρὸν ἢ μᾶλλον ξηρὸν ἢ ὑγρόν. εἰ μὲν δὴ μέσον ἀκριβῶς εἴη καθ' ἑκατέραν τῶν ἀντιθέσεων, ὡς μηδὲν μᾶλλον εἶναι θερμὸν ἢ ψυχρὸν ἢ ξηρὸν ἢ ὑγρόν, εὔκρατον ἁπλῶς τοῦτο λεχθήσεται· θατέρου δὲ πλεονεκτήσαντος ἤτοι κατὰ τὴν ἑτέραν ἀντίθεσιν ἢ κατ' ἀμφοτέρας, οὐκέτ' εὔκρατον. εἰ μὲν δὴ θερμὸν εἴη μᾶλλον ἢ ψυχρόν, ὃ μᾶλλόν ἐστι, τοῦτο λεχθήσεται.[38] κατὰ ταὐτὰ δὲ καὶ εἰ | ψυχρὸν εἴη μᾶλλον, ὀνομασθήσεται ψυχρόν, ὡσαύτως δὲ καὶ ξηρὸν καὶ ὑγρόν. εἰ δ' ἐξ ἑκατέρας τῆς ἀντιθέσεως ἐπικρατοίη θάτερον, ἤτοι θερμὸν ἅμα καὶ ὑγρὸν ἢ θερμὸν ἅμα καὶ ξηρὸν ἢ ψυχρὸν ἅμα καὶ ξηρὸν ἢ ψυχρὸν ἅμα καὶ ὑγρὸν ὀνομασθήσεται τὸ σῶμα.[39]

Ταύτας μὲν οὖν τὰς τέτταρας δυσκρασίας, ὡς καὶ πρόσθεν εἴπομεν, οἱ πλεῖστοι γιγνώσκουσιν ἰατροί τε καὶ φιλόσοφοι. τὰς δ' ἄλλας τέτταρας τὰς ἐξ ἡμίσεος τούτων γιγνομένας οὐκ οἶδ' ὅπως παραλείπουσιν, ὥσπερ καὶ τὴν πρώτην ἁπασῶν κρᾶσιν, τὴν ἀρίστην. ἀλλ' ὅτι γε δυνατὸν ἐπικρατοῦντος τοῦ θερμοῦ μηδὲν μᾶλλον ὑγρὰν ἢ ξηρὰν εἶναι τὴν κρᾶσιν, ὅσον ἐπὶ ταύτῃ τῇ συζυγίᾳ, πρόδηλον μὲν οἶμαι κἀκ τῶν ἤδη προειρημένων εἶναι. ῥᾷστον δέ, κἂν εἰ μηδὲν προείρητο, συλλογίσασθαι, συγχωρησάντων γ' ἅπαξ αὐτῶν ἑτέραν μὲν εἶναι κρᾶσιν ὑγρὰν καὶ θερμήν, ἑτέραν δὲ ξηρὰν καὶ θερμήν. εἰ γὰρ οὐκ ἀναγκαῖον

[38] post λεχθήσεται.: κατὰ ταὐτὰ δὲ καὶ εἰ ψυχρὸν εἴη μᾶλλον, ὀνομασθήσεται ψυχρόν, ὡσαύτως δὲ καὶ ξηρὸν καὶ

hot or cold, or more dry or wet. If, however, they are precisely median in relation to each of the antitheses, so as not to be more hot or cold, or more dry or wet, this will be called absolutely *eukratic*. On the other hand, when one predominates, it will relate to either one of the antitheses or to both and will no longer be *eukratic*. If, in fact, it is more hot or cold, it will be named the one that is more. In the same way too, if cold is more, it will be termed cold, and similarly dry and wet. If one from each antithesis predominates, it will be termed either hot and wet at the same time, or hot and dry at the same time, or cold and dry at the same time, or cold and wet at the same time—[that is, the body will be named according to what predominates.]²⁷

These then are the four *dyskrasias*, as I said previously. Most doctors and philosophers know these. However, they leave aside the other four, arising from half of these—how I don't know—just as they do the first *krasis* of all which is the best. But that, in fact, it is possible when hot predominates, for the *krasis* to be neither more wet nor more dry, as far as pertains to the this conjunction, is, I think, clear from what has been said previously. However, it would be very easy, even if nothing were said previously, to draw the conclusion, since we agree once and for all on these points—that one *krasis* is wet and hot, another dry

---

²⁷ Added following Kühn.

ὑγρόν. H; ὡσαύτως δὲ καὶ ξηρὸν καὶ ὑγρόν· κατὰ ταὐτὰ δὲ καὶ εἰ ψυχρὸν εἴη μᾶλλον, ὀνομασθήσεται ψυχρόν. K *(see H's note on this rearrangement, p. 29)*

³⁹ *post* τὸ σῶμα: κατὰ τὸ ἐπικρατοῦν *add.* K

ἐστι πάντως εἶναι ξηρὰν τὴν θερμήν, ἀλλ' ἐγχωρεῖ καὶ ὑγρὰν ὑπάρχειν αὐτήν, ἐγχωρήσει δηλονότι καὶ μέσην· ἐγγυτέρω γάρ ἐστιν ἡ μέση τῆς ξηρᾶς κράσεως ἤπερ ἡ ὑγρά. κατὰ δὲ τὸν αὐτὸν τρόπον ἐστί τις ἑτέρα ψυχρὰ κρᾶσις, ἐφ' ἧς ἐστιν ἰσχυρότερον τὸ ψυχρόν· οὐ μὴν οὔθ' ὑγρὰν οὔτε ξηρὰν ἀναγκαῖον εἶναι τὴν αὐτήν,[40] ἀλλ' ἐγχωρεῖ καὶ μέσην γενέσθαι· πάλιν γὰρ κἀνταῦθα τὸν αὐτὸν ἐπάξεις λόγον. ὥστ', εἴπερ οὐκ ἀναγκαῖόν ἐστιν ὑγρὰν εἶναι τὴν ψυχράν, ἀλλ' ἐγχωρεῖ καὶ ξηρὰν γενέσθαι, πρόδηλον, ὡς καὶ τὴν μέσην ἐγχωρήσει· ἐγγυτέρω γάρ ἐστιν αὕτη τῆς ὑγρᾶς ἤπερ ἡ ξηρά. ὡς οὖν αὗται αἱ δύο δυσκρασίαι κατὰ τὴν ἑτέραν ἀντίθεσιν ἐδείχθησαν, ἡ μὲν θερμὴ μόνον, ἡ δὲ ψυχρά, κατὰ τὸν αὐτὸν τρόπον ἄλλαι δύο γενήσονται κατὰ τὴν ἑτέραν ἀντίθεσιν, ἡ μὲν ξηρὰ μόνον, ἡ δ' ὑγρά, συμμέτρως ἐχόντων πρὸς ἄλληλα τοῦ θερμοῦ καὶ τοῦ ψυχροῦ. πάλιν γὰρ κἀνταῦθα φήσομεν, ὡς, εἴπερ οὐκ ἔστιν ἀναγκαῖον, εἴ τίς ἐστι ξηρὰ κρᾶσις, εὐθὺς ταύτην εἶναι καὶ θερμήν, ἀλλ' ἐνδέχεται καὶ ψυχρὰν ὑπάρχειν, οὐκ ἀδύνατον ἔσται καὶ τὸ μήτε ψυχρὰν εἶναί τινα μήτε θερμήν, ἀλλ' εὔκρατον μὲν κατὰ τοῦτο, ξηρὰν δὲ κατὰ τὴν ἑτέραν ἀντίθεσιν. ὡσαύτως δὲ καὶ τὴν ὑγρὰν κρᾶσιν οὐκ ἀναγκαῖον οὔτε θερμὴν οὔτε ψυχρὰν ὑπάρχειν, ἀλλ' ἐνδέχεται μέσην ἀμφοῖν εἶναι κατά γε ταύτην τὴν ἀντίθεσιν.

Εἰ τοίνυν οὐκ ἀναγκαῖον οὔτε τῇ κατὰ τὸ θερμὸν καὶ ψυχρὸν δυσκρασίᾳ τὴν ἐκ τῆς ἑτέρας ἀντιθέσεως

ON TEMPERAMENTS, BOOK I

and hot. For if it is not altogether necessary for the hot *krasis* to be dry; it is also possible for it to be wet. It will clearly be possible for it also to be median, for the median *krasis* is nearer the dry *krasis* than the wet is. On the same basis, there is another cold *krasis*, in which the cold is stronger, and this *krasis* is not of necessity wet or dry; it is possible for it to be median. Again, even here we deem the same argument right. As a consequence, if it is not necessary for the cold to be wet, but it is also possible for it to be dry, it is quite clear that a median will also be possible, for this is nearer to the wet than the dry is. Therefore, as these same two *dyskrasias* have been shown in relation to the one antithesis—the hot alone or the cold—by the same token, there will be two others in relation to the other antithesis—the dry only and the wet—having a moderation relating to each other of hot and cold. Again, even here we shall say that it is not necessary for what is a dry *krasis* to immediately also be hot, but it is possible for it also to be cold. It will also not be impossible for it to be neither cold nor hot, but *eukratic* with respect to this while dry in relation to the other antithesis. Similarly also, it is not necessary for the wet *krasis* to be either hot or cold; it is possible for it to be median in both in relation to this antithesis.

Accordingly, if it is not necessary either for the *dyskrasia* from the other antithesis to follow the *dyskrasia* relat-

---

⁴⁰ ἀναγκαῖον εἶναι τὴν αὐτήν H; ὑπάρχειν αὐτήν ἀνάγκη K

ἕπεσθαι οὔτ' ἐκείνῃ τὴν ἐκ ταύτης, ἐνδέχεταί ποτε καὶ
τὴν εὔκρατον ὅσον ἐπὶ θερμότητι καὶ ψυχρότητι φύσιν ἤτοι ξηρὰν ἢ ὑγρὰν γενέσθαι καὶ τὴν ἐν τούτοις
πάλιν εὔκρατον ἤτοι θερμὴν ἢ ψυχράν. ὥστ' εἶναι καὶ
ταύτας τέτταρας ἑτέρας ἐκείνων δυσκρασίας, ὡς οἱ
πρόσθεν ἰατροί τε καὶ φιλόσοφοι παρέδοσαν ἡμῖν,
καὶ μέσας γε ταύτας τετάχθαι τῶν εὐκράτων ἕξεων
καὶ τῶν κατ' ἀμφοτέρας τὰς ἀντιθέσεις δυσκράτων. ἡ
μὲν γὰρ ἄκρως εὔκρατος οὐδετέραν ἀντίθεσιν ἔχει
πλεονεκτοῦσαν, ἡ δ' ἐξ ὑπεναντίου τῇδε δύσκρατος
ἀμφοτέρας μοχθηράς. ἐν μέσῳ δ' ἑκατέρων⁴¹ ἐστὶν ἡ
κατὰ μὲν τὴν ἑτέραν εὔκρατος ὑπάρχουσα, κατὰ δὲ
τὴν ἑτέραν δύσκρατος,⁴² ἥτις ἐξ ἡμίσεος μὲν εὐκρα-
559K τος, ἐξ | ἡμίσεος δὲ δύσκρατος οὖσα, μέση δεόντως
ἂν εἶναι λέγοιτο τῆς ὅλης εὐκράτου τε καὶ δυσκράτου.
καὶ εἴπερ ἔχει ταῦθ' οὕτως, ὥσπερ οὖν ἔχει, θαρρού-
ντως ἤδη λέγομεν, ἐννέα τὰς πάσας εἶναι τῶν κράσεων διαφοράς, εὔκρατον μὲν μίαν, οὐκ εὐκράτους δὲ
τὰς ὀκτώ, τέτταρας μὲν ἁπλᾶς, ὑγρὰν καὶ ξηρὰν καὶ
ψυχρὰν καὶ θερμήν, ἄλλας δὲ τέτταρας συνθέτους,
ὑγρὰν ἅμα καὶ θερμὴν καὶ ξηρὰν ἅμα καὶ θερμὴν καὶ
ψυχρὰν ἅμα καὶ ὑγρὰν καὶ ψυχρὰν ἅμα καὶ ξηράν.

9. Ἐν ἑκάστῃ δὲ τῶν εἰρημένων κράσεων τὸ μᾶλλόν
τε καὶ ἧττον πάμπολυ κατά τε τὰς ἁπλῶς λεγομένας
κράσεις ἐπί τε τῆς ὅλης οὐσίας, ἤδη δὲ καὶ καθ' ἓν
ὁτιοῦν γένος. εἰ δή τις βούλεται διαγνωστικὸς εἶναι
κράσεων, ἄρχεσθαι τούτῳ προσήκει τῆς γυμνασίας

ing to cold and hot, or for the *dyskrasia* from this to follow that (i.e., vice versa), it is also possible at some time for the nature *eukratic* in the amount of hot and cold to become either dry or wet, and contrariwise the *eukrasia* in these to become either hot or cold. As a consequence, there are also these four other *dyskrasias* of those, as the prior doctors and philosophers handed down to us, and these have a median position between the *eukratic* states and the *dyskratic* states relating to both antitheses. The perfect *eukrasia* has neither antithesis predominant, while that from the opposite of this is *dyskratic* in respect of both abnormalities. In the middle of both, however, is *eukrasia*, existing in relation to one antithesis, while in relation to the other is *dyskrasia*, | this being what is half *eukratic* and half *dyskratic*, which may properly be called midway between complete *eukrasia* and complete *dyskrasia*. And if this is the case, as indeed it is, we may now confidently say there are nine *differentiae* of the *krasias* in all. One is *eukratic* while eight are not *eukratic*. Of the [latter], four are simple—wet, dry, cold, and hot—while the other four are compound—wet and hot at the same time, dry and hot at the same time, cold and wet at the same time, and cold and dry at the same time.

9. In each of the aforementioned *krasias*, the difference in respect to more and less is very great in relation to the *krasias* called absolute, in the case of the whole substance, while now also in relation to any one class (*genus*) whatsoever. Certainly, if someone wishes to be an able diagnostician of *krasias*, it is appropriate for him to

---

[41] ἑκατέρων om. K
[42] post δύσκρατος; ὑπάρχουσα add. K

ἀπὸ τῶν καθ' ἕκαστον γένος εὐκράτων τε καὶ μέσων φύσεων. ἐκείναις γὰρ τὰς ἄλλας παραβάλλων ῥᾳδίως ἐξευρήσει τὸ πλεονάζον ἢ λεῖπον ἐν ἑκάστῃ. περὶ πρώτων οὖν ῥητέον τῶν ἁπλῶς λεγομένων εὐκράτων τε καὶ δυσκράτων, ἃς ἐπὶ πάσης οὐσίας γεννητῆς, οὐκ ἐπὶ ζῴων μόνον ἢ φυτῶν, ἔφαμεν ἐξετάζεσθαι. πάλιν δὲ κἀνταῦθα τό γε τοσοῦτον χρὴ διαστείλασθαι περὶ τῶν ὀνομάτων, ὡς θερμὴ κρᾶσις ἄλλη μέν ἐστιν ἐνεργείᾳ, δυνάμει δ' ἄλλη, καὶ ὡς δυνάμει ταῦτ' εἶναι λέγομεν, ὅσα μήπω μέν ἐστιν ἃ λέγεται, ῥᾷστον δ' αὐτοῖς γενέσθαι φυσικήν τιν' ἐπιτηδειότητα κεκτημένοις εἰς τὸ γενέσθαι. περὶ πρώτων οὖν διέλθωμεν τῶν ἐνεργείᾳ θερμῶν καὶ ψυχρῶν καὶ ξηρῶν καὶ ὑγρῶν ἀπὸ τῆς συμπάσης οὐσίας ἀρξάμενοι κἄπειτα μεταβῶμεν ἐπὶ τὰ φυτὰ καὶ τὰ ζῷα. τελέως γὰρ ἂν οὕτως ἡμῖν ἀπειργασμένον εἴη τὸ προτεθέν.

Ἐπειδὴ τοίνυν τὸ μέσον ἐν ἅπαντι γένει καὶ μάλιστα κατὰ τὰς συμπάσας οὐσίας ἐκ τῆς τῶν ἄκρων μίξεως γίγνεται, χρὴ καὶ τὴν νόησιν αὐτοῦ καὶ τὴν διάγνωσιν ἐξ ἐκείνων συνίστασθαι. τὸ μὲν δὴ τῆς νοήσεως ῥᾷστον. ἀπὸ γὰρ τοῦ θερμοτάτου πάντων τῶν εἰς αἴσθησιν ἡκόντων, οἷον ἤτοι πυρὸς ἤ τινος ὕδατος ἄκρως ζέοντος, ἐπὶ τὸ ψυχρότατον καταντῶντες ἁπάντων ὧν ἴσμεν, οἷον ἤτοι κρύσταλλον ἢ χιόνα, νοήσαντές τι διάστημα, μέσον ἀκριβῶς τοῦτο τέμνο-

---

[28] The Greek term οὐσία is consistently rendered *substantia* in KLat. Of the five meanings listed in Glare's dictionary, three

ON TEMPERAMENTS, BOOK I

begin his training in this from the *eukrasias* in relation to each class and the median natures. Comparing the other *krasias* with those, he will readily discover | the predominant and the deficient in each [nature]. Therefore, one must speak first about *eukrasias* and *dyskrasias* called absolute, which we said are to be examined in the case of every generated substance and not only of animals or plants. Again even here, it is necessary to make such a distinction about names, that we say one hot *krasis* is so actually and another is so potentially, the latter referring to those that do not yet exist but very easily arise in those who have acquired a certain natural suitability for this to arise. Therefore, let us go over the first—those which are hot, cold, dry, and wet in actuality, beginning from the whole substance and then passing on to the plants and animals. In this way the task before us would finally be brought to completion.

Accordingly, since the median in every class, and particularly in relation to all substances (existing things),[28] arises from the mixture of the extremes and the conception, and recognition of this must be put together from those. That of the conception is very easy. | Proceeding from the hottest of all those coming to perception, (for example, either fire or boiling water) to the very coldest of all things we know, (for example, either ice or snow), and considering the separation between them, we divide

560K

561K

---

are relevant: the quality of being real or having an actual (corporeal) existence; underlying or essential nature; the material of which a thing is made. All these are indicated in the present context.

μεν. οὕτω γὰρ ἐξευρήσομεν τῇ νοήσει τὸ σύμμετρον, ὅπερ ἑκατέρου τῶν ἄκρων ἴσον ἀπέχει. ἀλλὰ καὶ κατασκευάσαι πως αὐτὸ δυνάμεθα τὸν ἴσον ὄγκον κρυστάλλου μίξαντες ὕδατι ζέοντι. τὸ γὰρ ἐξ ἀμφοῖν κραθὲν ἴσον ἑκατέρου τῶν ἄκρων ἀφέξει τοῦ τε καίοντος καὶ τοῦ νεκροῦντος διὰ ψύξιν. οὔκουν οὐδὲ χαλεπὸν ἔτι τοῦ κραθέντος οὕτως ἁψαμένους ἔχειν τὸ μέσον ἁπάσης οὐσίας ἐν τῇ κατὰ τὸ θερμόν τε καὶ ψυχρὸν ἀντιθέσει καὶ μεμνῆσθαι τούτου καὶ κρίνειν ἅπαντα τἆλλα καθάπερ τινὶ κανόνι παραβάλλοντας. καὶ μὲν δὴ καὶ ξηρὰν γῆν ἢ τέφραν ἤ τι τοιοῦτον ἕτερον ἀκριβῶς αὐχμηρὸν ἀναδεύσας ὕδατι κατὰ τὸν ὄγκον ἴσῳ τὸ μέσον ἐργάσῃ σῶμα τῆς κατὰ τὸ ξηρόν τε καὶ ὑγρὸν ἀντιθέσεως. οὔκουν οὐδ' ἐνταῦθα χαλεπὸν οὐδὲν ὄψει θ' ἅμα καὶ ἁφῇ τὸ τοιοῦτον σῶμα διαγνόντα παραθέσθαι τῇ μνήμῃ καὶ τούτῳ κανόνι τε καὶ κριτηρίῳ χρῆσθαι πρὸς τὴν τῶν ἐλλειπόντων ἢ πλεοναζόντων ὑγρῶν τε καὶ ξηρῶν διάγνωσιν. ἔστω δὲ δηλονότι τὸ κρινόμενον σῶμα συμμέτρως θερμόν. εἰ γὰρ εἰς ἄκρον ἤτοι θερμότητος ἢ ψύξεως ἄγοιτο τουτὶ τὸ μέσον ὑγροῦ καὶ ξηροῦ σῶμα, φαντασίαν ἐνίοτε παρέξει ψευδῆ καὶ δόξει ποτὲ μὲν ὑγρότερον εἶναι τοῦ συμμέτρου, ποτὲ δὲ ξηρότερον. εἰ μὲν γὰρ θερμανθείη πλέον ἢ δεῖ, τηκόμενόν τε καὶ ῥέον ὑγροτέρου φαντασίαν ἑαυτοῦ παρέξει· ψυχόμενον δὲ περαιτέρω τοῦ προσήκοντος ἵσταταί τε καὶ πήγνυται καὶ ἀκίνητον γίγνεται καὶ σκληρὸν ἁπτομένῳ φαίνεται κἀκ τούτου φαντασίαν προβάλλει ψευδῆ ξηρότη-

this precisely in the middle. In this way we shall discover conceptually the median (moderate) which is equally removed from each of the extremes. But we are also able to prepare this in a certain way when we mix an equal mass of ice with boiling water. For what is mixed from each of the two will be equally separate from each of the extremes, of what burns up and what is mortified by cold. There is no longer, therefore, any difficulty for those who touch the mixture to have the median of every substance in the antithesis relating to hot and cold, and having this in mind, to judge all the others, as if making a comparison by a certain rule. Furthermore, if you were to mix into a paste dry earth, or ashes, or some other such thing that is completely dry, with water in equal mass, you will make a body median of the antithesis relating to dry and wet. There will not, therefore, be any difficulty here | in recognizing such a body by sight and touch at the same time, and setting this aside in the memory, using it as a rule and criterion for the recognition of the deficiencies or excesses of the wet and the dry. Clearly the body being assessed should be moderately hot, for if it is brought to a peak of either hotness or coldness, this will sometimes provide a false impression in respect of a body median in terms of cold and dry, and will seem at one time to be wetter than the moderate and at another time drier. For if it is heated more than it ought to be, and is melting and flowing, this will provide an impression of being wetter than it is. On the other hand, if cooled beyond what is appropriate, it stands still and solidifies, and becomes immobile, also appearing hard to the touch, and from this presents a false impression of dryness. If, however, just as

562K

τος. εἰ δ' ὥσπερ ἴσον ὑγροῦ καὶ ξηροῦ μετέσχεν,[43] οὕτω καὶ θερμότητος καὶ ψύξεως εἴη μέσον, οὔτε σκληρὸν οὔτε μαλακὸν ἁπτομένῳ φανεῖται τὸ τοιοῦτον σῶμα.

Τὸ μὲν οὖν ὅλα δι' ὅλων αὐτὰ κεράσαι, τὸ θερμὸν λέγω καὶ τὸ ψυχρὸν καὶ τὸ ξηρὸν καὶ τὸ ὑγρόν, ἀδύνατον ἀνθρώπῳ. γῆ γὰρ ὑγρῷ | φυραθεῖσα μέμικται μέν, ὡς ἄν τῳ δόξειε, καὶ οὕτω κέκραται πᾶσα παντί, παράθεσις μὴν ἐστι τὸ τοιοῦτον κατὰ σμικρὰ καὶ οὐ δι' ὅλων κρᾶσις, ἀλλὰ δι' ὅλων ἄμφω κεράσαι θεοῦ καὶ φύσεως ἔργον, ἔτι δὲ μᾶλλον, εἰ καὶ τὸ θερμὸν καὶ τὸ ψυχρὸν ὅλα δι' ὅλων ἀλλήλοις κεραννύοιτο. τὸ μέντοι παράθεσιν ἐργάσασθαι τοιαύτην, ὡς ἐκφεύγειν τὴν αἴσθησιν ἕκαστον τῶν ἁπλῶν σωμάτων, οὐ φύσεως τοῦτό γε μόνης ἢ θεοῦ τοὖργον, ἀλλὰ καὶ ἡμέτερόν ἐστιν. οὐδὲν γὰρ χαλεπὸν ὑγροῦ καὶ ξηροῦ μέσον ἐργάσασθαι πηλὸν ἐκ τῆς τοιαύτης μίξεως, ὡσαύτως δὲ καὶ θερμοῦ καὶ ψυχροῦ, καί σοι φανεῖται τὸ τοιοῦτον σῶμα καὶ τῇ θερμότητι μὲν[44] εὔκρατον, ἀλλὰ καὶ σκληρότητος καὶ μαλακότητος ἐν τῷ μέσῳ.

Τοιοῦτον δ' ἐστὶ καὶ τὸ τῶν ἀνθρώπων δέρμα, μέσον ἀκριβῶς ἁπάντων τῶν ἐσχάτων, θερμοῦ καὶ ψυχροῦ καὶ σκληροῦ καὶ μαλακοῦ, καὶ τούτου μάλιστα τὸ κατὰ τὴν χεῖρα. γνώμων γὰρ αὕτη πάντων ἔμελλεν ἔσεσθαι τῶν αἰσθητῶν, ὄργανον | ἁπτικὸν ὑπὸ τῆς φύσεως ἀπεργασθεῖσα τῷ φρονιμωτάτῳ τῶν ζῴων

ON TEMPERAMENTS, BOOK I

it partakes equally of wet and dry, in this way also, it is median in terms of hotness and coldness, such a body will seem neither hard nor soft to someone touching it.

Therefore, a thoroughgoing mixing of all these throughout—I speak of hot and cold, dry and wet—is not possible for a person [to carry out]. When earth is mixed with water to make a paste so it would seem to someone that in this way there has been mixing of all with all, this sort of thing is a juxtaposition on a small scale and not a thoroughgoing *krasis*; a thoroughgoing mixing of both these is an action of God and Nature. Still more is this so if the hot and the cold are mixed with each other completely throughout. This, indeed, is to make such a juxtaposition that each of the simple bodies (components) escapes perception, which is not an action of Nature alone or of God, but is also our doing. For there is nothing difficult about a median of wet and dry, making mud from such a mixture, and similarly with [a median of] hot and cold, and for such a body to seem to you *eukratic* in terms of hotness and coldness but also in the middle of hardness and softness.[29]

The skin of humans is also such a thing, being exactly the median of all the extremes—hot, cold, hard, and soft—and this is especially so in relation to the hand. For this is destined to become the arbiter of all the perceptions, having been made by Nature the instrument of touch proper to the most intelligent of the animals. It must

---

[29] The Kühn text is followed here.

---

[43] *post* μετέσχεν: ἴσον *add.* K
[44] *post* μὲν: καὶ τῇ ψυχρότητι *add.* K

οἰκεῖον. ἴσον οὖν ἀπέχειν αὐτὴν ἐχρῆν ἁπάντων τῶν ἄκρων, θερμοῦ καὶ ψυχροῦ καὶ ξηροῦ καὶ ὑγροῦ. καὶ δὴ καὶ γέγονεν ἐκ τῆς τούτων ἁπάντων ἰσομοιρίας οὐ μιχθέντων μόνον, ἀλλὰ καὶ δι' ὅλων ἀλλήλοις κερασθέντων, ὅπερ οὐκέτ' οὐδεὶς ἡμῶν ἐργάσασθαι δυνατός ἐστιν, ἀλλὰ φύσεως τοὖργον. ὅσα μὲν οὖν σκληρότερα τοῦ δέρματός ἐστι μόρια, καθάπερ ὀστᾶ καὶ χόνδροι καὶ κέρατα καὶ τρίχες ὄνυχές τε καὶ σύνδεσμοι καὶ ὁπλαὶ καὶ πλῆκτρα, πλέον ἐν τούτοις ἐστὶ τὸ ξηρόν· ὅσα δὲ μαλακώτερα, καθάπερ αἷμα καὶ φλέγμα καὶ πιμελὴ καὶ στέαρ καὶ μυελὸς ἐγκέφαλός τε καὶ νωτιαῖος, ὑγροῦ πλέον ἐν τούτοις ἐστὶν ἢ ξηροῦ· καὶ μὲν δὴ καὶ ὅσῳ τὸ πάντων ξηρότατον ἐν ἀνθρώπῳ μόριον ὑπερβάλλει σκληρότητι[45] τοῦ δέρματος, τοσούτῳ πάλιν ἀπολείπεται τὸ ὑγρότατον. ἔοικε δέ πως ὁ λόγος ἤδη τῶν χρησιμωτάτων αὐτῶν ἐφάπτεσθαι καὶ διδάσκειν, ἅμα μὲν ὡς οὐ ζῴων μόνον, ἀλλὰ καὶ τῶν ἄλλων ἁπάντων σωμάτων εὐκρατότατός ἐστιν | ὁ ἄνθρωπος, ἅμα δ' ὡς τῶν ἐν αὐτῷ μορίων τὸ τῆς χειρὸς δέρμα τὸ ἔσωθεν ἁπάσας ἐκπέφευγεν ἀκριβῶς τὰς ὑπερβολάς.

Ἐπιστήσαντες οὖν πάλιν ἐνταῦθα τὸν λόγον ἐπισκεψώμεθα, τίς ἄριστα κέκραται πάντων ἄνθρωπος, ὃν καὶ τῆς ὅλης μὲν οὐσίας, ἔτι δὲ μᾶλλον ἀνθρώπων τε καὶ τῶν ἄλλων ζῴων ἐν τῷ μέσῳ χρὴ τάξαντας, καθάπερ τινὰ κανόνα καὶ γνώμονα, τοὺς ἄλλους ἅπαντας τούτῳ παραβάλλοντας θερμοὺς καὶ ψυχροὺς καὶ ξηροὺς καὶ ὑγροὺς ὀνομάζειν. δεῖ δὲ συνδραμεῖν

ON TEMPERAMENTS, BOOK I

therefore be equally distant from all the extremes—hot, cold, dry, and wet. And in particular it has become, from an equal share of all these, not only mixed, but compounded throughout with each other, which is not something any of us is able to create but is a work of Nature. Therefore, those parts that are harder than skin, such as bones, cartilages, horns, hair, nails, ligaments, hooves, and spurs have more dry in them, whereas those that are softer, like blood, phlegm, soft fat, hard fat, marrow, brain, and spinal cord, have more wet in them than dry. Furthermore, the amount by which the driest part of all in a human exceeds the dryness of the skin is in turn that by which the wettest part is deficient. The argument already seems somehow to touch on what is most useful, and to teach that | the human is at the same time not only the most *eukratic* of animals and of all other bodies, but also that of the parts in a human, the skin of the palmar surface of the hand completely escapes all the excesses.

565K

Stopping the argument again here, let us consider which person is best compounded (has the best temperament) of all—someone whom, in terms of the whole substance, we must place as the median of humans especially and of other animals, like some standard and criterion (*gnomon*), by comparison with which all others are termed hot, cold, dry, and wet. We need to bring together to this

---

45 σκληρότητι H; ξηρότητι K

ἐς ταὐτὸν ἐπὶ τοῦδε πολλὰ γνωρίσματα. καὶ γὰρ ὡς πρὸς τὴν ὅλην οὐσίαν ἐξετάζοντι μέσον χρὴ φαίνεσθαι τὸν τοιοῦτον, ἔτι δὲ μᾶλλον ὡς πρὸς ἀνθρώπους τε καὶ ζῷα. τὰ μὲν οὖν ἁπάσης τῆς οὐσίας κοινὰ γνωρίσματα προείρηται· τὰ δ' ὡς ἐν ζῴων εἴδεσιν ἐνεργείας τελειότητι κρίνεται τῆς ἑκάστῳ πρεπούσης. πρέπει δ' ἀνθρώπῳ μὲν εἶναι σοφωτάτῳ, κυνὶ δὲ πραοτάτῳ θ' ἅμα καὶ ἀλκιμωτάτῳ, λέοντι δ' ἀλκιμωτάτῳ μόνον, ὥσπερ γε καὶ προβάτῳ πραοτάτῳ. καὶ μέν γε καὶ ὡς τὰς τοῦ σώματος ἐνεργείας οἰκείας εἶναι προσήκει[46] τῷ τῆς ψυχῆς | ἤθει, δέδεικται μὲν καὶ πρὸς Ἀριστοτέλους ἐν τοῖς περὶ ζῴων μορίων, δέδεικται[47] δὲ καὶ πρὸς ἡμῶν ὑπὲρ αὐτῶν οὐδὲν ἧττον. ἡ μὲν δὴ μέθοδος αὕτη.

Τὸ δ' ἀσκῆσαι γνωρίζειν ἑτοίμως ἐν ἑκάστῳ γένει ζῴου καὶ κατὰ τὰ σύμπαντα τὸ μέσον οὐ τοῦ τυχόντος ἀνδρός, ἀλλ' ἐσχάτως ἐστὶ φιλοπόνου καὶ διὰ μακρᾶς ἐμπειρίας καὶ πολλῆς γνώσεως ἁπάντων τῶν κατὰ μέρος ἐξευρίσκειν δυναμένου τὸ μέσον. οὕτω γοῦν καὶ πλάσται καὶ γραφεῖς ἀνδριαντοποιοί τε καὶ ὅλως ἀγαλματοποιοὶ τὰ κάλλιστα γράφουσι καὶ πλάττουσι καθ' ἕκαστον εἶδος, οἷον ἄνθρωπον εὐμορφότατον ἢ ἵππον ἢ βοῦν ἢ λέοντα, τὸ μέσον ἐν ἐκείνῳ τῷ γένει σκοποῦντες. καί πού τις ἀνδριὰς ἐπαινεῖται Πολυκλείτου κανὼν ὀνομαζόμενος, ἐκ τοῦ πάντων τῶν

[46] προσήκει H; χρὴ K
[47] δέδεικται H; λέλεκται K

ON TEMPERAMENTS, BOOK I

the many signs in this case. For surely, to one examining the whole substance, such a person must appear to be median, especially as regards humans and animals. The signs common to the whole substance have been previously stated; those in species (kinds) of animals are judged by the perfection of function appropriate to each. Appropriate to a person is great cleverness, to a dog great docility along with great bravery, while to a lion great bravery alone, just as also to a sheep great docility. And further in fact that the functions of the body should be appropriate to the character of the soul has been shown also by Aristotle in the work, *On the Parts of Animals*.[30] No less have we also spoken about these. Certainly this itself is a method.

The practice is to readily recognize the median in each class of animal and in relation to all things. This is not for the ordinary man, but for one who is extremely diligent, and through long experience and much knowledge is able to discover the median of all things individually. Anyway, it is in this manner also that sculptors, painters, statue-makers, and image-makers in general draw and make things of the greatest beauty in relation to each kind, such as the most well-formed man, horse, ox, or lion, considering the median in that class. And, I suppose, a certain statue is praised, called the Canon of Polyclitus, which

---

[30] See particularly *Parts of Animals* 4 (LCL 323 [Peck], 232–47) where Aristotle describes the heart, considered by him to be the seat of the soul. On this, see his *De anima*, LCL 288 (W. H. Hett). Galen deals with the relation between body and soul in *Quod animi mores*, the third work in the present volume. There is extensive reference to Aristotle in that work.

μορίων ἀκριβῆ τὴν πρὸς ἄλληλα συμμετρίαν ἔχειν ὀνόματος τοιούτου τυχών. ἔστι μὲν οὖν ἐπὶ πλέον, ὃν νῦν ἡμεῖς ζητοῦμεν, ἢ ὁ κανὼν οὗτος. οὐ μόνον γὰρ ὑγρότητός τε καὶ ξηρότητος ἐν τῷ μέσῳ καθέστηκεν ὁ οὕτως εὔσαρκος ἄνθρωπος, ἀλλὰ καὶ διαπλάσεως ἀρίστης τετύχηκεν, ἴσως μὲν ἑπομένης τῇ τῶν τεττάρων στοιχείων εὐκρασίᾳ, τάχα δέ τινα θειοτέραν ἀρχὴν ἑτέραν ἐχούσης ἄνωθεν. ἀλλὰ τό γε πάντως εὔκρατον εἶναι τὸν τοιοῦτον ἐξ ἀνάγκης ὑπάρχει· τὸ γὰρ ἐν εὐσαρκίᾳ σύμμετρον εὐκρασίας ἐστὶν ἔκγονον. εὐθὺς δ' ὑπάρχει τῷ τοιούτῳ σώματι καὶ ταῖς ἐνεργείαις ἄριστα διακεῖσθαι καὶ σκληρότητός τε καὶ μαλακότητος ἔχειν μετρίως θερμότητός τε καὶ ψυχρότητος. καὶ ταῦθ' ὑπάρχει ἅπαντα τῷ δέρματι καὶ τούτου μάλιστα τῷ τῆς χειρὸς ἐντός, ὅταν γε μηδένα τύλον ἔχῃ τοιοῦτον, οἷος τοῖς ἐρέττουσί τε καὶ σκάπτουσι γίγνεται. διττῆς γὰρ ἕνεκα χρείας τῶν χειρῶν γεγενημένων, ἁφῆς καὶ ἀντιλήψεως, αἱ μαλακαὶ μὲν εἰς τὴν τῆς ἁφῆς ἀκρίβειαν, αἱ σκληραὶ δ' εἰς τὴν τῆς ἀντιλήψεως ἰσχὺν ἐπιτηδειότεραι.

Καὶ δὴ καὶ τὸ δέρμα τὸ μέσον οὐ μόνον ἁπάντων τῶν τοῦ ἀνθρώπου μορίων, ἀλλὰ καὶ τῆς ὅλης οὐσίας ἁπάντων τῶν ἐν γενέσει τε καὶ φθορᾷ σωμάτων οὐ τὸ τετυλωμένον ἐστὶ καὶ σκληρὸν καὶ λιθῶδες, ἀλλὰ τὸ κατὰ φύσιν ἔχον, ᾧ δὴ καὶ μάλιστά φαμεν ἀκρι-

happens to have such a name from the precise proportion of all the parts to each other.[31] It is, then, something more than this Canon which we are now seeking. For the man well-fleshed (*eusarkos*) in this way is not only to be placed in the median of wetness and dryness, but also must possess the best conformation, which perhaps follows the *eukrasia* of the four elements (elemental qualities), or perhaps has some other divine principle from above. But, in fact, such a person is of necessity *eukratic* in every way, for being well-fleshed (*eusarkos*) with a balanced symmetry is a product of *eukrasia*. In such a body there is immediately the best disposition in the functions and a balance between hardness and softness, hotness and coldness. And this is present in the skin entirely and particularly in the palmar surface of the hand whenever it has none of the kinds of callus formation as occur in those who row and dig. For the hands have been created for a twofold use—touching and holding. Soft hands are more suitable for the precision of touch while hard hands are more suitable for the strength of holding.

And actually, the skin which is the median not only of all the parts of the human, but also of the whole substance, and of all the bodies involved in generation and destruction, is not calloused, hard, and stony, but has an accord with nature, by which we also say it is made particularly

---

[31] Polyclitus was an Argive sculptor active during the middle to late third century BC. His most famous work was the Doryphorus (Spearbearer). He is said to have written a book (his *Canon*) detailing the principles of his art as exemplified by this statue. The important feature was the proportion of the sculpted parts.

βοῦσθαι τὴν ἁφήν. ὅτι μὲν οὖν σκληρότητός τε καὶ μαλακότητος ἐν τῷ μέσῳ καθέστηκεν ἁπάντων τῶν μορίων, ἱκανῶς ἐναργές· ὅτι δὲ καὶ θερμότητός τε καὶ ψυχρότητος, ἐκ τῆς οὐσίας ἂν αὐτοῦ μάλιστα καταμάθοις. ἔστι γὰρ οἷον ἔναιμόν τι νεῦρον, ἀκριβῶς μέσον ὑπάρχον νεύρου τε καὶ σαρκός, ὡς εἰ καὶ κραθέντων ἀμφοῖν ἐγένετο. ἀλλὰ νεῦρον μὲν ἅπαν ἄναιμόν τε καὶ ψυχρόν, σὰρξ δὲ πολύαιμός τε καὶ θερμή· μέσον δ' ἀμφοῖν τὸ δέρμα μήτ' ἄναιμον τὸ πάμπαν ὡς τὸ νεῦρον, ἀλλὰ μηδὲ πολύαιμον ὡς ἡ σὰρξ γενόμενον. εἰ δὴ τοῦτο κανόνα τε καὶ οἷον κριτήριον ἁπάντων τῶν τοῦ ζῴου μορίων προστησάμενος ἐξετάζοις τε καὶ παραβάλλοις αὐτῷ τἆλλα, τὰς ὀκτὼ διαφορὰς εὑρήσεις τῶν δυσκρασιῶν ἐν αὐτοῖς.

Καὶ δὴ καὶ κατὰ μέρος δίειμί σοι περὶ πάντων ἐφεξῆς. φλέγμα μέν ἐστιν ὑγρότατον καὶ ψυχρότατον, αἷμα δὲ θερμότατον, ἀλλ' οὐχ οὕτως ὑγρὸν ὡς τὸ φλέγμα. θρὶξ δὲ ψυχρότατον καὶ ξηρότατον· ἧττον δ' αὐτῆς ὀστοῦν ψυχρόν τ' ἐστὶ καὶ ξηρὸν καὶ τοῦδε χόνδρος ἧττον ξηρός, ἐφεξῆς δὲ χόνδρῳ σύνδεσμος, ἔπειτα τένων, εἶθ' ὑμὴν καὶ ἀρτηρία καὶ φλέψ, αὐτὰ δηλονότι τὰ σώματα τῶν ἀγγείων, εἶθ' ὅσα νεῦρα σκληρά. τὰ δὲ μαλακὰ νεῦρα κατὰ τὴν τοῦ δέρματος ὑπάρχει φύσιν ἐν τῇ καθ' ὑγρότητά τε καὶ ξηρότητα μεσότητι. κατὰ γὰρ τὴν ἑτέραν ἀντίθεσιν οὐκ ἔστι μέσον θερμοῦ καὶ ψυχροῦ τὸ νεῦρον τὸ μαλακόν, ἀλλὰ τοσοῦτον ἀπολείπεται θερμότητος, ὅσον καὶ αἵματος. οὕτω δὲ καὶ τἆλλα σύμπαντα τὰ πρόσθεν

ON TEMPERAMENTS, BOOK I

accurate in respect of touch. Therefore, that it is situated in the midpoint of hardness and softness of all the parts is sufficiently clear; that it is placed in the midpoint of hotness and coldness, you would learn from the substance itself. For it is like some blood-containing sinew (nerve), being precisely in the middle of sinew (nerve) and flesh, as if it were compounded from both. But a sinew (nerve) is entirely bloodless and cold, while flesh is full of blood and hot. The skin is intermediate between both, being neither entirely bloodless like sinew (nerve) or blood-filled like flesh. Certainly if you set up this canon and, as it were, criterion of all the parts of the animal, and you examine and compare the others with this, you will discover the eight *differentiae* of the *dyskrasias* in them.

And further, I shall go through all of these individually for you in order. Phlegm is wettest and coldest, while blood is hottest but not wet in the way phlegm is. Hair is very cold and very dry. Bone is less cold and dry than this, and cartilage less dry than this again. Next after cartilage is ligament, [which they also term cartilaginous ligament],[32] then tendon, thin membrane, artery and vein, these latter clearly being the actual bodies of the vessels, then those nerves that are hard. However, the soft nerves relate to the nature of the skin and are in the middle in terms of wetness and dryness. In relation to the other antithesis, the nerve that is soft is not a median of hot and cold, but lacks hotness to the degree that it also lacks blood. In this way too, all the other things (tissues) previously mentioned are

---

[32] Kühn has καὶ ὃν χονδροσύνδεσμον ὀνομάζουσιν. LSJ lists the term χονδροσύνδεσμος as meaning "cartilaginous connection" citing this passage and Aristotle, *History of Animals* 617b2.

εἰρημένα τοσούτῳ ψυχρότερα δέρματος ὅσῳ καὶ ἀναιμότερα. καὶ οἵ γε χιτῶνες αὐτοὶ τῶν ἐναιμοτάτων ἀγγείων, ἀρτηρίας λέγω καὶ φλεβός, ἄναιμοί τ' εἰσὶ καὶ ψυχροὶ φύσει. τῇ γειτνιάσει δὲ τοῦ αἵματος θερμαίνονταί τε καὶ εἰς μέσην ἀφικνοῦνται κατάστασιν κράσεως. τὸ δ' αἷμα πάλιν αὐτὸ παρὰ τῆς καρδίας ἔχει τὴν θερμασίαν. φύσει γὰρ ἐκεῖνο τὸ σπλάγχνον ἁπάντων ἐστὶ τῶν τοῦ ζῴου μορίων ἐναιμότατόν θ' ἅμα καὶ θερμότατον, ἐφεξῆς δ' αὐτῷ τὸ ἧπαρ. ἀλλ' ἡ μὲν καρδία βραχὺ δέρματος ἀποδεῖ σκληρότητι, τὸ δ' ἧπαρ πολύ. καὶ τοίνυν καὶ ὑγροτέρα τοσούτῳ δέρματός ἐστιν ὅσῳ μαλακωτέρα. καὶ μὲν δὴ καὶ ἡ σὰρξ ὑγροτέρα δέρματος, ἀλλ' αὕτη μὲν καὶ θερμοτέρα. νωτιαῖος δ' ὑγρότερος μέν, ἀλλὰ καὶ ψυχρότερος, καὶ τοῦδε μᾶλλον ἐγκέφαλος ὑγρότερος, ἔτι δὲ τοῦδε μᾶλλον ἡ πιμελή, καὶ ἡ πῆξις αὐτῆς διὰ τὴν τῶν ὑμένων γειτνίασιν· ἐλαίῳ γὰρ ἔοικε παχεῖ καὶ διὰ τοῦτο πήγνυται ψυχροῖς καὶ ἀναίμοις ὁμιλοῦσα μορίοις. οὔτε δ' ἥπατι περιπήγνυσθαι πιμελὴν οὔτ' ἀρτηρίαις καὶ φλεψὶν οὔτε καρδίᾳ δυνατόν, ἀλλ' οὐδ' ἄλλῳ τινὶ θερμῷ πάνυ μορίῳ. διότι δὲ πέπηγε ψυχρῷ, διὰ τοῦτο χεῖται θερμαινομένη τοῖς ἄλλοις ὁμοίως τοῖς πεπηγόσιν. οὐ μὴν ἐγκεφαλός γε θερμαινόμενος χεῖται καὶ διὰ τοῦθ' ἧττον ὑγρός ἐστι πιμελῆς. ἧττον δ' ὑγρὰ πιμελῆς ἐστι καὶ ἡ τοῦ πνεύμονος σάρξ, οὐδὲ γὰρ αὕτη χεῖται θερμαινομένη. πολὺ δ' ἔτι μᾶλλον ἡ τοῦ σπληνός τε καὶ τῶν νεφρῶν σὰρξ ἧττόν ἐστιν ὑγρὰ

colder than skin to the extent that they are more lacking in blood. In fact, the actual walls of the most blood-containing vessels—I speak of arteries and veins—are bloodless and cold in nature. They are, however, heated by the proximity of the blood and come to a median state of *krasis*. In turn, the blood itself has the heating from the heart, for in nature that viscus is the most blood-containing of all the parts of the animal and at the same time the hottest. Next in order to this is the liver, but the heart is lacking somewhat in hardness compared to the skin, while the liver is lacking much more. And accordingly, it is also wetter than the skin to the extent that it is softer. Furthermore, the flesh is wetter than skin but is itself also hotter, whereas spinal cord is wetter but also colder, and the brain is wetter than this to a greater extent, and fat still more so. The congelation of this is due to the proximity of the membranes, for it is like thick olive oil and because of this congeals when it comes into contact with cold and bloodless parts. On the other hand, it is impossible for fat to congeal around the liver, or arteries and veins, or heart, or any other very hot part. Because it is congealed by cold, it flows when heated, similar to other things that are congealed. Brain, however, does not in fact flow when heated, and because of this is less wet than fat. The flesh of the lung is also less wet than fat, for it does not flow when heated. The flesh of the spleen and kidneys is, to a much greater extent, less wet

πιμελῆς. ἅπαντα μέντοι ταῦτα δέρματός ἐστιν ὑγρότερα. τὰς δὲ τούτων ἀποδείξεις ἐν τῷ μετὰ ταῦτα λόγῳ διέξειμι καὶ μὲν δὴ καὶ ὅσα λείπει τῇ συμπάσῃ περὶ κράσεων πραγματείᾳ διὰ τῶν ἐφεξῆς δυοῖν ὑπομνημάτων εἰρήσεται.

than fat. And yet all these structures are wetter than skin. I shall go through the demonstrations of these matters in the course of the discussion that follows, and furthermore, those matters lacking a complete discussion of *krasias* will be spoken of in the two following books.

ΒΙΒΛΙΟΝ ΔΕΥΤΕΡΟΝ

1. Ὅτι μὲν δὴ τῶν πολλαχῶς λεγομένων ἐστὶν ὑγρόν τε σῶμα καὶ ξηρὸν καὶ ψυχρὸν καὶ θερμόν, ἐν τῷ πρὸ τούτου λόγῳ διήρηται. δέδεικται δὲ καί, ὡς ἐννέα διαφοραὶ τῶν κράσεών εἰσι, μία μὲν ἡ σύμμετρός τε καὶ εὔκρατος, αἱ λοιπαὶ δὲ πᾶσαι δύσκρατοι, τέτταρες μὲν ἁπλαῖ, μιᾶς ἐν ἑκάστῃ πλεονεκτούσης ποιότητος ἤτοι θερμότητος ἢ ψυχρότητος ἢ ξηρότητος ἢ ὑγρότητος, ἕτεραι δὲ τέτταρες, ἐπειδὰν ἐξ ἑκατέρας ἀντιθέσεως ἡ ἑτέρα κρατήσῃ δύναμις. λέγω δ' ἀντιθέσεις δύο, μίαν μὲν τὴν κατὰ τὸ θερμόν τε καὶ ψυχρόν, ἑτέραν δὲ τὴν κατὰ τὸ ξηρόν τε καὶ ὑγρόν.

Ἐφεξῆς δὲ τούτων ἐπὶ τὰ γνωρίσματα μεταβάντες ὑπὲρ τῆς εὐκράτου φύσεως ἐσκοπούμεθα, διότι πρώτη πασῶν ἥδε καὶ ἀρετῇ καὶ δυνάμει καὶ τάξει νοήσεώς ἐστιν. ἐπεὶ δ' εὔκρατον ἄλλο μὲν ἁπλῶς εὑρίσκεται λεγόμενον ἐν πάσῃ τῇ τῶν ὄντων φύσει, καθ' ἕκαστον δὲ γένος ἄλλο, περὶ πρώτου δεῖν ἐδόκει σκοπεῖσθαι τοῦ κοινῇ κατὰ πάσης φύσεως ἐξεταζομένου. κανὼν δ' ἦν αὐτοῦ καὶ κρίσις ἡ τῶν στοιχείων ἰσομοιρία, δι' ἣν καὶ τὸ τῶν ἐσχάτων ἁπάντων ἀκριβῶς μέσον ἀποτελεσθὲν εὔκρατόν τε καὶ σύμμετρον ὀνομάζεται. τὰ

BOOK II

1. It has surely been determined in the discussion prior to this, that a wet body, and a dry, cold and hot one are spoken of in many ways. It has also been shown that there are nine *differentiae* of the *krasias*; one which is in due proportion and *eukratic*, while the remainder are all *dyskratic*. Four are simple *dyskrasias*: one in each case when the prevailing quality is either hotness, coldness, dryness, or wetness. The other four occur when one or other power from each antithesis prevails—I speak of two antitheses, one is in relation to hot and cold, | while the other is in relation to dry and wet.

Next after these, having moved on to the signs, we gave consideration to the *eukratic* nature, since it is the first of all in excellence and potency, when considered in order. Since it was found that *eukratic* is in one way spoken of absolutely in every nature of existing things, and in another way in relation to each class, it seemed there was a need to consider the first examined in common in relation to every nature. A standard of this and a means of judgment was the equal distribution of the elements, as a result of which it was made the precise median of all the extremes and was called *eukratic* and well-balanced (mod-

δ᾽ ἄλλα τὰ καθ᾽ ἕκαστον γένος εὔκρατα ταῖς οἰκείαις τῶν σωμάτων ἐνεργείαις τε καὶ χρείαις κρίνεται καὶ διὰ τοῦτο ταὐτὸν σῶμα ζῴου τινὸς ἢ φυτοῦ μέσον μὲν εἶναι δύναται τῶν ὁμογενῶν ἁπάντων, τουτέστιν εὔκρατόν τε καὶ σύμμετρον ἐν ἐκείνῳ τῷ γένει, δύσκρατον δ᾽ ἑτέρῳ τινὶ παραβαλλόμενον ἢ φυτῶν ἢ ζῴων ἢ ἀψύχων γένει. τὸ μὲν γὰρ τοῦ ζῶντος σῶμα τῷ τοῦ νεκροῦ παραβαλλόμενον ὑγρότερόν ἐστι καὶ θερμότερον, οἷον, εἰ τύχοι, λέων ζῶν τεθνεῶτος λέοντος ἢ αὐτὸς ἑαυτοῦ τις ἢ ἕτερος ἑτέρου θερμότερός τ᾽ ἐστὶ καὶ ὑγρότερος, καὶ διὰ τοῦτ᾽ εἴρηται πρὸς τῶν παλαιῶν ὑγρόν τ᾽ εἶναι καὶ θερμὸν τὸ ζῷον, οὐχ ὡς ἢ τῆς ὑγρότητος ἐν αὐτῷ πλεονεκτούσης ἁπλῶς ἢ τῆς θερμότητος· οὕτω μὲν γὰρ εὑρεθήσεται πάμπολλα ζῷα ξηρὰ καὶ ψυχρά, καθάπερ ἐμπίδες τε καὶ κώνωπες καὶ μυῖαι καὶ μέλιτται καὶ μύρμηκες· ἀλλ᾽ ὡς τοῖς τεθνεῶσι παραβαλλόμενα. καὶ γὰρ καὶ μέλιττα ζῶσα τεθνεώσης μελίττης ὑγροτέρα τ᾽ ἐστὶ καὶ θερμοτέρα καὶ μύρμηξ μύρμηκος, ἀνθρώπῳ μέντοι παραβαλλόμενα καὶ ἵππῳ καὶ βοῒ καὶ τοῖς ἄλλοις ζῴοις τοῖς ἐναίμοις ἅπαντα τὰ τοιαῦτα ψυχρὰ καὶ ξηρὰ τὴν κρᾶσίν ἐστι. καὶ μὲν δὴ κἂν εἰ πρὸς τὴν ὅλην οὐσίαν ἀποβλέπων ἐξετάζοις, οὐδ᾽ οὕτως ἐκπέπτωκε τοῦ ξηρά τ᾽ εἶναι καὶ ψυχρά. ὥσπερ γὰρ καθ᾽ ἕκαστον γένος, ὅταν ἐξίστηταί τι τῆς μεσότητος, ἀπὸ τοῦ πλεονεκτοῦντος ὀνομάζεται, κατὰ τὸν αὐτὸν τρόπον ἐπὶ τῆς συμπάσης οὐσίας, ὅταν ὑπερβάλλῃ τι τὸ μέσον,

erate). The other *eukrasias*—those in relation to each class—are judged by the characteristic functions and uses of the bodies. And because of this, the same body, whether of an animal or a plant, is able to be the median of all things of the same class—that is to say, *eukratic* and well-balanced in that class, although it may be *dyskratic* when compared to some other class of either plants, or animals, or inanimate things. | For the body of the living [organism], when compared to that of the dead, is wetter and hotter. For example, as may happen, a living lion is hotter and wetter than a dead lion whether you compare the same lion with itself or one lion with another. And because of this the living animal was said by the ancients to be wet and hot, not because the wetness prevails in it absolutely, or the hotness, for in this way very many animals will be discovered to be dry and cold, like gnats, mosquitoes, flies, bees, and ants, but as compared to those that are dead. For surely a living bee is wetter and hotter than a dead bee, and a living ant than a dead ant, although when compared to a human, a horse, and an ox, and to the other sanguineous animals, all such (creatures) are cold and dry in *krasis*. And further certainly, even if you examine them, looking at the whole substance, they do not in this way cease to be dry and cold. For just as, in relation to each class, whenever something departs from the median position, | it is named from what predominates; in the same manner, in the case of the whole substance, whenever

οὐκέτ' εὔκρατον, ἀλλ' ἤτοι θερμὸν ἢ ψυχρὸν ἢ ξηρὸν ἢ ὑγρὸν ὀνομασθήσεται.

Δέδεικται γὰρ δὴ πρόσθεν, ὡς ἄνθρωπός ἐστιν οὐ τῶν ζῴων μόνον ἢ φυτῶν, ἀλλὰ καὶ τῶν ἄλλων ἁπάντων εὐκρατότατον. ἐπεὶ δ' ἐκ πολλῶν καὶ διαφερόντων σύγκειται μορίων, εὔδηλον, ὡς τὸ μέσον ἁπάντων τῇ κράσει τοῦτο καὶ ἁπλῶς ἐστιν εὔκρατον.[1] τὸ γὰρ τοῦ μέσου τῇ κράσει ζῴου μέσον μόριον ἁπάντων ἁπλῶς εὐκρατότατον ἔσται. ἐδείχθη δὲ τοῦτ' ἐν ἀνθρώπῳ τὸ καλούμενον δέρμα καὶ μάλιστα τοῦ δέρματος τὸ τῶν χειρῶν ἐντός,[2] ὅταν, οἷον ὑπὸ τῆς φύσεως ἀπειργάσθη, τοιοῦτον φυλάττηται. καὶ μὲν δὴ καὶ ὡς οὐ παντὸς ἀνθρώπου τὸ δέρμα μέσον ἁπλῶς ἐστιν ἁπάσης οὐσίας, ἐδείχθη[3] πρόσθεν, ἀλλ' ὅστις ἂν εὐκρατότατος ᾖ· πολλὴν γὰρ εἶναι καὶ αὐτοῖς τοῖς ἀνθρώποις πρὸς ἀλλήλους τὴν διαφοράν. εὐκρατότατος δ' ἐστίν, ὃς ἂν τῷ μὲν σώματι φαίνηται μέσος ἀκριβῶς ἁπάντων τῶν ἄκρων, ἰσχνότητός τε καὶ παχύτητος, μαλακότητός τε καὶ σκληρότητος, ἔτι δὲ θερμότητός τε καὶ ψυχρότητος. ἔστι γὰρ εὑρεῖν ἁψάμενον ἑκάστου τῶν ἀνθρωπίνων σωμάτων ἢ χρηστὴν καὶ ἀτμώδη θερμασίαν ἢ πυρώδη καὶ δριμεῖαν ἢ τούτων μὲν οὐδετέραν, ἐπικρατοῦσαν δέ τινα ψῦξιν. ἀκούειν δὲ χρὴ ψῦξιν ἐπικρατοῦσαν ὡς ἐν ζῴου σώματι καὶ ταῦτ' ἐναίμου τε καὶ ὑγροῦ ὄντος. τῷ μὲν δὴ σώματι τοιοῦτος ὁ εὐκρατότατος[4] ἄνθρωπος· ὡσαύτως δὲ καὶ τῇ ψυχῇ

---

[1] εὔκρατον H; εὐκρατότατον K

ON TEMPERAMENTS, BOOK II

something exceeds the median, it is no longer *eukratic* but will be termed either hot, cold, dry, or wet.

It has been shown before that a human is the most *eukratic*, not only of animals or plants, but also of all other things. However, since it is compounded from many and different parts, it is clear that this is the median of all of them in the *krasis* and this is [most] *eukratic* in an absolute sense, for the median part of the animals median in *krasis* will be the most *eukratic* of all in an absolute sense.[1] And in a human this was shown to be what is called the skin, and especially the skin on the palmar surface of the hands, whenever it preserves such a state as was fashioned by Nature. Furthermore also, that the skin of the whole person is not absolutely the median of every substance was shown before, but only one who is most *eukratic*, for the difference is great in humans themselves among one another. | The most *eukratic* is the one who appears in the body to be precisely midway between all the extremes— that is, thinness and thickness, softness and hardness, and further, hotness and coldness. For it is found when touched that in each of the human bodies, there is either useful and vaporous heat, or a fiery and sharp heat, or neither of these when a certain cooling prevails. It is necessary to understand prevailing cooling as being in an animal's body and one that is sanguineous and wet. Such, therefore, is the most *eukratic* person in the body. Similarly also he is pre-

576K

---

[1] Following the Kühn text.

---

[2] *post* ἐντός; ὅταν H; ἐὰν K
[3] ἐδείχθη H: ἐλέχθη K
[4] εὐκρατότατος H; εὔκρατος K

μέσος ἀκριβῶς ἐστι θρασύτητός τε καὶ δειλίας, μελλησμοῦ τε καὶ προπετείας, ἐλέου τε καὶ φθόνου. εἴη δ' ἂν ὁ τοιοῦτος εὔθυμος, φιλόστοργος, φιλάνθρωπος, συνετός.

Ἐκ τούτων μὲν οὖν ὁ εὐκρατότατος ἄνθρωπος γνωρίζεται πρώτως καὶ μάλιστα· προσέρχεται δ' αὐτοῖς οὐκ ὀλίγα τῶν ἐξ ἀνάγκης ἑπομένων· καὶ γὰρ ἐσθίει καὶ πίνει συμμέτρως καὶ πέττει καλῶς τὰς τροφὰς οὐκ ἐν γαστρὶ μόνον, ἀλλὰ κἂν ταῖς φλεψὶ καὶ καθ' ὅλην τὴν | ἕξιν τοῦ σώματος ἁπάσας τε συνελόντι φάναι τάς τε φυσικὰς καλουμένας ἐνεργείας καὶ τὰς ψυχικὰς ἀμέμπτους ἔχει. καὶ γὰρ καὶ ταῖς αἰσθήσεσιν ἄριστα διάκειται καὶ ταῖς τῶν κώλων κινήσεσιν εὔχρους τ' ἐστὶ καὶ εὔπνους ἀεὶ καὶ μέσος ὑπνώδους τε καὶ ἀγρύπνου καὶ ψιλοῦ τριχῶν καὶ δασέος καὶ μέλανος τὴν χρόαν καὶ λευκοῦ καὶ τρίχας ἔχει παῖς μὲν ὢν πυρροτέρας μᾶλλον ἢ μελαντέρας, ἀκμάζων δ' ἔμπαλιν.

2. Ἐπεὶ δὲ καὶ τῆς κατὰ τὰς ἡλικίας αὐτοῦ διαφορᾶς ἐπεμνήσθημεν, οὐδὲν ἂν εἴη χεῖρον ἤδη τι καὶ περὶ τούτων εἰπεῖν. ἐβουλόμην μὲν οὖν πρότερον ἑκάστου τῶν εἰρημένων γνωρισμάτων ἐπελθεῖν τὰς αἰτίας, ἀλλ' ἐπεὶ πρὸς τὰ παρόντα μᾶλλον ἢ περὶ τῶν ἡλικιῶν ἐπείγει σκέψις εὐπορωτέρους θ' ἡμᾶς πρὸς τὴν τῶν αἰτιῶν εὕρεσιν ἀπεργάζεται, πρώτην[5] ταύτην ἐνστησόμεθα.

Νοήσωμεν οὖν ἄρτι διαπλαττόμενον ἐν ταῖς μήτραις τῶν κυουσῶν[6] τὸ ζῷον, ἵνα γνῶμεν, ὅπως

ON TEMPERAMENTS, BOOK II

cisely median in the soul in terms of courage and timidity, hesitancy and rashness, compassion and envy. Such a person would be generous, affectionate, humane, and wise.

It is from these things, then, that the most *eukratic* person is recognized primarily and particularly. But it is necessary to add to these not a few of the following. For surely [such a person] also eats and drinks moderately and concocts nutriments well not only in the stomach, but also in the veins, and in the whole | state (*hexis*) of the body, and in summary, has all the so-called physical and psychical functions faultless. Further, he is in the best state as regards perceptions and the movements of the limbs, has a good color and good respiration, is midway between sleeping and wakefulness, bareness and abundance of hair, and blackness and whiteness in respect of color, and as a child, has hair that inclines more toward light red than dark, while in his prime the converse is the case.

2. Since we are reminded of the differences relating to the ages (stages of life) of this person, it would not be bad to say something about these now. I would normally wish to go over first the causes of each of the signs mentioned, but as the speculation on the present matters urges us more toward the ages and will make us more prepared for the discovery of the causes, we will put this in place first.

Let us, then, conceive now of the animal being formed in the wombs of those who are pregnant, so that we may

---

⁵ πρώτην *om.* K
⁶ τῶν κυουσῶν H; τῶν τικτουσῶν K

ὑγρότατόν τ' ἐστὶ καὶ θερμότατον. ἡ μὲν γὰρ πρώτη σύστασις ἐξ αἵματος αὐτῷ καὶ σπέρματος, ὑγρῶν καὶ θερμῶν χρημάτων. ἀεὶ δὲ καὶ μᾶλλον τούτων ξηρῶν γιγνομένων ὑμένες μὲν πρῶτα διαπλάττονται καὶ χιτῶνες καὶ σπλάγχνα καὶ ἀγγεῖα, τελευταῖα δ' ὀστᾶ καὶ χόνδροι καὶ ὄνυχες πηγνυμένης ἀποτελοῦνται τῆς οὐσίας· πρὶν γὰρ ἤτοι δύνασθαι τείνεσθαι τὴν ὑποβεβλημένην οὐσίαν ἢ πήγνυσθαι, τῶν εἰρημένων οὐδὲν ἐγχωρεῖ γενέσθαι. χιτῶνες μὲν οὖν καὶ ὑμένες ἀρτηρίαι τε καὶ φλέβες καὶ νεῦρα τεινομένης αὐτῆς, ὀστᾶ δὲ καὶ χόνδροι καὶ ὄνυχες καὶ ὁπλαὶ καὶ πλῆκτρα[7] πηγνυμένης ἀποτελοῦνται· τελειωθέντων δ' οὕτως ἐν τῇ κυούσῃ τίκτεται μὲν ἐφεξῆς, ἔτι δ' ὑγρὸν ἐσχάτως ἐστὶν ὥσπερ βρύον, οὐκ ἀγγείοις μόνον καὶ σπλάγχνοις καὶ σαρξίν, ἀλλὰ καὶ τοῖς ὀστοῖς αὐτοῖς, ἃ δὴ ξηρότατα τῶν ἐν ἡμῖν ὑπάρχει μορίων. ἀλλ' ὅμως καὶ ταῦτα καὶ ὅλα σὺν αὐτοῖς τὰ κῶλα διαπλάττουσιν αἱ τροφοὶ τῶν βρεφῶν ὥσπερ κήρινα. τοσαύτη τις ὑγρότης ἐστὶν ἐν ἅπαντι τῷ σώματι τῶν βρεφῶν.[8]

Ἀλλὰ καὶ νέον ἱερεῖον εἴτ' οὖν ἐσθίειν ἐθέλοις εἴτ' ἀνατεμὼν σκοπεῖσθαι, μυξώδη μὲν καὶ πλαδαρὰν εὑρήσεις τὴν σάρκα, τὸ δ' ὀστῶδες γένος ἅπαν ἄρτι πηγνυμένῳ τυρῷ ἐμφερές, ὥστε μηδὲ φαγεῖν ἡδέα δι' ὑπερβάλλουσαν ὑγρότητα τῶν νεογενῶν ζῴων εἶναι τὰ σώματα καὶ μάλιστά γε τοῦτο πέπονθε τὰ ὕεια καὶ

[7] πλῆκτρα H; σπλάγχνα K
[8] τῶν βρεφῶν H; τῶν παιδίων K

ON TEMPERAMENTS, BOOK II

know how very wet and very hot it is. For the first construction is from blood and semen in it, which are things that are wet and hot. And always, when these become more dry, membranes are formed first, and tunics, viscera and vessels, and finally when the substance congeals, they make bones, cartilages and nails. For it is not possible for any of the things mentioned to come into being before the underlying substance is able to be stretched or congealed. Thus, tunics and membranes, arteries and veins, and nerves are produced when this is stretched, while bones, cartilages, nails, hoofs and spurs when it is congealed. When they are completed in this way, next [the child] in the womb is born. However it is still extremely wet, like seaweed,[2] not only in vessels, viscera and flesh, but in the bones themselves, which in truth are the driest of parts in us. But nevertheless also, those who rear newborn infants mold these and the whole limbs with them, as if they were wax, so great is the wetness in the whole body of infants.

But also, if you wish either to eat a young animal[3] or to consider cutting it up, you will find the flesh mucoid and flabby, while the whole bony class resembles cheese just now congealed, so that eating it is not pleasant due to the body of newborn animals being excessively wet. And this in fact particularly affects pigs and sheep, because they are

---

[2] See Singer and van der Eijk, *Galen: Works on Human Nature*, on alternative readings of βρύον (109n33). I have opted for "seaweed" on the grounds of sense (KLat *maris alga*).

[3] The Greek is ἱερεῖον, a term generally used of sacrificial animals. KLat has *porcellum*.

τὰ προβάτεια, διότι καὶ μάλιστ' ἐστὶν ὑγρότατα· τὰ δ' αἴγεια, διότι ξηρότερα, βελτίω τ' ἐστὶ καὶ ἡδίω φαγεῖν. ἔμπαλιν δὲ τοῖς νέοις ἱερείοις τὰ γεγηρακότα ξηρὰ μὲν ἱκανῶς καὶ ἄνικμα καὶ ἄχυμα τά τ' ὀστᾶ σύμπαντα καὶ τοὺς συνδέσμους αὐτῶν ἔχει, νευρώδη δὲ καὶ σκληρὰν τὴν σάρκα καὶ τὰς ἀρτηρίας καὶ τὰς φλέβας καὶ τὰ νεῦρα⁹ δίκην ἱμάντων ἀηδῆ τε καὶ ἄχυμα. τὰ δ' ἐν τῷ μέσῳ τούτων καὶ τῶν ἄρτι γεγενημένων, ὅσα μὲν ἤδη προβέβηκε ταῖς ἡλικίαις, ὅσον ἀπολείπεται τοῦ γήρως, τοσοῦτον καὶ τῆς ἐσχάτης ξηρότητος· ὅσα δὲ νεώτερα καὶ ἔτ' αὐξανόμενα, τοσοῦτον καὶ ταῦτα τῆς τῶν ἐμβρύων ὑγρότητος ἀποκεχώρηκεν, ὅσον καὶ ταῖς ἡλικίαις προελήλυθεν. ἡ δ' ἀκμὴ μάλιστα πάντων τῶν ζῴων ἐν τῷ μέσῳ καθέστηκε τῶν ἀκροτήτων οὔτ' εἰς ἔσχατον ἥκουσα ξηρότητος, ὡς τὸ γῆρας, οὔτ' ἐν ὑγρότητι καὶ πλάδῳ πολλῷ καθεστῶσα, καθάπερ ἡ τῶν βρεφῶν ἡλικία.

Τί δὴ οὖν ἔνιοι τῶν ἐλλογίμων ἰατρῶν ὑγρὸν ἀποφαίνονται τὸ γῆρας; ἢ δηλονότι τῷ πλήθει τῶν περιττωμάτων ἐξαπατηθέντες; οἵ τε γὰρ ὀφθαλμοὶ δακρύουσιν αὐτοῖς αἵ τε ῥῖνες ἀναπίμπλανται κορύζης ἔν τε τῷ στόματι σιέλων πλῆθος ἀθροίζεται, ἀλλὰ καὶ βήττουσι καὶ ἀναπτύουσι φλέγμα, δηλοῦντες ἄρα καὶ τὸν πνεύμονα μεστὸν εἶναι τοῦ τοιούτου χυμοῦ· καὶ ἡ γαστὴρ δ' αὐτοῖς πεπλήρωται φλέγματος ἕκαστόν τε τῶν ἄρθρων ὑπόμυξον. ἀλλ' οὐδὲν τούτων ἐναντιοῦται

⁹ post τὰ νεῦρα: διὰ τὴν ἡλικίαν K om.

wetter to the greatest degree. Goats, because they are drier, are better and more pleasant to eat. Contrary to the young animals, those that have become old are dry and significantly without moisture and flavor, as are all their bones and ligaments, while the flesh is sinewy and hard, and the arteries, veins, and nerves, like leather straps due to the age, are unpleasant and without flavor. On the other hand, those midway between these and those which have just now been born, those who have already advanced through the stages of life, to the extent that they fall short of old age, also fall short of the extreme dryness. However, in those that are younger and are still growing, the extent to which they are removed from the wetness of the newborn is also related to how far they have advanced through the stages of life. For the most part, the highest point of all animals is that situated in the middle of the extremes, not having come to extreme dryness, as in old age, nor to have settled in wetness and great abundance of fluid, as in the stage of infancy.

Why then do some of the doctors of high repute declare old age to be wet? Were they obviously deceived by the excess of the superfluities? For in the aged, the eyes run, the nostrils fill with mucus, and a large amount of saliva collects in the mouth. But they also cough and expectorate phlegm, indicating that in fact the lungs are full of such a humor. And in them the stomach is filled with phlegm and each of the joints is somewhat charged with mucus.[4] But none of these things opposes the fact that the

---

[4] Hippocrates, *Joints* 8, LCL 149 (E. T. Wittington), 217ff.

τῷ ξηρὰ τῶν γερόντων εἶναι τὰ σώματα. τὰ μὲν γὰρ νεῦρα καὶ τὰς ἀρτηρίας καὶ τὰς φλέβας καὶ τοὺς ὑμένας καὶ τοὺς χιτῶνας ἁπάντων τῶν ὀργάνων ξηρότερα μὲν εὑρήσεις τῶν πρόσθεν πολύ, περιπεπλασμένον δ' αὐτοῖς ἔνδοθέν τε καὶ ἔξωθεν ἤτοι φλεγματώδη τινὰ χυμὸν ἢ ὑγρότητα μυξώδη.

Ἀλλὰ τοσούτου δεῖ τὰ τοιαῦτα σύμπαντα γνωρίσματα τὸ γῆρας ὑγρὸν ἀποφαίνειν, ὥστε καὶ μαρτυρεῖν μοι δοκεῖ τῇ ξηρότητι. δι' αὐτὸ γάρ τοι τοῦτο ξηρότερον ἕκαστον γίγνεται τῶν μορίων, ὅτι μηκέθ' ὁμοίως τρέφεται νῦν ὑπ' ἀρρωστίας τοῦ θερμοῦ. πλῆθος μὲν γὰρ ἔξωθεν αὐτὸ περιττωμάτων ὑγρῶν ἐπικλύζει, τὸ βάθος δ' αὐτὸ τοῦ σώματος ἑκάστου ξηρόν ἐστι μήθ' ἕλκειν εἴσω τὴν τροφὴν δυναμένου μήτ' ἀπαλαύειν ἱκανῶς. ὑγρὸς οὖν ὁ γέρων ἐστίν, οὐ τοῖς οἰκείοις μορίοις, ἀλλὰ τοῖς περιττώμασι, καὶ ξηρός, οὐ τοῖς περιττώμασιν, ἀλλὰ τοῖς μορίοις αὐτοῖς, ὥστ' ἄλλῳ μὲν ὑγρός, ἄλλῳ δὲ ξηρός. ἀλλ' οὐχ ὑπὲρ τῶν περιττωμάτων αὐτοῦ νῦν ὁ λόγος ἀλλὰ τῶν οἰκείων μορίων ἐστίν, ὧν αἱ κατὰ φύσιν ἐνέργειαι συμπληροῦσι τὴν ζωήν. τούτοις οὖν ξηρός ἐστιν ὁ γέρων, οἷς ὁ παῖς ἦν ὑγρός, αὐτοῖς τοῖς στερεοῖς μέρεσι τοῦ σώματος, ὀστοῖς καὶ συνδέσμοις καὶ ὑμέσι καὶ ἀρτηρίαις καὶ φλεψὶ καὶ νεύροις καὶ χιτῶσι καὶ σαρξί, καὶ καλῶς Ἀριστοτέλης εἰκάζει τὸ γῆρας αὐαινομένῳ φυτῷ. καὶ γὰρ οὖν καὶ τὰ φυτά, νέα μὲν ὄντα, μαλακά τ' ἐστὶ καὶ ὑγρά, γηρῶντα δ' ἀεὶ καὶ

bodies of old people are dry. For you will find the nerves, arteries, veins, membranes and the tunics of all the organs much drier than they were before, while what surrounds them inside and out is a certain phlegmatic humor or mucoid wetness.

But all such signs are lacking so much in indicating the wetness of old age that they also seem to me to be witness to the dryness. Because of this certainly, each of the parts becomes drier because it is no longer now similarly nourished due to weakness of the heat. For excess of wet superfluities swamp this externally, while the actual depths of each body are dry, and no longer able to draw the nutriment inward, nor derive sufficient benefit from it. Therefore, the old person is wet, not in the specific parts, but in the superfluities, and dry, not in the superfluities, but in the parts themselves, so they are wet in one respect but dry in another. However, the discussion is not now about the superfluities of the old person, but about the specific parts, of which the functions in accord with nature help fulfill the person's life. Thus, the old person is dry in those parts in which the young child is wet; that is, in the actual solid parts of the body, which are bones, ligaments, membranes, arteries, veins, sinews (nerves), tunics, and flesh. And Aristotle does well to liken old age to a withered plant.[5] For surely plants too, when they are young, are soft and wet, while as they grow old, they always seem to

---

[5] Aristotle, *On Respiration* 17.478b.28–29, LCL 288 (W. S. Hett), 470–71. "This phenomenon (i.e., death) is called withering in plants, and in animals, old age." Aristotle does not make any connection with drying, but focuses on loss of innate heat, which Galen also mentions below.

μᾶλλον φαίνεται ξηραινόμενα καὶ τελευτῶντα τελέως ἀποξηραίνεται καὶ τοῦτ' ἔστιν αὐτοῖς ὁ θάνατος.

Ὅτι μὲν δὴ ξηρότατον ὡς ἐν ἡλικίαις τὸ γῆρας, ἐκ τῶν εἰρημένων εὔδηλον· ὅτι δὲ καὶ ψυχρότατόν ἐστιν, ἔτ' ἐναργέστερον, ὥστ' οὐδ' ἠμφισβήτησεν οὐδεὶς ὑπέρ γε τούτου. καὶ γὰρ ἁπτομένοις οἱ γέροντες ψυχροὶ φαίνονται καὶ ῥᾳδίως ἀποψύχονται καὶ μελαίνονται καὶ πελιδνοῦνται καὶ τοῖς ψυχροῖς ἑτοίμως ἁλίσκονται νοσήμασιν, ἀποπληξίαις, παραλύσεσι, νάρκαις, τρόμοις, σπασμοῖς, κορύζαις, βράγχοις. ἀπόλωλε δ' αὐτῶν ὀλίγου δεῖν ἅπαν τὸ αἷμα, καὶ διὰ τοῦτο συναπόλωλεν ἡ τῆς χρόας ἐρυθρότης. ἀλλὰ καὶ πέψις αὐτοῖς καὶ ἀνάδοσις ἐξαιμάτωσίς τε καὶ πρόσθεσις καὶ θρέψις ὄρεξίς τε καὶ κίνησις καὶ αἴσθησις ἀμυδρὰ πάντα καὶ κακῶς διακείμενα. καὶ τί γὰρ ἄλλ' ἢ ὁδὸς ἐπὶ θάνατόν ἐστι τὸ γῆρας; ὥστ' εἴπερ ὁ θάνατος σβέσις ἐστὶ τῆς ἐμφύτου θερμασίας, εἴη ἂν καὶ τὸ γῆρας οἷον μαρασμός τις αὐτῆς.

Οὐ μὴν περί γε τῆς τῶν παίδων ἡλικίας καὶ τῆς τῶν ἀκμαζόντων οὔθ' ὡμολόγηταί τι τοῖς ἰατροῖς οὔτε κρῖναι τὴν διαφωνίαν αὐτῶν εὐπετές. πιθανοὶ γὰρ ἑκατέρων οἱ λόγοι τῶν τε τοὺς παῖδας ἀποφαινόντων θερμοτέρους εἶναι τῶν ἀκμαζόντων καὶ τῶν ἔμπαλιν τούτοις τοὺς ἀκμάζοντας τῶν παίδων. οἱ μὲν γὰρ ὅτι θερμότατος ἁπάντων ἐστὶ τῶν ἐν τῷ ζῴῳ κατὰ φύσιν ὑπαρχόντων ὁ τοῦ αἵματος χυμός, εἶθ' ὅτι τὰ κυούμενα τὸ μὲν πρῶτον ὀλίγου δεῖν αἷμα μόνον ἐστίν, ὕστερον δὲ διαπλαττομένων ἤδη τῶν μορίων τὸ μὲν

ON TEMPERAMENTS, BOOK II

dry out more, until finally they are dried up completely and for them this is death.

That old age is the driest in the stages of life is quite clear from the things said. Also that it is the coldest is even more clear, so that no one disputes this. Further, for those who touch them, old people seem cold and they easily become cold, dark colored and livid, and are readily seized by the cold diseases—apoplexies, paralyses, numbness, tremors, spasms (convulsions), catarrh, and sore throat. They have lost almost all their blood and because of this lose as well the redness of the complexion. But also in them concoction, distribution, blood formation, absorption, nourishing, appetite, movement, and perception are all feeble and in a bad state. For what is old age but the road to death? So that, if death is the quenching of the innate heat, old age would also be the wasting away, as it were, of this.

Concerning the | stage of life of children, and of those in their prime, there is no agreement among doctors, nor is the dispute about these easily resolved. For the arguments on both sides are credible: those who declare children are hotter than those in their prime, and conversely those who declare those in their prime are hotter than children. For there are those [who say] that the humor of blood is the hottest of all those existing in the animal in accord with nature, and that the fetus at first is almost only blood whereas later, when the parts are already formed,

ὀστοῦν γίγνεται, τὸ δ' ἀρτηρία, τὸ δὲ φλέψ, τὸ δ' ἄλλο τι, πάντα μὴν ἐρυθρὰ καὶ πλεῖστον αἵματος εἰλικρινεστάτου τε καὶ θερμοτάτου μετέχοντα, συλλογίζονται θερμότατον εἶναι τὸ κυούμενον· εἰ δὲ τοῦτο, καὶ τοὺς παῖδας ὅσωπερ ἐγγυτέρω τοῖς κυουμένοις, θερμοτέρους εἶναι τῶν ἀκμαζόντων. οἱ δ', ὅτι πολὺ μὲν κἂν τοῖς ἀκμάζουσι τὸ αἷμα καὶ πλέον ἢ ἐν τοῖς παισίν, ὥστε διὰ τοῦτο συνεχῶς αἱμορραγεῖν, ἀλλὰ καὶ ὁ τῆς ξανθῆς χολῆς χυμὸς | αἵματος πολὺ θερμότερος ὑπάρχων πλεῖστος αὐτοῖς ἐστι, διὰ τοῦτο θερμοτέρους ἀποφαίνουσι τῶν παίδων τοὺς ἀκμάζοντας.

Αὖθις δ' ἀπὸ τῶν ἐνεργειῶν οἱ μὲν ὅτι καὶ αὐξάνονται καὶ πλειόνων ἢ κατὰ τὴν ἀναλογίαν τοῦ σώματος ὀρέγονταί τε καὶ κρατοῦσιν ἐδεσμάτων, ἰσχυρὰν ἐν τοῖς παισὶν εἶναί φασι τὴν θερμασίαν· οἱ δὲ τὸ μὲν αὐξάνεσθαι διὰ τὴν ὑγρότητα μᾶλλον ἢ τὴν τοῦ θερμοῦ ῥώμην ὑπάρχειν αὐτοῖς φασι, ταῖς μέντοι πέψεσιν οὐχ ὅπως πλεονεκτεῖν ἀλλὰ καὶ πολὺ τῶν ἀκμαζόντων ἀπολείπεσθαι. ἐμέτους τε γὰρ ἀπέπτων αὐτοῖς γίγνεσθαι σιτίων καὶ διαχωρήσεις ὑγρῶν καὶ τραχέων καὶ ἀχυμώτων.[10] εἰ δ' ὀρέγονται πλειόνων, οὐδὲν εἶναι τοῦτό φασι πρὸς τὴν ῥώμην τοῦ θερμοῦ. πρῶτον μὲν γὰρ οὐδὲ πλεονεξίᾳ θερμότητος ὀρέγεσθαι τὰ ζῷα, τοὐναντίον [δ'] ἅπαν ἀποψυχομένων τῶν ὀρεκτικῶν μορίων· ἔπειτα δὲ διότι μὴ μόνον εἰς θρέψιν ἀλλὰ

[10] post ἀχυμώτων: αὐτῶν K om.

ON TEMPERAMENTS, BOOK II

and bone, artery, vein, and whatever else arise, they are all red and partake most of the purest and hottest blood, so they conclude the fetus is the hottest. However, if this is so, children are hotter than those in their prime to the extent they are nearer the fetuses. The others, however, say that there is much blood in those in their prime and more than in children, so they are continually hemorrhaging due to this, but also that the humor of yellow bile | is much hotter than blood, and is largest in amount in them, and because of this, they declare those in their prime are hotter than children.

In turn, from the functions, there are those who say the heat is strong in children because they grow and desire and have power over more foods than are in proportion to the body, while there are those who say growth is due more to the wetness than to the strength of the heat in them, and that not only do they not predominate in the concoctions in this matter, but also they are consistently lacking compared to those in their prime. For vomits of unconcocted foods occur in them as do excretions, since the foods themselves are watery, rough, and have not as yet been changed in flavor.[6] If, however, they desire more food, they say this is not related to the strength of the heat. For appetite in animals is not primarily due to the preponderance of hotness; on the contrary; it is generally when the appetitive parts are chilled. Therefore, because the nour-

584K

---

[6] The translation of this sentence follows KLat on the grounds of sense: *nam et vomitiones iis ex incoctis cibis accidere, et deiectiones, cum ipsi cibi adhuc humidi sunt asperique, nec adhuc in succum mutati.*

καὶ εἰς αὔξησιν αὐτοῖς ἡ τροφὴ διοικεῖται, διὰ τοῦτο πλειόνων ἐδεσμάτων προσδεῖσθαι. κατὰ μέντοι τὰς ἄλλας ἁπάσας ἐνεργείας καὶ πάνυ σαφῶς ἀπολείπεσθαι τοὺς παῖδας τῶν ἀκμαζόντων. οὔτε γὰρ βαδίζειν οὔτε θεῖν οὔτε βαστάζειν οὔθ' ὅλως οὐδὲν τῶν πρακτικῶν ἐνεργειῶν ὁμοίως ἐπιτελεῖν, ἀλλὰ καὶ τὰς αἰσθήσεις καὶ τὰς νοήσεις ἐν τοῖς ἀκμάζουσί φασιν εἰς ἄκρον ἥκειν ἀρετῆς. ὅλως δὲ τὸ μὲν ἀτελὲς ἔτι, τὸ δ' ἤδη τέλειον εἶναι ζῷον. ἐν δὲ τοῖς τελείοις εὔλογόν φασι τὸ πρακτικώτατόν τε καὶ ἀρχικώτατον τῶν στοιχείων ἐπικρατεῖν. ἀλλὰ καὶ τοὺς ὕπνους πλείστους μὲν ἐν τοῖς παισὶν ἰδεῖν ἔστι γιγνομένους, ἐλαχίστους δ' ἐν τοῖς ἀκμάζουσι. καίτοι τούτους γε, φασίν, οὐδὲ μανεὶς ἄν τις ἑτέρως ἡγήσαιτο γίγνεσθαι ἢ τοῦ θερμοῦ νικηθέντος πως καὶ βαρυνθέντος ὑπὸ πλήθους ὑγρότητος, ὡς ἔκ τε τῶν οἰνωθέντων ἔστιν ἰδεῖν ἔτι τε τῶν πλείω λουσαμένων. οὕτω δὲ καὶ μήκων ὑπνοποιός ἐστι καὶ μανδραγόρας καὶ θριδακίνη καὶ πάνθ' ὅσα τὴν κρᾶσιν ὑγρότερα καὶ ψυχρότερα.

Τοιαῦται μέν τινες αἱ ἑκατέρωθεν ἀμφισβητήσεις εἰσὶ περὶ τῶν προκειμένων ἡλικιῶν τῆς κράσεως. ἁπάσας γὰρ αὐτὰς ἐπεξέρχεσθαι περιττὸν εἶναί μοι δοκεῖ τοῦ τύπου τῶν ἐπιχειρημάτων ἤδη σαφῶς κἀξ ὧν εἰρήκαμεν ἐγνωσμένου. πόρρωθεν γὰρ ἑκάτεροι καὶ σχεδὸν ἀπὸ τῶν δευτέρων τὰ πρότερα συλλογίζονται καὶ ὥσπερ εἰδότων ἤδη τῶν ἀκροατῶν, ὅπως μὲν

ON TEMPERAMENTS, BOOK II

ishment in children provides not only for nutrition but also for growth, there is need for more food due to this. And further of course, in all the other functions, it is very clear that children are deficient compared to those in their prime. For they don't walk, run, and carry, nor in general perform any of the practical functions similarly. They also say the perceptive and intellectual functions come to a peak of excellence in those in their prime. In general terms, the one (the child) is still an incomplete animal while the other (adult in the prime of life) is already complete. They say in those who have come to completion, it is reasonable for the most effective and most dominant of the elements to prevail. But it is also seen that more sleep occurs in children and the least in those in their prime. And indeed, they say not even someone who is insane would think these things would occur otherwise than that the hotness is somehow overcome and weighed down by the excessive wetness, as seen from those drunk with wine, and further those who bathe excessively. In the same way, there is an hypnotic effect of poppy, mandrake,[7] wild lettuce, and all those that are wetter and colder in *krasis*.

Such are some of the disputes on either side concerning the *krasis* of the proposed stages of life. It seems to me to be superfluous to go over all these since the general character of the proofs[8] is already clearly known from the things we have said. For each group just concludes the primary things remotely from the secondary ones, making the argument as if their listeners already knew how growth,

---

[7] See Dioscorides 4.76 and Theophrastsus, *History of Plants* 9.8.8.   [8] ἐπιχειρημάτων is used here in a technical (Aristotelian) sense; KLat has *epicheirematum*.

αὔξησις, ὅπως δὲ πέψις, ὅπως δὲ θρέψις γίγνεται, ποιοῦνται τὸν λόγον. ὡσαύτως δ' ὑπὲρ αἰσθήσεως καὶ νοήσεως[11] καὶ πρακτικῶν καὶ φυσικῶν ἐνεργειῶν διέρχονται καὶ γενέσεως ὕπνου μνημονεύουσι καὶ σιτίων φύσεως, ὧν οὐδὲν ἁπλῶς ἐστι καὶ ῥᾳδίως γνωστόν, ἀλλὰ παμπόλλης μὲν τῆς ζητήσεως δεόμενον, ἴσως δ' οὐδ' εὑρεθῆναι δυνάμενον, εἰ μὴ πρότερόν τις εἰδείη γνωρίζειν ὑγρὰν καὶ ξηρὰν καὶ ψυχρὰν καὶ θερμὴν κρᾶσιν. ὅ τι γὰρ ἂν ἐκείνων ὡς γιγνώσκοντες ἤδη λέγωσιν, εἰ ἀναγκάσειέ τις αὐτοὺς ἀποδεικνύναι, πάντως δεήσονται τοῦ περὶ κράσεων λόγου τοῦ νῦν ἡμῖν ἐνεστῶτος. ὥστε δι' ἀλλήλων καὶ ἐξ ἀλλήλων αὐτοῖς γίγνεσθαι τὰς ἀποδείξεις, ἐκ μὲν τῶν νῦν ζητουμένων, ὡς ἤδη γιγνωσκομένων, ἐπειδὰν ὑπὲρ τῶν ἐνεργειῶν διαλέγωνται καὶ τὴν τῶν ἐδεσμάτων τε καὶ φαρμάκων ἐξευρίσκωσι δύναμιν ὕπνων τε πέρι καὶ τῶν ἄλλων τῶν τοιούτων ἐπισκέπτωνται. πάλιν δ' αὖ τὰ νῦν ἐνεστῶτα δι' ἐκείνων ὡς ἤδη προεγνωσμένων ἀποδεικνύουσιν. ἐγὼ δ' οὐκ ἐπαινῶ τὰς τοιαύτας ἀποδείξεις, ἀλλ', εἰ χρὴ τἀληθὲς εἰπεῖν, οὐδ' ἀποδείξεις εἶναι νομίζω, καθάπερ ἐπὶ πλέον δι' ἑτέρων ἐδήλωσα, καὶ βέλτιον εἶναί φημι κατὰ πᾶσαν διδασκαλίαν ὁρίσασθαι τὴν τάξιν τῶν νοημάτων.

Εἴπερ οὖν ἀρχὴ μὲν ἁπάσης ἐστὶ τῆς περὶ τῶν κράσεων πραγματείας ἡ περὶ τῶν στοιχείων ἐπίσκεψις, εἴτ' ἀπαθῆ καὶ ἀμετάβλητα τελέως ἐστὶν εἴτ' ἀλ-

---

[11] νοήσεως H; κινήσεως K

ON TEMPERAMENTS, BOOK II

concoction, and nutrition occur. And they go over perception and thought similarly, and practical and physical functions, and they make mention of the genesis of sleep and the nature of foods, knowledge of which is neither simple nor easy, but requires a prolonged inquiry, or perhaps cannot be discovered unless one knows first how to recognize a wet, dry, cold, and hot *krasis*. For on whichever of those things they speak, as if knowing them already, if compelled to provide a demonstration, they would in all cases need the discussion about *krasias*, which is now before us. As a result, the demonstrations arise for them through each other and from each other, from the matters now being investigated, as being already known, when they discourse about the functions, discover the potency of foods and medications, and consider sleep and other such things. Then again, they show by demonstration the present matters through those things as if they were already known beforehand. I, on the other hand, do not praise such demonstrations; if the truth must be told, I do not think they are demonstrations, just as I have shown at greater length through other works.[9] And I say it would be better, in relation to all teaching, to define the order of the concepts.

The starting point, then, of the whole subject concerning the *krasias* is the investigation concerning the elements, whether they are completely impassible and im-

---

[9] The main work referred to is presumably Galen's lost treatise, *On Demonstration*; see his *On My Own Books*, XIX.41K. For other references, both Galenic and more modern, see Singer and Van Der Eijk, *Galen: Works on Human Nature,* 114n55.

λοιοῦσθαί τε καὶ μεταβάλλεσθαι δυνάμενα, μετὰ δὲ
τὴν ἐκείνων ἐπιστήμην ἐφεξῆς ἐστι δεύτερος ὁ νῦν
ἡμῖν ἐνεστηκὼς λόγος, οὐ χρὴ λαμβάνειν αὐτῶν τὰς
πίστεις ἐκ τῶν μηδέπω γιγνωσκομένων, ἀλλ' ὥσπερ
ὀρθόν τ' ἐστι καὶ δίκαιον, ἤ τι τῶν ἐναργῶν εἶναι
προσήκει τὸ ληφθησόμενον εἰς τὴν ἀπόδειξιν ἤ τι
τῶν προαποδεδειγμένων. οὔτ' οὖν ὕπνου γενέσεως
588K μνημονευτέον | οὔτε πέψεως οὔτ' αὐξήσεως οὔτ' ἄλλου
τῶν τοιούτων οὐδενός, ἀλλ' ἀπὸ μόνης καὶ ψιλῆς τῆς
οὐσίας τῶν ὑποκειμένων πραγμάτων ἡ ζήτησις γιγνέ-
σθω, καθάπερ καὶ διὰ τοῦ πρώτου λόγου πεποιήμεθα.

Διελόμενοι γάρ, ὡς ἕτερον μέν ἐστι τὸ κατ' ἐνέργειαν, ἕτερον δὲ τὸ κατὰ δύναμιν, ὑπὲρ τοῦ κατ' ἐνέργειαν ὄντος ἤδη θερμοῦ καὶ ψυχροῦ καὶ ξηροῦ καὶ
ὑγροῦ τὸν λόγον ἔφαμεν χρῆναι ποιήσασθαι πρότερον, ἔπειθ' οὕτως ἐπὶ τὰ κατὰ δύναμιν ἀφικέσθαι.
πρόχειρος δ' ἐστὶ πᾶσι καὶ γνώριμος ἡ τῶν κατ' ἐνέργειαν ἤδη θερμῶν καὶ ψυχρῶν καὶ ξηρῶν καὶ ὑγρῶν
διάγνωσις· ἀφὴ γὰρ τά γε τοιαῦτα διακρίνειν πέφυκεν ἡ καὶ τὸ πῦρ αὐτὸ θερμὸν εἶναι διδάξασα καὶ τὸν
κρύσταλλον ψυχρόν. εἰ δ' ἄλλοθέν ποθεν ἔχουσιν ἔν-
νοιάν τε καὶ διάγνωσιν θερμοῦ καὶ ψυχροῦ, λεγέτωσαν ἡμῖν. ἀμήχανον γάρ τινα σοφίαν ἐπαγγέλλονται,
μᾶλλον δ', εἰ χρὴ τἀληθὲς εἰπεῖν, ἐμπληξίαν, εἰ πραγμάτων αἰσθητῶν ἕτερόν τι πρεσβύτερον αἰσθήσεως
ἔχουσι κριτήριον. καὶ μὴν εἰ μηδὲν ἄλλο τῶν ἐνεργείᾳ

mutable or are able to be changed and transformed. After the knowledge of those, second and next in order is the discussion now upon us. We must not take the beliefs of these from those things not yet known, but as is right and proper, something will be taken of what is clearly germane to the demonstration, or something of those that have previously been demonstrated. One must not, therefore, call to mind the genesis of sleep, or of concoction (digestion), or of growth, or any other of such things. Rather, the inquiry should be through the substance of the underlying matters alone and unadorned, as we have done in the first book.

For having made the division that one thing is that related to actuality (function) while the other is that related to potentiality (capacity), we said that we need to first make the discussion about what already is hot and cold, and dry and wet in actuality, and then in this way come to what relates to potentiality.[10] There is at hand and familiar to everyone the recognition of those things that are already hot, cold, dry, and wet in actuality. In fact, touch naturally distinguishes such things, teaching that fire itself is hot and ice cold. If, however, they have a concept and recognition of hot and cold from any other source whatsoever, let them tell us. For they proclaim a certain inconceivable wisdom, or rather, if the truth must be told, stupidity, if they have some other, more important criterion of perception of perceptible matters. And if no other criterion is possible of things that are hot in actuality,

---

[10] As these terms (ἐνέργεια and δύναμις) are used in an Aristotelian sense, they are rendered "actuality" and "potentiality," respectively; see particularly Aristotle, *Metaphysics* 9.5–9.

θερμῶν ἐγχωρεῖ κριτήριον | ὑπάρχειν, ἁπτέσθωσαν ἤδη πολλῶν ἐφεξῆς ἀνδρῶν καὶ γερόντων καὶ μειρακίων καὶ παίδων καὶ βρεφῶν. οὕτω γὰρ ἐξευρήσουσι τοὺς μᾶλλόν τε καὶ ἧττον θερμούς.

Εἰ δ' αἰσθητῶν πραγμάτων ἀποδείξεις λογικὰς ζητοῦσιν, ὥρα τι καὶ περὶ τῆς χιόνος αὐτῆς ἤδη σκοπεῖν, εἴτε λευκήν, ὡς ἅπασιν ἀνθρώποις φαίνεται, νομιστέον αὐτὴν εἴτε καὶ μὴ λευκήν, ὡς Ἀναξαγόρας ἀπεφήνατο. καὶ μὲν δὴ καὶ περὶ πίττης ὡσαύτως ἀνασκοπεῖν καὶ κόρακος ἁπάντων τε τῶν ἄλλων. οὐ γὰρ δὴ τὸ μὲν λευκὸν ἀπιστεῖσθαι χρὴ τοὺς ὀφθαλμοὺς ὁρῶντας, ἄνευ δ' ἀποδείξεως ἐπὶ τῶν μελάνων πιστεύεσθαι. ἅπαντ' οὖν ἤδη τὰ τῶν αἰσθήσεων ἄπιστα φάμενοι μήτε τὸν κύκνον λευκὸν εἶναι λεγόντων, ἐὰν μὴ πρότερον ἐπισκέψωνται λόγῳ, μήτε τὴν τίτανον ἢ τὴν ἡμέραν ἢ αὐτὸν τὸν ἥλιον. οὕτω δὲ καὶ περὶ τῶν φωνῶν ἀπιστησάτωσαν ἀκοῇ καὶ περὶ τῶν ὀσμῶν ταῖς ῥισὶ καὶ περὶ πάντων τῶν ἁπτῶν ἀφῇ. εἶτα ταῦτ' οὐ Πυρρώνειος ἀπορία καὶ λῆρος ἀπέραντος; οὐ μὴν δίκαιόν γ' ἦν τοὺς τὴν ἀρίστην αἵρεσιν ἑλομένους τῶν ἐν φιλοσοφίᾳ, | τὴν τὸ θερμὸν καὶ τὸ ψυχρὸν καὶ τὸ ξηρὸν καὶ τὸ ὑγρὸν ἀρχὰς καὶ στοιχεῖα τιθεμένην, εἰς τοσοῦτον ἀποπλανηθῆναι τῶν ταῦτα θεμένων ἀνδρῶν, ὡς μὴ γιγνώσκειν, ὅτι τε πάσης ἀποδείξεως ἀρχαὶ τὰ πρὸς αἴσθησίν τε καὶ νόησίν εἰσιν ἐναργῆ καὶ ὅστις περὶ τούτων ἀπορεῖ, μάτην ὑπὲρ τῶν ἄλλων ζητεῖ, μηδ' ὁπόθεν ἄρξηται καταλελοιπὼς ἑαυτῷ.

ON TEMPERAMENTS, BOOK II

then they should go ahead and touch many adult males in order, and old men, youths, children and infants, for in this way they will discover those who are more and less hot.

If, however, they seek logical demonstrations of perceptible matters, [let them] now look at and consider snow itself, whether it is white, as it appears to all people, or one must consider it not white, as Anaxagoras declared.[11] And indeed also, one could closely examine pitch similarly, and a raven similarly, and all the other things. For certainly we must not disbelieve our own eyes when they see white, but believe them without demonstration in the case of things that are black. Already then, since they say all the objects of perception are not to be trusted, unless first examined by reason, they should not say a swan is white, nor chalk, nor the day, nor the sun itself. In this way too, we should distrust the ears in hearing sounds, and the nose regarding odors, and touch regarding all things touched. So then are these not a Pyrrhonist problem and boundless nonsense? In fact, it is not right that those who have chosen the best sect in philosophy | which postulates the hot, cold, dry, and wet as principles and elements, should have been led astray to such an extent by the men proposing these things, as not to realize the starting points of every demonstration clearly pertain to perception and conception. And whoever is at a loss about these seeks in vain for the others, and has not left himself with anywhere to begin from.

---

[11] The essence of the argument is that snow is frozen water, and since water is black, it follows that snow should be black too; see DK, 59A97 (II.29), and Sextus Empiricus, *Outlines of Pyrrhonism* 1.33, LCL 273 (R. G. Bury), 22–23.

GALEN

Πόθεν οὖν εἰς μακρὰν οὕτως ἄλην ἐξετράποντο καὶ λόγῳ ζητεῖν ἐπεχείρησαν αἰσθητῶν πραγμάτων διάγνωσιν, ἐγὼ μὲν οὐδ' ἐπινοῆσαι δύναμαι, καὶ διὰ τοῦθ' ἁφῇ μὲν κρίνω τὸ κατ' ἐνέργειαν θερμόν· εἴ τι δ' οὔπω μέν ἐστι θερμόν, ἐπιτήδειον δὲ γενέσθαι τοιοῦτον, ὃ δὴ καὶ δυνάμει θερμὸν ὀνομάζεται, τοῦτ' ἐξευρίσκειν λόγῳ πειρῶμαι. τοῖς δ' οὐκ οἶδ' ὅπως ἀντέστραπται[12] πάντα καὶ ῥητορεία μακρὰ περὶ τῶν ἐπιχειρημάτων ἤσκηται. τούτους μὲν οὖν ἐάσωμεν, ἀναμνήσαντες δ' ἡμᾶς αὐτοὺς καὶ νῦν τό γε τοσοῦτον, ὡς ἀρχὴ δογμάτων μοχθηρῶν ἐστι μία τὸ μηδὲν ὑπὲρ ἀποδείξεως ἐσκέφθαι πρότερον, ἀλλ' ἅμα τά τε πράγματα ζητεῖν | καὶ ὡς εἰδότας, ὅ τί ποτ' ἐστὶν ἀπόδειξις, ἐπιχειρεῖν ἀποδεικνύειν, ἐπανελθόντες αὖθις ἐπὶ τὸ προκείμενον ἁφῇ κρίνωμεν πρώτως καὶ μάλιστα τὸ κατὰ τὰς ἡλικίας θερμόν. ἔσται δ' ἡ κρίσις ἡμῖν ἀρίστη καθ' ἓν καὶ ταὐτὸν σῶμα βρέφους ἑνός. οὐ γὰρ ἀδύνατον ὁποία τέ τις ἡ θερμασία διετεῖ τὴν ἡλικίαν ὑπάρχοντι προϋπῆρχεν αὐτῷ μεμνῆσθαι καὶ ὁποία νῦν ἐστι δυοῖν ἢ τριῶν ἐτῶν, εἰ τύχοι, μεταξὺ γενομένων. εἰ γὰρ ὅλως φαίνοιτο μεταβολή τις ἐπὶ τὸ θερμὸν ἢ ψυχρὸν γεγονέναι τῷ βρέφει, χαλεπὸν οὐδὲν ἔτι συλλογίζεσθαι τὴν ἕως τῆς ἀκμῆς ἐσομένην ὑπεροχήν. εἰ δὲ καὶ πλείω παιδία πολλοῖς ἀκμάζουσιν ἐθέλοις παραβάλλειν, ἰσχνὰ μὲν ἰσχνοῖς, εὔσαρκα δ' εὐσάρκοις καὶ παχέα παχέσι παράβαλλε· οὕτω δὲ καὶ χρόας ὡσαύτως ἔχοντα καὶ τῶν ἄλλων ἁπάντων ὡς

ON TEMPERAMENTS, BOOK II

From what source they are turned in this way to a prolonged wandering and attempt to seek by reason recognition of perceptible matters, I am unable to conceive, and because of this, I determine by touch what is hot in actuality. However, if something is not yet hot, but has a propensity to become so, this certainly is also called hot in potentiality. This I shall attempt to discover by reason. But everything is turned on its head by them—how I don't know—and they practice a long-winded rhetoric concerning their attempted proofs. Let us, therefore, allow them [to do so], while we ourselves bear in mind this much now at least: that one source of difficult theories is not to give consideration first to demonstration, but at the same time to investigate the matters, and as if knowing what demonstration is at some time, to attempt to demonstrate. Returning once again to what is before us, let us determine by touch first and foremost the heat related to the stages of life. The best means of judgment for us will be in relation to one and the same body—that of one infant. For it is not impossible to remember what kind of heat is required for one of that age, and advancing from this, what kind it now is after say two or three years that have occurred in between. For if in general terms some change toward hot or cold has manifestly occurred to the infant, it is not difficult to compute what the projected change will be up to the prime of life. If, however, you wish to, also compare more children with many in their prime, compare thin with thin, well-fleshed with well-fleshed, and fat with fat. In the same way too, complexions should be similar, and as many as possible of all the other things.

---

[12] ἀντέστραπται H; ἀνατέτραπται K

GALEN

οἷόν τε. τὴν γὰρ ἐν ταῖς ἡλικίαις διαφορὰν ἐξευρεῖν ζητῶν ἐπὶ τῶν ὁμοίων ὡς ἔνι μάλιστα φύσεων ἀσφαλέστερον ἂν ἐπισκέπτοιο.

Τὸ δ' ἐπὶ τῶν ἐναντίων ἐξετάζειν οὐ σμικρὸν ἔχει τὸν παραλογισμόν, οὐ διὰ τὴν ἡλικίαν ἐνίοτε τῆς τῶν δοκιμαζομένων σωμάτων διαφορᾶς ἀλλὰ διὰ τὴν φυσικὴν ὑπαρχούσης κρᾶσιν. ὡσαύτως δὲ καὶ διαίτῃ πάσῃ καὶ τοῖς καιροῖς, ἐν οἷς ἐξετάζεται, παραπλησίως ἔχοντα προαιρεῖσθαι τὰ σώματα, μὴ γεγυμνασμένον ἠργηκότι παραβάλλοντας ἢ λελουμένον ἀλούτῳ μηδ' ἄσιτον ἐδηδοκότι καὶ διψῶντα μεμεθυσμένῳ μηδὲ τὸν ἐν ἡλίῳ θαλφθέντα τῷ ῥιγώσαντι διὰ κρύος ἢ τὸν ἀγρυπνήσαντα τῷ κεκοιμημένῳ μηδ' ὅλως τοῖς ἐξ ἐναντίας φύσεως ἢ διαίτης ἢ περιστάσεως πραγμάτων ἡστινοσοῦν, ἀλλ' ὡς οἷόν τε πάνθ' ὡσαύτως ὑπαρχέτω τἆλλα πλὴν τῆς ἡλικίας αὐτῆς. οὕτω δὲ δηλονότι καὶ αὐτὸν τὸν ἕνα παῖδα παραβάλλων ἑαυτῷ τὰς ἔξωθεν ἁπάσας αὐτοῦ περιστάσεις ἀκριβῶς ὁμοίας φυλάξεις, ἵνα μὴ τὸ διά τινα τούτων ἐν θάλψει τε καὶ ψύξει διάφορον εἰς τὴν τῆς ἡλικίας ἀναφέρηται μεταβολήν. μακρὰν ἴσως σοι δόξω λέγειν τὴν ἐξέτασιν ἀλλ' ἀληθῆ γε παντὸς μᾶλλον ἐξ αὐτῆς τε τοῦ ζητουμένου τῆς οὐσίας λαμβανομένην, ὡς ἐν τοῖς Ὑπὲρ ἀποδείξεως ἐλέγετο. σὺ δ' ἴσως αἱρήσῃ τὴν ἐπίτομον οὐδὲν φροντίζων, εἰ ψευδὴς εἴη. ἴσθι τοίνυν οὐ μόνον ψευδῆ βαδίσων ἀλλὰ καὶ μακράν. οὐ γὰρ ἔτεσι τρισὶν ἢ τέτταρσιν ἐξευρήσεις τὸ

For if you are seeking to discover the difference in the stages of life, you would be safer if you were to examine the same natures as far as possible.

To examine those who are opposites holds not a little deception, not because sometimes the difference of the bodies being examined is due to the stage of life, but due to the natural *krasis*. Likewise too, choose beforehand bodies that are already similar in both the whole regimen and the times in which they are examined; don't compare one that has exercised with one that has been idle, or one that has bathed with one that has not, or one fasting with one who has eaten, or one who is thirsty with one who is intoxicated or one that has warmed himself in the sun with one shivering due to severe cold, or one who is sleepless with one who has slept well, or in general with those of an opposite nature, regimen, circumstances, or activities of whatever kind, but let everything be as similar as possible apart from the stage of life (age) itself. In this way too, it is clear that when comparing the same child with himself, you should preserve all his external circumstances exactly the same so that you do not, because of some difference of these in heating and cooling, refer the change to the stage of life. Perhaps I shall seem to you to be describing a protracted examination, but it is true above all, taken from the very essence of what is being investigated, as was said in the work, *On Demonstration*.[12] Perhaps, however, you would choose an epitome, not being concerned if it is false. If so, you should know that you are not only proceeding falsely, but also [embarking] on a long [road]. For you will not discover what you are seeking, even in

[12] See note 9 above.

ζητούμενον, ἀλλ' ἐν παντὶ τῷ βίῳ φυλάξεις τὴν ἄγνοιαν. ὅσον γὰρ ἐπὶ ταῖς ἀντιλογίαις τῶν ἀνδρῶν, οὐδὲν ἀποδειχθῆναι δύναται σαφῶς· οὐδὲ γὰρ εὔλογον ὅλως ἐκ τῶν ὑστέρων πιστοῦσθαι τὰ πρότερα.

Κρίνωμεν οὖν αἰσθήσει τὸ θερμὸν καὶ ψυχρὸν σῶμα τό γε κατ' ἐνέργειαν ἤδη τοιοῦτον καὶ μηκέτι δυνάμει, παρέντες τήν[13] γε πρώτην τὰ ἄλλα σύμπαντα γνωρίσματα. καὶ δὴ σὲ μὲν ὡς εὖ κρινοῦντα πρὸς τὴν πεῖραν ἀπολύω, τὴν δ' ἐμὴν αὐτὸς κρίσιν ἑρμηνεύσω. πολλῶν γὰρ ἐφεξῆς ἁπτόμενος σωμάτων ἐπιμελῶς οὐ παίδων μόνον ἢ βρεφῶν ἀλλὰ καὶ μειρακίων καὶ ἀκμαζόντων εὕρισκον οὐδετέρους ἀληθεύοντας οὔτε τοὺς θερμότερον ἁπλῶς οὔτε τοὺς ψυχρότερον εἰπόντας εἶναι τὸν ἀκμάζοντα τοῦ παιδός. εἰ γὰρ τὰς ἄλλας ἁπάσας τὰς ἔξωθεν ἀφελὼν ἀλλοιώσεις τὰς ἐκ τῆς ἡλικίας μόνης ἐπισκέπτοιο διαφοράς, οὐδέτερός σοι φανεῖται θερμότερος ἁπλῶς. ποιότητι γάρ τοι διαφέρουσιν αὐτῶν αἱ θερμότητες ἐπ' ἀνίσῳ τῇ διαπνοῇ, δι' ἣν καὶ σοφιζόμενοί τινες ἢ τοὺς πέλας ἢ σφᾶς αὐτοὺς οἱ μὲν τὴν τοῦ παιδός, οἱ δὲ τὴν τοῦ νεανίσκου θερμασίαν ἰσχυροτέραν εἶναι νομίζουσιν. ἔστι γὰρ ἡ μὲν τῶν παίδων ἀτμωδεστέρα τε καὶ πολλὴ καὶ ἡδεῖα τοῖς ἁπτομένοις, ἡ δὲ τῶν ἀκμαζόντων ὑπόδριμύ τι καὶ οὐχ ἡδὺ κέκτηται. τοῦτ' οὖν τὸ διάφορον τῆς προσβολῆς ἀναπείθει τοὺς πλείους ἀποφαίνεσθαι θερμότερον εἶναι τὸ τῶν ἀκμαζόντων σῶμα. τὸ δ' οὐχ οὕτως ἔχει. τῷ γὰρ ἀσκήσαντι τὴν

ON TEMPERAMENTS, BOOK II

three or four years, but will preserve your ignorance throughout your life. For, to the extent that it involves opposing arguments among men, nothing can be clearly demonstrated; it is not reasonable in general for prior things to be believed on the basis of those that are later.

Therefore, let us judge by perception the hot and cold body as it already is in actuality and no longer in potentiality, bypassing in respect of the primary judgment all the other signs. And further, I leave you to judge well from experience, while I explain my own judgment. Since I have carefully touched many bodies in succession, and not only those of children or infants but also of adolescents and people in their prime, I discovered neither of the parties was right—neither those who said that the person in his prime is absolutely hotter than the child, nor those who said he is absolutely colder. If you take away all the other external changes, and consider the differences from age alone, it would seem to you that neither is hotter in the absolute sense. For certainly the heats among them differ in quality as a result of unequal transpiration. This is also why they deceive either those around them or themselves—those who think the heat of the child is stronger or those who think the heat of the young man is stronger. For that of children is more vaporous and large in amount and more pleasant to those touching them, while of those in their prime, it is somewhat pungent and has not acquired any sweetness. This difference, then, of the sensory impact persuades many to declare the body of those in their prime to be hotter. But this is not the case. To one

---

[13] post παρέντες τήν: γε πρώτην τὰ H; τε κρίσιν πρώτην, τά τ' K

ἁφὴν ἐν διαφόροις ὕλαις διαγνωστικὴν εἶναι θερμότητος ἰσχυροτέρας τε καὶ ἀσθενεστέρας καὶ ἴσης εὖ οἶδ' ὅτι καὶ ἡ τῶν παίδων[14] ἴση γε φανεῖται κατὰ τὴν ἰσχὺν τῇ τῶν ἀκμαζόντων ἢ πλείων.

Ἡ δ' ἄσκησις ἥδε· χρὴ γὰρ ἀπὸ τῶν ἐναργεστάτων ἄρξασθαι. τῶν βαλανείων ἐνίοτε θερμὸς οὕτως ἐστὶν ὁ ἀήρ, ὡς μηδένα φέρειν αὐτὸν ἀλλὰ καίεσθαι δοκεῖν, ἐνίοτε δ' οὕτω ψυχρός, ὡς ἱδροῦν μὴ δύνασθαι. καὶ μὴν καὶ ὅτι τρίτη τις ἄλλη παρὰ τάσδε κατάστασίς ἐστιν, ἧς μάλιστα χρῄζομεν, ἡ εὔκρατος, οὐδὲν δέομαι λέγειν. αἱ δ' αὐταὶ τρεῖς καταστάσεις ἐν τῷ τῆς κολυμβήθρας ὕδατι φαίνονται. καὶ γὰρ θερμὸν οὕτως, ὡς καίεσθαι πρὸς αὐτοῦ, καὶ ψυχρόν, ὡς μηδὲ θερμαίνεσθαι, καὶ εὔκρατον, ὡς συμμέτρως θερμαίνεσθαι, πολλάκις εὑρίσκεται. εἰ τοίνυν ἐροίμην σε, πότερόν ἐστι θερμότερον, ἆρά γε τὸ ὕδωρ τὸ εὔκρατον ἢ ὁ ἀὴρ ὁ εὔκρατος, οὐκ ἂν ἔχοις εἰπεῖν οὐδέτερον. ἀμφοῖν γὰρ ὄντων ὁμοίως ἡδέων τε καὶ συμμέτρων τῷ σώματι τὸ μὲν θερμότερον εἶναι[15] λέγειν αὐτῶν, τὸ δὲ ψυχρότερον οὐδένα νοῦν ἔχειν ἡγοῦμαι. καὶ μὴν εἰ νοήσαις τὸ τῆς δεξαμενῆς ὕδωρ εἰς ἄκρον θερμότητος ἀφικνούμενον, ὡς ζεῖν, ἢ τὸν ἀέρα τελέως ἐκφλογούμενον, ὅτι πρὸς ἀμφοῖν ὡσαύτως καυθήσῃ, πρόδηλον. εἰ δὲ δὴ καὶ νοήσαις αὖθις ἢ τὸ ὕδωρ οὕτω ψυχρόν, ὡς ἐγγὺς ἤδη πήξεως ἥκειν,[16] ἢ τὸν ἀέρα τελέως ἐψυγμένον, ὡς ἐν τοῖς νιφετοῖς γίγνεται, δῆλον, ὡς καὶ πρὸς τούτων ἑκατέρων ὁμοίως ψυχθήσῃ τε καὶ ῥιγώ-

who has trained his touch to recognize hotness in different materials—whether it is stronger, weaker or equal—knows full well that the hotness of children will seem to him either equal in strength to that of those in their prime, or greater.

The training is as follows: It is necessary to start from those things that are very clear. Sometimes the air of bathhouses is so hot no one can bear it, but seems to be burning up, while sometimes it is so cold that no one is able to sweat. Further also, I don't need to say there is a third and other state besides these which we shall require particularly; this is the *eukratic*. These same three states are apparent in the water of the swimming pool. For it will often be found to be hot in such a way as to burn up the person and cold in such a way as not to heat, and *eukratic* in such a way as to heat moderately. Accordingly, if I were to ask you which is hotter—water that is *eukratic* or air that is *eukratic*—you would not to be able to say either one of these is hotter, for both are similarly very pleasant and moderate to the body, while I think no one would have it in mind to say one of these is the colder. And indeed, if you think the water of the tank comes to a peak of hotness, so as to boil, or the air is completely scorching, it is quite clear that you will be burned similarly by both. And certainly also, if you think conversely that the water is cold in such a way as to already come close to freezing, or the air is completely cooled, so that it becomes as in snowstorms, it is clear also that you will freeze and shiver similarly with

---

¹⁴ *post* τῶν παίδων: ἴση γε φανεῖται κατὰ H; αὐτῷ γ' ἴση φανεῖται K   ¹⁵ εἶναι *add.* H   ¹⁶ ἥκειν H; εἶναι K

σεις. οὐκοῦν καὶ θερμότητα καὶ ψῦξιν ἄκραν ὡσαύτως μὲν ἀέρι νοήσεις ἐγγιγνομένην, ὡς δ' αὔτως ὕδατι, καὶ τῶν ἄκρων ἑκατέρων τὸ μέσον ὁμοίως ἀμφοῖν ἐγγιγνόμενον. ὥστε καὶ τὸ μεταξὺ πάντων τῶν ἄκρων τε καὶ τοῦ μέσου κατά τε τὸ ὕδωρ καὶ τὸν ἀέρα τὰς αὐτὰς ὑπεροχάς τε καὶ διαστάσεις ἕξει, καὶ τοσούτῳ ποτὲ φήσεις εἶναι τοῦ μετρίου θερμότερον θάτερον, ὅσῳ θάτερον. οὕτω δὲ καὶ ψυχρότερον τοῦ μετρίου τοσούτῳ φήσεις εἶναί ποτε τὸ ὕδωρ, ὅσῳ καὶ τὸν ἀέρα, καίτοι τό γε τῆς προσβολῆς ἴδιον οὐ ταὐτὸν ἑκατέροις ἦν. οὐ γὰρ ὡσαύτως ὕδωρ εὔκρατον, ὡς ἀὴρ εὔκρατος προσπίπτει. καὶ τί δεῖ λέγειν ἐπὶ τῶν οὕτως ἀνομοίων; αὐτοῦ γὰρ τοῦ ἀέρος ὁμοίως ὄντος θερμοῦ διαφέρουσαι γίγνονται προσβολαὶ παρὰ τὸ ποτὲ μὲν οἷον ἀχλυώδη τε καὶ ἀτμώδη, ποτὲ δ' οἷον λιγνυώδη τε καὶ καπνώδη, ποτὲ δὲ καθαρὸν ἀκριβῶς ὑπάρχειν. ἐν πολλαῖς οὖν καὶ διαφερούσαις οὐσίαις ἰσότης γίγνεται | θερμότητος ἐξαπατῶσα τοὺς ἀσκέπτους ὡς ἄνισος, ὅτι γε μὴ κατὰ πᾶν ὁμοία φαίνεται. λελογισμένου μήν ἐστιν ἀνδρὸς τοὺς λογισμοὺς οὓς εἴρηκα καὶ γεγυμνασμένου τὴν αἴσθησιν ἐν πολλῇ τῇ τῶν κατὰ μέρος ἐμπειρίᾳ τὴν ἰσότητα τῆς θερμότητος ἐξευρεῖν ἔν τε τοῖς παισὶ καὶ τοῖς ἀκμάζουσι καὶ μὴ τῷ τὴν μὲν ἐφ' ὑγρᾶς οὐσίας φαίνεσθαι τὴν δ' ἐπὶ ξηρᾶς ἐξαπατᾶσθαι. καὶ γὰρ λίθος ὕδατι δύναταί ποτε τὴν ἴσην δέξασθαι θερμασίαν, οὐδὲν διαφέρον, εἰ ξηρὸς μὲν ὁ λίθος ἐστίν, ὑγρὸν δὲ τὸ ὕδωρ.

Οὕτως οὖν ἔμοιγε μυριάκις ἐπισκεψαμένῳ καὶ παι-

both of these. Accordingly, you should think that extreme heating and extreme cooling have occurred similarly in the case of air and water, and that the median of each of the extremes has occurred similarly in both. Consequently also, what is between all the peaks and the median in relation to both water and air, will have the same excesses and the same intervals, and sometimes you will say one is hotter than the median to the same extent as the other. In the same way too, you will sometimes say water is colder than the median to the same extent as the air. But of course, the specific characteristic of the sensory impact is not the same for each, for *eukratic* water does not fall on [perception] in the same way as *eukratic* air does. And what do I need to say in the case of things dissimilar in this way? For with air itself, when it is similarly hot, different impacts occur depending on whether sometimes it is, as it were, misty and vaporous, or sometimes, as it were, sooty and smoky, or sometimes completely clear. Therefore, in many different substances, equal | heat arises, deceiving those who are unreflecting as to inequality, because it does not seem the same in every respect. It is for the man who has taken into account the reckonings which I have stated, and has trained his perception in much experience of these individually to discover the equality of the hotness in children and those in their prime, and not to be deceived by this appearing in a wet substance in one case and in a dry substance in another. For stone can sometimes receive equal heating to water, there being no difference, although the stone is dry while the water is wet.

In this way, then, I myself, by countless examinations

δας καὶ νεανίσκους πολλοὺς καὶ μειράκια καὶ τὸν αὐτὸν παῖδα καὶ βρέφος καὶ μειράκιον γενόμενον, οὐδὲν μᾶλλον ἐφάνη θερμότερος οὔτε παιδὸς ἀκμάζων οὔτ' ἀκμάζοντος παῖς. ἀλλ', ὡς εἴρηται, μόνον ἐν μὲν τοῖς παισὶν ἀτμωδεστέρα τε καὶ πολλὴ καὶ ἡδεῖα, ἐν δὲ τοῖς ἀκμάζουσιν ὀλίγη καὶ ξηρὰ καὶ οὐχ ὁμοίως ἡδεῖα τῆς θερμασίας ἡ προσβολή. πολὺ μὲν γὰρ τῆς τῶν παίδων οὐσίας ὑγρᾶς οὔσης ἐκτὸς ἀπορρεῖ, βραχὺ δὲ τῆς τῶν ἀκμαζόντων ξηρᾶς ὑπαρχούσης. οὐδέτερος οὖν αὐτῶν ἁπλῶς φαίνεται θερμότερος, ἀλλ' ὁ μὲν τῷ πλήθει τῆς διαπνοῆς, ὁ δὲ τῇ δριμύτητι· τὸ γὰρ ἔμφυτον θερμὸν ὁ παῖς ἔχει πλέον, εἴ γ' ἐξ αἵματός τε καὶ σπέρματος ἡ γένεσις αὐτῷ, καὶ ἥδιον, ἐν δὲ τοῖς ἀκμάζουσιν ὀλίγη καὶ ξηρὰ καὶ οὐχ ὁμοίως ἡδεῖα τῆς θερμασίας ἡ προσβολή.

3. Θερμοῦ μὲν δὴ καὶ ψυχροῦ σώματος ἁφὴ μόνη γνώμων ἐστίν, ὑγροῦ δὲ καὶ ξηροῦ σὺν τῇ ἁφῇ καὶ λογισμός. εἰ μὲν γὰρ ξηρόν, πάντως καὶ σκληρόν. ἀλλὰ τοῦτο μὲν ἁφῇ αἰσθητόν· οὐ μὴν εἴ τι σκληρόν, εὐθὺς ἤδη καὶ ξηρόν. ἀχώριστος μὲν γάρ ἐστι ξηροῦ σώματος ἡ σκληρότης, οὐ μὴν ἰδία γε τούτου μόνου. καὶ γὰρ τὸ πεπηγὸς ὑπὸ ψύξεως σκληρὸν ὥσπερ ὁ κρύσταλλος. ὅθεν οὐδ' εὐθὺς ἐπιχειρεῖν δεῖ τῇ τοῦ ὑγροῦ τε καὶ ξηροῦ διαγνώσει πρὶν ἐπισκέψασθαι, πῶς ἔχει ψυχρότητος ἢ θερμότητος. οὔτε γάρ, εἰ μετὰ ψυχρότητος ἄκρας σκληρόν, ἤδη τοῦτο καὶ ξηρόν, οὔτ' εἰ μετὰ θερμότητος σφοδρᾶς μαλακόν, εὐθέως ὑγρόν. ἀλλ' ὅταν μετρίως θερμὸν ᾖ, σκοπεῖν ἐφεξῆς,

of children and many youths and young lads, and the same child both as an infant and when he has become a young lad, said a person in his prime is no hotter than a child nor a child than a person in his prime. But, as was said, only in children is the impact of the heating more vaporous, large in amount, and sweet, whereas in those in their prime it is small in amount, dry, and not similarly sweet. For much of the substance of children, being wet, flows away externally, whereas a small amount of the substance of those | in their prime, being dry, does so. Therefore, neither one of these seems absolutely hotter, but the one is due to the amount of the transpiration and the other to the acridity. For in respect of the innate heat, the child has a large amount, as in fact the genesis of this is from blood and sperm, and is sweet, whereas in those in their prime the impact of the heating is small in amount, dry, and not similarly sweet.

3. Certainly, touch is the only judge of a hot and a cold body, while reason along with touch is the judge of a wet and a dry body, for if it is dry, it is also altogether hard. But this is what is perceived by touch. However, if something is also hard, it is not already immediately dry. For hardness is inseparable from a dry body, but is not specific to this alone. What has been made solid by cooling is hard, as ice is, from which one must not immediately attempt the diagnosis of wet and dry before considering how it possesses coldness or hotness. For hardness, if associated with extreme coldness, is not already also dry, | nor is softness, if associated with strong hotness, immediately wet. But when hot is moderate, consider next, whether it is

εἰ μαλακόν ἐστιν ἢ σκληρόν. εἰ μὲν γὰρ μαλακόν, ὑγρόν, εἰ δὲ σκληρόν, ξηρόν. ἀλλ' εἴπερ ταῦθ' οὕτως ἔχει, τῶν ἐν ἀνθρώπου σώματι μορίων[17] σκληρὸν οὐδὲν ἂν εἴη [ὑγρόν]. οὐ γὰρ ἐγχωρεῖ τοσαύτην ἐν αὐτῷ γενέσθαι ψῦξιν, ὡς σκληρυνθῆναί τι διὰ πῆξιν. εἰς μὲν γὰρ σύστασίν τινα τὸ τέως ῥυτὸν ἀφίξεταί ποτε, καθάπερ ἡ πιμελή. τὸ γὰρ ἐλαιῶδες ἐν αἵματι καὶ λιπαρόν, ῥυτὸν ὄν, ὅταν ἐν ψυχρῷ γένηται χωρίῳ, πήγνυται, σκληρὸν μὴν οὐδ' οὕτω γίγνεται.

Δεόντως οὖν εἴρηται τοῖς παλαιοῖς, ὑγρότατον μὲν ἡ πιμελή, δεύτερον δ' ἐπ' αὐτῇ τὸ σαρκῶδες γένος, εἴδη δ' αὐτοῦ πλείω· πρῶτον μὲν ἡ κυρίως ὀνομαζομένη σάρξ, ἣν οὐκ ἂν εὕροις καθ' ἑαυτὴν οὐδαμόθι τοῦ σώματος, ἀλλ' ἔστιν ἀεὶ μόριον μυός. ἐφεξῆς δ' ἑκάστου τῶν σπλάγχνων ἡ ἰδίως οὐσία. καλοῦσι δ' αὐτὴν οἱ περὶ τὸν Ἐρασίστρατον παρέγχυμα | καὶ ὡς περὶ μικροῦ καὶ φαύλου διανοοῦνται πράγματος οὐκ εἰδότες, ὡς ἡ καθ' ἕκαστον σπλάγχνον ἐνέργεια τῆς σαρκὸς ταύτης ἐστίν. ἀλλὰ τούτων μὲν οὔπω νῦν ὁ καιρός.

Ὅτι δ' αὐτὸ τὸ ἴδιον ἐγκεφάλου σῶμα καὶ πνεύμονος ἐφεξῆς ἐστι τῇ πιμελῇ καθ' ὑγρότητα, τῇ μαλακότητι πάρεστι τεκμήρασθαι. οὐ γὰρ δὴ ὑπὸ ψυχροῦ γε πέπηγεν, ὅτι μηδὲ θερμῷ χεῖται. πλησίον δὲ τούτων ἐστὶ καὶ ὁ μυελὸς τὴν φύσιν, οὐ μὴν ὁμογενὴς ὁ καθ' ἕκαστον ὀστοῦν μυελὸς ἐγκεφάλῳ τε καὶ νωτιαίῳ. ἀλλ' ἐγκέφαλος μὲν καὶ νωτιαῖος ἐκ ταὐτοῦ γέ-

soft or hard, for if it is soft, it is wet, while if it is hard, it is dry. But if this should be the case, none of the parts in a human body that is hard would be wet. It is not possible for such a degree of cooling to occur in this, as to be hardened through some congelation. Sometimes what is at one time fluid may come to this consistency, like fat, while what is oily and greasy in blood, being fluid, congeals whenever it occurs in a cold place, but it does not become hard in this way.

It was, then, properly said by the ancients that fat is very wet (the wettest), and second to this the fleshy class, of which there are many kinds. First is the flesh termed properly, which you would not find anywhere in the body as itself, but is always part of a muscle. Next is the specific substance of each of the internal organs. The followers of Erasistratus call this *parenchyma*,[13] and regard it as a small and trivial matter, not realizing that the function related to each internal organ is of this flesh. But now is not the time for these matters.

600K

That the actual specific body of brain and lung is next after fat in respect of wetness is evidenced by the softness present. Of course, it is not congealed by cold because it is not liquefied by heat. And near to these in nature is the marrow, although the marrow in each bone is not of the same class as brain and spinal cord. But brain and

---

[13] This term remains in use, defined currently as follows: "The distinguishing or specific cells of a gland or organ, contained in or supported by the connective tissue framework or stroma" (S).

---

[17] *post* μορίων: σκληρὸν οὐδὲν ἂν εἴη [ὑγρόν] H (ὑγρόν, quod *om*. M, delendum); σκληρῶν οὐδὲν ἂν εἴη ὑγρόν K

νους, οἱ δ' ἄλλοι σύμπαντες μυελοὶ φύσεως ἑτέρας εἰσίν. ὑγρότερος μέν ἐστι καὶ θερμότερος ἐγκέφαλος νωτιαίου καὶ διὰ τοῦτο καὶ μαλακώτερος· καὶ μέντοι καὶ αὐτοῦ τοῦ ἐγκεφάλου τὰ πρόσθεν ὑγρότερα τοσοῦτον, ὅσονπερ καὶ μαλακώτερα. πάντα μὴν ταῦτα δέρματος οὐχ ὑγρότερα μόνον, ἀλλὰ καὶ ψυχρότερα καὶ ὅλως ἄναιμον πᾶν ἐναίμου ψυχρότερον. ἐγγυτάτω δ' ἐστὶ δέρματος ἡ τῶν μαλακῶν, νεύρων φύσις· ἡ δὲ τῶν σκληρῶν, οἷόνπερ αὐτὸ τὸ δέρμα, καθ' | ὑγρότητα δηλονότι καὶ ξηρότητα· θερμότητι γὰρ ἀπολείπεται τοσοῦτον, ὅσον εἰκὸς ἀπολείπεσθαι τὸ παντελῶς ἄναιμον ἐναίμου σώματος.

Ἡ δὲ τοῦ σπληνὸς καὶ τῶν νεφρῶν καὶ τοῦ ἥπατος σὰρξ ὑγροτέρα μὲν τοσούτῳ δέρματος, ὅσῳ καὶ μαλακωτέρα· θερμοτέρα δ' ὅσῳ καὶ πολυαιμοτέρα. καὶ μὴν καὶ ἡ τῆς καρδίας σὰρξ ἁπάντων μὲν τούτων ξηροτέρα τοσοῦτον, ὅσονπερ καὶ σκληροτέρα· θερμοτέρα δ' οὐ τούτων μόνων, ἀλλὰ καὶ πάντων ἁπλῶς τῶν τοῦ σώματος μορίων. καί σοι καὶ τοῦτο σαφῶς ἔνεστιν αἰσθήσει μαθεῖν ἐν ταῖς τῶν ζῴων ἀνατομαῖς ταῖς κατὰ τὸ στέρνον γιγνομέναις εἰς τὴν ἀριστερὰν κοιλίαν τῆς καρδίας καθέντι τοὺς δακτύλους· εὑρήσεις γὰρ οὐκ ὀλίγῳ τινὶ τὸ χωρίον τοῦτο τῶν ἄλλων ἁπάντων θερμότερον. ἀλλ' ἡ μὲν τοῦ ἥπατός τε καὶ τοῦ σπληνὸς καὶ τῶν νεφρῶν καὶ τοῦ πνεύμονος σὰρξ ἁπλῆ τὴν φύσιν ἐστὶ ταῖς καθ' ἕκαστον σπλάγχνον ἀρτηρίαις καὶ φλεψὶ καὶ νεύροις περιπεφυκυῖα. τῆς καρδίας δ' οὐχ ἁπλοῦν τὸ τῆς σαρκὸς εἶδος, ἀλλ'

spinal cord are of the same class. However, all the other marrows are of a different nature. Brain is wetter and hotter than spinal cord, and because of this also softer. And indeed also, the previous things are to such a degree wetter than the brain itself to the extent that they are also softer. All these are not only wetter than skin but also colder, and in general, every bloodless (nonsanguineous) part is colder than a sanguineous part. The nature of nerves that are soft is closest to that of skin, while the nature of those that are hard is like the skin itself, in relation to | wetness, clearly, and dryness. For they are lacking in heat by as much as the nonsanguineous body would seem to completely fall short of a sanguineous one.

601K

The flesh of the spleen and of the kidneys and liver is wetter than that of skin by as much as it is also softer, but hotter by as much as it is also more full of blood. Furthermore, the flesh of the heart is drier than all these by as much as it is also harder, while it is not only hotter than these, but also absolutely hotter than all the parts of the body. And it is possible for you to learn this clearly by perception in the dissections of animals involving the chest, when you insert your fingers into the left chamber of the heart, for you will discover this region to be hotter than all the others to no small degree. But the flesh of the liver, spleen, kidneys, and lungs is simple in nature, having developed around the arteries, veins, and nerves (sinews) in relation to each internal organ. However, the kind of flesh of the heart is not simple, but there are kinds of fibers

οἱαίπερ αἱ ἐν τοῖς μυσὶν ἶνές εἰσιν, αἷς ἡ σὰρξ περιπέπηγε, τοιαῦται κἀν τῇ καρδίᾳ. πλὴν οὐ ταὐτὸν γένος τῶν ἰνῶν, ἀλλ' αἱ μὲν ἐν τοῖς μυσὶν νεύρων εἰσὶ καὶ συνδέσμων μόρια. τῆς καρδίας δὲ τὸ τῶν ἰνῶν γένος ἴδιον ὥσπερ καὶ τοῦ τῆς ἀρτηρίας τε καὶ τῆς φλεβὸς χιτῶνος ἐντέρων τε καὶ γαστρὸς καὶ μήτρας καὶ τῶν κύστεων ἑκατέρων. ἔστι γὰρ οὖν δὴ κἀν τούτοις ἅπασι τοῖς ὀργανικοῖς τὴν οἰκείαν σάρκα περιπεπηγυῖαν ἰδεῖν ταῖς ἰδίαις αὐτῶν ἰσίν.

Αὗται μὲν οὖν αἱ σάρκες θερμότεραι τοῦ δέρματος ὑπάρχουσιν, αἱ δ' ἶνες, αἱ μὲν ὀλίγῳ τινὶ μᾶλλον, αἱ δ' ἧττον, βραχὺ δέρματος ψυχρότεραί τ' εἰσὶ καὶ ξηρότεραι, τινὲς δ' ὅμοιαι κατὰ πᾶν εἰσι τῇ τοῦ δέρματος οὐσίᾳ. πάντες δ' ὑμένες ἤδη ξηρότεροι δέρματος, ὥσπερ γε καὶ αἱ περὶ τὸν ἐγκέφαλόν τε καὶ τὸν νωτιαῖον μήνιγγες· ὑμένες γὰρ δὴ καὶ αἵδε. καὶ μὲν δὴ καὶ σύνδεσμοι πάντες, εἰς ὅσον σκληρότεροι δέρματος, εἰς τοσοῦτον καὶ ξηρότεροι. καὶ οἱ τένοντες δέ, κἂν εἰ τῶν συνδέσμων εἰσὶ μαλακώτεροι, δέρματος γοῦν ἐναργῶς ἤδη σκληρότεροι. χόνδροι δὲ μετὰ τοὺς συνδέσμους εἰσὶ καί τι μέσον ἀμφοῖν σῶμα· καλοῦσι δ' αὐτὸ νευροχονδρώδη σύνδεσμον ἔνιοι τῶν ἀνατομικῶν. ἔστι δὲ τοῦτο σύνδεσμος σκληρὸς καὶ χονδρώδης. ὀστοῦν δὲ τὸ πάντων σκληρότατον,[18] ὧν καλύπτει τὸ δέρμα, καὶ τῶν ἐξεχόντων αὐτοῦ ξηρότατον μὲν ἡ θρίξ, ἐφεξῆς δὲ κέρας, εἶτ' ὄνυχές τε καὶ ὁπλαὶ καὶ πλῆκτρα καὶ ῥάμφη καὶ ὅσα τοιαῦτα καθ' ἕκαστον τῶν ἀλόγων ζῴων ἐστὶ μόρια.

in the muscles around which the flesh has congealed, such as are in the heart, except they are not the same class of fibers. Rather, those in the muscles are parts of nerves (sinews) and ligaments. The class of the fibers of the heart is specific, just as it also is in the wall of the artery and vein, and of the intestines, and of stomach, uterus, and each of the bladders (gall and urinary). There is then, in all these organs, the characteristic flesh seen to have congealed around their specific fibers.

Therefore, these same fleshes are hotter than the skin, whereas the fibers are slightly colder and drier than skin, some a little more so and some less. Some, however, are the same in every respect to the substance of the skin. All membranes are already drier than skin, just as in fact the meninges around the brain and spinal cord also are, for these are certainly membranes. And furthermore, all ligaments are drier to the degree they are harder than skin, while the tendons, even though they are softer than the ligaments, are in fact already clearly harder than skin. Cartilages are after the ligaments, and there is a body midway between both, which some anatomists call "neurocartilaginous ligament."[14] This is a hard and cartilaginous ligament. Bone, however, is the hardest of all those [structures] the skin conceals. Of those that project from the skin, hair is the driest, then next horns, then nails, hooves, spurs, and beaks, and such things as are parts in each of the nonrational animals.

---

[14] On this, see Galen, *De usu partium*, VI.19; May, *Galen on the Usefulness*, 1.326–27.

---

[18] σκληρότατον H; ξηρότατον K

Τῶν δὲ χυμῶν ὁ μὲν χρηστότατός τε καὶ οἰκειότατός ἐστι τὸ αἷμα. τούτου δ' οἷον ὑπόστασίς τις καὶ ἰλὺς ἡ μέλαινα χολή· ταῦτ' ἄρα καὶ ψυχροτέρα τ' ἐστὶ καὶ παχυτέρα τοῦ αἵματος· ἡ δέ γε ξανθὴ θερμοτέρα μακρῷ, ψυχρότατον δὲ καὶ ὑγρότατον ἁπάντων τῶν ἐν τῷ ζῴῳ τὸ φλέγμα. κριτήριον δὲ καὶ τῆς τούτου διαγνώσεως ἡ ἁφή, καθάπερ καὶ Ἱπποκράτης ἐν τῷ Περὶ φύσεως ἀνθρώπου πεποίηται.[19] ἀλλ' ὅτι μὲν ψυχρόν, ἡ ἁφὴ μόνη διαγιγνώσκει· τὸ δ' ὅτι καὶ ὑγρόν, ἁφή θ' ἅμα καὶ ὄψις καὶ λογισμός, ἁφή μὲν καὶ ὄψις, ὅτι τοιοῦτον ἑκατέρα φαίνεται, λογισμὸς δὲ διορισάμενος, ὡς οὐ πλήθει θερμότητος, ἀλλ' ὑγρότητι συμφύτῳ τοιοῦτον ἐγένετο. περὶ μὲν οὖν τῶν κατὰ τὸ σῶμα μορίων τε καὶ χυμῶν ὧδ' ἔχει.

4. Περὶ δὲ τῶν ἑπομένων ταῖς κράσεσιν ἐφεξῆς χρὴ διελθεῖν. ἕπεται μὲν οὖν καὶ τὰ προειρημένα, μᾶλλον δ' ἀχώριστα τελέως ἐστί, ξηρῷ μὲν σκληρότης, ὑγρῷ δὲ μαλακότης, ὅταν γε μετὰ χλιαρᾶς ᾖ θερμότητος. ἀλλὰ καὶ παχύτητες ἕξεως καὶ λεπτότητες ἕπονται κράσεσιν, οὐ ταῖς συμφύτοις μόνον, ἀλλὰ κἂν ἐξ ἔθους μακροῦ τις ἐπίκτητος γένηται. πολλοὺς γὰρ καὶ τῶν φύσει λεπτῶν ἐθεασάμην παχυνθέντας καὶ τῶν παχέων λεπτυνθέντας τοὺς μὲν ἀργίᾳ τε καὶ τῷ ἁβροδιαίτῳ τὴν ὅλην κρᾶσιν ὑπαλλάξαντας ἐπὶ τὸ ὑγρότε-

[19] *post* πεποίηται: λόγῳ K

[15] Added following Kühn.

ON TEMPERAMENTS, BOOK II

Of the humors, the most useful and most familiar is the blood. The black bile is a kind of sediment and dregs of this. This, then, is colder and thicker than the blood. The yellow bile is in fact much hotter, while the phlegm is the coldest and wettest of all the humors in the animal. Touch is judge of the recognition of this, just as Hippocrates has also established [by reason][15] in the work, *On the Nature of Man*.[16] But that it is cold, the touch alone recognizes, whereas that it is wet, touch along with sight and reason recognize; touch and sight because it seems so to each of them, while reason since it distinguishes that it is not by the abundance of hotness, but by innate wetness, that it has become like this. This is enough, then, about the parts and the humors in the body.

4. Next we must go over those things that follow the *krasias*. Thus, the things previously mentioned also follow, and particularly those that are completely inseparable. Hardness follows dryness while softness follows wetness, whenever in fact hotness is associated with warmth. But also thickness and thinness of bodily state follow *krasias*, and not only those that are innate, but also even those that occur as something acquired from long custom. For we have seen many of those who are thin in nature made thick, and also those who are thick made thin—the former when they change the whole *krasis* to wetter by idleness

---

[16] Hippocrates, *De natura hominis*: "Phlegm increases in a man in winter; for phlegm, being the coldest constituent of the body, is closest akin to winter. A proof that phlegm is very cold is that if you touch phlegm, bile and blood, you will find phlegm the coldest" (translation after W. H. S. Jones, *Hippocrates* IV, LCL 150, 18–19).

ρον, τοὺς δ' ἐν ταλαιπωρίαις πλείοσι καὶ φροντίσι καὶ διαίτῃ λεπτῇ καταξηρανθέντας. εἰρήσεται δὲ καὶ τούτων τὰ γνωρίσματα. κάλλιον γὰρ ἡμᾶς αὐτοὺς ἔκ τινων σημείων ὁρμωμένους, πρὶν παρ' ἑτέρου πυθέσθαι, δύνασθαι γνωρίζειν, εἰ φύσει τοιοῦτος ἦν ὁ ἄνθρωπος ἢ ἐξ ἔθους ἐγένετο. διδάσκαλος δὲ καὶ τούτων τῶν γνωρισμάτων, | ὥσπερ οὖν καὶ τῶν ἄλλων ἁπάντων, ὁ θαυμάσιος Ἱπποκράτης.

Ὅσοι μὲν οὖν εὐρυτέρας ἔχουσι τὰς φλέβας, θερμότεροι φύσει, ὅσοι δὲ στενοτέρας, ψυχρότεροι. τοῦ θερμοῦ γὰρ ἔργον ἀνευρῦναί τε καὶ διαφυσῆσαι ταύτας· ὥστ' εὐλόγως εἰς ταὐτὸν ὡς τὸ πολὺ συντρέχει στενότης μὲν φλεβῶν ἕξει πιμελώδει τε καὶ παχυτέρᾳ, λεπτότης δ' ἕξεως εὐρύτητι φλεβῶν. εἰ δ' ἅμα τις εἴη πιμελώδης τε καὶ παχὺς καὶ τὰς φλέβας εὐρείας ἔχοι, δι' ἔθος οὗτος, οὐ φύσει πιμελώδης ἐγένετο· ὥσπερ εἰ καὶ στενὰς μὲν ἔχοι τὰς φλέβας, εἴη δ' ἰσχνός, οὐδ' οὗτος ἐξ ἀνάγκης φύσει τοιοῦτος. κἂν τοῖσι λιμαγχικοῖσι,[20] φησί, τὰς μετριότητας ἀπὸ τούτων σκεπτέον, τῆς τῶν ἀγγείων εὐρύτητος δηλονότι καὶ στενότητος, οὐ τῆς ἄλλης ἕξεως ὅλου τοῦ σώματος. οἱ μὲν γὰρ στενὰς ἔχοντες τὰς φλέβας ὀλίγαιμοί τ' εἰσὶ καὶ μακρὰς ἀσιτίας οὐ φέρουσιν. ὅσοις δ' εὐρεῖαι καὶ πλῆθος αἵματος, τούτοις ἔνεστι καὶ χωρὶς βλάβης ἀσιτῆσαι. αἱ δ' αἰτίαι τῶν εἰρημένων ἤδη μὲν | δῆλαι, κἂν ἐγὼ μὴ λέγω, τοῖς γε προσέχουσι

[20] λιμαγχικοῖσι H; λιμαγχονικοῖσι K

and an over-delicate regimen, and the latter when they are dried up by many hardships and anxieties and by a thinning regimen. The signs of these will also be stated. For it is better if we ourselves, starting from certain signs, are able to recognize whether a person is such as he is by nature or has become so from custom, before learning this from someone else. And a teacher of these signs, as also of all the others, is the wondrous Hippocrates.

Thus, those who have more dilated (broader) veins are hotter by nature, whereas those who have more constricted (narrower) veins are colder. An action of the hot is to dilate and distend these (veins), so it is reasonable in regard to this that for the most part narrowness of veins goes along with a fatty and thicker bodily state (*hexis*), while thinness of bodily state goes along with broadness of veins. If, however, someone is at the same time fat and thick, and has veins that are broad, he has become fat due to custom not by nature, just as also, if someone were to have narrow veins, while being thin, he is not of necessity like this by nature. Even in those weakened by hunger, he says,[17] one must consider the moderate from these—that is, clearly of the broadness and narrowness of the vessels and not of the other state of the whole body. For those who have narrow veins are oligaemic and do not tolerate prolonged fasts, whereas in those with broad veins and abundant blood, it is possible for them to fast without harm. The causes of the things stated should already be clear, even if I do not say so, to those who are paying attention.

---

[17] Hippocrates, *Epidemics* 2.1.8, LCL 477 (W. D. Smith), 26–27.

τὸν νοῦν. ἐπεὶ δ' οὐ πάντες προσέχουσιν, ἀναγκαῖον[21] ἤδη καὶ δι' ἐκείνους εἰπεῖν, ὡς ὅσον ἐν αἵματι πῖόν τ' ἐστὶ καὶ κοῦφον καὶ λεπτόν, ἐν μὲν τοῖς θερμοτέροις σώμασι τροφή τις γίγνεται τοῦτο τῷ θερμῷ, κατὰ δὲ τὰ ψυχρότερα διασώζεται, καὶ τῶν φλεβῶν ἔξω διηθούμενον, ἐπειδὰν μὲν ψυχροῖς περιπέσῃ μορίοις, οἷοίπερ οἱ ὑμένες εἰσίν, ἐκείνοις περιπήγνυται, κατὰ δὲ τὰ φύσει θερμότερα[22]—τοιαῦτα δ' ἐστὶ δηλονότι τὰ σαρκώδη—δαπανᾶταί τε πρὸς τοῦ θερμοῦ καὶ διαφορεῖται, πλὴν εἴ ποτε πρὸς τῷ ψυχροτέρῳ τῆς κράσεως ἔτι καὶ τὸ τῆς διαίτης ἀταλαίπωρον ἐπιθρέψειέ τι καὶ αὐτοῖς τοῖς σαρκώδεσι μορίοις πιμελῆς. οὕτω τοι καὶ τὰ φωλεύοντα ζῷα πολλάκις εὑρίσκεται πιμελωδέστερα καὶ γυναῖκες ἀνδρῶν, ὅτι καὶ τῇ κράσει ψυχρότερον ἄρρενος τὸ θῆλυ καὶ οἰκουρεῖ τὰ πολλά.

Ὅσαι μὲν οὖν ἕξεις σωμάτων εὔκρατοί τ' εἰσὶ φύσει καὶ πονοῦσι τὰ μέτρια, ταύτας ἀναγκαῖον εὐσάρκους γίγνεσθαι, | τουτέστι πάντῃ συμμέτρους· ὅσοις δὲ τὸ μὲν ὑγρὸν αὔταρκες, ἀπολείπεται δ' οὐ πολλῷ τῆς ἄκρας συμμετρίας τὸ θερμόν, οὗτοι πολύσαρκοι γίγνονται. πολύσαρκοι δὲ καὶ ὅσοι φύσει μὲν εὔκρατοι, ῥᾳθύμως δὲ καὶ ἀπόνως βιοῦσιν. εἴρηται γὰρ δὴ καὶ τοῦτο κάλλιστα πρὸς τῶν παλαιῶν, ὡς ἐπίκτητοι φύσεις εἰσὶ τὰ ἔθη, καὶ οὐδὲν ἴσως δεήσει[23] τοῦθ' ἅπαξ εἰρηκότας νῦν μηκέτι διορίζεσθαι καθ' ἕκαστον κεφάλαιον, εἴτε φύσει ψυχρότερος εἴτ' ἐξ ἔθους ὅδε τις,[24] ἀλλὰ τοῦτο μὲν ἀπολιπεῖν τοῖς ἀναγιγνώσκου-

ON TEMPERAMENTS, BOOK II

Since, however, not everyone is paying attention, it is perhaps necessary now to also say something for them: that as much as is rich, light, and thin in blood becomes nourishment for the heat in the hotter bodies, while in those that are colder, it is preserved, and is filtered through the veins outwardly, then falling around cold parts, as the membranes are, congeals around those, whereas in the parts that are hotter by nature (the fleshy parts are clearly such), it is consumed by the heat and dispersed, unless sometimes in the colder *krasis* and the laxity of the regimen, it nourishes fat in the fleshy parts themselves. Certainly, in this way too, hibernating animals are often found to be more fatty, as are women compared to men, because the female is colder in *krasis* than the male, and for the most part stays at home.

Therefore, those bodily states that are *eukratic* by nature and labor in due measure, become of necessity well-fleshed; that is to say, are well-balanced in every way. Those where the wetness is sufficient while the hotness is lacking, but not by much, of the peak balance, become very fleshy. The very fleshy and the *eukratic* in nature live easily and idly. For certainly this was said best by the ancients—that the customs are acquired natures—and perhaps, having said this now once and for all, it will require no further distinction under each heading, whether this particular person has become colder due to nature, or due to habit, but to leave this to the readers, while for the sake

---

[21] post ἀναγκαῖον: ἤδη H; ἴσως ἐστί τι K
[22] post θερμότερα: add. μόρια K   [23] δεήσει H; διοίσει K
[24] post τις: add. ἐγένετο K

σιν, αὐτὸν δὲ βραχυλογίας ἕνεκα τὰς οἰκείας ἑκάστῃ τῶν κράσεων ἕξεις τοῦ σώματος ἐπελθεῖν. εἰσὶ δή τινες ἰσχνοί θ' ἅμα καὶ φλέβας ἔχοντες μικράς, ἀλλ' εἰ τέμοις ἐξ αὐτῶν ἡντινοῦν, προπίπτει πιμελή, δῆλον ὡς ὑποπεφυκυῖα τῷ δέρματι κατὰ τὸν ἔνδον ὑμένα. σπάνιον μὲν οὖν ἐπ' ἀνδρῶν τὸ τοιοῦτον, ἐπὶ δὲ γυναικῶν καὶ πάνυ πολλάκις εὑρισκόμενον. ἐστὶ γὰρ καὶ φύσεως ψυχροτέρας καὶ ἀργοτέρου βίου τὸ τοιοῦτον γνώρισμα. πιμελὴ μὲν γὰρ ἀεὶ διὰ ψῦξιν ἕξεως γίγνεται· πολυσαρκία δὲ πλήθους | αἵματος ἔκγονος, εὐσαρκία δὲ φύσεως εὐκράτου γνώρισμα. πάντως μὲν οὖν οἱ πολύσαρκοι καὶ τὴν πιμελὴν εὐθέως ἔχουσι πλείονα τῶν εὐσάρκων.[25] οὐ μὴν ἀνάλογον ἀεὶ ταῖς σαρξὶν ἡ πιμελὴ συναύξεται, ἔστι γὰρ ἰδεῖν τῶν παχέων τοὺς μὲν τὴν σάρκα πλείονα, τοὺς δὲ τὴν πιμελὴν ἔχοντας, ἐνίοις δ' ὁμοίως ἄμφω συνηυξημένα. οἷς μὲν οὖν ὁμοίως δ, ἄμφω συνηύξηται, τοσούτῳ πλέον ὑγρόν ἐστιν ὑπὲρ τὴν εὔκρατον φύσιν, ὅσῳ καὶ ψυχρόν. οἷς δ' ἡ πιμελὴ πλείων, τὸ ψυχρὸν ἐπὶ τούτων πλέον ἐστὶν ἤπερ τὸ ὑγρόν, ὥσπερ οἷς ἡ σὰρξ πλείων, ὑγρότης μέν ἐστι τοῦ δέοντος πλείων, οὐ μὴν καὶ ψῦξις. ὅταν γὰρ ἐν τοῖς οἰκείοις ὅροις μένοντος τοῦ θερμοῦ προσγένηταί τις αἵματος χρηστοῦ περιουσία,[26] πολυσαρκίαν ἀναγκαῖον ἀκολουθῆσαι.

Τὸ δ' ὅσῳ χρὴ πλέον εἶναι τοῦ συμμέτρου τὸ αἷμα, μέτρῳ μὲν οὐχ οἷόν τε μηνῦσαι καὶ σταθμῷ, λόγῳ δ' ἐγχωρεῖ διελθεῖν. ὡς ἐπειδὰν μήπω νοσῶδες μηδὲν

ON TEMPERAMENTS, BOOK II

of brevity, I myself recount the characteristic states of the body in each of the *krasias*. Certainly there are some who are thin and have at the same time small veins, but if you cut any one of these whatsoever, fat falls out which has clearly grown under the skin in relation to the internal membrane. Such a thing is found rarely, then, in men but very often in women, for as such it is a sign of both a colder nature and a more idle life. Fat always arises due to cooling of a bodily state, whereas much flesh (*polysarkia*) is the product of a large amount of blood; *eusarkia* is a sign of a *eukratic* nature. Altogether then, those who are very fleshy (*polysarkos*) straightaway have more fat than those who are well-fleshed (*eusarkos*). The fat does not always increase together in proportion to the flesh, for it is seen among those who are fat that some have a large amount of flesh while some have a large amount of fat. In those in whom both are increased together similarly, the wet is greater than the *eukratic* nature to the same extent as the cold. However, in those in whom the fat is greater, the cold is more or the wet is, just as in those in whom the flesh is more, the wetness is more than it should be but not also cooling. For whenever the hot remains within the proper limits, but some larger share of useful blood is added, of necessity excess flesh will be a consequence.

The amount by which it is necessary for the blood to be greater than the moderate cannot be made known and measured quantitatively, whereas it is possible to go over it by reason. So when nothing morbid has yet arisen in the

---

²⁵ εὐσάρκων H; εὐκράτων K
²⁶ περιουσία H; πλεονεξία K

γένηται τῷ τοῦ ζῴου σώματι σύμπτωμα παχυνομένῳ, τὸ πλῆθος τῆς ὑγρότητος τηνικαῦτα τῶν τῆς ὑγιείας ὅρων ἐντός ἐστιν. | ἐπιδέδεικται γὰρ ἡμῖν καὶ δι᾽ ἄλλων, ὡς ἀναγκαῖόν ἐστιν οὐ σμικρὸν ὑποθέσθαι πλάτος τῆς ὑγιεινῆς καταστάσεως· ἀλλὰ καὶ νῦν φαίνεται σχεδὸν ἐν ὅλῳ τῷ λόγῳ τὴν μὲν εὔκρατόν τε καὶ μέσην φύσιν οἷον κανόνα τινὰ τῶν ἄλλων ἀεὶ τιθεμένων ἡμῶν,[27] ὅσαι δ᾽ ἐφ᾽ ἑκάτερα τῆσδε, δυσκράτους ἀποφαινόντων· ὅπερ οὐκ ἂν ἦν, εἰ μὴ τὸ μᾶλλόν τε καὶ ἧττον ἡ ὑγιεινὴ κατάστασις ἐδέχετο. ἄλλη μὲν γάρ ἐστιν ἡ ὑγιεινή, ἄλλη δ᾽ ἡ νοσώδης δυσκρασία· νοσώδης μὲν ἡ ἐπὶ πλεῖστον ἀποκεχωρηκυῖα τῆς εὐκράτου, ὑγιεινὴ δ᾽ ἡ ἐπ᾽ ὀλίγον. ὁρίσαι δ᾽ οὐδ᾽ ἐνταῦθα μέτρῳ καὶ σταθμῷ τὸ ποσὸν ἐγχωρεῖ, ἀλλ᾽ ἱκανὸν γνώρισμα τῆς ὑγιεινῆς δυσκρασίας τὸ μηδέπω μηδεμίαν ἐνέργειαν τοῦ ζῴου βεβλάφθαι σαφῶς. ὅσον δ᾽ οὖν μεταξὺ τοῦ τ᾽ ἄκρως ἐνεργεῖν καὶ τοῦ βεβλάφθαι σαφῶς ἐνέργειαν ὑπάρχει, τοσοῦτον καὶ τῆς ὑγιείας τὸ πλάτος ἐστὶ καὶ τῆς κατ᾽ αὐτὴν δυσκρασίας. τούτῳ δ᾽ ἐφεξῆς ἐστιν ἡ νοσώδης δυσκρασία, ὅταν γε διὰ δυσκρασίαν νοσῇ | τὸ ζῷον· οὐ γὰρ δὴ διὰ ταύτην γε μόνην ἀλλὰ καὶ κατ᾽ ἄλλας διαθέσεις οὐκ ὀλίγας, ὑπὲρ ὧν ἐν τοῖς Περὶ τῆς τῶν νοσημάτων διαφορᾶς λογισμοῖς ἐπὶ πλέον εἰρήσεται.

Νυνὶ δὲ πάλιν ἀναληπτέον τὸν ἐξ ἀρχῆς λόγον. ὡς γὰρ τοῦ συμφύτου θερμοῦ τὴν ἀρίστην εὐκρασίαν

---

[27] τιθεμένων ἡμῶν H; τιθεμένοις K

ON TEMPERAMENTS, BOOK II

body of the animal that is being thickened, the quantity of the wetness is, under these circumstances, within the limits of health. | For we have shown through other things[18] that it is necessary for no small latitude to be assumed for the healthy condition, but it is now also apparent in almost the whole discussion, since we always set up the *eukratic* and median nature as a kind of standard for the others, that those on either side of this are called *dyskratic*. This would not be so unless the healthy condition admitted of more and less, for on one side there is the healthy *dyskrasia* and on the other the morbid *dyskrasia*; that which deviates to the greatest degree from the *eukratic* is morbid, while that which deviates to a slight degree is healthy. Here, however, it is not possible to define the amount by measure and rule, but it is a sufficient sign of the healthy *dyskrasia* for no function of the animal to have yet been clearly damaged. Therefore as much as exists between functioning perfectly and function clearly having been harmed is the extent of the latitude of the healthy and of the *dyskrasia* in relation to this. Next in order to this is the morbid *dyskrasia*, whenever, due to a *dyskrasia*, | the organism is diseased. For certainly it is not due to this alone but also in relation to quite a number of other conditions, which will be spoken about to a greater extent in the considerations in *On the Differentiae of Diseases.*[19]

Now, however, we must again take up the argument from the beginning. For just as, when the innate heat

---

[18] It is not specified what these other things are, but see Galen's *Hygiene*, VI.11–21K.   [19] *De morborum differentiis*, VI.836–80; see particularly sections iv.2 and v.2–5, Johnston, *On Diseases and Symptoms*, 138, 141–43.

φυλάττοντος αὐξηθὲν τὸ ὑγρὸν ἐν ὅροις ὑγιεινοῖς οὐ πιμελώδη τὸν ἄνθρωπον, ἀλλὰ πολύσαρκον ἀποδείκνυσι συναυξανομένης μὲν ἐπ᾽ ὀλίγον καὶ τῆς πιμελῆς, ἀλλὰ πολλῷ πλείονι μέτρῳ τῆς σαρκός, οὕτως αὖ πάλιν, εἰ καὶ τὸ ὑγρόν τε καὶ ξηρὸν ἀκριβῆ τὴν πρὸς ἄλληλα φυλάττοι συμμετρίαν, ἧττον δ᾽ εἴη θερμὸς ὁ ἄνθρωπος, ἀνάγκη πιμελῶδες μᾶλλον ἢ πολύσαρκον γενέσθαι τούτῳ τὸ σῶμα. εἰ δ᾽ αὖ πάλιν αὐξηθείη τὸ θερμόν, ἐν συμμετρίᾳ μενούσης τῆς ἑτέρας ἀντιθέσεως, πλέον ἀπολείπεται τούτῳ τὸ σῶμα[28] πιμελῆς ἢ σαρκός· ὥσπερ εἰ καὶ κρατῆσαι ποτὲ τὸ ξηρόν, ἐν συμμετρίᾳ μενούσης τῆς ἑτέρας ἀντιθέσεως, ἰσχνότερόν θ᾽ ἅμα καὶ σκληρότερον ἔσται τὸ σῶμα. ταῦτ᾽ εἴρηταί μοι καὶ ἤδη δῆλον, ὡς οὐ λόγῳ μόνον ἐδείχθησαν αἱ ἁπλαῖ δυσκρασίαι τοῖς τῶν ζῴων ὑπάρχουσαι σώμασιν, ἀλλὰ καὶ τὰ γνωρίσματα σαφῆ πασῶν ἐστιν οὐκ ἐν θερμότητι καὶ ψυχρότητι καὶ μαλακότητι καὶ σκληρότητι μόνον, ἀλλὰ κἀν ταῖς ἄλλαις ἁπάσαις τῆς ἕξεως ὅλου τοῦ σώματος διαφοραῖς, ὧν ὑπὲρ μὲν τῆς κατὰ λεπτότητά τε καὶ πάχος εἴρηται νῦν, ὑπὲρ δὲ τῶν ἄλλων ἤδη λεγέσθω.

5. Δασεῖα μὲν ἡ θερμὴ καὶ ξηρὰ κρᾶσίς ἐστιν, ἀλλ᾽ αὕτη μὲν ἐσχάτως· μετρίως δ᾽ ἡ θερμὴ μέν, σύμμετρος δὲ κατὰ τὴν ἑτέραν ἀντίθεσιν, ὥσπερ γε καὶ ἡ ξηρὰ μέν, εὔκρατος δὲ κατὰ τὸ θερμόν τε καὶ ψυχρόν· ἔστι γὰρ καὶ ἥδε μετρίως δασεῖα. ψιλαὶ δὲ τριχῶν αἱ ψυχραὶ πᾶσαι κράσεις, εἴτ᾽ οὖν ἀμέτρως ἔχοιεν ὑγρότητος εἴτε μετρίως. ἀλλ᾽ ἐσχάτως μὲν ἄτριχος ἡ

preserves the best *eukrasia*, the wet is increased within healthy limits, and the person does not exhibit fatness but is excessively fleshy (*polysarkos*), the fat having increased quite a lot but the flesh by a very much greater measure. Conversely, in the same way, if the wet and dry are preserved in a precise balance with each other, the person would be less hot, and of necessity the body would become more fatty than excessively fleshy in this. If, on the other hand, the hot is increased, while the other antithesis remains in balance, in this the body is wanting more in fat than in flesh. This is also so sometimes if the dry predominates, while the other antithesis remains in balance, the body will be thinner and at the same time harder. I have already said these things clearly, that not by reason alone the simple *dyskrasias* were shown to exist in the bodies of the animals, but also the signs of all these are clear, not in hotness and coldness, softness and hardness alone, but also in all the other differences of the state (*hexis*) of the whole body. Now that we have spoken about those which relate to thinness and thickness, let us forthwith speak about the others.

5. The hot and dry *krasis* is hirsute, but this is extreme. However, if it is hot but well-balanced in the other antithesis, there is moderate hirsutism, just as, in fact, if it is dry but *eukratic* in relation to hot and cold, there is also moderate hirsutism. All the cold *krasias* are bare of hair, whether wetness is either immoderate or moderate. But

---

[28] *post* ἀπολείπεται; τούτῳ τὸ σῶμα H; τούτου τὸ ζῷον K

ψυχρὰ καὶ ὑγρὰ κρᾶσις [ἐστίν], ἔλαττον δὲ ταύτης ἡ ψυχρά θ' ἅμα καὶ κατὰ τὴν ἑτέραν ἀντίθεσιν εὔκρατος, ἔτι δ' ἔλαττον ἡ ψυχρά θ' ἅμα καὶ ξηρά. καίτοι δόξει τις, ὡς ἐν γῇ ξηρᾷ ταῖς πόαις ἀδύνατόν ἐστι καὶ φῦναι καὶ τραφῆναι καὶ αὐξηθῆναι, κατὰ τὸν αὐτὸν λόγον κἂν τῷ δέρματι ταῖς θριξίν. ἔχει δ' οὐχ οὕτως· γῆ μὲν γὰρ ὡς γῆ ξηρὰ λέγεται, δέρμα δ' ὡς δέρμα· καὶ τοίνυν τὸ μὲν ἐν γῇ ξηρὸν ἄνικμον ἐσχάτως ἐστί, τὸ δ' ἐν ἀνθρώπου σώματι καὶ τῶν ὁμοίων ἀνθρώπῳ ζῴων οὔτ' ἄνικμον ἐπιτήδειόν τε καὶ μάλιστα πάντων εἰς γένεσιν τριχῶν. ἐκ μὲν γὰρ τῶν ὀστρακοδέρμων τε καὶ μαλακοστράκων, οἷον ὀστρέων καὶ καράβων καὶ καρκίνων, ὅσα τε φολιδωτὰ τῶν ζῴων ἐστίν, ὥσπερ οἱ ὄφεις, ἢ λεπιδωτά, καθάπερ οἱ ἰχθύες, οὐκ ἂν δύναιτο φύεσθαι θρίξ. ὄντως γάρ ἐστι τὰ τούτων δέρματα τελέως ξηρὰ δίκην ὀστράκου τινὸς ἢ πέτρας. ἐκ μέντοι τῶν μαλακοδέρμων, οἷόνπερ καὶ ὁ ἄνθρωπός ἐστιν, ὅσῳπερ ἂν ξηρότερόν τε καὶ θερμότερον ᾖ τὸ δέρμα, τοσούτῳ μᾶλλον ἐγχωρεῖ φύεσθαι τρίχας. ἵνα γάρ, ὡς ἐκεῖνοι προκαλοῦνται, τῷ τῆς γῆς ἑπώμεθα παραδείγματι, τὰς πόας οὔτ' ἐν ξηρᾷ πάνυ καὶ αὐχμώδει[29] φύεσθαι δυνατὸν οὔτ' ἐν ὑγρᾷ καὶ τελματώδει, ἀλλ' ἐπειδὰν μὲν ἄρχηται δαπανᾶσθαι τὸ περιττὸν τῆς ὑγρότητος, ἐκφύονται τῆς γῆς· αὐξάνονται δ' ἐπὶ πλέον, ὅταν καὶ ἥδε ξηραίνηται, μετρίως μὲν ἐν τῷ ἦρι, τάχιστα δὲ καὶ μέχρι πλείστου κατὰ τὴν ἀρχὴν τοῦ θέρους, ἀποξηραίνονται δὲ τελέως αὐανθείσης τῆς γῆς ἐν μέσῳ τῷ θέρει. καί σοι πάρε-

ON TEMPERAMENTS, BOOK II

the cold and wet *krasis* is extremely hairless. Less than this is that which is cold and at the same time *eukratic* in the other antithesis. Still less is the cold and dry. And indeed, someone will think that, as in dry earth, it is impossible for grasses to grow and be nourished and increased, the same argument applies | even for the hairs in the skin. However, this is not the case. For earth is said to be dry as earth while skin is said to be dry as skin. And accordingly what is dry in earth is without moisture to an extreme degree, whereas that in the body of a human, and of the animals similar to a human, is not without moisture and is the most suitable of all for the generation of hair. For in the case of hard-shelled and soft-shelled creatures, like oysters, crayfish, and crabs, and those animals that are scaly like snakes, or covered with scales as fish are, it would not be possible to grow hair. For truly the skins of these creatures are completely dry, like some earthen vessel or rock. Of course, in the case of those who are soft-skinned, as the human is, the drier and hotter the skin is, the more it is possible to grow hair. So then, as those men invite us to do, let us follow the example of the earth—grasses are unable to grow in either very dry and parched earth or in wet and marshy earth, but whenever the excess of wetness begins to be used up, the grasses grow out | from the earth. However, they grow still more whenever the earth is dried out—moderately in the spring and very quickly and to the greatest extent at the beginning of summer. They, on the other hand, are dried out when the earth is completely dried up in the middle of summer. And it is allowable for

612K

613K

---

²⁹ *post* αὐχμώδει: *add.* πάνυ γῇ K

στιν, εἰ βούλει, καὶ νῦν, ὥσπερ που κἀν τῷ πρὸ τούτου λόγῳ δέδεικται, τὸ μὲν ἔαρ, ὅτι τῶν ὡρῶν εὐκρατότατόν ἐστιν, εἰκάζειν εὐκράτου δέρματος φύσει καὶ μάλιστά γε τὰ μέσα τῆς ὥρας τῆσδε. τηνικαῦτα γὰρ οὖν καὶ ἡ γῆ μέση πως ὑγρότητός τε καὶ ξηρότητός ἐστιν. ὅσα δὲ τῷ θέρει συνάπτει τῆς ἠρινῆς ὥρας, ταῦτ᾽ ἤδη ξηροτέραν ἔχει τοῦ συμμέτρου τὴν γῆν, ἔτι δὲ μᾶλλον ἀρχομένου θέρους.

Ὃ τοίνυν λέγω δέρμα θερμὸν καὶ ξηρόν, εἰκάζοις ἂν μάλιστα τῇ τῆς γῆς διαθέσει τῇ γιγνομένῃ τελευτῶντος ἦρος ἢ ἀρχομένου θέρους. μεσοῦντος γὰρ θέρους ἄκρως ξηρὰ γίγνεται τοῖς τῶν ὀστρακοδέρμων[30] ὁμοίως, οὐ μὴν ἀνθρώπων γ᾽ ἢ συῶν ἢ ὄνων ἢ ἵππων ἢ ἄλλου του τῶν τριχωτῶν ζώων. ὥστ᾽, εἴπερ τῇ γῇ βούλονται παραβάλλειν τὸ δέρμα, καὶ κατὰ τοῦτο τὸν λόγον ὁμολογοῦντα τοῖς πρὸς ἡμῶν ἔμπροσθεν εἰρημένοις εὑρήσουσιν. αὐτοὶ δὲ σφᾶς αὐτοὺς ὑπὸ τῆς ὁμωνυμίας σοφισθέντες παραλογίζονται. ἐν γὰρ τῷ θερμῷ καὶ ξηρῷ δέρματι πολλὰς καὶ μεγάλας ἐλέγομεν φύεσθαι τρίχας, ὡς ὑπὲρ ἀνθρώπου δηλονότι ἢ ζῴου τριχωτοῦ τὸν λόγον, οὐχ ὑπὲρ ὀστρέων τε καὶ καρκίνων ποιούμενοι. διαπνεῖται μὲν γὰρ ἀεί τι καθ᾽ ἕκαστον δέρμα [ὑπὸ] τοῦ θερμοῦ συναπάγοντος ἑαυτῷ τῆς ἔνδοθεν ὑγρότητος οὐκ ὀλίγον. ἀλλ᾽ ἐν οἷς μὲν ὑγρόν ἐστι τὸ δέρμα καὶ ἀκριβῶς μαλακόν, οἷος ὁ νεωστὶ πηγνύμενος τυρός, οὐχ ὑπομένουσιν αἱ τῶν διεκπεσόντων ὁδοί, τῶν τέως διεστηκότων αὐτοῦ μορίων αὖθις ἀλλήλοις ἑνουμένων· ἐν οἷς

ON TEMPERAMENTS, BOOK II

you, if you so wish also now, as has been shown in the argument before this, to liken the spring, because it is the most *eukratic* of the seasons, to the nature of *eukratic* skin, and this is particularly so in the middle of this season. For at that time the earth is in some way at the midpoint of wetness and dryness. In that part of the spring season which connects to the summer the earth is already drier than the moderate, while it is still more so when the summer begins.

Accordingly, I say skin is hot and dry, especially if you should likeness it to the condition of the earth which occurs at the end of spring or the beginning of summer. In the middle of summer it becomes maximally dry, similar to the hard-shelled animals but not to humans, pigs, donkeys, horses, and other hairy animals. As a consequence, if they wish to compare the skin to the earth, they will also discover in this argument agreement with the things we said previously. However, these same people lead themselves astray, being deceived by the homonymy. When we said many and large hairs grow in the hot and dry skin, we made the argument obviously about a human or a hair-growing animal, not about oysters and crabs. For there is always some transpiration in each skin by the heat gathering to itself not a little of the wetness within. But in them the skin is wet and entirely soft, like recently coagulated cheese, and the pathways of the things escaping do not survive, since the parts of this which are separated for a time are united with each other once again. However, in

614K

---

30 *post* ὀστρακοδέρμων: *add.* ζώων K

δ' ἤδη σκληρὸν ὑπάρχει πεπηγότι παραπλήσιον τυρῷ, κατατιτρᾶται μὲν ὑπὸ τῆς ῥύμης τῶν ἐξιόντων, ἑνωθῆναι δ' ὑπὸ ξηρότητος οὐ δυνάμενον ὑπομένοντας ἴσχει τοὺς πόρους ἀεὶ καὶ μᾶλλον συριγγουμένους ταῖς συνεχέσι πληγαῖς τῶν διαρρεόντων. ἐὰν μὲν οὖν τὸ διαρρέον ἢ ἀτμὸς ἢ ὑγρὸν εἰλικρινὲς ᾖ, τῷ μὲν ἀτμῷ ταχεῖά τ' ἐστὶ καὶ ἀκώλυτος ἡ | φορά· τὸ δ' ὑγρὸν ἴσχεται πολλάκις ἐν τοῖς μικροτέροις πόροις καί τι καὶ παλινδρομεῖν αὖθις πρὸς τὸ βάθος ἀναγκάζεται. εἰ δ' οἷον αἰθαλώδης τε καὶ παχεῖα καὶ γεώδης ἡ ἀναθυμίασις εἴη, κίνδυνος αὐτῇ πολλάκις ἐν ταῖς στεναῖς τῶν διεξόδων σφηνωθείσῃ μήτ' εἴσω ῥᾳδίως ὑπονοστεῖν ἔτι μήτε κενοῦσθαι δύνασθαι. ταύτην οὖν ἑτέρα τοιαύτη πάλιν ἐκ τοῦ βάθους ἀναφερομένη πλήττει τε καὶ ὠθεῖ πρόσω καὶ ταύτην αὖθις ἑτέρα κἀκείνην ἄλλη καὶ πολλὰς αἰθαλώδεις οὕτω μοι νόει σφηνουμένας ἐπ' ἀλλήλαις ἀναθυμιάσεις ἐν τῷ χρόνῳ περιπλέκεσθαί τε καὶ συνάπτεσθαι καί τι ποιεῖν ἓν σῶμα τοιοῦτον, οἷον ἐκτὸς ἡ λιγνύς ἐστι, πλὴν ὅσῳ πεπύκνωται, τοσούτῳ καὶ ἀκριβῶς ἔσφιγκται τῇ τῆς διεξόδου στενότητι πιληθέν. ἐπειδὰν δὲ τὸν πόρον ἀποφράξῃ πάντα τὸ τοιοῦτον σῶμα, τοὐντεῦθεν ἤδη βιαίως πληττόμενον ὑπὸ τῶν ὁμοίων ἑαυτῷ περιττωμάτων οὐκ ἐχόντων διέξοδον ὠθεῖται πρόσω σύμπαν ἐν τῷδε, ὥστε καὶ προκύπτειν ἀναγκάζεται τοῦ δέρματος ἱμαντῶδες ἤδη γεγονός. | ἔοικε δ' αὐτοῦ τὸ μὲν ἐν τῷ πόρῳ σφηνωθὲν οἷον ῥίζῃ τινὶ πόας ἢ φυτοῦ, τὸ δ' ἐξέχον ἤδη τοῦ δέρματος οἷόνπερ αὐτὸ τὸ φυτόν.

those where there is already hardness similar to congealed cheese, it is opened up by the force of those things going out, while it is not able to be brought together due to dryness, the pores preserving themselves always unchanged and more pipe-like by the continuous blows of those things flowing out. Therefore, if what flows out is either vapor or pure fluid, | the passage for the vapors is rapid and unobstructed, while the fluid is often held back in the smaller pores, and some is forced to run back inward to the depths. If, however, the exhalation is, as it were, sooty, thick, and earthy, there is often a danger to it of being obstructed in those outflow channels that are narrow such that it is no longer easily able to go down inward or be evacuated. Therefore, another such exhalation carried up in turn from the depths strikes it and pushes it onward, and then another, and another after that. And in this way it seems to me many sooty exhalations wedged together with each other over time are interwoven and join together and are made into a sort of single body, like the thick soot that is external, except to the extent it has thickened, so it is entirely pressed together by the narrowness of the outflow channel and compressed. Whenever such a body blocks up the pore altogether, it is henceforth already forcibly propelled by the superfluities similar to itself, and not having an outflow channel, it is pushed forward as a whole in this, so when it is compelled to emerge through the skin, it has already become fibrous. | It seems that what of this is plugged up in the pore is like some root of a grass or plant, whereas what already projects from the skin is like the plant itself.

615K

616K

Ἡ θρὶξ δ' ἐστὶ μέλαινα μέν, ὅταν ὑπὸ ῥώμης τοῦ θερμοῦ συγκαυθείσης τῆς ἀναθυμιάσεως ἀκριβὴς λιγνὺς γένηται τὸ περίττωμα, ξανθὴ δ', ὅταν ἧττον κατοπτηθῇ. ξανθῆς γὰρ χολῆς τηνικαῦτ' ἐστὶν ἰλυῶδες περίττωμα τὸ σφηνωθέν, οὐ μελαίνης. ἡ δὲ λευκὴ θρὶξ ἔκγονος φλέγματος, ἡ πυρρὰ δ', ὥσπερ τῇ χρόᾳ μεταξὺ ξανθῆς ἐστι καὶ λευκῆς, οὕτω καὶ τῇ γενέσει μεταξὺ φλεγματώδους τε καὶ χολώδους ἰλύος. οὖλαι δὲ τρίχες ἢ διὰ τὴν ξηρότητα τῆς κράσεως ἢ διὰ τὸν πόρον, ἐν ᾧ κατερρίζωνται, γίγνονται, διὰ μὲν τὴν ξηρότητα παραπλησίως ἱμᾶσι τοῖς ἐπὶ πλέον ὑπὸ πυρὸς ξηρανθεῖσι. καίτοι τί δεῖ τῶν ἱμάντων μνημονεύειν αὐτὰς τὰς τρίχας ὁρῶντας, εἰ πλησιάσειαν πυρί, παραχρῆμα διαστρεφομένας; οὕτω μὲν οὖν Αἰθίοπες ἅπαντες οὖλοι. τῇ δὲ τῶν πόρων ἐν οἷς ἐρρίζωνται φύσει κατὰ τάδε. πολλάκις μὲν ἡ ἀναθυμίασις εὐθυπορεῖν ὑπ' ἀρρωστίας ἀδυνατοῦσα, καθ' ὃν ἂν αὐτὴ τρόπον ἑλίττηται, καὶ τὸν πόρον οὕτως ἐτύπωσεν. ἐνίοτε δ' ἡ μὲν ἀναθυμίασις εὔρωστός ἐστιν, ὑπὸ δὲ τῆς τοῦ δέρματος φύσεως σκληροτέρας τοῦ προσήκοντος οὔσης εἰργομένη φέρεσθαι κατ' εὐθὺ πρὸς τὸ πλάγιον ἐπιστρέφεται, καθάπερ γε κἀκτὸς ἰδεῖν ἔστιν οὐ μόνον ἀτμὸν ἢ καπνὸν ἀλλὰ καὶ τὰς φλόγας αὐτάς, ὅταν ἀποκλεισθῶσι τῆς ἄνω φορᾶς, λοξὰς ἐφ' ἑκάτερα μέρη σχιζομένας. οὕτως οὖν καὶ ἐκ τοῦ σώματος ἀναθυμίασις, ὅταν εἰρχθῇ κατά τι καὶ κωλυθῇ φέρεσθαι πρόσω,[31] λοξὴν ὑπὸ τὸ δέρμα διέξοδον ὁδοποιεῖται, μέχρι περ ἂν ἀθροισθεῖσα

ON TEMPERAMENTS, BOOK II

The hair is black whenever, being burned up by the strength of the heat of the exhalation, the superfluity becomes entirely smoky, whereas it is yellow (fair) when it is less overheated. Under these circumstances what is plugged up is the muddy superfluity of yellow bile, not black bile. White hair is the product of phlegm, while red hair, as it is intermediate in color between yellow and white, is also in this way intermediate in genesis between phlegmatous and bilious sediment. Curly hair arises due to the dryness of the *krasis* or due to the pore in which it is rooted. It is due to the dryness similar to leather straps dried up to a great extent by a fire. Indeed, what need is there to call to mind such straps when we have seen hairs themselves, which if brought close to a fire, immediately become twisted? In this way then all Ethiopians are curly haired. On the other hand, in the nature of the pores in which the hairs are rooted, it is as follows: often the exhalation is unable to go straight due to weakness, in relation to which it would coil around in this manner, and in this way affect the pore. Sometimes, however, the exhalation is strong, but due to the nature of the skin, which is harder than appropriate, it is prevented from going straight, being turned to the side, as seen in the external world when not only vapor or smoke but also the flames themselves are turned back in their upward passage, and are then divided into oblique paths in either direction. In this way too, the exhalation from the body, whenever it is relatively hindered and is prevented from being carried to the outside, makes an oblique passage under the skin until, having

617K

---

31 πρόσω H; πρὸς τοὐκτός K

χρόνῳ πλείονι βιάσηταί τε καὶ ἀναπνεύσῃ πρὸς τοὐκτός. ἐνίοτε δ' ἅμ' ἀμφοῖν συνδραμόντων, ἀρρώστου τε τῆς πρώτης ἀναθυμιάσεως, ἣ τὸν πόρον ἐδημιούργησε, καὶ ξηροῦ τοῦ δέρματος, ἡ λοξότης γίγνεται ταῖς ῥίζαις τῶν τριχῶν. οἷαι δ' ἂν ἐν τῇ ῥιζώσει διαπλάττωνται, τοιαύτας εἰκὸς ὑπάρχειν αὐτὰς ἕως παντός. οὐδὲ γὰρ οὐδ' ἄλλο τι τῶν σκληρῶν καὶ ξηρῶν σωμάτων εὐθῦναι δυνατὸν | ἄνευ τοῦ μαλάξαι πρότερον. αὕτη μὲν οὖν ἡ γένεσις τῶν τριχῶν.

Ἐφεξῆς δ' ἂν εἴη λέγειν τὰς αἰτίας ἁπάντων τῶν συμβεβηκότων ταῖς κράσεσιν ἐν ταῖς τῶν τριχῶν διαφοραῖς καθ' ἡλικίαν καὶ χώραν καὶ φύσιν σώματος. Αἰγύπτιοι μὲν οὖν καὶ Ἄραβες καὶ Ἰνδοὶ καὶ πᾶν τὸ ξηρὰν καὶ θερμὴν χώραν[32] ἐποικοῦν ἔθνος μελαίνας τε καὶ δυσαυξεῖς καὶ ξηρὰς καὶ οὔλας καὶ κραύρας ἔχουσι τὰς τρίχας. ὅσοι δ' ἔμπαλιν τούτοις ὑγρὰν καὶ ψυχρὰν χώραν ἐποικοῦσιν, Ἰλλυριοί τε καὶ Γερμανοὶ καὶ Δαλμάται καὶ Σαυρομάται καὶ σύμπαν τὸ Σκυθικὸν εὐαυξεῖς μετρίως καὶ λεπτὰς καὶ εὐθείας καὶ πυρράς· ὅσοι δ' ἐν τῷ μεταξὺ τούτων εὔκρατον νέμονται γῆν, εὐαυξεστάτας τε καὶ ἰσχυροτάτας καὶ μελαίνας μετρίως καὶ παχείας συμμέτρως καὶ οὔτ' ἀκριβῶς οὔλας οὔτ' ἀκριβῶς εὐθείας. οὕτω δὲ κἀν ταῖς ἡλικίαις βρέφεσι μέν, οἷαίπερ αἱ τῶν Γερμανῶν, ἀκμάζουσι δὲ [οἷαι] τοῖς Αἰθίοψιν, ἐφήβοις δὲ καὶ παισὶ τοῖς εὔκρατον ἐποικοῦσι γῆν ἔθνεσιν ἀνάλογον αἱ τρίχες ἔχουσιν ἰσχύος τε καὶ πάχους καὶ μεγέθους καὶ χρόας.

been gathered together over a longer time, it is forced and sent forth toward the outside. Sometimes, however, when both these [factors] concur—that is, weakness of the primary exhalation which caused the formation of the pore, and dryness of the skin—there is obliqueness in the roots of hairs. Those hairs that are formed and take root are likely to be such as they are forever. For it is not possible to straighten any other of the hard and dry bodies without their being softened first. This then is the genesis of the hair.

Next would be to state the causes of all the contingent things in the *krasias* in the differences of the hairs in relation to age, place, and nature of a body. And so Egyptians and Arabs and Indians and every race living in a hot, dry region have hair that is black, grows with difficulty, and is dry, curly, and brittle. Conversely those who live in a wet and cold region—Illyrians, Germans, Dalmatians, Sauromatians, and every Scythian—have hair that grows moderately well and is fine, straight, and red. Those who are in between these, inhabiting a *eukratic* land have hair that grows very well and is very strong, moderately black, moderately thick, and is not entirely curly or entirely straight. It is the same too in the ages: in infants the hair is like that of the Germans; in those in their prime, the hair is like that of the Ethiopians; in youths and children, it is like that of the races living in a *eukratic* land, the hair being proportionate in strength, thickness, size, and color.

---

[32] χώραν add. H

Οὕτω δὲ καὶ κατ' αὐτὰς τῶν σωμάτων τὰς φύσεις ἀνάλογον ἡλικίαις τε καὶ χώραις αἱ τρίχες διάκεινται. παῖδες μὲν οἱ πάνυ σμικροὶ ψιλοὶ τριχῶν, ὅτι μήπω μήτε πόρος αὐτοῖς ἐστι μηδεὶς κατὰ τὸ δέρμα μήτε λιγνυώδη περιττώματα. προσάγοντες δὲ τῷ ἡβάσκειν ὑποφύουσι σμικρὰς καὶ ἀσθενεῖς, ἀκμάζοντες δ' ἰσχυροτέρας καὶ πολλὰς καὶ μεγάλας καὶ μελαίνας ἴσχουσιν, ὅτι τε πλῆθος ἤδη πόρων ἐν αὐτοῖς ἐγένετο καὶ ὅτι μεστοὶ τῶν αἰθαλωδῶν εἰσι περιττωμάτων ὑπὸ ξηρότητός τε καὶ θερμότητος· αἱ δ' ἐν τῇ κεφαλῇ τε καὶ ταῖς ὀφρύσι καὶ κατὰ τὰ βλέφαρα καὶ παισὶν οὖσιν ἡμῖν ὑπάρχουσιν ἤδη. γένεσις γὰρ δὴ ταύταις οὐχ οἵα ταῖς πόαις, ἀλλ' οἵα τοῖς φυτοῖς κατὰ πρῶτον λόγον ὑπὸ τῆς φύσεως ἀπειργασμέναις, οὐκ ἐξ ἀνάγκης ἑπομέναις ταῖς κράσεσιν, ὡς κἀν τοῖς Περὶ χρείας μορίων δείκνυται. ἀλλά τοι κἀν ταῖσδε τὸ μὲν εἶναι διὰ τὴν τῆς φύσεως τέχνην, τὸ δ' ἤτοι μελαίναις ἢ πυρραῖς ἤ τιν' ἄλλην ἐχούσαις διαφορὰν ἐξ ἀνάγκης ἕπεται τῇ κράσει τῆς ἡλικίας. ὑπόπυρροι μὲν γάρ εἰσι τοὐπίπαν, ὅτι καὶ τὸ σφηνούμενον ἐν τοῖς πόροις οὐδέπω μέλαν ἐστίν·[33] ἥ τε γὰρ ὑγρότης πολλὴ καὶ ἡ διέξοδος ῥᾳδία καὶ ἡ σύγκαυσις ἀσθενής· εὐαυξεῖς δὲ καὶ παχεῖαι συμμέτρως τῇ τῶν τρεφόντων αὐτὰς περιττωμάτων ἀφθονίᾳ. τὸ μὲν γὰρ μόριον αὐτὸ τοῦ σώματος, ἐν ᾧ γίγνονται, ξηρόν· ὅλον γὰρ τὸ κρανίον ὀστέϊνόν ἐστι· τὸ δὲ περικείμενον αὐτῷ δέρμα ξηρότερον τοσούτῳ τοῦ κατὰ τὸ λοιπὸν σῶμα δέρματος ἅπαντος, ὅσωπερ καὶ σκληρότερον. ἀνα-

ON TEMPERAMENTS, BOOK II

In this way too, in relation to the actual natures of the bodies, the hair is disposed analogously to ages and regions. Very small children are devoid of hairs, in that none of them yet have a channel for them in the skin, nor do they have smoky superfluities. As they move toward puberty, they grow hairs that are small and weak, while in their prime, they have hairs that are stronger, copious, large, and black, because already a greater number of pores have arisen in them and because they are full of the sooty superfluities due to dryness and hotness. However, the hair on the head and the eyebrows and in relation to the eyelids is already present in us as children. For genesis in these is not like the grasses but like the plants established by Nature, in the first generation, not necessarily following the *krasias*, as was also shown in the work *On the Use of the Parts*.[20] But certainly, even in these, they are due to the art of Nature, while whether they are black or red or have some other difference necessarily follows the *krasis* of the age. For they are on the whole reddish because what is plugged up in the pores is not yet black altogether, for the wetness is abundant, the outward passage easy, and the burning weak. They grow well and are moderately thick due to the abundance of the superfluities nourishing them. The actual part of the body in which they arise is dry, for the entire cranium is bony, while the skin surrounding it is drier compared to the skin of the whole of the rest of the body to the extent that it is harder. And

---

[20] *De usu partium*; see particularly XI.14, III.889–911K; May, *Galen on the Usefulness*, 1.531–36.

---

33 *post ἐστίν: add.* ὅλον K

φέρεται μέντοι πλῆθος οὐκ ὀλίγον ἔκ τε τῶν κατὰ τὸν ἐγκέφαλον, ἤδη δὲ κἀξ ὅλου τοῦ σώματος αἰθαλωδῶν περιττωμάτων, ὥσθ᾽ οἷον τοῖς ἀκμάζουσιν ὅλον γίγνεται τὸ σῶμα, τοιοῦτον ἤδη τὸ δέρμα τῆς κεφαλῆς ἐστι τοῖς βρέφεσιν.

Εὐλόγως οὖν ἔνιοι φαλακροῦνται τοῦ χρόνου προϊόντος, οἷς ἐξ ἀρχῆς ἦν ξηρότερον τὸ δέρμα. δέδεικται γὰρ ἔμπροσθεν, ὡς τῶν γηρασκόντων ἅπαντα ξηραίνεται τὰ μόρια. γίγνεται δὲ πολλοῖς ὀστρακῶδες τὸ δέρμα πλέον ἢ δεῖ ξηρανθέν. ἐν τοιούτῳ δ᾽ οὐδὲν φύεσθαι δύναται, καθότι καὶ διὰ τῶν ἔμπροσθεν ὡμολόγηται. καὶ γὰρ δὴ καὶ τῶν χειρῶν τὰ ἔνδον ὥσπερ γε καὶ τὰ κάτω τῶν ποδῶν ἄτριχα καὶ ψιλὰ ξηρότητί τε καὶ πυκνότητι τοῦ κατ᾽ αὐτὰ τένοντος, ὃς ὑποτέτακται τῷ δέρματι. ὅσοις δ᾽ εἰς τέλος ξηρότητος οὐκ ἀφικνεῖται τὸ δέρμα τῆς κεφαλῆς, ἄρρωστοι τούτοις γίγνονται καὶ λευκαὶ πάντως αἱ τρίχες, ἃς ὀνομάζουσιν οἱ ἄνθρωποι πολιάς, ἄρρωστοι μὲν ἐνδείᾳ τῆς οἰκείας τροφῆς, λευκαὶ δέ, διότι καὶ τὸ τρέφον αὐτὰς τοιοῦτον οἷον εὐρώς τις φλέγματος ἐν χρόνῳ διασαπέντος. ὅταν γὰρ ὁ μὲν πόρος ἔτι μένῃ, τὸ περίττωμα δ᾽ ὀλίγον ᾖ καὶ γλίσχρον, ἀρρώστως δ᾽ ὑπὸ τῆς θερμασίας ὠθῆται πρόσω, πάσχει τι παραπλήσιον ἐν τῷδε σηπεδόνι· καὶ δὴ φαλακροῦνται μὲν μᾶλλον τὸ βρέγμα γηρῶντες οἱ ἄνθρωποι, πολιοῦνται δὲ τοὺς κροτάφους μᾶλλον, ὅτι τὸ μὲν ξηρότατόν ἐστι τῶν μορίων ἁπάν-

indeed, no small amount of sooty superfluities is brought up from those in the brain and already from the whole body. As a consequence, the skin of the head in infants is already such that it is like the whole body becomes in those in their prime.

It is, therefore, with good reason that some people become bald as time progresses, the skin in them being drier from the beginning, for it was shown previously that all the parts of those who are aging are dried out. In many the skin becomes shell-like being dried out more than it should be. In such skin, nothing is able to grow, as was also agreed upon through those things [said] previously. Furthermore also, the skin on the palms of the hands, just as that on the soles of the feet is hairless and bare due to dryness and thickness of the tendon in these which lies under the skin. The hair in those in whom the skin of the head does not reach complete dryness becomes weak and altogether white, which men term gray-haired; weakness is due to lack of proper nourishment, while whiteness is because what does nourish them is as if decayed, some phlegm coming to putrefaction over time. For whenever the pore still remains, while the superfluity is small in amount and viscous, and is pushed onward weakly by the heat, it suffers something similar in the putrefaction, and certainly men growing old become bald more in respect of the bregma,[21] whereas they become white-haired more at the temples. The former is the

---

[21] The term *bregma* is now applied to the point on the skull corresponding to the junction of the coronal and sagittal sutures, or more generally where the two frontal bones meet the two parietal bones.

των τῆς κεφαλῆς· ἐπὶ γὰρ ὀστῷ τὸ δέρμα ταύτῃ ψιλῷ· οἱ κρόταφοι δ' ὑγρότεροι· μύες γὰρ ἐνταῦθ' εἰσὶν ὑπὸ τῷ δέρματι μεγάλοι, σαρκώδης δὲ πᾶς μῦς, ἡ δὲ σὰρξ ὀστοῦ καὶ δέρματος ὑγροτέρα. |

6. Ἀκριβῶς δὲ χρὴ προσέχειν τῷ λεγομένῳ τὸν νοῦν, ὅπως μὴ λάθωμεν ἡμᾶς αὐτοὺς παρακούσαντές τι καὶ σφαλέντες, οἷα δὴ πολλοὶ τῶν πάνυ δοκούντων ἀρίστων ἰατρῶν εἶναι σφάλλονται, εἴ τίς ἐστι φαλακρός, εὐθὺς τοῦτον οἰόμενοι ξηρὰν ἔχειν ἅπαντος τοῦ σώματος τὴν κρᾶσιν. οὐ γὰρ δὴ ἁπλῶς οὕτως εἰκάζειν ἐχρῆν, ἀλλὰ διορίζεσθαι πρότερον ἄμεινον ἦν, ὡς τῶν ἀνθρώπων τὸ σῶμα τῶν μὲν ὁμαλῶς κέκραται σύμπαν, ἐνίων δὲ καὶ οὐκ ὀλίγων τούτων ἀνωμάλως διάκειται. τὰ μὲν γάρ τινα[34] τῶν μορίων αὐτοῖς ὑγρότερα τοῦ συμμέτρου τε καὶ προσήκοντός ἐστι, τὰ δὲ ψυχρότερα, τὰ δὲ ξηρότερα, τὰ δὲ θερμότερα, τὰ δὲ καὶ παντελῶς εὐκρατά τε καὶ σύμμετρα. δεῖ δὲ προσέχειν μάλιστα τούτῳ τὸν νοῦν, ἐπειδὰν ἐπισκέπτῃ σώματος κρᾶσιν. εἰ μὲν γὰρ ὁμαλῶς εὔρυθμον ὅλον ἐστὶν ἁπάσας τε τῶν μορίων ἀποσῶζον τὰς πρὸς ἄλληλα συμμετρίας ἐν μήκει καὶ πλάτει καὶ βάθει, δύναιτ' ἂν ὅλον ὁμοίως κεκρᾶσθαι τὸ τοιοῦτον. εἰ δέ τι σῶμα θώρακα μὲν ἔχει καὶ τράχηλον καὶ ὤμους μεγίστους, ἰσχνὰ δὲ καὶ σμικρὰ τὰ κατ' ὀσφὺν | καὶ σκέλη λεπτά,[35] πῶς ἂν ὁμοίως εἴη τοῦτο διακείμενον ἅπασι τοῖς μορίοις; οὐ μὴν οὐδ' εἰ τὰ μὲν

[34] τινα add. H [35] post λεπτά: καὶ ξηρὰ add. K

driest of all the parts of the head, for the skin there is attached to bare bone.[22] The temples, however, are wetter, for the muscles under the skin are large here, while the whole muscle is fleshy, the flesh being wetter than bone and skin.

6. It is necessary to direct attention precisely to what is being said, so we do not unwittingly hear something carelessly and erroneously, as certainly many of the seemingly very best doctors do when they fall into the error of thinking if a person is bald, he automatically has a dry *krasis* of the whole body. For it is certainly necessary not to conjecture absolutely in this way; it is better to distinguish first that in respect of people's bodies, some have been compounded entirely evenly, while others—and there are not just a few of these—are compounded unevenly (non-uniformly). For some of their parts are wetter than is balanced and appropriate, some colder, some drier, some hotter, and some that are altogether *eukratic* and well-balanced. It is particularly necessary to pay attention to this, when you consider the *krasis* of a body. For if the whole is evenly well-proportioned and all the parts preserve the due proportions with each other in length, breadth, and depth, it would be possible for such a body to have been similarly compounded throughout. If, however, some body has a very large chest, neck, and shoulders, while the parts in relation to the loins are thin and small and has legs that are thin and dry, how would this be disposed similarly in all the parts? Nor again, if the legs

---

[22] Following KLat (*haeret enim cutis illic ossi nudo*) on grounds of sense.

σκέλη παχέα καὶ τὰ κατ' ὀσφὺν εὐρέα, τὸν θώρακα δ' ἔχει στενόν, οὐδὲ τοῦτ' ἂν εἴη κεκραμένον ὁμαλῶς τοῖς μορίοις. ἕτερα δὲ σώματα μεγίστην ἔχει τὴν κεφαλήν, ἕτερα δὲ σμικράν, οἷανπερ οἱ στρουθοί. καὶ τοῖς σκέλεσι τὰ μὲν βλαισά, τὰ δὲ ῥαιβά, καὶ ἄκροις τοῖς κώλοις τὰ μὲν ἰσχνά, τὰ δὲ παχέα, καὶ θώραξ τοῖς μέν, ὡς εἴρηται πρόσθεν, εὐρύς, ἐνίοις δὲ στενὸς οὕτως ὡς σανίς, οὓς δὴ καὶ σανιδώδεις ὀνομάζουσιν. ὅταν δὲ καὶ τὰ κατ' ὠμοπλάτας αὐτοῖς ἄσαρκα τελέως ᾖ καὶ γυμνὰ καὶ προπετῆ δίκην πτερύγων, ὀνομάζονται μὲν αἱ τοιαῦται φύσεις ὑπὸ τῶν ἰατρῶν πτερυγώδεις, εἰς ὅσον δ' ἥκουσι κακίας ἀπολωλεκότος τοῦ θώρακος ὀλίγου δεῖν ἅπασαν τὴν ἐντὸς εὐρυχωρίαν, ἐν ᾗ πνεύμων τε καὶ καρδία τέτακται, πρόδηλον παντί. μυρίαι δ' ἄλλαι τῶν τοῦ σώματός εἰσι μορίων προδήλως αἱ διαθέσεις, ὅταν ἐκτραπόμενον τῆς φυσικῆς ἀναλογίας εἰς ἀνώμαλόν τινα δυσκρασίαν εὐθὺς ἐν τῷ κυΐσκεσθαι μεταπέσῃ. οὔκουν ἐπὶ τῶν τοιούτων ἐξ ἑνὸς χρὴ μορίου τεκμαίρεσθαι περὶ τοῦ παντός.

Οὐδὲ γὰρ οἱ φυσιογνωμονεῖν ἐπιχειροῦντες ἁπλῶς ἀποφαίνονται περὶ πάντων, ἀλλ' ἐκ τῆς πείρας καὶ

---

23 See Aretaeus, *On Causes and Signs of Chronic Diseases* 1.8, on *phthisis* (F. Adams, *The Extant Works of Aretaeus the Cappadocian* [London, 1886], 309–12).

24 A winged scapula (scapula alata) is a rare condition with a number of causes; for a modern review, see S. G. Lee et al., "Scapular Winging: Anatomical Review, Diagnosis and Treat-

ON TEMPERAMENTS, BOOK II

are thick and the parts in relation to the loins broad, while the chest is narrow, this would not have been compounded evenly in the parts. Some bodies have a very large head, while others have a small head, like birds, while in the legs there are bow legs and bandy legs, and in the distal parts of the legs, some are thin, whereas others are thick. And the chest in some, as was said before, is broad, whereas in others it is narrow as a plank, and people in fact call them "plank-like."[23] However, when the parts in relation to the scapulae in them are also completely without flesh, and are bare and incline forward like wings, such natures are called "wing-like"[24] by doctors, and they come to such a degree of abnormality that it is clear to everyone the chest has lost almost the whole internal space in which the lungs and heart are situated. However, there are clearly countless other conditions of the parts of the body when a deviation from the natural proportion to some non-uniform *dyskrasia* is fallen into immediately on conception. | In such cases one must not therefore derive evidence concerning the whole from one part.

624K

Physiognomists[25] do not attempt to make an absolute statement about all the features; they too have learned

---

ment," *Current Reviews in Musculoskeletal Medicine* 1(1) 2008: 1–11. Early reference to emaciation as a cause is found in Hippocrates, *Epidemics* 3.14 and 6.3.10. LSJ also refers to the present passage.

[25] The term *physiognomia* appears in Hippocrates and is the subject of a pseudo-Aristotelian treatise. Physiognomy became a separate *techne*. A detailed account of physiognomics in Roman times is provided by T. S. Barton, *Power and Knowledge: Astrology, Physiognomics and Medicine under the Roman Empire* (Ann Arbor, 1994).

οἴδε διδαχθέντες. εἰ μέν τις ἱκανῶς εἴη δασὺς τὰ στέρνα, θυμικὸν ἀποφαίνονται, μηροὺς δ' εἴπερ εἴη τοιοῦτος, ἀφροδισιαστικόν· οὐ μὴν τήν γ' αἰτίαν προστιθέασιν. οὐδὲ γὰρ ὅτι λέοντι μὲν ἐμφερὴς τὰ στέρνα, τράγῳ δὲ τὰ κατὰ μηροὺς ἐπειδὰν φῶσι, τὴν πρώτην αἰτίαν ἐξευρήκασι. διὰ τί γὰρ ὁ μὲν λέων θυμικός, ὁ δὲ τράγος ἀφροδισιαστικός, ὁ λόγος ἐξευρεῖν ἐπιζητεῖ. μέχρι γὰρ τοῦδε τὸ μὲν γιγνόμενον εἰρήκασι, τὴν δ' αἰτίαν αὐτοῦ παραλελοίπασιν. ἀλλ' ὁ φυσικὸς ἀνὴρ ὥσπερ τῶν ἄλλων ἁπάντων οὕτω καὶ τὰς τούτων αἰτίας ἐξευρίσκειν ἐπιχειρεῖ. διότι γὰρ ἀνωμάλως διάκεινται κατὰ τὰς τῶν μορίων κράσεις οὐ λέων μόνον καὶ τράγος ἀλλὰ καὶ τῶν ἄλλων ζῴων πάμπολλα, διὰ τοῦτ' ἄλλα πρὸς ἄλλας ἐνεργείας ἑτοίμως ἔχει. περὶ μὲν δὴ τούτων Ἀριστοτέλει καλῶς ἐπὶ πλεῖστον εἴρηται. τὸ δ' οὖν εἰς τὰ παρόντα χρήσιμον ἤδη φαίνεται, διότι χρὴ σκοπεῖσθαι τῶν ἀνθρώπων τὰς κράσεις, ἕκαστον τῶν μορίων ἐξετάζοντα καθ' ἑαυτό, καὶ μὴ νομίζειν, εἴ τῳ δασὺς ὁ θώραξ, ὅλον ἐξ ἀνάγκης τούτῳ τὸ σῶμα ξηρότερόν τε καὶ θερμότερον ὑπάρχειν, ἀλλ' ἐν τῇ καρδίᾳ τὸ θερμὸν εἶναι πλεῖστον, διὸ καὶ θυμικόν· δύνασθαι δ' ἐνίοτε δι' αὐτὸ τοῦτο μὴ ὁμοίως ἅπαν αὐτοῖς τὸ σῶμα θερμὸν καὶ ξηρὸν ὑπάρχειν, ὅτι πλεῖστον ἀνέπνευσεν ἐνταῦθα καὶ πρὸς τὸ περιέχον ἐξεκενώθη τὸ θερμόν. εἰ μὲν γὰρ ὅλου τοῦ σώματος ἡ κρᾶσίς ἐστιν ὁμαλή, εὐθὺς ἂν

from experience. If someone is very hairy on the chest, they declare him to be "high-spirited" (irascible), while if on the thighs, "lecherous." But they still do not add the cause, for they would not, when they refer to the chest in a lion or the parts in relation to the thighs in a goat, say they have discovered the primary cause. Why it is that the lion is high-spirited, while the goat is lecherous, reason seeks to discover. Up to this point, they have stated what has occurred, having left the cause of this untold. But the man who is a natural scientist attempts to discover, as of all other things, so too the causes of these. It is because they are disposed non-uniformly in relation to the *krasias* of the parts—not only lion and goat, but also a very large number of the other animals, since they are suited to different functions in different ways. Certainly, most has been well said by Aristotle about these matters.[26] What is useful, therefore, for present purposes is already apparent—that it is necessary in considering the *krasias* of people, to scrutinize closely each of the parts in itself and not to think, if the chest is hairy, the whole body in this case is necessarily drier and hotter, but the heat in the heart is very great, on which account the person is high-spirited. It is, however, possible sometimes, due to this very thing, for the whole body in them not to be similarly hot and dry, because the greater part of the heat has been exhaled here and evacuated to the surroundings. For if the *krasis* of the whole body is uniform, straightaway in those

[26] Presumably, a reference to the pseudo-Aristotelian work referred to in the previous note (*Physiognomica*); for an English translation, see *The Complete Works of Aristotle*, ed. Jonathan Barnes (Princeton, 1984), 1.1237–50.

εἴη τούτοις αὐτός τε σύμπας ὁ θώραξ εὐρύτατος αἵ τε φλέβες εὐρεῖαι καὶ αἱ ἀρτηρίαι μεγάλαι θ' ἅμα καὶ μέγιστον καὶ σφοδρότατον σφύζουσαι καὶ τρίχες πολλαὶ καθ' ὅλον τὸ σῶμα καὶ αἱ τῆς κεφαλῆς εὐαυξέσταται μὲν καὶ μέλαιναι καὶ οὖλαι κατὰ τὴν πρώτην ἡλικίαν, ἐπὶ δὲ προήκοντι τῷ χρόνῳ φαλάκρωσις ἀκολουθήσει. καὶ μὲν δὴ καὶ σύντονον καὶ διηρθρωμένον καὶ μυῶδες, ἐπειδὰν ὁμαλῶς ἔχωσι τῆς κράσεως, ὅλον ἔσται τοῖς τοιούτοις ἀνθρώποις τὸ σῶμα καὶ τὸ δέρμα σκληρότερόν τε καὶ μελάντερον ὥσπερ καὶ δασύτερον. οὕτω δὲ κἂν εἰ τἀναντία περὶ τὸν θώρακα συμπέσοι, τῆς κράσεως ὁμαλῆς ὑπαρχούσης ἐν ὅλῳ τῷ σώματι, τουτέστιν ὑγροτέρων τε καὶ ψυχροτέρων ἁπάντων τῶν μορίων γενομένων, ὁ μὲν θώραξ αὐτοῖς στενὸς καὶ ἄτριχος ἔσται, καθάπερ οὖν καὶ σύμπαν τὸ σῶμα ψιλὸν τριχῶν ἁπαλόν τε καὶ λευκὸν τὸ δέρμα καὶ ὑπόπυρρον ταῖς θριξὶ καὶ μάλιστ' ἐν νεότητι καὶ οὐ φαλακροῦνται γηρῶντες, εὐθὺς δὲ καὶ δειλοὶ καὶ ἄτολμοι καὶ ὀκνηροὶ καὶ σμικρὰς καὶ ἀδήλους ἔχοντες τὰς φλέβας καὶ πιμελώδεις καὶ νεύροις καὶ μυσὶν ἄρρωστοι καὶ ἀδιάρθρωτοι τὰ κῶλα καὶ βλαισοὶ γίγνονται. διαφόρου μέντοι τῆς κράσεως ἐν τοῖς μορίοις ἀπεργασθείσης, οὐκέτ' ἐξ ἑνὸς αὐτῶν οἷόν τε περὶ τοῦ σύμπαντος ἀποφαίνεσθαι σώματος, ἀλλ' ἄμεινον ἐφ' ἕκαστον ἰέναι καὶ σκοπεῖσθαι, πῶς μὲν ἡ γαστὴρ ἔχει κράσεως, ὅπως δ' ὁ πνεύμων, ὅπως δ' ὁ ἐγκέφαλος ἕκαστόν τε τῶν ἄλλων ἰδίᾳ καὶ καθ' ἑαυτό.

people the chest itself as a whole would be very broad, the veins dilated, and the arteries large and at the same time the pulses very large and very strong, the hair plentiful over the whole body, with that on the head growing very well, and being black and curly during the first stage of life, but with the progression of time baldness will follow. And furthermore, if they have uniformity of the *krasis*, the whole body in such people will be strained tight, well articulated, and muscular, the skin will be harder and darker, just as it will also be more hairy. In this way too, even if the opposite should happen involving the chest, when the *krasis* is uniform in the whole body—that is to say, if all the parts become wetter and colder—the chest will be narrow and hairless in them, just as also the whole body will be bare of hair, the skin soft and white, the hair reddish in color, especially in a youth, and when old they will not become bald. At once they will also be cowardly, wanting in spirit and hesitant, have small veins that are hard to detect; they are fatty, weak in the sinews and muscles, and the limbs are disjointed and become distorted. Of course, when a different *krasis* is produced in the parts, it is no longer still possible to make a statement about the whole body from one of these parts, but it is better to go to each one and consider what kind of *krasis* the stomach has, and what kind the lungs, and what kind the brain, and each of the others individually, in and of itself.

Ταυτὶ μὲν οὖν ἐκ τῶν ἐνεργειῶν γνωρίζειν· οὐ γὰρ ἔστιν οὔθ' ἁψάμενον[36] οὔτ' ὀφθαλμοῖς θεασάμενον ἐξευρεῖν τὴν κρᾶσιν αὐτῶν. προσεπισκέπτεσθαι δὲ καὶ τὰς τῶν περιεχόντων αὐτὰ μορίων διαθέσεις, ὧν ἁπάντων ἔξωθέν ἐστι τὸ δέρμα, κατὰ μὲν τὴν ἡμετέραν οἴκησιν εὔκρατον οὖσαν ἐνδεικνύμενον τῶν ὑποκειμένων μορίων τὴν φύσιν οὐδ' ἐν ταύτῃ πάντων ἁπλῶς, ἀλλ' ὅσα ταῖς κράσεσιν ὡσαύτως ἔχει τῷ δέρματι. κατὰ δὲ τὰς ὑπὸ ταῖς ἄρκτοις τε καὶ τῇ μεσημβρίᾳ χώρας, ἐπειδὴ τῶν μὲν εἰς τὸ βάθος ἀπελήλαται τὸ θερμὸν ὑπὸ τοῦ περιέχοντος ἔξωθεν κρύους νικώμενον, τῶν δ' εἰς τὸ δέρμα προελήλυθεν ὑπὸ τοῦ περιέχοντος θάλπους ἑλκόμενον, οὐκ ἐκ τῆς κατὰ τὸ δέρμα διαθέσεως οἷόν τε γνῶναι σαφῶς ὑπὲρ τῆς τῶν ἐντὸς μορίων κράσεως. ἀνώμαλος γὰρ ἡ τοῦ σώματος κρᾶσις ἐν ταῖς δυσκράτοις χώραις οὐχ ὡσαύτως ἐχόντων τῶν τ' ἔξωθεν μορίων καὶ τῶν ἐντός. Κελτοῖς μὲν γὰρ καὶ Γερμανοῖς καὶ παντὶ τῷ Θρακίῳ τε καὶ Σκυθικῷ γένει ψυχρὸν καὶ ὑγρὸν τὸ δέρμα καὶ διὰ τοῦτο μαλακόν τε καὶ λευκὸν καὶ ψιλὸν τριχῶν· ὅσον δ' ἔμφυτον θερμόν, εἰς τὰ σπλάγχνα καταπέφευγεν ἅμα τῷ αἵματι κἀνταῦθα κυκωμένου τε καὶ στενοχωρουμένου καὶ ζέοντος αὐτοῦ θυμικοὶ καὶ θρασεῖς καὶ ὀξύρροποι ταῖς γνώμαις ἀποτελοῦνται. Αἰθίοψι δὲ καὶ Ἄραψι καὶ ὅλοις τοῖς κατὰ μεσημβρίαν ἡ μὲν τοῦ δέρματος φύσις, ὡς ἂν ὑπό τε τοῦ περιέχοντος θάλπους καὶ τῆς ἐμφύτου θερμασίας ἔξω φερομένης διακεκαυμένη, σκληρὰ καὶ ξηρὰ καὶ μέλαινα. τὸ δ' ὅλον

ON TEMPERAMENTS, BOOK II

These, then, are made known from the functions, for it is not possible to discover their *krasis* through what is touched, or what is seen by eyes. Also to be considered besides are the conditions of the parts surrounding them, of which the skin is external to all. In our place of habitation, which is *eukratic*, the skin displays the nature of the underlying parts, although not in this absolutely of all, but those that are similar to the skin in their *krasias*. In the northern and southern regions, it is not possible to recognize clearly the *krasis* of the parts within from the condition in the skin; in the former this applies if the heat has departed to the depths, overcome by the icy cold of the external surroundings, while in the latter it applies if the heat is brought near to the skin, drawn by the ambient warmth. Thus the *krasis* of the body is non-uniform in the *dyskratic* places, not being similar in the external and internal parts. For in Celts and Germans, and in every Thracian and Scythian race the skin is cold and wet, and due to this soft, pale, and devoid of hair. However much innate heat there is has fled for refuge to the internal organs, along with the blood, which here is agitated, confined in narrow spaces and itself seething, rendering them high-spirited, rash, and quick to change their opinions. In Ethiopians and Arabs and all those in the south, the nature of the skin, as it would be roasted by the surrounding warmth and by the innate heat being carried outward, is hard, dry, and dark. The whole body partakes

---

36 *post* ἀψάμενον: ἀφῇ add. K

GALEN

σῶμα τῆς μὲν ἐμφύτου θερμότητος ἥκιστα μετέχει, θερμὸν δ' ἐστὶν ἀλλοτρίῳ τε καὶ ἐπικτήτῳ θερμῷ.

Καὶ γὰρ δὴ καὶ τοῦτο κάλλιστα πρὸς Ἀριστοτέλους ἐπὶ πολλῶν διώρισται. καὶ χρὴ προσέχειν αὐτῷ τὸν νοῦν, εἴπερ τῳ καὶ ἄλλῳ, καὶ σκοπεῖσθαι καθ' ἕκαστον σῶμα, πότερον οἰκείῳ θερμῷ θερμόν ἐστιν ἢ ἐπικτήτῳ. πάντα γοῦν τὰ σηπόμενα θερμὰ μὲν ἐπικτήτῳ θερμῷ, ψυχρὰ δ' οἰκείῳ καὶ τὰ τῶν ἐποικούντων τὴν μεσημβρινὴν χώραν σώματα[37] θερμὰ μὲν ἐπικτήτῳ θερμῷ, ψυχρὰ δ' οἰκείῳ, καὶ παρ' ἡμῖν δὲ κατὰ μὲν τὸν χειμῶνα τὸ φύσει θερμὸν πλέον, τὸ δ' ἐπίκτητον ἔλαττον, ἐν δὲ τῷ θέρει τὸ μὲν ἐπίκτητον πλέον, ἔλαττον δὲ τὸ σύμφυτον. ἅπαντ' οὖν ταῦτα διορίζεσθαι χρὴ τὸν μέλλοντα καλῶς διαγνώσεσθαι κρᾶσιν. οὐ γὰρ δὴ ἁπλῶς, εἰ τὸ δέρμα μελάντερον, ἤδη θερμότερος ὁ ἄνθρωπος ὅλος, ἀλλ' εἰ τῶν ἄλλων ἁπάντων ὡσαύτως ἐχόντων. εἰ γὰρ ὁ μὲν ἐν ἡλίῳ θερμῷ διέτριψεν ἐπὶ πλέον, ὁ δ' ἐν σκιᾷ, τῷ μὲν ἔσται μελάντερον τὸ χρῶμα τῷ δὲ λευκότερον. ἀλλ' οὐδὲν τοῦτο πρὸς τὴν τῆς ὅλης κράσεως ὑπάλλαξιν. αὐτὸ γὰρ τὸ δέρμα ξηρότερον μὲν ἡλιούμενον, ὑγρότερον δ' ἔσται σκιατραφούμενον. ἡ φυσικὴ δ' οὐκ εὐθὺς ὑπαλλαχθήσεται κρᾶσις οὔθ' ἥπατος οὔτε καρδίας οὔτε τῶν ἄλλων σπλάγχνων.

Ἄριστον οὖν, ὡς εἴρηται καὶ πρόσθεν, ἑκάστου τῶν μορίων ἴδια πεπορίσθαι τῆς κράσεως τὰ γνωρίσματα,

---

[37] σώματα add. H

least of the innate heat but is hot due to alien and acquired heat.

Furthermore, this was distinguished best by Aristotle in many cases.[27] And it is necessary to direct attention to this, if to anything, and to consider in relation to each body whether it is hot by its own heat or by acquired heat. Anyway, all things putrefying are hot by an acquired heat but cold by their own cold as the bodies of those settled in a southern region are hot by an acquired heat, and cold by their own cold. And among us there is greater natural heat in the winter, while that which is acquired is less. Conversely, in the summer, acquired heat is more while innate heat is less. All these, then, must be distinguished by someone who is going to properly recognize *krasis*. If the skin is already darker, the whole person is already hotter, not absolutely, but only if all the other parts are in a similar state. For if one person has spent more time in hot sun, and another in shade, in the former the complexion will be darker, while in the latter it will be paler, but this has no relevance to the change of the whole *krasis*. For the skin itself is drier when exposed to the sun but will be wetter when a person is brought up in the shade. The natural *krasis* will not automatically be changed—neither that of the liver, nor the heart, nor of the other internal organs.

And so it is best, as was also said previously, to have provided the signs of the *krasis* specific to each of the

---

[27] Aristotle, *Parts of Animals* 2.2, LCL 323 (A. L. Peck), 116–29.

οἷον τῆς μὲν γαστρός· εἰ πέττει καλῶς, εὔκρατος, εἰ δ' οὐ πέττει καλῶς, δύσκρατος. ἀλλ' εἰ μὲν κνισσώδεις τινὰς ἢ καπνώδεις ἐργάζοιτο τὰς ἐρυγάς, ἄμετρον αὐτῇ καὶ πυρῶδες[38] τὸ θερμόν, εἰ δ' ὀξείας, ἄρρωστόν τε καὶ ἀσθενές. οὕτω δὲ καὶ τὰ μὲν βόεια κρέα καὶ πάντα τὰ δυσκατέργαστα τῶν καλῶς πεττόντων ἄμετρον τὸ θερμόν, ἀσθενὲς δὲ τῶν ταῦτα μὲν ἀπεπτούντων, ἰχθῦς δὲ πετραίους ἤ τι τοιοῦτον πεττόντων. ἐπισκέπτεσθαι δ' ἐνταῦθα πάλιν, εἰ μὴ διά τινα χυμὸν ἑτέρωθεν ἐπιρρέοντα τὸ σύμπτωμα γίγνεται τῇ γαστρί. φλέγμα μὲν γὰρ ἐνίοις ἐκ τῆς κεφαλῆς, ξανθὴ δ' ἄλλοις ἐξ ἥπατος εἰς τὴν γαστέρα καταρρεῖ χολή, σπάνιον μὲν δὴ τοῦτο καὶ ὀλιγίστοις συμβαίνει. παμπόλλοις δ' ἐκ τῆς κεφαλῆς κατέρχεται φλέγμα καὶ μάλιστ' ἐν Ῥώμῃ τε καὶ τοῖς ὑγροῖς οὕτω χωρίοις. ἀλλά τοι καὶ τὸ σπάνιον ἐπιβλέπειν χρὴ καὶ μηδὲν ἐν παρέργῳ τίθεσθαι μηδ' ἀμελεῖν. οἶδα γὰρ ἐγώ τισιν ἱκανῶς φλεγματώδεσιν ἀνθρώποις ἀθροιζομένην ἐν τῇ γαστρὶ χολὴν παμπόλλην ξανθήν, ἣν ἐδέοντο πρὸ τῶν σιτίων ὕδωρ πολὺ πιόντες ἢ οἶνον ἐξεμεῖν. εἰ δ' ἥψαντό ποτε σιτίων, πρὶν ἐμέσαι, διέφθειρόν τε ταῦτα καὶ τὴν κεφαλὴν ὠδυνῶντο καὶ τούτους ᾤοντό τινες εἶναι χολώδεις φύσει. καίτοι μαλακοί τ' ἦσαν ὅλον τὸ σῶμα καὶ λευκοὶ καὶ ἄτριχοι καὶ πιμελώδεις καὶ ἄφλεβοι καὶ ἄμυοι καὶ ἄναιμοι καὶ ἁπτομένοις οὐ λίαν θερμοί. ἄλλους δέ τινας οἶδα

---

[38] post πυρῶδες: ὑπάρχει add. K

parts. Take, for example, the stomach; if it concocts well it is *eukratic,* whereas if it doesn't concoct well, it is *dyskratic.* But if it were to create certain eructations that are steamy and smoky, the heat in it is disproportionate and fiery, whereas if they are sharp (pungent), | the heat is weak and feeble. In this way too, in the case of ox flesh and all those things that are hard to digest, those who concoct this well will have a heat that is disproportionate, while it is weak in those who do not concoct these but do concoct fish of the rocks or some such thing. Here again it is necessary to consider whether the symptom in the stomach does not occur due to a certain humor flowing in from somewhere else. In some cases phlegm flows down to the stomach from the head, while in others yellow bile flows down from the liver to the stomach, although this is rare and happens in very few cases. In very many cases, however, phlegm comes down from the head, and particularly in Rome and in wet places such as this. But certainly also what is rare must be taken into consideration, and not put to one side or neglected. For I have seen, in some very phlegmatic people, a great quantity of yellow bile collected in the stomach, which they need to vomit up by drinking much water or wine before food. If, however, they took the foods sometime before they vomit, they destroy these and suffer pain in the head. Some think these people are bilious in nature. And indeed, they are soft in the whole body, and pale, hairless and | fatty, lacking in veins and muscles, bloodless, and not very hot to touch. I have seen certain others who have

630K

631K

μηδεπώποτε μὲν ἐμέσαντας χολὴν ξανθήν, ἰσχνοὺς δὲ καὶ δασεῖς καὶ μυώδεις καὶ μέλανας καὶ φλεβώδεις καὶ θερμοὺς ἱκανῶς, εἴ τις ἄψαιτο, φαινομένους, οἷος καὶ ὁ φιλόσοφος Εὔδημος.

Ἀλλ' ἐνταῦθα[39] καὶ ἀνατομικόν τι θεώρημα συνέπεσεν, ὃ μὴ γιγνώσκοντες ἔνιοι τῶν ἰατρῶν ἀποροῦνται δεινῶς ἐπὶ τῇ διαφωνίᾳ τῶν συμπτωμάτων, ἀγνοοῦντες, ὡς ὁ πόρος, ᾧ τὴν χολὴν εἰς τὴν γαστέρα τὸ ἧπαρ ἐξερεύγεται, τοῖς μὲν διπλοῦς ἐστι, τοῖς δ' ἁπλοῦς, ὡς κἂν ταῖς τῶν τετραπόδων ζῴων ἀνατομαῖς ἔνεστι θεάσασθαι. τὰ πολλὰ μὲν οὖν ἁπλοῦς ἐστιν εἰς τὸ μεταξὺ πυλωροῦ τε καὶ νήστεως ἐμφυόμενος, ὃ δὴ γαστρὸς ἔκφυσιν ὀνομάζουσιν, ἢ διπλοῦς γιγνόμενος εἰς μὲν τὴν ἔκφυσιν ἐμβάλλει θατέρῳ στόματι τῷ μείζονι, θατέρῳ δὲ τῷ μικροτέρῳ κατὰ τὸν πυθμένα μικρὸν ἄνω τοῦ πυλωροῦ. σπανιώτατα δέ ποτε τὸ μὲν ἄνω μέρος αὐτοῦ μεῖζον εὑρίσκεται, τὸ κάτω δ' ἔλαττον. ἀλλ' ἐφ' ὧν γε μεῖζόν ἐστιν, ἐπὶ τούτων ἡ γαστὴρ ἐφ' ἡμέραν ἐμπίπλαται χολῆς οὐκ ὀλίγης, ἣν ἐμεῖν τε δέονται πρὸ τῶν σιτίων καὶ βλάπτονται κατασχόντες. οἷς δὲ παντελῶς ἁπλοῦς ἐστιν ὁ πόρος, εἰς τὴν νῆστιν ἡ χολὴ τούτοις σύμπασα καταρρεῖ. πῶς οὖν χρὴ διαγιγνώσκειν αὐτούς; οὐ γὰρ δὴ ἀνατεμεῖν γε ζῶντας τοὺς ἀνθρώπους ἀξιῶ. πρῶτον μὲν ὅλῃ τοῦ σώματος τῇ κράσει, καθότι καὶ μικρὸν ἔμπροσθεν ἐλέγετο, δεύτερον δὲ τοῖς ὑπιοῦσι κάτω.

---

[39] post ἐνταῦθα: καὶ ἀνατομικόν τι θεώρημα συνέπεσεν H; μὲν ἐμπέπτωκε καὶ ἀνατομικόν τι θεώρημα, K

never vomited yellow bile and were thin, hirsute, muscular, dark, with prominent veins and seemed very hot, if someone were to touch them—the philosopher Eudemus is an example.

But here a certain anatomical speculation has intruded; some doctors, if they are not aware of this, are greatly at a loss when faced with the discordance of the symptoms, being unaware that the channel by which the liver discharges the bile to the stomach is double in some [animals] and single in others, as can be observed also in the dissections of quadrupeds.[28] In the majority, then, it is single and is implanted between the pylorus and the jejunum, which is what they call the outgrowth of the stomach, while if it is double, it inserts into the outgrowth by the larger opening and by the other and smaller opening into the base (of the stomach) a little above the pylorus. Very rarely the upper part of this is found to be larger, and the lower part smaller. | But in fact, in those in whom it is larger, the stomach is filled by day with not a little bile, which they need to vomit before foods and causes harm if retained. However, in those in whom the duct is entirely single, all the bile flows down to the jejunum. How then must we recognize these cases? I certainly don't think to carry out a dissection while people are living! First, it is from the whole *krasis* of the body, which I also spoke about a little earlier, and second, it is from what is excreted below (the stools).

632K

---

[28] This is taken to be a reference to inter- rather than intraspecies variation, but it is not altogether clear; see the detailed note in Singer and van der Eijk, *Galen: Works on Human Nature*, 143n163.

Χολώδη μὲν γὰρ ἄκρατα διὰ τῆς γαστρὸς ἐξεκενοῦτο συνεχῶς Εὐδήμῳ, διότι καὶ πολλὴν ἤθροιζε χολὴν καὶ οὐδὲν αὐτῆς εἰς τὴν ἄνω κοιλίαν ἀφικνεῖτο. τοῖς δ' ἄλλοις, οἷς ἡ μὲν ἕξις φλεγματώδης, ἐμοῦσι δὲ χολήν, ἥκιστα διαχωρεῖται χολώδη· καὶ γὰρ ὀλίγον γεννᾶται τῆς ξανθῆς χολῆς αὐτοῖς καὶ πλεῖστον εἰς τὴν ἄνω κοιλίαν ἀφικνεῖται. τρίτον ἐπὶ τοῖς εἰρημένοις εἶδος γνωρισμάτων ἐν αὐτοῖς τοῖς ἐμουμένοις[40] ἐστίν. οἷς μὲν γὰρ ἐν τῇ γαστρὶ θερμοτέρᾳ τὴν κρᾶσιν ὑπαρχούσῃ[41] τὸ χολῶδες γεννᾶται περίττωμα, πρασοειδὲς φαίνεται· ξανθὸν δ' ἀκριβῶς ἐστιν ἢ ὠχρόν γε πάντως, οἷς ἐξ ἥπατος ὑπέρχεται, καὶ οἷς μὲν ἐν γαστρὶ τὸ πρασοειδὲς τοῦτο γεννᾶται, χρὴ πάντως αὐτοῖς τὸ ἐδηδεσμένον σιτίον οὐκ ἄρτον οὐδὲ κρέας εἶναι χοίρειον ἤ τι τοιοῦτον παραπλήσιον, ἀλλὰ θερμότερόν τι τῶνδ' ἐξ ἀνάγκης καὶ οὐκ εὔχυμον. οἷς δ' ἐξ ἥπατος εἰς αὐτὴν ἀφικνεῖται, ξανθὸν ἢ ὠχρὸν ἐμεῖται, κἂν εὐχυμώτατον ᾖ τὸ ληφθὲν σιτίον κἂν ἄκρως πεφθῇ, καὶ μᾶλλόν γε τοῖς ἀκριβῶς πέψασιν ἐμεῖται τὰ ξανθὰ καὶ πολλῷ μᾶλλον ἔτι τοῖς ἐπὶ πλέον αὐτῶν ἀσιτήσασι. τὰ δὲ πρασοειδῆ μόνοις τοῖς κακῶς πέψασιν ἐν τῇ κοιλίᾳ γεννᾶται. καὶ μὲν δὴ καὶ φροντίδες καὶ θυμοὶ καὶ λῦπαι καὶ πόνοι καὶ γυμνάσια καὶ ἀγρυπνίαι καὶ ἀσιτίαι καὶ ἔνδειαι πλείονα τὸν τῆς ξανθῆς χολῆς τούτοις ἀθροίζουσι χυμόν, ὅτι καὶ πλείονα γεννῶσιν ἐν ἥπατι. ταῦτά τ' οὖν ἀκριβῆ τὰ γνωρίσματα καὶ πρὸς τούτοις ἔτι, τῷ μὲν αὐχμηρῷ

ON TEMPERAMENTS, BOOK II

In Eudemus, unmixed bilious material was continuously emptied out through the stomach because he collected a lot of bile and none of this came to the abdomen above. In others, however, in whom the bodily state (*hexis*) is phlegmatous, they vomit bile, their stools being least bilious, for little of the yellow bile is generated in them and most of it comes to the upper abdomen. There is a third kind of sign in addition to those mentioned in the evacuations themselves. For in those in whom the bilious superfluity is generated in the stomach which is hotter in *krasis*, it appears green. | It is, however, completely yellow or altogether ocher in those in whom it comes from the liver. And in those in whom this greenness is generated in the stomach, the food eaten by them must at all events not be bread or pork or some such similar thing, but of necessity something hotter than these and not *euchymous*. In those in whom it comes from the liver to the stomach, the vomitus is yellow or ocher, even if the food taken is perfectly *euchymous* and completely concocted. The yellowness of the vomitus is more in those who have concocted perfectly, and much more still in those of them who have fasted for a longer time. The green color is generated only in those who have concocted badly in the abdomen. And certainly also in those who are anxious, angry, sad, have worked hard, exercised, have not slept, and have fasted, and are lacking, the humor of yellow bile collects more because they also generate more in the liver. These, then, are the precise signs, and in addition to these further, since

---

[40] ἐμουμένοις H; κενουμένοις K
[41] *See H's note, p. 77, on the addition of* θερμοτέρᾳ τὴν κρᾶ-τιν ὑπαρχούσῃ *post* τῇ γαστρὶ.

καὶ πυρώδει τοῦ κατὰ τὴν γαστέρα θερμοῦ τῆς εἰς τὸ χολῶδες τροπῆς ἑπομένης, ἄρτοι καὶ κρέα χοίρεια καὶ βόεια κάλλιον πεφθήσονται τῶν πετραίων ἰχθύων. εἰ δ' ἐξ ἥπατος καταρρέοι, παρὰ τὴν τῶν ἐδεσμάτων ὑπάλλαξιν οὐδεμία τῆς πέψεως ἔσται διαφορά. ἐν τούτοις μὲν δὴ διώρισται τὸ διά τι ἄλλο καὶ μὴ διὰ τὴν κρᾶσιν γιγνόμενον.

Κατὰ δὲ τὸν αὐτὸν τρόπον, εἰ κἀκ τῆς κεφαλῆς εἰς τὴν γαστέρα φλέγμα καταρρέον ὀξυρεγμίας αἴτιον γίγνοιτο, χρὴ κἀνταῦθα κατὰ τὰς ὁμοίας μεθόδους ἀποχωρίζειν αὐτὸ τοῦ τῆς γαστρὸς ἰδίου παθήματος. οὕτω δὲ καὶ τὰς τῆς κεφαλῆς ὀδύνας, εἰ διὰ τὴν οἰκείαν αὐτῆς δυσκρασίαν ἢ διὰ τὰ τῆς γαστρὸς περιττώματα <γίγνονται>. καὶ μὲν δὴ καὶ τὸν ἐγκέφαλον ἐπισκέπτεσθαι καθ' ἑαυτόν, ὁποίας ἐστὶ κράσεως, ἄμεινον, οὐκ ἐκ τῆς τοῦ παντὸς σώματος διαθέσεως. αὐτοῦ δὲ καθ' ἑαυτὸν ὁποίας ἐστὶ κράσεως ἐπίσκεψις ἥ τε πολίωσίς ἐστιν οἵ τε κατάρροι καὶ βῆχες καὶ κόρυζαι καὶ σιέλων πλῆθος. ἅπαντα γὰρ ταῦτα ψυχρότερον αὐτὸν ἐμφαίνει καὶ ὑγρότερον ὑπάρχειν, ἔτι δὲ μᾶλλον, εἰ ἐπὶ ταῖς τυχούσαις προφάσεσιν εἰς τοιαύτας ἔρχεται διαθέσεις. ἡ δέ γε φαλάκρωσις ἐπὶ ξηρότητι καὶ ἡ τῶν μελαινῶν τε καὶ πολλῶν γένεσις τριχῶν εὐκρασίας ἐγκεφάλου γνώρισμα.

Κατὰ τοῦτον οὖν τὸν τρόπον ἀεὶ χρὴ σκοπεῖσθαι περὶ κράσεως, ἕκαστον ἰδίᾳ μόριον ἐξετάζοντα, καὶ μὴ περὶ πάντων ἀποφαίνεσθαι τολμᾶν ἐξ ἑνός, ὥσπερ ἐποίησαν ἔνιοι, τοὺς μὲν σιμοὺς ὑγροὺς εἶναι φάμε-

the change to what is bilious follows the parched and fiery heat in the stomach, bread, pork | and ox flesh will be concocted better than the fish of the rocks. If, however, it flows down from the liver, there will not be any difference of concoction apart from the alteration of the foods. Undoubtedly in these there is differentiation of what occurs due to something else and not to the *krasis*.

In the same manner, if phlegm flows down to the stomach from the head, it may become a cause of acid indigestion (heartburn), and even here this must be separated by similar methods from the specific affection of the stomach. And the same applies to the pains of the head that arise due either to the specific *dyskrasia* of this or to the superfluities of the stomach. Furthermore also, it is better to consider the brain in and of itself, as to which kind of *krasis* there is, and not (to judge it) from the condition of the whole body. Inspection of the kind of *krasis* of this in itself involves the hair turning gray, catarrhs, coughs, coryzas, and abundant saliva. All these show a head that is colder and wetter, and more so, if it comes to such conditions following contingent causes. If, in fact, baldness follows dryness and also the genesis of dark and thick hair | it is a sign of *eukrasia* of the brain.

In the same way, then, examination of *krasis* must always involve investigating each part individually and not making a bold declaration about all from one, as some did, saying those with snub noses are wet while those with

νοι, τοὺς δὲ γρυποὺς ξηροὺς καὶ οἷς μὲν οἱ ὀφθαλμοὶ μικροί, ξηρούς, οἷς δὲ μεγάλοι, ὑγρούς. τοῦτο μέν γε καὶ διαπεφώνηται πρὸς αὐτῶν. οἱ μὲν γάρ τινες ὑποθέμενοι τῶν ὑγρῶν εἶναι μορίων τοὺς ὀφθαλμούς, ἐν οἷς ἂν μείζους εὑρίσκωσιν, ὑγρότητα κράσεως ἐν τούτοις κρατεῖν ὑπολαμβάνουσιν. ἔνιοι δὲ τῇ ῥώμῃ τοῦ θερμοῦ κατὰ τὴν πρώτην διάπλασιν ἀναπνεύσαντος ἀθροωτέρου τε καὶ πλείονος οὐκ ὀφθαλμοὺς μόνον ἀλλὰ καὶ στόμα καὶ τοὺς ἄλλους ἅπαντας πόρους γενέσθαι φασὶ μείζονας, ὅθεν οὐχ ὑγρότητος, ἀλλὰ θερμότητος εἶναι γνώρισμα. ἀμφότεροι δὲ διαμαρτάνουσι τῆς ἀληθείας ἑνὶ μὲν καὶ κοινῷ λόγῳ, διότι περὶ παντὸς τοῦ σώματος ἐξ ἑνὸς ἀποφαίνεσθαι τολμῶσι μορίου· κατὰ δεύτερον δὲ τρόπον, ὅτι τῆς διαπλαστικῆς ἐν τῇ φύσει δυνάμεως οὐ μέμνηνται τεχνικῆς τ' οὔσης καὶ τοῖς τῆς ψυχῆς ἤθεσιν ἀκολούθως διαπλαττούσης τὰ μόρια. περὶ ταύτης γάρ τοι καὶ ὁ Ἀριστοτέλης ἠπόρησε, μή ποτ' ἄρα θειοτέρας τινὸς ἀρχῆς εἴη καὶ οὐ κατὰ τὸ θερμὸν καὶ ψυχρὸν καὶ ξηρὸν καὶ ὑγρόν. οὔκουν ὀρθῶς μοι δοκοῦσι ποιεῖν οἱ προπετῶς οὕτως ὑπὲρ τῶν μεγίστων ἀποφαινόμενοι καὶ ταῖς ποιότησι μόναις ἀναφέροντες τὴν διάπλασιν. εὔλογον γὰρ ὄργανα μὲν εἶναι ταύτας, τὸ διαπλάττον δ' ἕτερον. ἀλλὰ καὶ χωρὶς τῶν τηλικούτων ζητημάτων ἐνὸν ἐξευρίσκειν, ὡς ἔμπροσθεν ἐδείξαμεν, ὑγρὰν καὶ ξηρὰν καὶ ψυχρὰν καὶ θερμὴν κρᾶσιν ἁμαρτάνουσιν οἱ τῶν οἰκείων μὲν ἀμελοῦντες γνωρισμάτων, ἐπὶ δὲ τὰ πόρρω τε καὶ ζητήσεως ἱκανῆς τετυχηκότα καὶ μέ-

hooked noses are dry, and in them small eyes [indicate they are dry] and large eyes that they are wet. This, in fact, is also a matter of disagreement among them. For some suppose the eyes to be among the wet parts, and they would discover more in them, and they assume the wetness of *krasis* prevails in them. Some, however, say that, due to the strength of the hot in the first conformation when exhalation is more concentrated and greater, not only the eyes but also the mouth and all the other channels become larger, which is why it is a sign not of wetness but of hotness. Both, however, depart from the truth, due to one common reason: because they boldly make a declaration about the whole body from one part, while in a second way, because they do not call to mind the formative power in Nature, | which, being demiurgical, also fashions the parts to follow character traits of the soul. Regarding this even Aristotle doubted whether it might be of some more divine origin and not related to hot and cold, dry and wet. Those who make precipitate declarations in this way on the greatest matters, and refer the formation to the qualities alone, do not therefore seem to me to act correctly. For it is reasonable that the instruments (i.e., bodily components) are these but the conformation is something else. But also, apart from such great inquires, it is possible to discover, as we showed previously, wet, dry, cold, and hot *krasis*, so those who neglect the characteristic signs err in regard to those things which happen to have been subjects of a considerable investigation, and up to the present

636K

χρι τοῦ δεῦρο καὶ παρ' αὐτοῖς τοῖς ἀρίστοις φιλοσόφοις ἀπορούμενα μεταβαίνοντες. οὐδὲ γὰρ οὐδ' ὅτι τὰ μὲν παιδία σιμότερα, γρυπότεροι δ' οἱ παρακμάζοντες, εὔλογον ὑγροὺς μὲν νομίζειν τοὺς σιμοὺς ἅπαντας, ξηροὺς δὲ τοὺς γρυπούς. ἀλλ' ἐνδέχεται μὲν καὶ τῆς διαπλαστικῆς δυνάμεως ἔργον εἶναι τὸ τοιοῦτον μᾶλλον ἢ τῆς κράσεως. εἰ δ' ἄρα καὶ τῆς κράσεως εἴη γνώρισμα, τῆς ἐν τῇ ῥινὶ μόνης ἂν εἴη, οὐ τῆς ἐν ὅλῳ τῷ σώματι. μάτην οὖν ὑπ' αὐτῶν κἀκεῖνο λέγεται τὸ ῥῖνά τ' ὀξεῖαν γίγνεσθαι καὶ ὀφθαλμοὺς κοίλους καὶ κροτάφους συμπεπτωκότας ἐν ταῖς ξηραῖς φύσει κράσεσιν, ὅτι κἀν τοῖς πάθεσιν οὕτω συμπίπτει τοῖς συντήκουσί τε καὶ κενοῦσι πέρα τοῦ μετρίου τὰ σώματα. πολλάκις μὲν γὰρ οὕτω συμπίπτει, πολλάκις δ' οὐχ οὕτως. ἀλλ' ἔστιν ἰδεῖν καὶ μαλακὴν καὶ πιμελώδη καὶ λευκὴν καὶ πολύσαρκον ὅλου τοῦ σώματος τὴν ἕξιν ἐπὶ σμικροῖς ὀφθαλμοῖς ἢ ὀξείᾳ ῥινὶ καὶ ξηρὰν καὶ ἄσαρκον καὶ μέλαιναν καὶ δασεῖαν ἐπὶ μεγάλοις ὀφθαλμοῖς σιμῇ τε ῥινί. βέλτιον οὖν, εἴπερ ἄρα, τῆς ῥινὸς μόνης ὑγρότητα μὲν τῇ σιμότητι, ξηρότητα δὲ τῇ γρυπότητι τεκμαίρεσθαι καὶ μὴ περὶ τῆς ἅπαντος τοῦ ζῴου κράσεως ἐντεῦθεν ἀποφαίνεσθαι.

Κατὰ ταυτὰ δὲ καὶ τῶν ὀφθαλμῶν καὶ παντὸς οὑτινοσοῦν ἑτέρου μορίου τὴν ἰδίαν κρᾶσιν ἐκ τῶν οἰκείων ἐπισκοπεῖσθαι γνωρισμάτων ἄμεινον, οὐ περὶ τῆς ὅλου τοῦ σώματος κράσεως ἀφ' ἑνὸς μέρους ἔνδειξιν λαμβάνειν. εἴτε γὰρ ὑγρότητος εἴτε θερμότητος

day continue to be problematic among the best philosophers themselves. For it is not reasonable that just because children are more snub-nosed while those past their prime are more hook-nosed to think that all who are snub-nosed are wet, whereas those who are hook-nosed are dry. But it is possible that such a thing is more the work of the formative power than of the *krasis*. If, then, it is also a sign of the *krasis*, it would only be of that in the nose and not of that in the body as a whole. Therefore, it is unreasonable for them to say that a sharp nose and sunken eyes and temples occur together in the naturally dry *krasias* because this is what happens in those affected when the bodies waste away and are emptied beyond the moderate. It does often happen like this, but it often does not. It is also possible to see a soft, fatty, pale, and excessively fleshed state of the whole body along with small eyes or a sharp nose, and conversely a dry, fleshless, dark, and hirsute state along with large eyes and a snub nose. It is better, therefore, to take a snub nose as evidence of wetness of the nose alone and a hook nose of dryness and not here to make a statement about the *krasis* of the whole animal (organism).

In the same way too, it is better, concerning the eyes and any other part whatsoever, to observe the specific *krasis* from the characteristic signs, and not to take an indication concerning the *krasis* of the whole body from one part. If we must set up blue eyes as a sign of dominance

ἐπικρατούσης εἴτε καὶ ἀμφοτέρων ὀφθαλμοὺς γλαυκοὺς τίθεσθαι γνώρισμα χρή, τῆς οἰκείας ἂν εἶεν οὕτω γε, οὐ τῆς ἁπάντων τῶν τοῦ σώματος μορίων ἐνδεικτικοὶ κράσεως. οὐδὲ γάρ, εἰ ξηρὰ καὶ ἄσαρκα τὰ σκέλη, ξηρὰ πάντως καὶ ἡ κρᾶσις ὅλου τοῦ σώματος. ἔνιοι γὰρ ἱκανῶς εὔσαρκοι καὶ πιμελώδεις καὶ παχεῖς καὶ προγάστορες καὶ μαλακοὶ καὶ λευκοὶ γίγνονται μετὰ τοιούτων σκελῶν. ἀλλ' εἰ μὲν ὁμαλῶς ἅπαν ἔχει τὸ σῶμα τῆς κράσεως, οἷς μὲν ἰσχνὰ τὰ σκέλη, ξηροὶ πάντως εἰσίν, ὑγροὶ δ', οἷς παχέα. καὶ οἷς μὲν ἡ ῥὶς ὀξεῖα καὶ γρυπή, ξηροί, σιμῆς δ' οὔσης ὑγροί· κατὰ ταὐτὰ δὲ καὶ περὶ τῶν ὀφθαλμῶν τε καὶ τῶν κροτάφων ἁπάντων τε τῶν ἄλλων μορίων. οἷς δ' ἀνώμαλος ἡ κρᾶσις καὶ οὐχ ἡ αὐτὴ πάντων τῶν μερῶν, ἄτοπον ἐπὶ τούτων ἐξ ἑνὸς μορίου φύσεως ὑπὲρ ἁπάντων ἀποφαίνεσθαι. τοιοῦτον δέ τι τοὺς πλείστους αὐτῶν ἠπάτησεν, οὐχ ὑπὲρ τῶν ἀνθρώπων μόνον, ἀλλὰ καὶ περὶ τῶν ἄλλων ζῴων ἀποφήνασθαι τολμήσαντας ὑπὲρ ὅλης τῆς κράσεως ἐκ μόνων τῶν κατὰ τὸ δέρμα γνωρισμάτων. οὔτε γάρ, εἰ σκληρὸν τὸ δέρμα, ξηρὸν ἐξ ἀνάγκης τὸ ζῷον, ἀλλ' ἐγχωρεῖ τὸ δέρμα μόνον, οὔτ' εἰ μέλαν οὔτ' εἰ δασύ. κατὰ δὲ τὸν αὐτὸν τρόπον οὐδ' εἰ μαλακὸν ἢ λευκὸν ἢ ψιλὸν τριχῶν, ὑγρὸν ἐξ ἀνάγκης ὅλον τὸ ζῷον. ἀλλ' εἰ μὲν ὁμαλῶς κέκραται σύμπαν, εὔλογόν ἐστιν, οἷόνπερ τὸ δέρμα, τοιοῦτον εἶναι καὶ τῶν ἄλλων ἕκαστον μορίων, εἰ δ' ἀνωμάλως, οὐκέτι. τῶν γοῦν ὀστρέων ὑγρότατον μὲν ὅλον τὸ σῶμα, ξηρότατον δὲ τὸ δέρμα· τὸ γὰρ δὴ

ON TEMPERAMENTS, BOOK II

of either wetness or hotness or both, they would in fact be indicators of their own *krasis* and not that of all the parts of the body. Nor would the *krasis* of the whole body be altogether dry, if the legs were dry and fleshless. For some people become quite well-fleshed and fatty, and thick and potbellied, and soft, and pale along with such legs. But if the whole body has a uniformity of the *krasis*, those in whom the legs are thin are altogether dry, while those in whom they are thick are wet. And those in whom the nose is sharp or hooked are dry, while those in whom it is snub are wet. It is the same too regarding the eyes and the temples and all the other parts. On the other hand, in those in whom the *krasis* is non-uniform, and is not the same in all the parts, it is unwonted to make a statement about all the parts from the nature of one part. Such a thing deceives the majority of them, not only about humans, but also about the | other animals, when they boldly make a statement about the whole *krasis* from the signs pertaining to the skin alone. For the animal is not necessarily dry, if the skin is hard; it is possible for the skin alone [to be hard]. [The same applies] if it is black or hirsute. In the same way, if the skin is soft or pale or devoid of hair, the whole animal is not necessarily wet. But if the whole has been compounded uniformly, it is reasonable for each of the other parts, like the skin, to be so too. On the other hand, if the *krasis* is non-uniform this is no longer the case. Anyway, the whole body of oysters is very wet while the skin is very dry, for certainly the shell in them is like the

639K

ὄστρακον αὐτοῖς οἱόνπερ ἡμῖν τὸ δέρμα. καὶ ἡ προσηγορία δ' ἐντεῦθεν, ὀστρακοδέρμων ἁπάντων τῶν τοιούτων ζῴων ὀνομασθέντων ἐκ τοῦ τὸ δέρμα παραπλήσιον ἔχειν ὀστράκῳ. καὶ τὰ μαλακόστρακα δέ, καθάπερ ἀστακοί τε καὶ κάραβοι καὶ καρκῖνοι, ξηρὸν μὲν ἔχει τὸ δέρμα, τὴν δ' ἄλλην ἅπασαν κρᾶσιν ὑγράν. καὶ αὐτό γε τοῦτο πολλάκις αἴτιον ὑπάρχει τοῖς ζῴοις τῆς ἐν ταῖς σαρξὶν ὑγρότητος, ὅτι πᾶν αὐτοῖς τὸ ξηρὸν καὶ γεῶδες ἡ φύσις ἀποτίθεται πρὸς τὸ δέρμα. μὴ τοίνυν μήθ' ὅτι ξηρὸν τοῦτο τοῖς ὀστρέοις, εὐθέως καὶ τὴν σάρκα νομιστέον ὑπάρχειν ξηρὰν μήθ', ὅτι πλαδαρὰ καὶ μυξώδης ἥδε, τοιοῦτον ὑποληπτέον εἶναι καὶ τὸ δέρμα. δίκαιον γὰρ ἕκαστον τῶν μορίων ἐξ ἑαυτοῦ γνωρίζεσθαι.

Ταῦτά τ' οὖν ἁμαρτάνουσιν οἱ τὰ περὶ κράσεων ἡμῖν ὑπομνήματα καταλελοιπότες ἔτι τε πρὸς τούτοις, ὅτι μηδὲ μέμνηνται τοῦ πρὸς Ἱπποκράτους ὀρθότατα παρῃνημένου τοῦ δεῖν ἐπισκέπτεσθαι τὰς μεταβολὰς ἐξ οἵων εἰς οἷα γίγνονται. πολλάκις γὰρ τὰ παρόντα γνωρίσματα τῆς ἔμπροσθεν κράσεώς ἐστιν, οὐ τῆς νῦν ὑπαρχούσης τῷ σώματι. φέρε γὰρ εἴ τις ἔτη γεγονὼς ἑξήκοντα δασὺς ἱκανῶς εἴη, μὴ διότι νῦν ἐστι θερμὸς καὶ ξηρός, ἀλλ' ὅτι πρόσθεν μὲν ἐγένετο τοιοῦτος, ὑπομένουσι δ' αἱ τότε γεννηθεῖσαι τρίχες, ὥσπερ ἐν τῷ θέρει πολλάκις αἱ κατὰ τὸ ἔαρ ἀναφυεῖσαι βοτάναι. τισὶ μὲν γὰρ ἐν τῷ χρόνῳ προϊόντι κατὰ βραχὺ συνέβη τῆς ἄγαν ἐκείνης ἀπαλλαγῆναι δασύτητος ἐκπιπτουσῶν ὑπὸ ξηρότητος ἄκρας τῶν τριχῶν,

ON TEMPERAMENTS, BOOK II

skin in us. And the appellation of ostrakoderms here, when applied to all such animals, comes from the skin in them being similar to an *ostrakon* (earthenware vessel). And soft-shelled animals[29] like smooth lobsters, crayfish, and crabs have dry skin but all the other *krasis* is wet. And this itself is often a cause for the animals of the wetness in the flesh; in them all | the dryness and earthiness, Nature puts aside to the skin. Accordingly, it is not because there is this dryness in oysters that one must immediately think the flesh is dry, nor, because the flesh is flabby and mucoid, must one suppose the skin is also like this. It is proper for each of the parts to be recognized in its own right.

640K

In these respects, then, those who have left us treatises about the *krasias* are mistaken, and further, in addition to these, failed to mention what was very correctly advised by Hippocrates: the need to consider from what and to what the changes occur.[30] For often the present signs are of the previous *krasis* and not of what exists in the body presently. So then, if some sixty-year-old man is notably hirsute, it is not because he is now hot and dry, but because he was so previously and the hairs generated at that time remain, just as often plants from the spring grow up again in the summer. In some people what happens over the course of time and gradually, is that they lose the marked hairiness when the hairs fall out due to extreme dryness.

---

[29] On the term μαλακόστρακος (soft-shelled) applied to the animals listed, see Aristotle, *History of Animals* 490b11, LCL 437 (A. L. Peck), 30–31. As Peck remarks in his note, the term is applied to crustaceans.

[30] I am unable to determine to whom Galen is referring here or to locate the specific Hippocratic reference.

ἐνίοις δ' ἄχρι πλείστου παραμένουσιν, οἷς ἂν μήτ' ἀποξηρανθῶσιν ἱκανῶς ἐπὶ προήκοντι τῷ χρόνῳ καὶ τὴν πρώτην ἔκφυσιν αἱ τρίχες αὐτοῖς ἰσχυρὰν ποιήσωνται δίκην φυτῶν ἀκριβῶς ἐνερριζωμένων τῇ γῇ. μὴ τοίνυν, εἰ δασύς τις ἱκανῶς ἐστιν, εὐθὺς τοῦτον οἰώμεθα μελαγχολικὸν ὑπάρχειν, ἀλλ' εἰ μὲν ἀκμάζων, οὔπω τοιοῦτον· εἰ δὲ παρακμάζων, ἤδη μελαγχολικόν· εἰ δὲ γέρων, οὐκέτι. γίγνονται μὲν γὰρ αἱ μελαγχολικαὶ κράσεις ἐκ συγκαύσεως αἵματος. οὐ μήν, ἐπειδὰν ἄρξηται τοῦτο πάσχειν, εὐθὺς καὶ κατώπτηται τελέως. ἀλλ' ἐν τάχει μὲν ἱκανῶς ἔσται δασὺς ὁ θερμὸς καὶ ξηρός, εἴ τι μεμνήμεθα τῶν ἔμπροσθεν λόγων, οὐκ εὐθέως δὲ μελαγχολικός. ἡ γὰρ τοῦ δέρματος πύκνωσις εἴργουσα τῶν παχυτέρων περιττωμάτων τὴν διέξοδον ἀναγκάζει συγκαίεσθαι κατὰ τὰς ἄκρως θερμὰς κράσεις, ὥστε τοιοῦτον αὐτοῖς ὑπάρχειν ἤδη τὸ περίττωμα τὸ φύον τὰς τρίχας, οἷον ἐν τοῖς ἀγγείοις ἔσεσθαι μέλλει προελθόντος τοῦ χρόνου.

Καὶ ταῦτ' οὖν ἠμέληται τοῖς ἔμπροσθεν ἔτι τε πρὸς τούτοις, ἐπειδὰν ἐκ τῆς φύσεως τῶν περιττωμάτων ἀδιορίστως ὑπὲρ τῶν κράσεων ἀποφαίνωνται. νομίζουσι γὰρ ἀνάλογον ἔχειν τὰς κράσεις τῶν μορίων τῇ φύσει τῶν περιττωμάτων. τὸ δ' οὐχ ὅλως ἀληθές ἐστιν, ἀλλ' ἐγχωρεῖ ποτε περίττωμα μὲν ἀθροίζεσθαι φλεγματῶδες, ὑγρὸν δ' οὐκ εἶναι τὸ μόριον, ἀλλὰ ψυχρὸν μὲν ἐξ ἀνάγκης, οὐ γὰρ δὴ ἄλλη γέ τις ἡ τοῦ

In others, however, the hair remains for a very long time because they have not been dried out significantly with the passage of time, and the hair in them made the first outgrowth strong, like plants which are completely rooted in the earth. Moreover, if someone is markedly hirsute, we should not immediately think this person is melancholic, but if in his prime, not yet like this. If, however, he is past his prime, he is already melancholic, whereas if old, no longer so. For the melancholic *krasias* occur from burning up of blood. When this begins to be affected, it is not also immediately and completely overheated.[31] But the hot and dry person will rapidly become significantly hirsute, if we recall something of the previous discussions; he will not, however, immediately become melancholic. For the condensation of the skin impedes the outward passage of the thicker superfluities and compels the burning-up as well in the extremely hot *krasias*, so the existing superfluity that creates the hairs in them will come to be in the vessels with the passage of time.

These things, then, were neglected by those before us, and further, besides these, whenever they make pronouncements about the *krasias* loosely from the nature of the superfluities. For they think the *krasias* of the parts are proportional to the nature of the superfluities. This is not, however, wholly true. Rather, it is sometimes possible for a phlegmatic superfluity to be collected, but the part not to be wet. But it is cold of necessity, for the generation

---

[31] On the verb κατοπτάω meaning "overheating of the blood," see Galen, *De symptomatum causis*, VII.246K.

φλέγματος γένεσις,[42] ὑγρὸν δ' οὐκ ἐξ ἀνάγκης· ἐγχωρεῖ γὰρ καὶ ξηρὸν εἶναι. τὸ δ' ἀπατῆσαν αὐτοὺς εὐφώρατον. οὐ γὰρ ἐνενόησαν, ὡς ἐκ τῶν σιτίων, οὐκ ἐξ αὐτοῦ τοῦ σώματος ἡμῶν γίγνεται τὸ φλέγμα. θαυμαστὸν οὖν οὐδέν, εἰ μὴ κρατῆσάν ποτε τὸ σῶμα τῶν προσενεχθέντων σιτίων, ὑγρῶν, εἰ τύχοι, τὴν φύσιν ὑπαρχόντων, ὅμοιον αὐτοῖς ἀποτελέσει καὶ τὸ περίττωμα. μὴ τοίνυν ὑπολαμβανέτωσαν ὥσπερ τὸ σῶμα ξηρὸν οὕτω καὶ τὸ περίττωμα δεῖν ξηρὸν εἶναι. εἰ γάρ τις εὐθὺς ἐξ ἀρχῆς ἐγένετο τῇ κράσει ψυχρότερός τε καὶ ξηρότερος, οὐ μελαγχολικὸς ὁ τοιοῦτος, ἀλλὰ φλεγματικός ἐστι τοῖς περιττώμασιν. εἰ δ' ἐκ μεταπτώσεως ἐγένετο ψυχρὸς καὶ ξηρός, ἐξ ἀνάγκης ὁ τοιοῦτος εὐθὺς ἤδη καὶ μελαγχολικός ἐστιν, οἷον εἴ τις ἔμπροσθεν ὑπάρχων[43] θερμὸς καὶ ξηρὸς ἐκ συγκαύσεως τοῦ αἵματος πλείστην ἐγέννησε τὴν μέλαιναν χολήν. οὗτος γάρ ἐστιν ὁ πρὸς τῷ ξηρός εἶναι καὶ ψυχρὸς εὐθὺς καὶ μελαγχολικὸς ὑπάρχων. εἰ δ' ἀπ' ἀρχῆς εἴη ψυχρὸς καὶ ξηρός, ἡ μὲν ἕξις τοῦ σώματος τούτῳ λευκὴ καὶ μαλακὴ καὶ ψιλὴ τριχῶν, ἄφλεβος δὲ καὶ ἄναρθρος καὶ ἰσχνὴ καὶ ἁπτομένοις ψυχρὰ καὶ τὸ τῆς ψυχῆς ἦθος ἄτολμον καὶ δειλὸν καὶ δύσθυμον, οὐ μὴν μελαγχολικά γε τὰ περιττώματα.

Ταῦτ' οὖν ἅπαντα διαμαρτάνουσιν οἱ πολλοὶ τῶν ἰατρῶν ἐκ τοῦ τῶν οἰκείων μὲν ἀποχωρῆσαι γνωρισμάτων, ἐπὶ δὲ τὰ συμβεβηκότα μὴ διὰ παντός, ἀλλ' ὡς ἐπὶ τὸ πολὺ μεταβῆναι. ταύτῃ τοι καὶ τὸ θερμαί-

of the phlegm does not occur otherwise, although not wet of necessity as it is also possible for it to be dry. What deceived them is easy to detect. They had not realized that phlegm arises from the foods and not from our body itself. It is no wonder then, if the body sometimes does not prevail over the proffered foods, and they are wet in nature as may be the case, they will also make the superfluity like themselves. Let them not assume, therefore, that as the body is dry, so too the superfluity must also be dry. For if someone immediately from the beginning became colder and drier in *krasis*, such a person is not melancholic but phlegmatic from the superfluities. If, however, he became cold and dry from change, of necessity such a person is immediately already also melancholic, as for example someone who was previously hot and dry from overheating of the blood generated very much black bile. This person, in addition to being dry and cold, is also immediately melancholic. If, however, he were cold and dry from the start, the state (*hexis*) of the body in him would be pale, soft, and devoid of hairs, without visible veins and joints, thin and cold to those touching it, and the disposition of the soul cowardly, craven and dispirited, although the superfluities are not in fact melancholic.

Thus the majority of doctors mistake all these things by departing from the characteristic signs, moving over to those things that are contingent, not present forever but for the most part. In this way they think what heats also

---

[42] οὐ γὰρ δὴ ἄλλη γέ τις ἡ τοῦ φλέγματος γένεσις, *add.* H *(see note, p. 83)* [43] ὑπάρχων *add.* H

νον ἡγοῦνται πάντως ξηραίνειν. ἔγνωκα γὰρ ἔτι τοῦτο προσθεὶς οἷον κορωνίδα τε καὶ κεφαλήν τινα τῷ λόγῳ παντὶ καταπαύειν ἤδη τὸ δεύτερον γράμμα. θερμὸν γοῦν ὕδωρ ἐπαντλοῦντες ἑκάστοτε τοῖς φλεγμαίνουσι μορίοις, εἶθ᾽ ὁρῶντες αὐτῶν ἐκκενουμένην τὴν ὑγρότητα προφανῶς οἴονται δείκνυσθαι τὸ ξηραίνεσθαι πάντα πρὸς τῆς θερμασίας, οὐκ εἰ μετὰ ξηρότητος μόνον, ἀλλ᾽ εἰ καὶ μεθ᾽ ὑγρότητος εἴη. ἔστι δ᾽ οὐ ταὐτὸν ἢ κενῶσαί τινος ὑγρότητα παρεσπαρμένην ἔν τισι χώραις ἢ ξηροτέραν ἀπεργάσασθαι τὴν οἰκείαν κρᾶσιν. ἀνώμαλος γάρ τις ἐν τοῖς φλεγμαίνουσι μορίοις γίγνεται δυσκρασία, τῶν μὲν ὁμοιομερῶν σωμάτων οὔπω τῆς οἰκείας ἐξεστηκότων φύσεως, ἀλλ᾽ ἔτι μεταβαλλομένων τε καὶ ἀλλοιουμένων, ἐμπεπλησμένων δὲ τοῦ ῥεύματος ἁπασῶν τῶν μεταξὺ χωρῶν. ἅπαντ᾽ οὖν, ὅσα θερμὰ καὶ ὑγρὰ τὴν κρᾶσιν ἐστί, προσαγόμενα τοῖς οὕτω διακειμένοις ἐκκενοῖ μὲν τὸ περιττὸν ἐκεῖνο τὸ τὰς μεταξὺ χώρας τῶν ὁμοιομερῶν κατειληφός· αὐτὰ δὲ τὰ σώματα τοσοῦτον ἀποδεῖ τοῦ ξηραίνειν, ὥστε καὶ προσδίδωσιν αὐτοῖς ὑγρότητος. τὸ μὲν οὖν ἀληθὲς ὧδ᾽ ἔχει. δεῖ δὲ τοῖς εἰρημένοις ἀποδείξεως, ἣν μακροτέραν τ᾽ εἶναι νομίζων ἢ ὥστε προσγράφεσθαι κατὰ τόνδε τὸν λόγον ἔτι τ᾽ ἀκροατοῦ δεομένην ἐπισταμένου περὶ φαρμάκων δυνάμεως, ἀναβάλλομαι τό γε νῦν διελθεῖν. ἀλλ᾽ ἐπειδὰν τὸν τρίτον λόγον περὶ κράσεων ἅπαντα διέλθω καὶ δείξω περὶ τῶν κατὰ δύναμιν ὑγρῶν καὶ ξηρῶν καὶ ψυχρῶν

always dries. I have determined to add this as a kind of completion and heading to the whole argument and now to finish the second book. Anyway, when they pour hot water over the inflamed parts on each occasion, if they see the wetness of these emptied out, they think this quite clearly demonstrates that the drying is altogether from the heating, not | only if it is accompanied by dryness, but also if it is accompanied by wetness. However, to empty out of something wetness dispersed in the regions and to make drier the characteristic *krasis* are not the same. For some non-uniform *dyskrasia* arises in the inflamed parts of which the *homoiomerous* bodies have not yet departed from the characteristic nature, but are still changing and altering, while all the spaces between have been filled with the flux. Therefore, all those things that are hot and wet in terms of *krasis*, when taken by those in this state, evacuate the superfluity there, which has taken over the spaces between the *homoiomeres*. But these same bodies lack the dryness to such a degree that as a result they also increase the wetness in themselves. Really and truly this is so. However, there needs to be a demonstration for the things stated which I think would be too long, or that to write besides this particular discussion would require of the listener knowledge concerning the potency of medications, which in fact I now put off | going over. But when I come to the third book about *krasias*, I shall show the whole methodology relating to potency of all those things that are

καὶ θερμῶν ἅπασαν τὴν μέθοδον, ἐφεξῆς οὕτω βιβλίον ὅλον ὑπὲρ ἀνωμάλου δυσκρασίας ἔγνωκα γράψαι. τελειωθήσεται γὰρ ἅπας ἡμῖν ὁ περὶ κράσεων λόγος εἴς τε τὴν θεραπευτικὴν μέθοδον οὐ σμικρὰς ἀφορμὰς παρέξει.

wet, dry, cold and hot. I have decided next to write a whole book about the non-uniform *dyskrasia*, so my whole discussion about *krasias* will be completed and will provide no small resource for the therapeutic method.

ΒΙΒΛΙΟΝ ΤΡΙΤΟΝ

1. Ὅτι μὲν οὖν ἕκαστον τῶν ἐνεργείᾳ θερμῶν καὶ ψυχρῶν καὶ ξηρῶν καὶ ὑγρῶν ἢ τῷ τὴν ἄκραν δεδέχθαι ποιότητα τοιοῦτον εἶναί φαμεν ἢ ἐπικρατήσει τινὸς ἐξ αὐτῶν ἢ πρὸς τὸ σύμμετρον ὁμογενὲς παραβάλλοντες ἢ πρὸς ὁτιοῦν τῶν ἐπιτυχόντων, ἔμπροσθεν εἴρηται. δέδεικται δὲ καί, ὡς ἄν τις μάλιστα δύναιτο διαγιγνώσκειν ἀκριβῶς αὐτά. λοιπὸν δ' ἂν εἴη περὶ τῶν δυνάμει τοιούτων διελθεῖν αὐτὸ πρότερον ἐξηγησαμένους τοὔνομα τί ποτε σημαίνει τὸ δυνάμει.

Σύντομος δὲ καὶ ῥᾴστη καὶ σαφὴς ἡ ἐξήγησις. ὃ γὰρ ἂν | ὑπάρχῃ μὲν μήπω τοιοῦτον, οἷον λέγεται, πέφυκε δὲ γενέσθαι τοιοῦτον, δυνάμει φαμὲν ὑπάρχειν αὐτό, λογικὸν μὲν τὸν ἄρτι γεγενημένον ἄνθρωπον, πτηνὸν δὲ τὸν ὄρνιν, καὶ θηρατικὸν μὲν τὸν κύνα, ταχὺν δὲ τὸν ἵππον, ὅπερ ἔσεσθαι πάντως ἕκαστον αὐτῶν μέλλει μηδενὸς τῶν ἔξωθεν ἐμποδὼν αὐτῷ γενομένου, τοῦθ' ὡς ὂν ἤδη λέγοντες. ὅθεν οἶμαι καὶ δυνάμει ταῦτα πάντα φαμὲν ὑπάρχειν, οὐκ ἐνεργείᾳ. τέλειον μὲν γάρ τι καὶ ἤδη παρὸν ἡ ἐνέργεια·

---

[1] The terms ἐνεργείᾳ and δυνάμει are rendered "in actuality/

BOOK III

1. It was said before that each of the things that are hot, cold, dry, and wet in actuality, we say are so either by having received such a quality to an extreme degree, or through one of them dominating, or by comparison with the median of the same class, or whatever of these has happened. It has also been shown how someone is most able to recognize these accurately. It now remains to go over such things "in potentiality" (potentially),[1] having first explained this term and what it sometimes signifies.

The explanation is brief, very easy, and clear. For if something | is not yet such a thing as it is called, but is of a nature to become such a thing, we say this exists "in potentiality"—a human that has just come into being is "rational" [in this sense], a bird "flying," a dog "hunting," or a horse "swift." What each of these will, in all likelihood, become in the future, providing nothing external arises to prevent it, we speak as already existing. This is why, I think, we say all these things exist potentially but not actually. For "actuality" is said of something that is complete

---

actually" and "in potentiality/potentially," respectively. The three terms ἐνεργεία, δυνάμις, and ὕλη are rendered in line with Aristotle's usage in *Metaphysics* 8 and 9, the last being understood as "matter."

τὸ δυνάμει δ' ἀτελές τι καὶ μέλλον ἔτι καὶ οἷον ἐπιτήδειον μὲν εἰς τὸ γενέσθαι, μήπω δ' ὑπάρχον ὃ λέγεται. οὔτε γὰρ τὸ βρέφος ἤδη λογικόν, ἀλλ' ἔσεσθαι μέλλει, οὔθ' ὁ γεγενημένος ἄρτι κύων ἤδη θηρατικός, ὅς γε μηδὲ βλέπει μηδέπω, τῷ δύνασθαι δ', εἰ τελειωθείη, θηρᾶν οὕτως ὀνομάζεται. κυριώτατα μὲν οὖν ἐκεῖνα μόνα δυνάμει λέγομεν, ἐφ' ὧν ἡ φύσις αὐτὴ πρὸς τὸ τέλειον ἀφικνεῖται μηδενὸς τῶν ἔξωθεν ἐμποδὼν αὐτῇ γενομένου, ἤδη δὲ καὶ ὅσαι προσεχεῖς ὗλαι τῶν γιγνομένων εἰσίν. οὐδὲν δὲ διοίσει προσεχῆ λέγειν ἢ οἰκείαν ἢ ἰδίαν· ἐξ ἁπάντων γὰρ αὐτῶν δηλοῦται τὸ πλησίον καὶ μὴ διὰ μέσης ἄλλης μεταβολῆς, οἷον εἰ τὸ αἷμα δυνάμει σάρκα προσαγορεύοις ἐλαχίστης μεταβολῆς δεόμενον εἰς σαρκὸς γένεσιν. οὐ μὴν τό γ' ἐν τῇ γαστρὶ πεπεμμένον σιτίον ὕλη προσεχὴς σαρκὸς ἀλλὰ διὰ μέσου τοῦ αἵματος· ἔτι δὲ μᾶλλον ἡ μᾶζα καὶ ὁ ἄρτος ἐπὶ πλέον ἀποκεχώρηκε· τριῶν γὰρ δεῖται μεταβολῶν εἰς σαρκὸς γένεσιν. ἀλλ' ὅμως καὶ ταῦτα δυνάμει λέγεται σὰρξ καὶ πρὸ τούτων ἀὴρ καὶ πῦρ καὶ ὕδωρ καὶ γῆ καὶ ἡ τούτων αὐτῶν ὕλη κοινή. ταυτὶ μὲν οὖν ἅπαντα καταχρωμένων ἢ μᾶλλον ἢ ἧττον λέγεται. ὁ δὲ πρῶτος τρόπος ὁ κυριώτατός ἐστι τῶν δυνάμει τόδε τι λεγομένων ὑπάρχειν, ἐφεξῆς δ' ὁ κατὰ τὴν οἰκείαν ὕλην, οἷον εἰ τὴν ἀναθυμίασιν τὴν καπνώδη φλόγα λέγοις ὑπάρχειν ἢ τὸν ἀτμὸν

and already present, whereas "in potentiality" is said of what is incomplete and still about to be, and of a kind appropriate to come about, but doesn't yet exist. For the infant is not already rational but will become so in the future, nor is a dog that has just been born already a hunting creature—in fact, it doesn't yet see. But because it has the potential to hunt if it is brought to completion, it is termed "hunting" in this way. Therefore, we say "potentially" most properly only of those things which Nature herself brings to completion when none of those things occurring externally prevents this coming about, while also enough materials are already at hand for those things to come into being. It makes no difference whether we say close at hand, characteristic, or specific, for from all of these the close proximity is indicated, and is not due to another mediating change. For example, if you say the blood is potentially flesh, the genesis to flesh requires the least change. In fact, the food that has been concocted in the stomach is a material closely connected to flesh, but through the mediation of the blood. Still more, the barley cake and bread are to the greatest extent removed, for they require three changes for the genesis of flesh. But nevertheless, these things are said to be flesh in potentiality, and before these air, fire, water, and earth and the common material[2] of these things themselves. These things, then, are all said more or less catachrestically. The first way, and the most important, is of saying something exists potentially. Next is in relation to the characteristic material—for example, if you say the smoky flame rising up is

[2] See Galen's *De elementis secundum Hippocratem* 20, I.444K.

ἀέρα δυνάμει. λέγεται δέ ποτε δυνάμει καὶ τὸ τῷ κατὰ συμβεβηκὸς ἀντιδιαιρούμενον, οἷον εἰ τὴν ψυχρολουσίαν ἐπ' εὐσάρκου νέου θερμαίνειν τις φαίη τὸ σῶμα κατὰ συμβεβηκός, οὐκ οἰκείᾳ δυνάμει.

Κατὰ τοσούτους δὴ τρόπους καὶ τὰ δυνάμει θερμὰ καὶ ψυχρὰ καὶ ξηρὰ καὶ ὑγρὰ λεχθήσεται καὶ ζητηθήσεται δεόντως, τί δή ποτε καστόριον ἢ εὐφόρβιον ἢ πύρεθρον ἢ στρουθίον ἢ νίτρον ἢ μίσυ θερμὰ λέγομεν ἢ θριδακίνην ἢ κώνειον ἢ μανδραγόραν ἢ σαλαμάνδραν ἢ μήκωνα ψυχρά· πότερον τοῖς εἰρημένοις τρόποις ὑποπέπτωκεν ἢ κατ' ἄλλον τινὰ λέγεται μηδέπω διῃρημένον.[1] ἄσφαλτος μὲν γὰρ καὶ ῥητίνη καὶ στέαρ ἔλαιόν τε καὶ πίττα δυνάμει θερμά, διότι ῥᾳδίως ἐνεργείᾳ γίγνεται θερμά· καὶ γὰρ ἐκφλογοῦται τάχιστα καὶ τοῖς σώμασιν ἡμῶν προσαγόμενα θερμαίνει σαφέστατα. χαλκῖτις δὲ καὶ μίσυ καὶ νᾶπυ καὶ νίτρον ἄκορόν τε καὶ μῆον καὶ κόστος καὶ πύρεθρον ἡμῖν μὲν προσαγόμενα θερμαίνειν φαίνεται τὰ μὲν μᾶλλον αὐτῶν, τὰ δ' ἧττον, οὐ μὴν ἐκφλογοῦσθαι πέφυκεν. ἢ παραλογίζονται σφᾶς αὐτοὺς οἱ τοῦτο μόνον ἐπισκοποῦντες, εἰ μὴ ῥᾳδίως ἐκφλογοῦνται; ἐχρῆν γὰρ οὐχ οὕτως, ἀλλ' εἰ μηδ' ἀνθρακοῦνται σκοπεῖν, ὡς οὐδέν γ' ἧττον φλογὸς ὁ ἄνθραξ πῦρ. ἀλλ' ἡ μὲν φλὸξ ἀέρος ἐκπυρωθέντος ἤ τινος ἀερώδους σώματος, ὁ δ' ἄνθραξ γῆς ἤ τινος γεώδους γίγνεται. καὶ δὴ μέχρι τοῦδε συμφωνεῖν ὁ λόγος ἔοικεν ἑαυτῷ πάντῃ·

[1] διῃρημένον H; δ' εἰρεμένον K

fire, or the vapor is air potentially. Sometimes, however, in potentiality is also said in contradistinction to the contingent—for example, if someone were to say bathing in cold water would heat the body of a well-fleshed youth, he would be using it contingently and not in reference to the characteristic potential [of the cold bath].

In just so many ways, certainly, hot, cold, dry, and wet will be said "potentially" and will be looked into as they ought to be, as to why we ever call castor, spurge, pellitory, soapwort, nitron (sodium carbonate), or copper ore[3] hot, or lettuce, hemlock, mandragora, salamander, or poppy cold, and whether they fall under the ways stated or are said in some other manner which has not yet been determined. For asphalt, pine resin, hard fat (suet), olive oil, and pitch are hot potentially because they readily become hot actually. Also they scorch very quickly, and when applied to our bodies, very clearly heat. Rock alum, copper ore, mustard, nitron, yellow flag, spignel, *kostos*,[4] and pellitory when applied seem to heat us, some of them more and some less, but are not naturally scorching. Or do they mislead themselves by false reasoning, those who consider this alone, if it does not scorch easily? For it is necessary not to consider it in this way, but whether or not it is turned into charcoal, for charcoal is fire no less than flame is. But flame is when air is burned, or some airy body, whereas charcoal arises from earth or something earthy. And certainly up to this point, the argument seems to

[3] μίσυ is listed in LSJ as "a copper ore found in Cyprus" but also as "truffle" (*Tuber cinereum*). [4] Κόστος is listed in LSJ as a root used as a spice (*Sausurrea Lappa*); see Theophrastus, *History of Plants* 9.7.3, and Dioscorides 1.16.

φαίνεται γὰρ ὅσα πυρὸς ἁπτόμενα φάρμακα ῥᾳδίως ἐκπυροῦται, ταῦτα καὶ ἡμᾶς θερμαίνοντα, πλὴν εἴ τι διὰ τὸ παχυμερὲς εἶναι μὴ παραδέχεται ῥᾳδίως εἴσω τὸ σῶμα. διορισθήσεται γὰρ ἐπὶ πλέον ὑπὲρ τούτων ἐν τοῖς περὶ φαρμάκων δυνάμεως. ὅσα μέντοι τὸ σῶμα τὸ ἡμέτερον φαίνεται θερμαίνοντα, ταῦθ' ἑτοίμως ἐκπυροῦται. πῶς οὖν, φασίν, ἁπτομένοις οὐ φαίνεται θερμά; τοῦτο δ' οὐκ οἶδα, τίνος ἕνεκα λέγουσιν. εἰ μὲν γὰρ ἐνεργείᾳ τε καὶ ἤδη θερμὸν ἐλέγομεν ἕκαστον τῶν εἰρημένων ὑπάρχειν, ἦν ἂν δήπου θαυμαστόν, ὅπως ἁπτομένοις οὐ φαίνεται θερμά. νυνὶ δὲ τῷ δύνασθαι γενέσθαι ῥᾳδίως θερμὰ δυνάμει προσαγορεύομεν τὰ τοιαῦτα. θαυμαστὸν οὖν οὐδὲν πέπονθεν, εἰ μήπω θερμαίνει τοὺς ψαύοντας αὐτῶν. ὡς γὰρ οὐδὲ τὸ πῦρ αὔξει τὰ ξύλα πρὶν ὑπ' αὐτοῦ νικηθέντα μεταβληθῆναι καὶ τοῦτο πάντως ἔν τινι χρόνῳ γίγνεται, κατὰ τὸν αὐτὸν τρόπον οὐδὲ τὴν ἐν τοῖς ζῴοις θερμασίαν τὰ φάρμακα, εἰ μὴ πρότερον ὑπ' αὐτῆς ἐκείνης μεταβληθείη. καθ' ἕτερον μὲν γὰρ τρόπον ὁ παρὰ πυρὶ θαλπόμενος ἢ ἐν ἡλίῳ θερμαίνεται, καθ' ἕτερον δ' ὁ ὑφ' ἑκάστου τῶν εἰρημένων φαρμάκων· ἐκεῖνα μὲν γὰρ ἐνεργείᾳ θερμά, τῶν φαρμάκων δ' οὐδέν. οὔκουν οὐδὲ θερμαίνειν ἡμᾶς δύναται πρὶν ἐνεργείᾳ γενέσθαι τοιαῦτα, τὸ δ' ἐνεργείᾳ παρ' ἡμῶν αὐτῶν λαμβάνει, καθάπερ οἱ ξηροὶ κάλαμοι παρὰ τοῦ πυρός. οὕτω δὲ καὶ τὰ ξύλα ψυχρὰ μὲν ἅπαντα κατά γε τὴν ἑαυτῶν φύσιν, ἀλλὰ τὰ μὲν ξηρότερά τε καὶ σμι-

agree with itself completely, for it appears those medications readily burned when in contact with fire also heat us, unless due to being thick-particled they are not readily taken up into the body. There will be further distinctions made about these in the writings *On the Powers of Medications*.[5] Of course, those things that seem heating for our own body are readily burned. How then, they say, do they not appear hot to those touching them? For what purpose they say this, I do not know. If we were saying each of the things mentioned is already hot in actuality, it would obviously be surprising if they did not seem hot when touched. For the present, however, because they are easily able to become hot, we call such things potentially hot. No one is affected with surprise if they don't yet heat those touching them, as the wood does not increase the fire before being overcome by it, and has undergone a change; this at all events occurs in a certain time. In the same way, medications do not increase heat in animals unless beforehand they are changed by the heat itself. A person is heated in one way when warmed beside a fire or in the sun, but in another way by each of the medications mentioned. Thus, the former are hot in actuality, whereas none of the medications are. Such things are not therefore able to heat us before they become hot "in actuality," while they take this "in actuality" from us ourselves, just as dry reeds do from the fire. In like manner too, all wood is cold in its own nature, but the

---

[5] Presumably, *De simplicium medicamentorum temperamentis et facultatibus*, XI.369–892K and XII.1–377K; see particularly XI.398–410K.

κρὰ ῥᾳδίως εἰς πῦρ μεταβάλλει, τὰ δ' ὑγρότερά τε καὶ μεγάλα χρόνου δεῖται πλείονος.

Οὐδὲν οὖν θαυμαστόν, εἰ καὶ τὰ φάρμακα πρῶτον μὲν εἰς λεπτὰ καὶ σμικρὰ καταθραυσθῆναι δεῖται, δεύτερον δὲ χρόνῳ τινὶ κἂν ἐλαχίστῳ τοῖς σώμασιν ἡμῶν ὁμιλῆσαι πρὸς τὸ γενέσθαι θερμά. σὺ δ', εἰ μήτε καταθραύσας αὐτὰ μήτε θερμήνας πρότερον ἀξιοῖς ἤδη φαίνεσθαι[2] θερμά, τί ποτε σημαίνει τὸ δυνάμει θερμόν, ἐπιλελῆσθαί μοι δοκεῖς· ὡς ἐνεργείᾳ γοῦν θερμὰ βασανίζεις αὐτά. καὶ μὴν οὐδ' ἐκεῖνο θαυμαστόν, εἰ θερμανθῆναι δεῖται πρότερον, ἵν' ἀντιθερμήνῃ. γίγνεται γὰρ δὴ τοῦτο καὶ κατὰ τὴν τῶν ξύλων εἰκόνα. τὴν γοῦν ἀποσβεννυμένην φλόγα διασῴζει θ' ἅμα καὶ αὔξει θερμαινόμενα πρότερον ὑπ' ἐκείνης αὐτῆς. οὔκουν ἀπεικός ἐστιν οὐδὲ τὴν ἐν τοῖς ζῴοις θερμασίαν οἷον τροφῇ τινι χρῆσθαι τοῖς τοιούτοις φαρμάκοις ὡς τὸ πῦρ τοῖς ξύλοις. οὕτω γὰρ δὴ καὶ φαίνεται γιγνόμενον. εἰ δὲ κατεψυγμένῳ σώματι περιπάττοις ὁτιοῦν αὐτῶν ἀκριβῶς λεπτὸν ἐργασάμενος, οὐδ' ὅλως θερμαίνεται καὶ διὰ τοῦτο τρίβομεν ἐπὶ πλεῖστον τὰ κατεψυγμένα μόρια τοῖς τοιούτοις φαρμάκοις, ἅμα μὲν ἀνάπτοντες τῇ τρίψει θερμασίαν, ἅμα δ' ἀραιὸν ἐργαζόμενοι τὸ τέως ὑπὸ τῆς ψύξεως πεπυκνωμένον, ἵν' εἴσω τε δύῃ τὸ φάρμακον ὁμιλοῦν τε τῷ συμφύτῳ τοῦ ζῴου θερμῷ μεταβάλληταί τε καὶ θερμαίνηται. καὶ γὰρ εἰ μόριον αὐτοῦ τι σμικρότατον ἐνεργείᾳ κτήσαιτο τὴν θερμασίαν, εἰς ἅπαν οὕτω διαδίδωσι κατὰ τὸ συνεχές, ὡς εἰ καὶ τῆς δᾳδὸς ἅψαις

drier and smaller pieces change easily to fire, whereas pieces that are wetter and larger require a longer time.

It is no surprise then, if also the medication will first need to be broken into fine and small pieces, and second a certain time, even very short, to come into association with our own bodies to become hot. You, however, if you expect these to already seem hot without being broken into pieces or heated beforehand, seem to me to have forgotten what in the world hot "potentially" signifies, as in fact you are focusing closely on those things that are hot in actuality. And that is not surprising, if it needs to be heated first, so that it heats in turn. This certainly occurs in relation to the example of wood. Anyway, the flame that is being quenched is preserved and at the same time increases when previously heated by that fire itself. It is not therefore unreasonable that the heat in living creatures uses such medications like some nutrition, as the fire uses the wood. For surely this is what manifestly occurs. If, however, you sprinkle these (medications) around in any way whatsoever on a body that has been cooled, making them extremely fine, it is not heated at all, and for this reason, we massage to the greatest extent the cooled parts with such medications, stirring up heat with the massage at the same time, while making loose-textured what up to that time had been condensed by the cold, so that the medication may sink inward and, associating with the innate heat of the animal, be changed and heated. For surely also, if the smallest part of it should acquire the heat in actuality, it distributes it to the whole in this way by virtue of the continuity, just as if you touch the end of a

---

² φαίνεσθαι H; γενέσθαι K

τὸ ἄκρον ἀπὸ σμικροῦ σπινθῆρος· ἅπασαν γὰρ καὶ ταύτην ἐπινέμεται ῥᾳδίως τὸ πῦρ οὐδὲν ἔτι τοῦ σπινθῆρος δεόμενον. ἕκαστον οὖν τῶν δυνάμει θερμῶν οὔπω μὲν ἐν τῇ φύσει πλεονεκτοῦν ἔχει τὸ θερμὸν τοῦ ψυχροῦ, πλησίον δ' ἤδη τοῦ πλεονεκτεῖν ἐστιν, ὥστε βραχείας τῆς ἔξωθεν ἐπικουρίας δεῖσθαι πρὸς τὸ κρατῆσαι, καὶ ταύτην αὐτῷ ποτὲ μὲν ἡ τρῖψις ἱκανὴ παρασχεῖν ἐστι, ποτὲ δ' ἤτοι τὸ πῦρ ἤ τι τῶν φύσει θερμῶν σωμάτων ἁπτόμενον τούτου.[3]

Οὐδὲν οὖν θαυμαστὸν οὐδὲ[4] διὰ τί τὰ μὲν εὐθὺς ἅμα τῷ ψαῦσαι τοῦ σώματος ἡμῶν ἀντιθερμαίνειν αὐτὸ πέφυκε, τὰ δ' ἐν πλείονι χρόνῳ δρᾷ τοῦτο. καὶ γὰρ καὶ τῶν πλησιαζόντων τῷ πυρὶ τὰ μὲν εὐθὺς ἐξάπτεται, καθάπερ ἡ θρυαλλὶς καὶ ἡ δὰς ἡ λεπτὴ ἥ τε πίττα καὶ ὁ κάλαμος ὁ ξηρός, τὰ δ', εἰ μὴ πολλῷ χρόνῳ πλησιάσειεν, οὐ νικᾶται, καθάπερ τὸ ξύλον τὸ χλωρόν. ἀλλὰ μᾶλλον ἐκεῖνο διελέσθαι δικαιότερον, οὗ τὴν μὲν ἀπόδειξιν ἐν τοῖς Περὶ φυσικῶν δυνάμεων ἐροῦμεν, ἐξ ὑποθέσεως δ' ἂν ἕνεκα τῶν παρόντων καὶ νῦν αὐτῷ χρησαίμεθα, τέτταρας μὲν εἶναι παντὸς σώματος δυνάμεις,[5] ἑλκτικὴν μὲν τῶν οἰκείων μίαν, ἑτέραν δὲ τὴν τούτων αὐτῶν καθεκτικὴν καὶ τρίτην τὴν ἀλλοιωτικὴν καὶ τετάρτην ἐπ' αὐταῖς τὴν τῶν ἀλλοτρίων ἀποκριτικήν, εἶναί τε ταύτας τὰς δυνάμεις ὅλης τῆς οὐσίας ἑκάστου τῶν σωμάτων, ἣν ἐκ θερμοῦ καὶ ψυχροῦ καὶ ξηροῦ καὶ ὑγροῦ κεκρᾶσθαί φαμεν.

[3] τούτου add. H (see his note, p. 90)

pine torch with a small spark, the fire is readily distributed to the whole torch, and has no further need of the spark. Therefore, each of the things that are hot potentially doesn't yet have in its nature the hot prevailing over the cold, although it is already close to prevailing so that it needs little external aid in order to prevail. And sometimes massage is sufficient to provide this for it, whereas in other instances, it is either the fire or one of the naturally hot bodies contacting it.

There is nothing remarkable about why some things that touch our body immediately and naturally heat this in turn, while some do this over a longer time. Furthermore, of the things coming near to the fire, some are kindled into a flame immediately, like a plantain wick, thin pinewood, pitch, and dry reeds, while some, if not previously in contact for a long time, are not overcome, like green wood. But that was defined more strictly, and we shall speak of the demonstration of this in the works *On the Natural Faculties*.[6] Now, however, we shall also use it *ex hypothesi* for our present purposes, saying there are four capacities (faculties, powers) of every body: the first is the capacity to attract suitable things; the second is the capacity of retaining these same things; the third is the transformative; and the fourth after these is the capacity of separating alien things. And these capacities are of the whole substance of each of the bodies, which we say is compounded from hot, cold, dry, and wet. However, when the body

---

[6] *De naturalibus facultatibus*, I.2, LCL 71 (A. J. Brock), 4–13.

---

[4] οὐδὲν οὖν θαυμαστὸν οὐδὲ H; οὔκουν οὐδὲν θαυμαστὸν εἰπεῖν K   [5] δυνάμεις H; δυνάμεις οἰκείας K

ἐπειδὰν δὲ κατὰ μίαν ἡντινοῦν τῶν ἐν αὐτῷ ποιοτήτων τὸ σῶμα μεταβάλλῃ τὸ πλησιάζον, οὔτε καθ' ὅλην ἐνεργεῖν αὐτοῦ τηνικαῦτα τὴν οὐσίαν ὑποληπτέον οὔτ' ἐξομοιωθῆναι δύνασθαι [ποτὲ] τὸ μεταβαλλόμενον· ὥστ' οὐδὲ θρέψειεν ἄν ποτε τὸ οὕτω μεταβληθὲν οὐδὲ τῶν μεταβαλλόντων. εἰ δ' ἱκανῶς[6] μεταβάλλοι, τουτέστι καθ' ὅλην αὑτοῦ τὴν οὐσίαν ἐνεργῆσαν, ἐξομοιώσειεν ἂν οὕτως ἑαυτῷ καὶ τραφείη πρὸς | τοῦ μεταβληθέντος· οὐδὲ γὰρ ἄλλο τι θρέψις ἐστὶ παρὰ τὴν τελείαν ὁμοίωσιν.

2. Ἐπειδὴ δὲ ταῦτα διώρισται, πάλιν ἐκεῖθεν ἀρκτέον τοῦ λόγου. τῶν ζῴων ἕκαστον οἰκείαις τρέφεται τροφαῖς· οἰκεία δ' ἐστὶν ἑκάστῳ τροφὴ πᾶν ὅ τι ἂν ἐξομοιωθῆναι δύνηται τῷ τρεφομένῳ σώματι. χρὴ τοίνυν ὅλην τὴν οὐσίαν τοῦ τρέφοντος ὅλῃ τῇ τοῦ τρεφομένου φύσει κοινωνίαν τέ τινα καὶ ὁμοιότητα κεκτῆσθαι, πάντως οὐκ ὀλίγης οὐδ' ἐνταῦθα τῆς κατὰ τὸ μᾶλλόν τε καὶ ἧττον ὑπαρχούσης διαφορᾶς· τὰ μὲν γὰρ μᾶλλον οἰκεῖά τ' ἐστὶ καὶ ὅμοια, τὰ δ' ἧττον, ὥστε καὶ τῆς κατεργασίας τὰ μὲν ἰσχυροτέρας τε καὶ πολυχρονιωτέρας, τὰ δ' ἀσθενεστέρας θ' ἅμα καὶ ὀλιγοχρονιωτέρας προσδεῖσθαι, τὸ μὲν ὀρνίθειον ἐλάττονος, τὸ δὲ χοίρειον πλείονος, τὸ δὲ βόειον ἔτι πλείονος. ἐλαχίστης δὲ δεῖται μεταβολῆς εἰς ἐξομοίωσιν ὁ οἶνος, ὅθεν καὶ τρέφει καὶ ῥώννυσι τάχιστα. πάντως μὲν οὖν καὶ τοῦτον ὁμιλῆσαι χρὴ τοῖς πεπτικοῖς ὀργάνοις, γαστρί τε καὶ ἥπατι καὶ φλεψίν, ἐν οἷς ἂν προκατεργασθεὶς | τρέφειν ἤδη δύναιτο τὸ σῶμα·

changes what comes near in any one of the qualities in it whatsoever, one must not suppose that under the circumstances it acts in relation to its whole substance, nor that what has changed is able at some time to be assimilated. As a consequence, nothing that is changed in this way would ever nourish those things effecting the change. If, on the other hand, it were to effect a significant change, that is, in relation to its whole existing function, it would assimilate to itself in this way and be nourished by what is changed. For nourishing is nothing else apart from complete assimilation.

2. Since these are distinguished, we must in turn begin the discussion from that fact. Each of the animals is nourished by suitable nutriments. A suitable nutriment for each is any that would be able to be assimilated by the body being nourished. Accordingly, it is necessary for the whole substance of what is nourishing to have acquired a certain commonality and similarity with the whole nature of what is being nourished. At all events, no small difference exists here in relation to more and less, for some are those that are more suitable and similar, while some are less, so that in the working up, some require this to be stronger and over a long time, while others require it to be weaker and over a shorter time. Thus, the flesh of birds requires less, pork more, and beef more still. Wine, however, needs the least change for assimilation, for which reason it very quickly nourishes and strengthens. By all means then, this too must come into contact with the digestive organs—stomach, liver and veins—in which, when it is worked up

---

6 ἱκανῶς H; ἐκεῖνο K

πρὶν δὲ τῆς ἐν τούτοις μεταβολῆς ἐπιτυχεῖν οὐχ οἷόν τ᾽ αὐτῷ τροφὴν ζῴου γενέσθαι, κἂν εἰ δι᾽ ὅλης ἡμέρας καὶ νυκτὸς ἐπικείμενος εἴη ἔξωθεν τῷ χρωτί.[7] πολὺ δὲ δὴ μᾶλλον ἄρτος ἔξωθεν ἐπικείμενος ἢ τεῦτλον ἢ μάζα τρέφειν ἀδύνατα.

Τὰ μὲν οὖν ὁμοιούμενα πάντη τροφαί, τὰ δ᾽ ἄλλα σύμπαντα φάρμακα καλεῖται. διττὴ δὲ καὶ τούτων ἡ φύσις· ἢ γὰρ οἷάπερ ἐλήφθη διαμένοντα νικᾷ καὶ μεταβάλλει τὸ σῶμα, καθ᾽ ὃν τρόπον ἐκεῖνο τὰ σιτία, καὶ πάντως ταῦτα τὰ φάρμακα δηλητήριά τε καὶ φθαρτικὰ τῆς τοῦ ζῴου φύσεώς ἐστιν, ἢ μεταβολῆς ἀρχὴν παρὰ τοῦ σώματος λαβόντα σήπεται τοὐντεῦθεν ἤδη καὶ διαφθείρεται κἄπειτα συνδιασήπει τε καὶ συνδιαφθείρει τὸ σῶμα· δηλητήρια δ᾽ ἐστὶν ἔτι καὶ ταῦτα. τρίτον δ᾽ ἐπ᾽ αὐτοῖς ἐστιν εἶδος φαρμάκων τῶν ἀντιθερμαινόντων μὲν τὸ σῶμα, κακὸν δ᾽ οὐδὲν ἐργαζομένων· καὶ τέταρτον, ὅσα καὶ ποιοῦντά τι καὶ πάσχοντα νικᾶται τῷ χρόνῳ καὶ τελέως ἐξομοιοῦται. συμπέπτωκε δὲ τούτοις ἅμα τε φαρμάκοις εἶναι καὶ τροφαῖς.

Θαυμαστὸν δ᾽ οὐδέν, εἰ βραχείας ἀφορμῆς ἔνια λαβόμενα μεγίστην ἐκτροπὴν ἴσχει τῆς ἀρχαίας φύσεως. ὁρᾶται γοῦν καὶ τῶν ἔξω πολλὰ τοιαῦτα. κατὰ μέν γε τὴν ἐπὶ τῆς Ἀσίας Μυσίαν οἰκία ποτὲ κατεκαύθη τρόπῳ τοιῷδε· κόπρος ἀπέκειτο περιστερῶν ἤδη σεσηπυῖα καὶ τεθερμασμένη καὶ ἀτμὸν ἀναπέ-

---

[7] χρωτί H; σώματι K

beforehand, it is now able to nourish the body. However, before the change in these happens, it is not possible for it to become nourishment for an animal, even if placed externally on the skin for a whole day and night. Much more so, certainly, if bread is placed externally, or beet, or barley cake, they would be unable to nourish.

Therefore, those things that are assimilated in every way are nutriments, whereas all the others are called medications. And the nature of the latter is twofold: there are those that remain as they were when taken and which overcome and change the body in the way foods do—these medications are in every way deleterious and destructive to the nature of the animal—or taking a beginning of change from the body, and from there, being putrefied and already destroyed, the body is then putrefied and destroyed at the same time. There is a third kind of medications in addition to these—those which heat the body in turn while doing no harm. And there is a fourth kind—those that both act and are acted on | and in time are overcome and completely assimilated. What happens with these is that they are at the same time medications and nutriments.

It is no surprise then, if some taking a slight starting point have a very large deviation from the original nature. Anyway, many such things are seen in the external world. In fact, in the Mysian region of Asia,[7] a house was once burned down in the following manner. The excrement of pigeons was left lying already putrefied and heated, and

[7] Galen here refers to a region north of Pergamum.

μπουσα καὶ ἁπτομένοις ἱκανῶς θερμή. ταύτης δὲ πλησίον ἦν, ὡς καὶ ψαύειν ἤδη, θυρὶς ἔχουσα ξύλα νεωστὶ καταληλιμμένα ῥητίνῃ πολλῇ. θέρους οὖν μέσου λάβρος ἥλιος προσβαλὼν ἐξῆψε τὴν ῥητίνην τε καὶ τὰ ξύλα. κἀντεῦθεν ἤδη θύραι τινὲς ἕτεραι πλησίον ὑπάρχουσαι καὶ θυρίδες ἔναγχος ἐξαληλιμμέναι ῥητίνῃ ῥᾳδίως τε διεδέξαντο τὸ πῦρ καὶ μέχρι τῆς ὀροφῆς ἐξέτειναν. ἐπεὶ δ' ἅπαξ ἐκείνης ἡ φλὸξ ἐλάβετο, ταχέως ἐπὶ πᾶσαν ἐνεμήθη τὴν οἰκίαν. οὕτω δέ πως, οἶμαι, καὶ τὸν Ἀρχιμήδην φασὶ διὰ τῶν πυρείων ἐμπρῆσαι τὰς τῶν πολεμίων τριήρεις. ἀνάπτεται δ' ἑτοίμως ὑπὸ πυρείου καὶ ἔριον καὶ στυπεῖον καὶ θρυαλλὶς καὶ νάρθηξ καὶ πᾶν ὅ τι ἂν ὁμοίως ᾖ ξηρόν τε καὶ χαῦνον. ἐξάπτουσι δὲ φλόγα καὶ λίθοι παρατριβόμενοι καὶ μᾶλλον ἢν θείου τις αὐτοῖς ἐπιπάσσῃ. καὶ τὸ τῆς Μηδείας δὲ φάρμακον τοιοῦτον ἦν. πάντα γοῦν ἀνάπτεται προσβαλλούσης θερμασίας οἷς ἂν ἐπαλειφθῇ. σκευάζεται δ' ἐκεῖνο διά τε θείου καὶ τῆς ὑγρᾶς ἀσφάλτου. καὶ μὲν δὴ καὶ ὡς θαῦμά τις ἐδείκνυ· ἀποσβεννὺς λύχνον αὖθις ἧπτε τοίχῳ προσφέρων· ἕτερος δὲ λίθῳ προσέφερεν· ἐτεθείωτο δ' ἄρα καὶ ὁ τοῖχος καὶ ὁ λίθος. καὶ ὡς ἐγνώσθη τοῦτο, θαυμαστὸν οὐκέτ' ἦν τὸ γιγνόμενον.

Πάντ' οὖν ταῦτα τὰ φάρμακα θερμὰ μὲν οὔπω τελέως ἐστίν, ἐπιτηδειότατα μέντοι πρὸς τὸ γενέσθαι

giving off a vapor was also notably hot to those touching it. Near this, so close as to touch it, was a wooden window frame newly covered with much resin. Since it was the middle of summer and a fierce sun assailed it, it set fire to the resin and the wood. And here, since there were already some other doors and window frames lately having been coated with resin, the fire readily spread until it reached the roof. When the fire completely took hold of that, it quickly set fire to the whole house. Somewhat in the same way, I think, they say Archimedes, using fire sticks, set fire to the triremes of the enemy.[8] | And wool, flax, plantain, fennel, and anything that is similarly dry and porous is easily kindled to flames by a fire stick. However, stones also kindle flames when rubbed together, and particularly if some brimstone is sprinkled over them. And the medication of Medea was of this sort.[9] At all events, anything that it is smeared over is kindled to flame when heat is applied to it. And that is prepared from brimstone and the liquid of asphalt. Furthermore also, what is shown as a mystery is an extinguished lamp that is set alight again when brought together with a wall, and another when brought together with a stone. The wall and the stone were in fact covered with brimstone. And when this was known, the occurrence was no longer surprising.

Thus all these medications are not yet completely hot, but are of course very suited to becoming hot, and because

658K

---

[8] See the detailed note in Singer and van der Eijk on this matter: *Galen: Works on Human Nature*, 163n43.

[9] This was a "gift" sent by Medea to the woman for whom her husband had abandoned her. It was to burst into flames when first worn; see Euripides, *Medea* 995ff.

θερμὰ καὶ διὰ τοῦτο δυνάμει θερμὰ λέγεται. περὶ μὲν δὴ τούτων οὐδὲν ἄπορον, ἀλλ' οὐδὲ διὰ τί πινόμενος μὲν ὁ οἶνος ἱκανῶς θερμαίνει τὸ σῶμα, κατὰ δὲ τοῦ δέρματος ἐπιτιθέμενος οὐ θερμαίνει. δέδεικται γὰρ ὀλίγῳ πρότερον οὐχ ἁπλῶς ὡς θερμὸν φάρμακον ἀλλ' ὡς οἰκεία τροφὴ θερμαίνων τὸ ζῷον. ὡς γὰρ αἱ τοῦ πυρὸς ἐπιτήδειοι τροφαὶ τὸ πῦρ αὔξουσιν, οὕτω καὶ τῶν φύσει θερμῶν σωμάτων ὅ τί περ ἂν οἰκεία τε καὶ σύμφυτος ὑπάρχῃ τροφὴ ῥώσει τε πάντως αὐτὰ καὶ τὴν ἔμφυτον αὐξήσει θερμασίαν. καὶ τοῦτο μὲν ἁπάσης τροφῆς κοινόν· οἴνῳ δ' ἴδιον ἐξαίρετον ὑπάρχει τὸ τάχος τῆς μεταβολῆς ὡς δᾳδὶ καὶ θρυαλλίδι καὶ στυπείῳ καὶ πίττῃ. καὶ δὴ καὶ τῆς εἰκόνος ἐχόμενοι τοῦ πυρὸς ἀναμνησθῶμεν αὖθις ξύλων ὑγρῶν, ἃ τροφὴ μέν ἐστι καὶ αὐτὰ τοῦ πυρός, ἀλλ' οὐκ εὐθὺς οὐδ' ἐκ τοῦ παραχρῆμα, καὶ διὰ τοῦτ' ἐπιβληθέντα πολλάκις τῷ πυρὶ κατακρύπτει τε τὴν φλόγα καὶ μάλιστ' ἂν ἀσθενὴς ὑπάρχῃ καὶ σμικρά, καὶ κίνδυνον ἐπάγει φθορᾶς. οὕτως οὖν κἂν τοῖς ζῴοις ὅσα τῶν ἐδεσμάτων, ἵν' ἐξομοιωθῇ τελέως καὶ θρέψῃ τὸ σῶμα, χρόνου δεῖται, ψῦχος μᾶλλον ἐπάγειν ἢ θάλπος ἐν τῷ παραυτίκα φαίνεται. θερμαίνει μὴν ἐν τῷ χρόνῳ καὶ ταῦτα, παραπλησίως τοῖς ἄλλοις ἐδέσμασιν, εἰ μόνον αὐτοῖς προσγένοιτο τὸ θρέψαι τὸ σῶμα. τροφὴ γὰρ ἅπασα κατὰ τὸν ἑαυτῆς λόγον αὔξει τὸ τοῦ ζῴου θερμόν. εἰ δὲ καταποθείη μὲν ὡς τροφή, μὴ μέντοι κρατηθείη μηδ' ἐξομοιωθείη,[8] τοῦτ' ἐκεῖνο τὸ πρὸς Ἱπποκράτους εἰρημένον, εἴη ἂν ὄνομα τροφῆς, ἔργον δ'

of this are said to be hot potentially. There is certainly nothing strange about these, but nor is there as to why wine heats the body significantly when drunk, but doesn't heat when applied to the skin. It has been shown a little earlier that it is not | simply as a hot medication but as a suitable nutriment that it heats the organism, for as the nutriments suitable for the fire increase the fire, in the same way too for the bodies hot in nature, what is a suitable and natural nutriment will strengthen these throughout and increase the innate heat. And this is common to every nutriment. The rapidity of the change is peculiarly remarkable to wine, as to pinewood, plantain, flax, and pitch. Furthermore, since they have a likeness to the fire, we shall call to mind again wood that is wet which is itself a nutriment for the fire, but not immediately or straightaway, and because of this, when it is thrown on the fire, on many occasions it conceals the flame and particularly should that be weak and small, there is danger of it leading to extinction. Therefore, in this way, even in animals, those of the foods, in order for them to be completely assimilated and nourish the body, need time, and seem to bring in more of cold than of hot in the immediate term. But these things also heat in time, similar to the other foods, | only if nourishing the body occurs with them. For every nutriment, by reason of itself, increases the heat of the animal. If, however, it is swallowed as a nutriment and of course is neither overcome nor assimilated, it would, as was said by Hippocrates, have the name of a nutriment but

---

[8] μηδ' ἐξομοιωθείη add. H

οὐχί. τριχῶς γὰρ τῆς τροφῆς λεγομένης, ὡς καὶ τοῦτ' αὐτὸς ἐδίδαξεν εἰπών· "τροφὴ δὲ τὸ τρέφον, τροφὴ καὶ τὸ οἷον τροφὴ καὶ τὸ μέλλον," ἡ μὲν ἤδη τρέφουσα καὶ προστιθεμένη καὶ μηκέτι μέλλουσα κυρίως ὀνομάζεται τροφὴ καὶ θερμαίνει πάντως τὸ τρεφόμενον σῶμα, τῶν δ' ἄλλων οὐδετέρα, διότι μηδὲ τροφὴ κυρίως ἐστίν, ἀλλὰ τὸ μὲν οἷον τροφή, τὸ δ' ὅτι μέλλει τοιοῦτον γενήσεσθαι. ταῦτ' ἄρα καὶ αὐτὸς ὁ οἶνος οὐκ ἀεὶ θερμαίνει τὸ ζῷον, ὥσπερ οὐδὲ τοὔλαιον ἀνάπτει τὴν φλόγα, καίτοι γ' οἰκειοτάτη τροφὴ πυρὸς ὑπάρχον, ἀλλ' ἐὰν ἀσθενεῖ καὶ σμικρᾷ φλογὶ καταχέῃς ἔλαιον ἀθρόον καὶ πολύ, καταπνίξεις καὶ τελέως ἀποσβέσεις αὐτὴν μᾶλλον | ἢ αὐξήσεις. οὕτως οὖν καὶ ὁ οἶνος, ἐπειδὰν πολὺς ὡς μὴ κρατεῖσθαι πίνηται, τοσοῦτον ἀποδεῖ τοῦ θερμαίνειν τὸ ζῷον, ὥστε καὶ πάθη ψυχρότατα γεννᾷ· ἀποπληξίαι γοῦν καὶ παραπληξίαι καὶ κάροι καὶ κώματα καὶ παραλύσεις ἐπιληψίαι τε καὶ σπασμοὶ καὶ τέτανοι ταῖς ἀμέτροις οἴνου πόσεσιν ἕπονται, ψυχρὰ σύμπαντα πάθη. καθόλου γὰρ ὅσα τῶν εἰς τὸ σῶμα λαμβανομένων ὡς τροφὴ θερμαίνει, ταῦθ' εὕροις ἄν ποτε καὶ ψύχοντα, καθότι καὶ τὴν φλόγα πρὸς τῆς αὐτῆς ὕλης οὐκ αὐξανομένην μόνον, ἀλλὰ καὶ σβεννυμένην ἐνίοτε. ταυτὶ μὲν οὖν ἔοικεν ὁμολογεῖν ἅπαντα τοῖς τε περὶ τῶν στοιχείων καὶ τοῖς περὶ τῶν κράσεων λογισμοῖς.

---

[10] Hippocrates, *Nutriment* 21, LCL 147 (*Hippocrates* I,

not the action. For nutriment is said in three ways, as he himself taught, when he said: "Nutriment is what nourishes; nutriment is also what is like nutriment; and nutriment is what will nourish in the future."[10] What has already nourished and is added, and is no longer going to do so, is properly called nutriment and altogether heats the body being nourished. However, neither of the other two do, because they are not properly nutriment, but what in one case is like nutriment, and in the other what is going to become nutriment. And wine itself doesn't always heat the animal, just as olive oil does not stir up the flame, even though it is in fact the most suitable nutriment for fire, but if oil is poured all at once onto a weak and small flame, and in a large amount, it will suffocate and completely quench the flame rather than increase it. Therefore, in this way too, wine, when so much is drunk that it is not overcome, it is deficient in the heating of the animal to such an extent that it generates very cold affections. Apoplexies, at any rate, hemiplegias, torpors, hypersomnias, paralyses, epilepsies, convulsions, and tetanus follow immoderate drinking of wine. All are cold affections. In general, although those things, when taken into the body as nutriment, heat, you would find they are sometimes also cooling in the way in which the flame is not only increased in relation to the actual material, but is also sometimes quenched. All these things, then, seem to agree with the theories concerning the elements and the *krasias*.

W. H. S. Jones), 348–49. The text differs somewhat from Galen. Jones' translation reads: "Nutriment not nutriment if it have not its power. Not nutriment if it can nourish. Nutriment in name, not in deed; nutriment in deed, not in name."

3. Ἐκεῖνο δ' ἂν ἴσως δόξειε διαφέρεσθαι τό τινα τῶν ἐσθιομένων ἐν τροφῆς χρείᾳ κατὰ τοῦ δέρματος ἐπιτιθέμενα διαβιβρώσκειν τε καὶ ἑλκοῦν αὐτό, καθάπερ νᾶπυ καὶ τάριχος σκόροδά τε καὶ κρόμυα. καίτοι καὶ τοῦτο συμφωνεῖ τοῖς ἐξ ἀρχῆς ὑποκειμένοις· ἅμα μὲν γὰρ ὅτι μεταβάλλεται καὶ ἀλλοιοῦται κατά τε τὴν γαστέρα πεττόμενα κἀν ταῖς φλεψὶν αἱματούμενα, πρὸς δὲ τούτοις καὶ διότι μὴ μένει καθ' ἕνα τόπον, ἀλλ' εἰς πολλὰ μερίζεται πάντῃ φερόμενα, καὶ πρὸς τούτοις διότι μίγνυται καὶ χυμοῖς πολλοῖς καὶ τοῖς ἅμ' αὐτοῖς λαμβανομένοις σιτίοις, ἔτι τε πρὸς τούτοις ὅτι διὰ ταχέων ἥ τε πέψις αὐτῶν γίγνεται καὶ ἡ διάκρισις, ὡς τὸ μὲν οἰκεῖον ἐξομοιωθῆναι, τὸ δὲ περιττὸν ἐν αὐτοῖς καὶ δριμὺ διὰ γαστρός θ' ἅμα καὶ οὔρων καὶ ἱδρώτων ἐκκριθῆναι, διὰ ταῦτα πάντα τὸ ἔξωθεν ἑλκοῦν ἐσθιόμενον οὐχ ἑλκοῖ. καίτοι κἂν εἰ τῶν εἰρημένων ἓν ὁτιοῦν ὑπῆρχεν αὐτοῖς, ἱκανὸν ἂν ἦν δήπουθεν ἀβλαβῆ φυλάξαι τὰ ἐντός,[9] οἷον, εἰ τύχοι, τὸ μεταβάλλειν πρῶτον. εἰ γὰρ οὐ μένει τὸ νᾶπυ τοιοῦτον, οἷον ἔξωθεν ἐλήφθη, δῆλον, ὡς οὐδὲ τὴν δύναμιν αὐτοῦ μένειν ἀξιώσεις· εἰ δὲ καὶ διακρίνεται καὶ καθαίρεται, πολὺ δὴ καὶ μᾶλλον. ἤρκει δὲ καὶ τὸ μὴ χρονίζειν ἐν ἑνὶ χωρίῳ. φαίνεται γὰρ οὐδὲ περὶ τὸ δέρμα δρᾶσαί τι δυνάμενον ἄνευ χρόνου πλείονος. ἀλλὰ καὶ τὸ μίγνυσθαι πολλοῖς ἑτέροις σιτίοις οὐ σμικρὸν οὐδ' αὐτό. γνοίης δ' ἄν, εἰ μόνον αὐτὸ προσενέγκοις χωρὶς τῶν ἄλλων σιτίων, ὅσην ἀνίαν τε καὶ

---

[9] ἐντός H; ἐντὸς K

ON TEMPERAMENTS, BOOK III

3. What would perhaps seem to be different is that some of the things eaten in the service of nutrition, when applied to the skin, erode and ulcerate it—things like mustard, pickled meat, garlic, and onions. And indeed, this also agrees with the things assumed from the beginning. For at the same time, because things concocted in the stomach are changed and transformed and are even changed to blood in the veins, while in addition to these things also, because they do not remain in one place, but are divided into many pieces and carried everywhere, and in addition to these again, because they are mixed both with many humors, and with the actual foods taken at the same time, and further still, because their concoction occurs rapidly, as does the breakdown to what is suitable to be assimilated, while what is superfluous in them and pungent is expelled through the stomach, urine, and sweat, due to all these things, they cause ulceration externally but do not ulcerate when eaten. And indeed, if any one whatsoever of the things mentioned exists with them, it would obviously be sufficient to preserve the internal structures unharmed, if for example, as may happen, it is changed first. For if the mustard does not remain as such when received, as it were, externally, it is clear that you would not expect its potency to remain as it was. However, if it is broken down and reduced, this is certainly much more so. And it is sufficient also for it not to spend time in one place. For it is apparent that it is not able to do anything to the skin without a longer time. But also, not being mixed with many other foods is no small matter. You would know, if you were to take this alone without other foods,

662K

663K

δῆξιν ἐπιφέρει τῇ γαστρί. καὶ μὲν δὴ κἂν εἰ πολλοῖς γλυκέσι χυμοῖς ἀναμίξας αὐτὸ κατὰ τοῦ δέρματος ἐπιθείης, οὐδὲν ἐργάσεται φαῦλον.

Ὁπότ' οὖν ἕκαστον τῶν εἰρημένων ἱκανόν ἐστι καθ' ἑαυτὸ κωλῦσαι τὸ νᾶπυ δρᾶσαί τι τοιοῦτον ἐντός, οἷον ἐκτὸς ἕδρα περὶ τὸ δέρμα, πολὺ δήπου μᾶλλον οἶμαι πάνθ' ἅμα συνελθόντα. καὶ γὰρ ἀλλοιοῦται πεττόμενον καὶ διακρίνεται καὶ καθαίρεται καὶ πολλοῖς ἑτέροις ἀναμίγνυται καὶ μερίζεται πολλαχῇ καὶ πάντῃ φέρεται καὶ χρονίζει κατ' οὐδὲν τῶν μορίων. ὅτι δ', εἴπερ ἔμενε δριμύ, πάντως ἂν ἥλκωσε καὶ τὰ ἐντός, ἐκ τῶν αὐτομάτων ἑλκῶν ἐπιγνώσῃ. γίγνεται γὰρ πολλοῖς πολλάκις τοῖς μὲν ἐξ ἐδεσμάτων μοχθηρῶν, τοῖς δ' ἔκ τινος ἐν αὐτῷ τῷ σώματι διαφθορᾶς καὶ σηπεδόνος ἡ καλουμένη κακοχυμία καὶ τούτοις ἐνίοτε μὲν ἑλκοῦταί τι καὶ τῶν ἐντός, ὡς τὰ πολλὰ δὲ τῷ τὴν φύσιν ἀποτρίβεσθαι τὰ κατὰ τὴν ἕξιν περιττώματα πρὸς τὸ δέρμα τοῦθ' ἑλκοῦται πολλοῖς καὶ συνεχέσιν ἕλκεσι. καρκῖνός τε γὰρ καὶ φαγέδαινα καὶ ἕρπης ὁ ἀναβιβρωσκόμενος ἄνθρακές τε καὶ τὰ χειρώνεια καὶ τηλέφεια καλούμενα καὶ ἄλλαι μυρίαι

---

11 Cheiron was one of the centaurs. He is described as a healer at *Iliad* 4.19, where he is said to have treated Menelaus' wound. He is regarded as the teacher of Asclepius. Telephus was the son of Hercules and Auge. He suffered a wound when fleeing from Achilles. This became chronic but was ultimately healed by Achilles' spear. See Galen's *De methodo medendi*, X.84 and 1006K. See also Pliny, *Natural History* 34.45, where the cure is attributed to

how much distress and biting it brings to the stomach. And certainly, if you were to mix this with many sweet juices, when you apply it to the skin, it will do nothing bad.

Therefore, since each of the things mentioned is sufficient in itself to prevent the mustard doing the sort of thing internally that it does externally when lying around on the skin, I think this is perhaps much more so, when all these things come together at the same time. For also, when it is concocted, it is changed, broken down, reduced, and mixed with many other things, and divided up in many ways and carried in all directions, not spending time in any one of the parts. That however, if it remained pungent, it would also altogether ulcerate the parts within, you would realize from the spontaneous ulcers. For so-called *kakochymia* occurs often in many people. In some this is from foods that are abnormal and in some from a certain destruction and putrefaction in the body itself. And in these sometimes, there is also some ulceration of internal structures, as in many instances, by virtue of their nature, the superfluities in relation to the bodily state rub against the skin, ulcerating this with many and chronic ulcers. For cancer, *phagedaina* (cancerous sores), eroding herpes, malignant pustules (carbuncles), and the so-called Cheironian and Telephian [ulcers][11] and countless other classes

664K

---

rust of iron. The terms Cheironian and Telephian for chronic ulcers remained in use well into the second millennium; see "A Treatise on Ulcers" in *The Works of that Famous Physitian Dr. Alexander Read* (London, 1650), 161–66. Read concludes his discussion of these ulcers as follows: "maligne ulcers who are not easily cured, are called *Chironia*, because *Chiron* was able to cure them; and *Telephia* because *Telephus* was troubled with such an one."

γενέσεις ἑλκῶν ἔκγονοι τῆς τοιαύτης εἰσὶ κακοχυμίας. οὔτ' οὖν τῶν τοιούτων οὐδὲν ἄπορον οὔτε διὰ τί τῶν φαρμάκων ἔνια μὲν οὐδὲν ἡμᾶς ἔξωθεν ἀδικοῦντα μέγα τι κακὸν ἐργάζεται καταποθέντα. τινὰ δὲ πολλάκις μὲν ἔβλαψεν εἴσω ληφθέντα, πολλάκις δ' ὠφέλησεν· ἔνια δ' οὐ μόνον ἔσωθεν ἀλλὰ καὶ ἔξωθεν ἀδικεῖ. συλλήβδην δ' εἰπεῖν οὐδὲν ὁμοίως ἔσωθέν τε καὶ ἔξωθεν ἐνεργεῖν πέφυκεν. οὔτε γὰρ ὁ τοῦ λυττῶντος κυνὸς ἀφρὸς οὔθ' ὁ τῆς ἀσπίδος οὔθ' ὁ τῆς ἐχίδνης ἰός, οἳ δὴ καὶ χωρὶς ἕλκους ἔξωθεν προσπεσόντες ἀδικεῖν πεπίστευνται, τὴν ἴσην ἔχουσι δύναμιν ἢ τῷ δέρματι μόνον ὁμιλήσαντες ἢ εἴσω μεταληφθέντες. οὐ μὴν οὐδ' ἐκεῖνο θαυμάζειν ἄξιον, εἰ τινων φαρμάκων οὐκ ἐξικνεῖται πρὸς τὸ βάθος ἡ δύναμις· οὐ γὰρ ἀναγκαῖον ἅπαντα τὴν αὐτὴν ἔχειν ἰσχύν.

Εἰ δὲ πολλὰ τῶν εἴσω λαμβανομένων ἐν μὲν τῷδε τῷ καιρῷ καὶ μετὰ τοσῆσδε ποσότητος καὶ τῆς πρὸς τάδε μίξεως ὠφέλησεν, ἀκαίρως δὲ καὶ πολλὰ καὶ ἄμικτα ληφθέντα βλάβην ἤνεγκεν, οὐδὲν οὐδ' ἐντεῦθεν ἀπόρημα τῷ λόγῳ. καὶ γὰρ καὶ τοῖς σιτίοις ὑπάρχει τοῦτό γε καὶ τῷ πυρὶ καὶ πᾶσιν ὡς οὕτω φάναι τοῖς προσπίπτουσι τῷ σώματι. συμμέτρου γοῦν φλογὸς ἔστιν ὅτε δεόμεθα καὶ χρώμενοι μεγάλως πρὸς αὐτῆς ὀνινάμεθα καίτοι τῆς ἀμέτρου καιούσης ἡμᾶς. οὕτω δὲ καὶ ψυχροῦ πόσις ἡ μὲν σύμμετρος ὀνίνησιν, ἡ δ' ἄμετρος ἐσχάτως βλάπτει. τί τοίνυν θαυμαστὸν εἶναί τι φάρμακον οὕτω δυνάμει θερμόν, ὡς, εἰ μὲν πολύ τε

ON TEMPERAMENTS, BOOK III

of ulcers are products of this kind of *kakochymia*. Therefore, there is nothing problematic about such things, as to why there are some medications that do no harm to us externally, but create great harm when swallowed. However, there are some that often cause damage when taken internally, but often benefit, while some not only cause harm externally but also internally. Anyway, to sum up, there is nothing of such a nature that it functions similarly internally and externally. For neither the saliva of a rabid dog, nor the venom of an asp or viper, which in fact are believed to cause harm when falling upon [someone] externally, apart from a wound (ulcer) have equal potency when associated with the skin alone or when taken internally. And it is not worth remarking that the potency of some medications does not reach the depths, for all do not necessarily have the same strength.

If, however, many of these [medications] taken internally at the appropriate time and in a certain quantity, and mixed with others, bring benefit, whereas if they are taken unmixed, in a large amount and at an inappropriate time, bring harm, this should not henceforth present any problem for the argument. This is also the case with foods, and with fire, and may be said similarly with all things that fall upon the body. Anyway, there are times when we need a moderate fire and when we use this we are benefitted greatly, and yet an immoderate fire would burn us. In this way too, a drink of cold water, when it is moderate, benefits, but when immoderate is extremely harmful. What then is surprising, if there is a medication hot in potentiality in this way, that this, when taken in large amounts

λαμβάνοιτο καὶ κενῷ τῷ σώματι προσφέροιτο, διαβιβρώσκειν τε καὶ κατακαίειν αὐτό, εἰ δὲ παντελῶς ὀλίγον εἴη ἢ καὶ σὺν τοῖς κολάζουσι τὴν ἰσχὺν αὐτοῦ, πρὸς τῷ βλάπτειν μηδὲν ἔτι καὶ θερμαῖνον ὠφελεῖν; ὀπὸν γοῦν ἤτοι τὸν Κυρηναϊκὸν ἢ τὸν Μηδικὸν ἢ τὸν Παρθικὸν αὐτὸν καθ' ἑαυτὸν οὐκ ἔνεστιν ἀλύπως λαβεῖν· ἀλλ' εἰ παντελῶς ὀλίγος ἢ σὺν ἄλλοις ἐν καιρῷ προσήκοντι ληφθείη, μεγάλως ὠφελεῖ.

Ταυτὶ μὲν οὖν ὅσα θερμαίνει τὸ σῶμα, μεταβολῆς ἀρχὴν ἐν αὐτῷ λαβόντα, καθότι πρόσθεν ἐρρέθη, πάλιν ἀντιθερμαίνειν αὐτὸ πέφυκεν· ὅσα δὲ ψύχει, καθάπερ ὀπὸς μήκωνος, οὐ μεταβάλλεται πρὸς τοῦ σώματος οὐδ' ἐπ' ὀλίγον, ἀλλ' εὐθὺς αὐτὸ νικᾷ καὶ μεταβάλλει, κἂν εἰ θερμήνας αὐτὰ δοίης. ἔστι γὰρ ἡ σφῶν αὐτῶν φύσις ψυχρά, καθότι καὶ τὸ ὕδωρ. ὀρθῶς οὖν καὶ τοῦτο σὺν πολλοῖς ἄλλοις ὑπ' Ἀριστοτέλους εἴρηται, τὸ τῶν θερμῶν καὶ ψυχρῶν καὶ ξηρῶν καὶ ὑγρῶν σωμάτων τὰ μὲν εἶναι καθ' ἑαυτὰ τοιαῦτα, τὰ δὲ κατὰ συμβεβηκός, οἷον καὶ τὸ ὕδωρ καθ' ἑαυτὸ μὲν ψυχρόν, κατὰ συμβεβηκὸς δέ ποτε καὶ θερμόν. ἀλλ' ἡ μὲν ἐπίκτητος αὐτοῦ θερμασία ταχέως ἀπόλλυται, μένει δ' ἡ σύμφυτος ψυχρότης. ὡς οὖν ὕδωρ θερμὸν ἐπιβληθὲν φλογὶ κατασβέννυσιν αὐτήν, οὕτω καὶ τὸ μηκώνιον εἰ καὶ ὅτι μάλιστα θερμήνας δοίης, ἀποψύξεις τε τὴν ἐν τῷ ζῴῳ θερμασίαν καὶ κίνδυνον ἐπάξεις θανάτου. πάντ' οὖν τὰ τοιαῦτα φάρμακα βραχέα τε διδόμενα καὶ σὺν τοῖς κολάζειν αὐτῶν τὸ σφοδρὸν τῆς ψύξεως δυναμένοις ἔστιν ὅτε χρείαν τινὰ παρέχει

and presented to an empty body, erodes and burns this, whereas if it is altogether small in amount and with those things that curtail its strength, there is nothing harmful from it and the heating still benefits? Anyway, | to take a juice, either the Kurēnaikos, Mēdikos or Parthikos,[12] by itself alone is not without harm, but if it is altogether small in amount, or is taken with other things at an appropriate time, it benefits greatly.

In the same way, then, those medications heat the body, taking an origin of change in it, as was said previously, naturally heating it in turn. On the other hand, those that cool, like the juice of poppy, are not changed by the body, not even a little, but immediately overcome and change the body, even if you give them heated. For the nature of these themselves is cold, just as water also is. And this matter, along with many others, was correctly stated by Aristotle—that of bodies which are hot, cold, dry, and wet, there are those that are so of themselves and those that are so contingently. An example is water which is cold of itself, and yet also sometimes hot contingently. But its acquired heat is quickly destroyed, while the innate cold remains. Therefore, as hot water thrown on a fire quenches it, in the same way too, opiates, even if you give them as heated as possible, | will cool the heat in the animal and will bring danger of death. Thus, all such medications, when given in a small amount, and with those things able to curtail the strength of their cooling, sometimes

---

[12] There is some uncertainty about the juices or saps of plants being referred to here; see the introductory section on medications.

τοῖς σώμασιν ἡμῶν, ὡς ἐν τοῖς περὶ φαρμάκων εἰρήσεται. καὶ γὰρ δὴ καὶ τὸ διὰ τῶν κανθαρίδων φάρμακον ἱκανῶς ὀνίνησι τοὺς ὑδερικούς, καίτοι τοὐπίπαν ἑλκοῖ τὴν κύστιν ἡ κανθαρίς· ἀλλ᾽ ἐπειδὰν ὑπό τε τῶν μιχθέντων αὐτῇ κολασθῇ καὶ σώματι προσάγηται παμπόλλην ὑγρότητα περιέχοντι, κενοῖ διὰ τῶν οὔρων αὐτό. μάλιστ᾽ οὖν χρὴ προσέχειν τὸν νοῦν ἐν ἅπασι τοῖς δυνάμει θερμοῖς ἢ ψυχροῖς εἶναι λεγομένοις, εἴτε τῆς φύσεώς ἐστι τῶν τρέφειν δυναμένων εἴτε μόνην ἀφορμὴν ἀλλοιώσεως λαμβάνοντα κἄπειτα κατὰ τὴν οἰκείαν φύσιν ἀλλοιούμενα διατίθησί πως τὸ σῶμα καὶ τρίτον ἐπὶ τούτοις, εἰ μηδ᾽ ὅλως ἀλλοιοῦται πρὸς αὐτοῦ κατὰ μηδέν. εἰ μὲν γὰρ ἐκ τοῦ γένους εἴη τῶν τρεφόντων, εἰ μὲν κρατηθείη, θερμαίνει, μὴ κρατηθέντα δὲ ψύχει· εἰ δὲ τῶν ἐπ᾽ ὀλίγον ποσὸν ἀλλοιουμένων, θερμαίνει πάντως, εἰ δὲ τῶν μηδ᾽ ὅλως, ψύχει μάλιστα.

4. Προσέχειν δ᾽, ὡς εἴρηται, μάλιστα καὶ διορίζεσθαι τὰ καθ᾽ ἑαυτὰ τῶν κατὰ συμβεβηκός, οὐκ ἐπὶ θερμῶν καὶ ψυχρῶν μόνον, ἀλλ᾽ οὐδὲν ἧττον αὐτῶν ἐφ᾽ ὑγρῶν τε καὶ ξηρῶν. ἔνια γὰρ τῶν τοιούτων ξηρὰ ταῖς οὐσίαις ὑπάρχοντα πολλῷ θερμῷ τακέντα φαντασίαν ὑγρότητος ἔλαβεν ὡς χαλκὸς καὶ σίδηρος, ἢ καθ᾽ ἑαυτὰ μέν ἐστιν ὑγρὰ καθάπερ ὁ κρύσταλλος,

---

[13] Galen's three major works on medications are as follows: *De simplicium medicamentorum temperamentis et facultatibus*, XI.379–892K and XII.1–377K; *De compositione medicamento-*

provide a certain use for our bodies, as will be described in the works on medications.[13] For certainly also, the medication derived from the cantharides[14] will significantly benefit those with dropsy, and yet in general the cantharis ulcerates the bladder. But when curtailed by the things mixed with it and administered to a body containing a great deal of moisture, it evacuates this through the urine. Therefore, it is particularly necessary to direct your attention, in all the medications described as heating or cooling potentially, to whether it is of the nature of the potencies to nourish, or whether they take only the origin of change, and then, in relation to the characteristic nature, change how the body is disposed, and third, in addition to these, nothing at all is changed in any way regarding the body itself. For if it is from the class of those things that nourish, if it is overcome, it heats, whereas if it is not overcome, it cools. | If, however, it is among those that change to a small amount, it altogether heats; if it is among those not changed at all, it particularly cools.

4. Attend, however, as was said, particularly to distinguishing those things that are of themselves from those things that are contingent (*per accidens*), not only in the case of those that are hot and cold, but no less than these, those that are wet and dry. For some of such things are dry in their substances but when melted by great heat take the appearance of wetness, such as copper and iron. Or they are wet in themselves, like ice, but when brought into

rum secundum locos, XII.378–1007K and XIII.1–361K; *De compositione medicamentorum per genera*, XIII.362–1058K.

[14] Cantharis is the blister beetle or Spanish fly (*Lytta vesicatoria*).

ἀκράτῳ δὲ ψύξει πλησιάσαντα ξηρὰ φαίνεται. χρὴ τοίνυν ἁπάντων τούτων τὴν κρίσιν οὐχ ἁπλῶς ποιεῖσθαι, καθότι κἂν τοῖς ἔμπροσθεν ἐλέγομεν, ἀλλὰ μετὰ τοῦ συνεπισκέπτεσθαι, πῶς ἔχει θερμότητος ἢ ψυχρότητος· εἰ μὲν γὰρ ὀλίγης μετέχοντα θερμότητος ὅμως ὑγρὰ φαίνοιτο, διὰ τὴν οἰκείαν φύσιν ἐστὶ τοιαῦτα, κἂν εἰ μετὰ δαψιλοῦς θερμασίας, ξηρά. τὰ δ' ἤτοι ῥέοντα μετὰ θερμότητος ζεούσης ἢ πεπηγότα διὰ | ψῦξιν ἄκρατον, οὐ καθ' ἑαυτὰ νομίζειν εἶναι τὰ μὲν ὑγρά, τὰ δὲ ξηρά. ταύτῃ τ' οὖν διαιρεῖσθαι τὰ καθ' ἑαυτὰ τῶν κατὰ συμβεβηκὸς ἀναφέροντά τε πρὸς αὐτὰ ταῦτα τὴν κρίσιν οὕτω ποιεῖσθαι τῶν δυνάμει θερμῶν ἢ ψυχρῶν ἢ ξηρῶν ἢ ὑγρῶν. οὐ γὰρ πρὸς τὸ κατὰ συμβεβηκὸς ἀλλὰ πρὸς τὸ καθ' ἑαυτὸ κρίνεσθαι χρὴ τὸ δυνάμει. κοινὴ δ' ἐπὶ πάντων ἡ κρίσις καὶ μία τὸ τάχος τῆς ἀλλοιώσεως. ὁμωνύμως δὲ λεγομένου τοῦ θερμοῦ καὶ ψυχροῦ καὶ ξηροῦ καὶ ὑγροῦ· τὸ μὲν γὰρ κατ' ἐπικράτησιν οὕτως ὀνομάζεται, τὸ δ' ὡς ἄκραν ἔχον ἧς παρονομάζεται ποιότητα· εἰς ὁπότερον ἂν αὐτῶν ἑτοίμως μεθίσταται τὸ κρινόμενον, ἔσται δυνάμει τοιοῦτον. ἔλαιον γοῦν ἐστι δυνάμει θερμόν, ὅτι ῥᾳδίως φλὸξ γίγνεται, κατὰ ταῦτα δὲ καὶ ῥητίνη καὶ πίττα καὶ ἄσφαλτος, οἶνος δέ, διότι ῥᾳδίως αἷμα γίγνεται, κατὰ ταὐτὰ δὲ καὶ μέλι καὶ κρέας καὶ γάλα.

Ταυτὶ μὲν οὖν ὅλαις ἀλλοιούμενα ταῖς οὐσίαις τροφαὶ | τῶν ἀλλοιούντων εἰσί· τὰ δὲ κατὰ μίαν ἡντι-

contact with unmixed cooling seem dry. Accordingly, it is necessary to make the judgment of all these not absolutely, just as we said in those beforehand, but along with a joint examination as to how there is hot or cold. For if such things partake of a little heat, but nevertheless appear wet, it is because their own nature is such, and even if [they appear wet] with abundant heat, they are dry. Those that either flow with boiling heat, or are solidified due to strong cooling, should not be thought of as wet of themselves in the former case, or dry in the latter. This, then, is how those things "of themselves" are distinguished from those that are "contingent,"[15] and referring to these same things, and in this way make the judgment of those things potentially hot, cold, dry, or wet. For one must not judge the potential by those things that are contingent but by those that are so of themselves. In all cases, the criterion is common and single—the rapidity of the change. For hot, cold, dry, and wet are said homonymously, in one instance using the term in this way in relation to prevailing and in another as being the peak of the quality paronomastically. When the judgment is readily changed to one or other of these, it will be such in potentiality. At all events, olive oil is hot potentially because it easily becomes a flame, and on the same basis, pine resin, pitch, and asphalt are so. Wine, however, is hot potentially because it easily becomes blood, and the same also applies to honey, flesh, and milk.

These things then, which change in their whole substances, are nutriments for those things effecting the

[15] The distinction is signified by $\kappa\alpha\theta$' $\dot{\epsilon}\alpha\upsilon\tau\alpha$ (*per se*) and $\kappa\alpha\tau\dot{\alpha}$ $\sigma\upsilon\mu\beta\epsilon\beta\eta\kappa\acute{o}\varsigma$ (*per accidens*).

νοῦν ποιότητα τὴν ἀλλοίωσιν ἴσχοντά τε καὶ παρέχοντα[10] φάρμακα μόνον, ὥσπερ γε καὶ ὅσα ταῖς ὅλαις ἑαυτῶν οὐσίαις ἄτρεπτα μένοντα διατίθησί πως[11] τὸ σῶμα, φάρμακα μέν ἐστι καὶ ταῦτα, χαλεπὰ δὲ καὶ φθαρτικὰ τῆς τοῦ ζῴου φύσεως. ὅθενπερ οἶμαι τὸ γένος αὐτῶν ἅπαν ὀνομάζεται δηλητήριον. οὐ γὰρ δὴ διὰ τοῦτό γε λεκτέον οὐκ εἶναι τῷ γένει δηλητήρια τὰ τοιαῦτα, διότι παντελῶς ἐλάχιστα προσαχθέντα βλάβην οὐδεμίαν αἰσθητὴν ἐπιφέρει· οὕτω γὰρ ἂν οὐδὲ τὸ πῦρ εἴη θερμὸν οὐδ᾽ ἡ χιὼν ψυχρά, ὡς καὶ τούτων γε τὰ παντάπασι σμικρὰ σαφὲς οὐδὲν ἀποτελεῖ περὶ τοῖς σώμασιν ἡμῶν πάθος. σπινθῆρος γὰρ ἑνὸς ἑκατοστὸν μέρος ἐστὶ μὲν πάντως τῷ γένει πῦρ, ἀλλ᾽ οὐ μόνον οὐκ ἂν ἡμᾶς καύσειεν ἢ θερμήνειεν, ἀλλ᾽ οὐδ᾽ ἂν αἴσθησίν τινα παράσχοι προσπεσόν. οὕτω δὲ καὶ τὸ τῆς ψυχρᾶς ῥανίδος ἑκατοστὸν μέρος οὐ μόνον οὐδ᾽ ἂν βλάψειεν ἢ ψύξειεν, ἀλλ᾽ οὐδ᾽ ἂν αἴσθησίν τινα παράσχοι. μὴ τοίνυν οὕτω μηδὲ τὰ δηλητήρια κρίνειν, ἀλλὰ τῇ τῆς ὅλης φύσεως ἐναντιώσει, τὴν δ᾽ ἐναντίωσιν ἐκ τῆς ἐν μέσῳ[12] κρίνειν μεταβολῆς. οἷον εὐθέως ἀπὸ τῶν στοιχείων οὔθ᾽ ὕδωρ εἰς πῦρ οὔτε πῦρ εἰς ὕδωρ μεταβάλλειν πέφυκεν, ἀλλ᾽ εἰς ἀέρα μὲν ἄμφω, ἐκεῖνος δ᾽ εἰς ἑκάτερα, εἰς ἄλληλα

[10] παρέχοντα H; τρεπόμενα K
[11] πως add. H
[12] ἐν μέσῳ H; ἐμμέσου K. H has: ἐν μέσῳ scripsi; ἐμμέσου codd. For the whole sentence, KLat has: porro judicabitur contrarietas ex ea quae media intercedit mutatione.

ON TEMPERAMENTS, BOOK III

change. However, those things that both undergo the change in relation to any one quality whatsoever and produce it are medications only, just as in fact are those also that remain unchanged in the whole substances of themselves when applied to the body in some way. These latter are also medications, but are troublesome and destructive of the nature of the organism. Whence, I think, the whole class of these is termed noxious (deleterious).[16] For certainly because of this, one must not speak of such destructive things as not being in the class because, when administered in the smallest possible dose, they bring no perceptible harm. For in this way, fire would not be hot while snow would not be cold, as with these it is clear to everyone that in small amounts they do not affect our bodies. A hundredth part of a single spark is altogether fire in terms of class, but not only would it not burn or heat us, it would not reveal itself to any sense modality it impinged upon. The same would apply to a hundredth part of a drop of cold water; not only would it not cause harm or cool, but it would not reveal itself to any sense modality. Accordingly, do not judge in this way those things that are noxious, but by the opposition of the whole nature, evaluating the opposition from the change in an intermediate. For example, among the elements, water is not of a nature to immediately change to fire, nor fire to water, but both change to air, and the latter into either one of them but in

671K

---

[16] On the definition of a "noxious medication" (δηλητήριον φαρμάκον), see *De naturalibus facultatibus*, III.7, LCL 71 (Brock), 250–51. Brock has: "the latter (a deleterious drug) masters the forces of the body whereas the former (nourishing food) is mastered by them."

GALEN

δὲ ταῦτ' οὐδαμῶς. ἄμεσος μὲν οὖν ἡ εἰς ἀέρα μεταβολὴ τοῦ ὕδατος, ὡσαύτως δὲ καὶ τοῦ πυρός· οὐκ ἄμεσος δ' ἡ πυρὸς καὶ ὕδατος εἰς ἄλληλα. ταῦτ' οὖν ἀλλήλοις ἐστὶν ἐναντία καὶ πολέμια. κατὰ δὲ τὸν αὐτὸν τρόπον καὶ ὀπὸς μήκωνος ἀνθρωπείῳ σώματι τελέως ἐστὶν ἐναντίος οὐδὲν εἰς αὐτὸν οὔτε κατὰ μίαν ποιότητα δρᾶσαι δυναμένῳ οὔτε πολὺ δὴ μᾶλλον ἔτι καθ' ὅλην τὴν οὐσίαν.

Ἓν μὲν δὴ τὸ γένος τοῦτο τῶν δηλητηρίων. ἕτερον δὲ τῶν ἀφορμὴν μέν τινα μεταβολῆς λαμβανόντων ἐκ τῆς ἐν ἡμῖν θερμασίας, εἰς πολυειδεῖς δ' ἐντεῦθεν ἐκτρεπομένων ἀλλοιώσεις, ὑφ' ὧν διαφθείρεσθαι τὴν φύσιν ἡμῶν συμβέβηκε.[13] πάντα γὰρ τὰ τοιαῦτα τῷ γένει δηλητήρια, κἂν διὰ σμικρότητά | ποτε μηδὲν αἰσθητὸν ἀπεργάζηται. τὰ μὲν δὴ διαβιβρώσκοντα καὶ σήποντα καὶ τήκοντα τὴν τοῦ σώματος ἡμῶν φύσιν εἰκότως ὀνομάζεται δυνάμει θερμά, τὰ δ' αὖ ψύχοντα καὶ νεκροῦντα[14] ψυχρά. τὰ μὲν οὖν πρότερα παράλογον οὐδὲν οὔτ' αὐτὰ πάσχειν οὔτε τοῖς σώμασιν ἡμῶν ἐναπεργάζεσθαι δοκεῖ. θερμῷ γὰρ σώματι πλησιάσαντα καί τινα ῥοπὴν ἀλλοιώσεως ἐντεῦθεν λαβόντα τὰ μὲν εἰς ἐσχάτην ἀφικνεῖται θερμότητα, τὰ δ' εἰς σηπεδόνα. διατίθησιν οὖν εἰκότως καὶ τὰ τῶν ζῴων σώματα κατὰ τὴν ἑαυτῶν διάθεσιν. ὅσα δὲ ψύχει τὸ σῶμα, κἂν θερμήνας αὐτὰ προσενέγκῃς, ἀπορίαν οὐ σμικρὰν ἐπιφέρει, τίνος ποτὲ φύσεώς ἐστιν. εἰ γὰρ ἅπαξ ἐνεργείᾳ θερμὰ γέγονε, τί οὐ θερμαίνει τὸ ζῷον; ἢ εἰ μήπω τεθέρμανται, πῶς φαίνεται

no way to each other. Thus the change of water to air is without an intermediate, and similarly also of fire. However, fire and water do not [change] into each other without an intermediate. These, then, are opposite and inimical to each other. In the same way too, the juice of the poppy is completely opposite to a human body which is unable to do anything to this, either in relation to one quality, or, much more, in relation to the whole substance.

Undoubtedly there is this one class of things that are noxious. Another is of those taking a certain origin of change from the heating in us and from that source turning aside to many kinds of changes from which our nature happens to be destroyed. For all such things are deleterious in class, even if in a very small [dose] | they sometimes bring about nothing perceptible. Certainly, those things that erode, putrefy, and dissolve the nature of our body are reasonably termed hot potentially. Others in turn that are cooling and necrosing are cold. The former, then, do not seem to do anything unexpected, neither themselves being affected nor acting on our bodies. When they are brought near to a hot body and take from that source a certain tendency to change, some come to an extreme hotness and some to putrefaction. Therefore, it is likely they also dispose the bodies of the animals according to their own condition. Those that cool the body, even if you administer them heated, carry no small doubt as to what their nature is. For if something has once become hot in actuality, why doesn't it heat the animal? Or if they have not yet been heated, how do they seem hot? Resolu-

---

[13] συμβέβηκε H; συμβαίνε K
[14] νεκροῦντα H; ναρκοῦντα K

θερμά; λύσις δὲ τῆς ἀπορίας, εἰ διορισθείη τὸ καθ᾽ αὑτὸ ψυχρὸν τοῦ κατὰ συμβεβηκός, ὡς Ἀριστοτέλης ἐδίδαξεν. ἀπόλλυται γὰρ ἐν τάχει τῶν κατὰ συμβεβηκὸς θερμῶν ἡ ἐπίκτητος διάθεσις, ὥστ᾽ εἰς τὴν ἀρχαίαν αὐτὰ φύσιν ἐπανέρχεσθαι ῥᾳδίως. ἐν δὲ τῷ πλησιάζειν ἡμῖν τὰ φύσει μὲν ψυχρά, κατὰ συμβεβηκὸς δὲ θερμά, δύο ταῦτ᾽ ἐξ ἀνάγκης γίγνεται· τὸ μὲν ἐπίκτητον αὐτῶν ἀπόλλυται θερμόν, ἡ δ᾽ οἰκεία κρᾶσις οὐδὲν ὑπὸ τῆς ἡμετέρας πάσχουσα μένει ψυχρά. καὶ τί θαυμαστόν, εἰ μήκωνος ὀπὸς ἢ μανδραγόρας ἢ κώνειον ἤ τι τῶν τοιούτων, εἰ καὶ θερμανθέντα προσενεχθείη, μικρὸν ὕστερον γίγνεται ψυχρά, πτισάνης καὶ γάλακτος καὶ χόνδρου καὶ ἄρτου ταὐτὸν τούτοις[15] πασχόντων, ἐπειδὰν εἰς ἄρρωστον ἐμπεσόντα γαστέρα μὴ κρατηθῇ πρὸς αὐτῆς. ἐμεῖται γοῦν πολλάκις ἱκανῶς ψυχρά. καὶ τό γε τούτου μεῖζον, ὃ δὴ καὶ Ἱπποκράτης ἐπεσημήνατο, καίτοι χυμὸς ὂν ἤδη τὸ φλέγμα κἀκ τῶν σιτίων τῶν μὴ πεφθέντων ἐν τῇ γαστρὶ γεννώμενον ὅμως ψυχρὸν ἁπτομένοις φαίνεται οὐ μόνον ἐν τῇ γαστρὶ συστὰν ἀλλὰ κἀκ τῶν φλεβῶν αὐτῶν ὑπό τινος τῶν καθαιρόντων φαρμάκων ἑλχθέν. καίτοι γε γλισχρότατόν ἐστι φύσει[16] καὶ βιαίως ἄγεται, ἀλλ᾽ ὅμως οὐδ᾽ ἡ βία τῆς ὁλκῆς ἐκθερμαίνειν αὐτὸ δύναται. τί οὖν θαυμαστόν, εἰ καὶ τὸ μηκώνιον, οὕτως ἐναντίον ἡμῶν τῇ φύσει φάρμακον, ἀποψύχεται μὲν αὐτίκα μάλα, κἂν θερμὸν ποθῇ, συγκαταψύχει δ᾽ ἑαυτῷ τὸ σῶμα; τὴν μὲν γὰρ ἐπίκτητον οὐ φυλάττει θερμασίαν, ὅτι φύσει ψυχρὸν ἦν· τῷ δὲ

tion of the problem [is achieved] if the distinction is made
between cold in itself and cold *per accidens* (contingently),
as Aristotle taught. For the acquired condition of those
things hot *per accidens* is quickly lost, so that these are
easily returned to their original nature. However, when
those things that are cold in nature but contingently hot
come into contact with us, these two things necessarily
occur: their acquired heat is lost, while their own proper
krasis is not affected by us and remains cold. And what
is so remarkable, if poppy juice, mandragora, hemlock,
or some such thing, administered heated becomes cold
slightly later, when there is this same effect, if ptisane, milk,
gruel, or bread, falling upon a weak stomach, are not overcome by it? At any rate, sufficiently cold things are often
vomited. And, in fact, what is more than this, which of
course Hippocrates also pointed out, is that phlegm, even
though already being a cold humor and generated from
foods that have not been concocted in the stomach, nevertheless seems cold to the touch, not only when formed in
the stomach but also from the veins themselves, drawn by
one of the purging medications. And yet, in fact, it is very
viscid in nature and is led forcibly, but nevertheless, the
force of the attraction is not able to heat it. Is it any wonder then, if opium also, in this way a medication opposite
to us in nature, is immediately very much cooled, even if
drunk hot, and while it cools the body is cooled along with
it? For it does not preserve the acquired heat because it
is cold in nature. But because its own substance is not

---

15 τούτοις add. H *(see his note to line 26, p. 102)*
16 φύσει H; σφόδρα K

μηδ' ἀλλοιοῦσθαι τὴν οὐσίαν αὐτοῦ πρὸς ἡμῶν ἀλλὰ μᾶλλον ἀλλοιοῦν τε καὶ μεταβάλλειν ἡμᾶς οὔτ' ἐκθερμαίνεταί που πρὸς ἡμῶν αὐτό τε διατίθησιν ἡμᾶς καθ' ἑαυτό· ψυχρὸν οὖν ὑπάρχον φύσει ψύχει δήπου καὶ ἡμᾶς· οὔκουν οὐκέτ' ἄπορον οὐδὲν ὑπόλοιπον ἐν τῷ λόγῳ. καὶ γὰρ δὴ καὶ ὅτι τούτων ἁπάντων τῶν φύσει ψυχρῶν ὅ τι ἂν ἐπὶ πλέον ἐκθερμήνῃς, ἐξίσταται τῆς ἰδίας φύσεως, πρὸς τῷ μηδὲν ἄπορον ἔχειν ἔτι καὶ μαρτυρεῖ τοῖς προειρημένοις. ὡς γὰρ καὶ ἡ σαλαμάνδρα μέχρι μέν τινος οὐδὲν ὑπὸ πυρὸς πάσχει, κατακαίεται δ', εἰ[17] πλείονα χρόνον πλησιάσειεν, οὕτω καὶ μανδραγόρας καὶ κώνειον καὶ μηκώνιον καὶ ψύλλιον ἐπὶ βραχὺ μὲν ὁμιλήσαντα πυρὶ τὴν οἰκείαν ἔτι διαφυλάττει κρᾶσιν, ἐπὶ | πλέον δὲ θερμανθέντα διαφθείρεται παραχρῆμα καὶ δρᾶν οὐκέτ' οὐδὲν ὧν πρότερον ἐπεφύκει δύναται.

Τῶν μὲν δὴ τοιούτων ἁπάντων ἡ φύσις ἐναντιωτάτη τοῖς ἀνθρώποις ἐστί· φύσιν δ' ὅταν εἴπω, τὴν ὅλην οὐσίαν τε καὶ κρᾶσιν λέγω τὴν ἐκ τῶν πρώτων στοιχείων, θερμοῦ καὶ ψυχροῦ καὶ ξηροῦ καὶ ὑγροῦ· τῶν δέ γε τάχιστα τρεφόντων οἰκειοτάτη. τὰ δ' ἄλλα πάντα μεταξὺ τούτων ἐστί, τὰ μὲν μᾶλλον, τὰ δ' ἧττον δρᾶν καὶ πάσχειν ὑπὸ τοῦ σώματος ἡμῶν δυνάμενα, καστόριον μὲν καὶ πέπερι δρᾶν μᾶλλον ἢ πάσχειν [ὑπὸ τοῦ σώματος ἡμῶν δυνάμενα],[18] οἶνος δὲ καὶ μέλι καὶ πτισάνη πάσχειν μᾶλλον ἢ δρᾶν.

---

[17] post εἰ: πλείονα χρόνον H; πλείονι χρόνῳ τῷ πυρὶ K

changed by us but rather changes and alters us and is not heated in any way by us, it disposes us by virtue of itself; being cold in nature, it also clearly cools us. There is not, therefore, any difficulty remaining in the argument. For of a surety also, all these things cold in nature which, should you heat one of them to the greatest extent, change the specific nature, present no difficulty in this matter, but are further evidence of what has previously been said. Thus, as the salamander,[17] up to a certain point, suffers nothing from fire, but is burned up if it is closer to the fire over a longer time, in this way too mandragora, hemlock, opium, and fleawort, if brought into contact with the fire for a short time, still preserve their characteristic *krasis*, whereas if they are heated to a greater extent, are immediately destroyed and are no longer able to do what they had been of a nature to do previously.

Certainly, the nature of all such things is most inimical to humans. When I say "nature," I speak of the whole substance and *krasis*—that from the primary elements, hot, cold, dry, and wet—which is most suitable of those things that nourish most rapidly. All the others are, however, between these, being more or less in terms of being able to act upon or be acted upon by our body. For castor and pepper are able to act more than be acted upon by our body, whereas wine, honey, and ptisane are acted upon

---

[17] On the medicinal actions of salamanders, see Aristotle, *History of Animals* 332b16, and Dioscorides 2.67.

[18] H *has:* vocabula ὑπὸ τοῦ σώματος ἡμῶν δυνάμενα *om.* L¹T *add.* L² *in marg. (p. 104)*

ἅπαντα γοῦν ταῦτα καὶ πάσχει τι καὶ δρᾷ περὶ τὸ σῶμα· καθ' ὅλου γὰρ ἐπειδὰν εἰς ταὐτὸν ἀλλήλοις ἥκοντα δύο σώματα διαμάχηταί τε καὶ στασιάζῃ πρὸς ἄλληλα περὶ τῆς ἀλλοιώσεως ἐν χρόνῳ πλείονι, δρᾶν καὶ πάσχειν ἑκάτερον αὐτῶν ἀναγκαῖόν ἐστιν. ἴσως δέ, κἂν μὴ πάνυ πολλῷ χρόνῳ γίγνηται τοῦτο, δρᾷ μέν τι καὶ τότε τὸ νικώμενον εἰς τὸ κρατοῦν, ἀλλ' οὕτω σμικρὸν ὡς λανθάνειν τὴν αἴσθησιν. οὐδὲ γὰρ ὁ τμητικώτατος σίδηρος εἰ τὸν μαλακώτατον τέμνοι κηρὸν δι' ὅλης ἡμέρας καὶ νυκτός, δυνατὸν[19] αὐτῷ μὴ οὐκ ἀμβλεῖ γενέσθαι σαφῶς. οὕτω δήπου κἀκεῖνο καλῶς εἰρῆσθαι δοκεῖ

Πέτρην κοιλαίνει ῥανὶς ὕδατος ἐνδελεχείῃ.

καὶ γὰρ καὶ φαίνεται γιγνόμενον οὕτως· ἀλλ' ἐπὶ μιᾶς ἢ καὶ δευτέρας προσβολῆς οὐδὲν οὔπω σαφὲς ἐν τοῖς τοιούτοις ἰδεῖν ἔστιν. ὅθεν οἶμαι καὶ τὸ μηδ' ὅλως πάσχειν ἔνια πρὸς τῶν ὁμιλούντων ἐδοξάσθη τισὶ καὶ συγχωρητέον γε τοῖς οὕτω λέγουσι πολλάκις καὶ αὐτοὺς ὁμοίως ἐκείνοις λεκτέον ἐστὶ τὰ πολλά, πλὴν εἴ ποτε μέχρι τῆς ἐσχάτης ἀκριβείας ἀνάγοιμεν τὸν λόγον, ὥσπερ ἐν τῷ παρόντι ποιοῦμεν. οὕτως οὖν καὶ τὸ τῆς ἀειπαθείας δόγμα τῷ λόγῳ μὲν αὐτῷ μόνῳ

[19] δυνατὸν H; ἀδύνατον K

[18] This is apparently a line from a lost epic poem attributed to Cherylus of Samos (*Ep. Gr.* 3.271, frag. 10).

ON TEMPERAMENTS, BOOK III

more than they act. Anyway, all these are both acted on and act with regard to the body. In general, when two bodies come to the point of contending with each other and being at variance with one another regarding the change in a longer time, each of them is of necessity acting and acted upon. Perhaps, however, even if this occurs over a not very long time, what acts would still at that time be prevailing toward what is prevailed upon, but in such a small way as to be hidden to perception. For the sharpest iron, if it were to cut the softest cheese for a whole day and night, cannot but become clearly dulled in itself. Of course, this seems to be said well, as follows:

Constant dripping of water hollows out a stone.[18]

For it manifestly occurs like this, but after one or two impacts, there is not yet anything clear to see in such cases. This is why, I think, it seemed to some that some things are not at all affected by things contacting them. And one must agree with those who often speak in this way, and we ourselves must speak similarly to those men in many instances, except we should sometimes elevate the argument to the extreme of accuracy, as we are doing in the present instance. Therefore, in this way too, they consider the dogma of perpetual affection[19] has a strong dem-

---

[19] The term ἀειπάθεια is defined in LSJ as "perpetual passivity," but this is not an appropriate reading here. Galen considers the matter in a number of other places (*On the Constitution of the Art of Medicine*, I.256–7K; *The Art of Medicine*, I.317K; and *Hygiene*, VI.18K). Rather, it seems an almost necessary consequence of his theory of bodily *krasis*. See also Boudon's note given in Johnston, LCL 523, 175.

σκοποῦσιν ἰσχυρὰν ἔχει τὴν ἀπόδειξιν. οὐ μὴν χρεία γ' αὐτοῦ τίς ἐστιν πρὸς τὰς κατὰ μέρος πράξεις. ἂν γὰρ οὕτω σμικρά τινα περὶ ἡμᾶς ᾖ | πάθη διὰ παντός, ὡς μηδεμίαν αἰσθητὴν καὶ σαφῆ βλάβην ἐνεργείας ἐργάζεσθαι μηδεμιᾶς, εὐκαταφρόνητα δήπουθέν ἐστι καὶ τῷ μηδ' εἶναι φάσκοντι τὰ τοιαῦτα συγχωρητέον. οὕτως οὖν ἔχει κἀπὶ τῶν τρεφόντων ὀλίγου δεῖν ἁπάντων. ἐργάζεται μὲν γάρ τι καὶ αὐτὰ περὶ τὸ σῶμα τῶν ἀνθρώπων, ἀλλ' οὐκ αἰσθητὸν οὐδὲ σαφὲς εἰσάπαξ. ἡ μέντοι πολυχρόνιος αὐτῶν προσφορὰ μεγάλως ἀλλοιοῖ καὶ μεταβάλλει σαφῶς ἤδη τὰ σώματα. ἔνια μέν γε καὶ κατὰ τὴν πρώτην χρῆσιν εὐθὺς ἐναργῶς ἐνδείκνυται τὴν ἀλλοίωσιν, οἷον καὶ ἡ θριδακίνη τοὺς μὲν ἐγκαιομένους τὴν γαστέρα σαφῶς ἐμψύχουσά τε καὶ ἀδίψους ἐργαζομένη, τοὺς δὲ κατεψυγμένους ἐναργῶς βλάπτουσα. συντελεῖ δ' οὐ σμικρὰ καὶ τοῖς ὕπνοις οὐ κατ' ἄλλον τινὰ λόγον, ἀλλ' ἢ ὅτι ψυχρά τ' ἐστὶ καὶ ὑγρὰ τὴν κρᾶσιν, ἀλλ' οὕτω ψυχρὰ καὶ ὑγρὰ πρὸς ἄνθρωπόν τε καὶ τἆλλ' ὅσα τρέφειν πέφυκεν ὡς καὶ τὰ ξύλα τὰ χλωρὰ πρὸς τὸ πῦρ. ὥστ' εὐλόγως ἄμφω τοῖς τοιούτοις ἐδέσμασιν ὑπάρχει | τό θ' ὡς φαρμάκοις διατιθέναι τὰ σώμαθ' ἡμῶν καὶ τὸ τρέφειν, παρ' ὅλον μὲν τὸν χρόνον τῆς πέψεως ὡς φαρμάκοις· ἡνίκα δ' ἤδη τρέφει τε καὶ τελέως ὁμοιοῦται, τότ' οὐκέτ' οὐδὲν ἡμᾶς ἀντιδρῶντα τὴν ἔμφυτον αὔξει θερμασίαν, ὡς καὶ πρόσθεν εἴρηται. κοινὸν γὰρ δὴ τοῦτο τῶν τρεφόντων ἁπάντων ἐστὶ καὶ οὐ χρὴ θαυμάζειν, εἴ τι, πρὶν μὲν ἐξομοιοῦσθαι καὶ τρέφειν, ἔτι

onstration by reason itself alone. In fact, there is no use of this in regard to actions individually, for in this way, there would be certain small | affections involving us continuously, such as to produce no perceptible and clear harm to any functions, and are, I presume, to be ignored. As a result one must agree with someone who says such things don't exist. It is the same in the case of all nutriments apart from a few. That is to say, these also produce something involving the body of humans, but not something perceptible or clear once and for all. Nevertheless, the administration of these over a long period of time greatly alters and already clearly changes the bodies. Some, in fact, at the first use immediately and clearly display the change; an example is lettuce, which in some cases clearly cools the stomach when it is burning and produces lack of thirst, while in others who are cooled, it is obviously injurious. And it has no small effect on those who are sleeping, and not for any other reason than that it is cold and wet in *krasis*, but in this way being cold and wet toward a person and the other things it naturally nourishes, as green wood nourishes a fire. Consequently, both are to be expected with such foods; | to be applied to our bodies as medications and to nourish through the whole time of concoction as medications, but when it has already nourished and is completely assimilated, at that time it no longer acts against us in any way, but increases the innate heat, as was also said previously. For this is a common feature of all those things that are nourishing, and it is not necessarily surprising, if something, before it is assimilated and nourishing, while still being concocted,

πεττόμενον ἔψυξεν, ἐξομοιωθὲν δὲ καὶ θρέψαν ἐθέρμηνε, μεμνημένους ἀεὶ τοῦ τῶν χλωρῶν ξύλων παραδείγματος. ὥστε καὶ ἡ χρεία τῶν τοιούτων ἁπάντων διττὴ τοῖς ἰατροῖς ἐστι καὶ ὡς σιτίων καὶ ὡς φαρμάκων. φέρε γὰρ ὑπηλλάχθαι τινὶ τὴν ἀρίστην ἐν τῇ γαστρὶ κρᾶσιν ἐπὶ τὸ θερμότερον. οὗτος ἄχρι μὲν οὗ πέττει τὴν θριδακίνην, ἐμψυχθήσεται καὶ συμμετρίαν κτήσεται κράσεως· ἐπειδὰν δ' ἐξ αὐτῆς ἤδη τρέφηται, τὴν οὐσίαν αὐξήσει τῆς ἐμφύτου θερμασίας.

Ἐν τούτῳ δὴ καὶ μάλιστα δοκοῦσί μοι παραλογίζεσθαι σφᾶς αὐτοὺς οἱ πολλοὶ τῶν νεωτέρων ἰατρῶν ἀγνοοῦντες, ὡς ἐνίοτε μὲν ἡ ποιότης ἐπιτείνεται τῆς ἐν ἡμῖν θερμασίας, ἐνίοτε δ' ἡ οὐσία παραύξεται καὶ ὡς ἑκατέρως οἱ παλαιοὶ θερμότερον γεγονέναι φασὶ τὸ ζῷον. καὶ γὰρ καὶ γίγνεται θερμότερον, ἄν τ' ἐπιτείνῃς αὐτοῦ τὴν θερμασίαν ἄν τ' αὐξήσῃς τὴν οὐσίαν, ἐν ᾗ πρώτῃ περιέχεται. φέρε γὰρ εἶναι τὸ αἷμα τὸ ἐν τῷ σώματι τοῦ ζῴου καθ' ἑαυτὸ θερμὸν ἢ καὶ νὴ Δία, εἰ βούλει, τὴν ξανθὴν χολήν, ἅπαντα δὲ τἆλλα κατὰ συμβεβηκὸς ὑπάρχειν θερμὰ τῷ τούτων μετέχειν. ἆρ' οὐκ ἀναγκαῖον ἔσται διττῶς γίγνεσθαι θερμότερον τὸ ζῷον ἤτοι τῷ πλείονας ἐπικτήσασθαι τοὺς θερμοὺς χυμοὺς ἢ τῷ θερμοτέρους ἔχειν ἢ πρότερον; ἐμοὶ μὲν καὶ πάνυ φαίνεται. κατὰ δὲ τὸν αὐτὸν οἶμαι τρόπον καὶ ψυχρότερον ἔσται διττῶς ἢ τῷ πλείονας ὑποτραφῆναι τοὺς ψυχροὺς χυμούς, οἷον τό τε φλέγμα καὶ τὴν μέλαιναν χολὴν ἢ τῷ τῆς αὐτῆς ἁπάντων μενούσης συμμετρίας ὑπαλλαχθῆναι μόνην

has cooled, whereas when it has been assimilated and has nourished, it has heated, bearing in mind always the example of green wood. As a consequence, the use of all such things by doctors is twofold—as foods and as medications. Suppose then, the best *krasis* in the stomach in someone has been changed to hotter. This person, while he has still not concocted the lettuce, will be cooled and will acquire a balance of *krasis*, whereas when he is already nourished from this, he will increase the substance of the innate heat.

In this, surely and particularly, many of the younger doctors seem to me to have misled themselves, being unaware that sometimes the quality | of the heat in us is increased, while sometimes the substance is augmented and, as the ancients say, the animal has become hotter in both ways. For truly it also becomes hotter, whether you increase its heat or augment the substance contained in the first formation. Suppose then the blood in the body of the animal is hot in and of itself, or also, by Zeus, if you wish, the yellow bile, while all the other things are hot contingently by partaking of the heat of these. Will it not then be necessary for the animal to become hotter in two ways; that is, either by more hot humors being acquired, or by the humors being hotter than they were before? To me this is very obvious. In the same way, I think, it will also be colder in two ways: because it is nourished by more cold humors, such as phlegm and black bile, or because the balance of all these remaining the same, the quality alone is changed. Why, then, is it surprising

679K

τὴν ποιότητα. τί δὴ οὖν θαυμαστόν, ἄχρι μὲν ἂν πέττηται τὸ ψυχρὸν τῇ φύσει σιτίον, οἷον ἀνδράχνη τε καὶ θριδακίνη, ψυχρᾶς ποιότητος ἀναπίμπλασθαι τὸ σῶμα, πεφθέντων δ' ἀκριβῶς καὶ γενομένων αἵματος χρηστοῦ θερμότερον αὖθις ἑαυτοῦ γίγνεσθαι τὸ σῶμα τῇ τοῦ θερμοῦ χυμοῦ γενέσει;

Καὶ μὴν εἰ μηδὲν τούτων μήτ' ἀδύνατόν ἐστι μήτε θαυμαστὸν ἔτι, παυσάσθωσαν οἱ μὴ συγχωροῦντες ἓν καὶ ταὐτὸν ἔδεσμα καὶ τὴν ὡς τροφῆς καὶ τὴν ὡς φαρμάκου χρείαν τῷ ζώῳ παρέχειν. ὡς γὰρ εἰ καὶ μηδ' ὅλως ἐπέφθη, διὰ παντὸς ἂν ἐφυλάχθη φάρμακον, οὕτω πεφθὲν ἄμφω γίγνεται. φέρε γὰρ μηδ' ὅλως πεφθῆναι τὴν θριδακίνην ἢ νὴ Δία τὸν χυλὸν αὐτῆς, ἐπειδὴ καὶ παραπλήσια τῷ τῆς μήκωνος ὀπῷ δρᾷ τὸν ἄνθρωπον, εἰ πάμπολυς ληφθείη, ἆρ' οὖν οὐ φάρμακον ἔσται τηνικαῦτα μόνον, ἄλλο δ' οὐδέν; οὐκ οἶμαί τινα περί γε τούτου διαμφισβητήσειν· ὥστ' ἔχει πάντως καὶ τὴν τοῦ φαρμάκου δύναμιν ἡ θριδακίνη. ἀλλὰ μὴν εἶχε καὶ τὴν τῆς τροφῆς· ἔθρεψε γὰρ πολλάκις· ὥστ' ἀμφοτέρας μὲν ἅμα τὰς δυνάμεις ἐν ἑαυτῇ περιέχει, δείκνυσι δ' οὐχ ὁμοίως ἀμφοτέρας, ἀλλ' ἐπειδὰν μὲν αὐτὴ πλέον ἐνεργῇ περὶ τὸν ἄνθρωπον ἢ πάσχῃ, τὴν ὡς φαρμάκου μᾶλλον ἐπιδείκνυται δύναμιν, ἐπειδὰν δὲ πάσχῃ πλέον ἢ ποιῇ, τὴν ὡς σιτίου. καὶ τί θαυμαστόν, εἰ τῇ θριδακίνῃ καὶ δρᾶν καὶ πάσχειν συμβέβηκεν, ὅπου γε καὶ τῷ ξίφει, καθότι καὶ μικρὸν ἔμπροσθεν ἐλέγετο, μὴ μόνον δρᾶν εἰς τὸν κηρὸν ἀλλὰ καὶ πάσχειν ὑπάρχει. τῷ δ' εἶναι πολὺ πλέον ὃ

ON TEMPERAMENTS, BOOK III

that the body is filled with a cold quality, when the food currently being concocted is cold in nature, like purslane and lettuce, whereas when it is concocted completely | and has become useful blood, the body in turn becomes hotter than it was by the genesis of the hot humor?

And in fact, if none of these is impossible or even surprising, they should refrain from disagreeing that this one food provides a use to the animal as both a nutriment and a medication. For if it were not altogether concocted, it would be preserved as a medication throughout, so in this way when concocted it becomes both. Let us suppose then that the lettuce is not concocted at all or, by Zeus, the juice of this is not, since in what it does to the human it is similar to the juice of the poppy. If taken in a large amount, will it not then, under these circumstances, be only a medication but nothing else? I do not think anyone would in fact dispute this. As a consequence, lettuce has, at all events, the potency of the medication, but it would also have that of the nutriment, for it frequently nourishes. Consequently, it provides in itself both potencies at the same time. It does not, however, display both similarly, but when it acts more on the person than it is acted upon, | it displays the potency as a medication more, whereas, when it is acted upon more than it acts, it displays the potency of a food. It is no surprise then, if both acting and being acted upon happen due to the lettuce, when in fact with a sword also, as we said a little earlier, not only is there acting on the wax but there is also being acted on [by the wax]. However, it is because its nature is very much more to act

πέφυκε δρᾶν τοῦ πάσχειν λανθάνει θάτερον. ἀλλ' εἰ σκληρότατον αὐτῷ παραβάλλοις σίδηρον, ἔμπαλίν σοι φανεῖται πάσχειν μᾶλλον ἢ δρᾶν, καίτοι δρᾷ μέν τι καὶ τότε, παρορᾶται δ' ἡ δύναμις αὐτοῦ διὰ σμικρότητα.

Θαρροῦντες οὖν ἐπὶ πάντων μὲν ἁπλῶς ἀποφαινόμεθα τῶν σιτίων, ὡς οὐ μόνον πάσχειν ὑπὸ τοῦ σώματος ἡμῶν ἀλλὰ καὶ δρᾶν εἰς αὐτὸ πέφυκεν· ἤδη δὲ καὶ περί τινων, οἷς ἐναργῶς καὶ σαφῶς ὑπάρχει τὸ δρᾶν, ὡς οὐ σιτία μόνον ἐστὶν ἀλλὰ καὶ φάρμακα. θριδακίνη μὲν οὖν καὶ τροφὴ καὶ φάρμακον ψυχρόν, εὔζωμον δὲ καὶ τροφὴ καὶ φάρμακον θερμόν. εἰ δὲ καὶ καστόριον ἐν τῷ χρόνῳ πέττεται, εἴη ἂν καὶ τοῦτο καὶ τροφὴ καὶ φάρμακον θερμόν· οὕτω δὲ καὶ νᾶπυ καὶ πέπερι καὶ τῶν βοτανῶν ἄνηθόν τε καὶ πήγανον ὀρίγανόν τε καὶ γλήχων καὶ καλαμίνθη καὶ θύμβρα καὶ θύμος. πάντα γὰρ ταῦτα καὶ τροφαὶ καὶ φάρμακα θερμά, πρὶν μὲν εἰς αἷμα μεταβαλεῖν, ἔτι γε πεττόμενα, φάρμακα, μεταβληθέντα δ' οὐκέτι μὲν φάρμακα, τροφαὶ δ' ἤδη κατὰ τὸ δεύτερον δηλονότι τῆς τροφῆς σημαινόμενον, ὃ οὔπω μέν ἐστι τροφή, οἷον δὲ τροφή.

Ὡς οὖν ἔμπροσθεν ἐπὶ τῆς θριδακίνης ὑπεθέμεθα μίαν μὲν κοιλίαν θερμοτέραν τοῦ δέοντος, ἑτέραν δὲ ψυχροτέραν, οὕτω καὶ νῦν ἐπὶ πάντων τῶν δυνάμει θερμῶν ὑποκείσθωσαν αἱ δύο κοιλίαι. τὴν μὲν οὖν ψυχροτέραν τοῦ δέοντος, ἄχρι περ ἂν ᾖ ἐν αὐτῇ περιεχόμενα καὶ πεττόμενα τὰ τοιαῦτα σύμπαντα, θερμα-

than to be acted upon, that the latter goes unnoticed. But if you were to expose very hard iron to it, it would seem to you conversely to be acted upon more than it acts, and yet at that time it does also act to some extent but the potency of this is overlooked due to being small.

Let us then be so bold as to declare, in the case of all foods absolutely, that not only are they acted upon by our body but they also naturally act upon this. But already also concerning some in which the acting is clear and obvious, they are not only foods but also medications. Lettuce, then, is both a nutriment and a cold medication, while rocket is both a nutriment and a hot medication. And if castoreum is concocted over time, this would also be a nutriment and a hot medication, as in same way too are mustard, pepper, and of the herbs, dill, rue, oregano, pennywort, catmint, savory, and thyme. For all these are both nutriments and hot medications before the change into blood; while they are still in fact being concocted, they are medications, whereas, when they are changed, they are no longer medications but are already nutriments, in relation obviously to the second signification of "nutriment"—what is not yet a nutriment, but is like a nutriment.

682K

Therefore, as we postulated before in the case of lettuce that one abdomen was hotter than it ought to be while another was colder, in this way now the two abdomens are assumed in the case of all these things hot potentially. Thus, in respect of the abdomen colder than it should be, while all such things are being contained and concocted

νεῖ τε καὶ εἰς ἰσότητα κράσεως ἐπανάξει καὶ ὠφελήσει λόγῳ φαρμάκων, τὴν δ' ἑτέραν τὴν θερμὴν ἐκπυρώσει τε καὶ μεγάλως βλάψει, καὶ ταύτας μὲν τὰς ἀλλοιώσεις ἐργάσεται κατὰ ποιότητα. πεφθέντα δ' ἀκριβῶς καὶ μεταβληθέντα καὶ χρηστὸν αἷμα γενόμενα κατ' οὐσίαν αὐξήσει τὸ ἔμφυτον τοῦ ζῴου θερμόν, οὐ κατὰ ποιότητα. καθ' ὅλου γάρ, ἄν τε ψυχρὸν ἄν τε θερμὸν ᾖ τῇ δυνάμει τὸ σιτίον, ἐπειδὰν αἱματωθῇ, τὴν ἔμφυτον ὡσαύτως αὔξει θερμασίαν· ἄχρι δ' ἂν ἄγηται μὲν εἰς αἵματος ἰδέαν, οὔπω δὲ τελέως αἷμα <γένηται>,[20] ψύχει καὶ θερμαίνει τὸ σῶμα δίκην φαρμάκου. ἅπας δ' οὗτος ὁ λόγος ἐκ μιᾶς, ὡς ἔοικεν,[21] ἀρχῆς ἤρτηται. διὸ καὶ φυλάττειν αὐτὴν ἀεὶ χρὴ καὶ μεμνῆσθαι διὰ παντός, ὡς ἕκαστον τῶν σωμάτων ἰδιότητά τινα κέκτηται κράσεως, οἰκείαν μὲν τῇδέ τινι τῇ φύσει, διαφερομένην δὲ τῇδέ τινι, καὶ ὡς εἰ μὲν ἀλλοιώσειε τὸ οἰκεῖον εἰς τὴν ἑαυτοῦ φύσιν, αὐξήσει τὴν οὐσίαν οὕτω τῆς ἐν αὐτῷ θερμασίας· εἰ δ' ἀλλοιωθείη, δυοῖν θάτερον αὐτῷ συμβήσεται, θερμαίνοντος μὲν τοῦ μεταβάλλοντος ἐπικτήσασθαί τινα θερμασίαν, μὴ θερμαίνοντος δὲ τὴν οἰκείαν ἀπολέσαι. δῆλον οὖν ἐκ τούτων, ὡς ἐν τῷ πρός τι σύμπαντ' ἐστὶ τὰ τοιαῦτα. πρὸς γὰρ τὴν ἰδιότητα τῆς ἀλλοιούσης φύσεως ἕκαστον τῶν προσφερομένων ἢ τροφῆς ἢ φαρμάκου λόγον ἢ ἀμφοτέρων ὑφέξει, οἷον τὸ κώνειον τῷ ψαρὶ μὲν τροφή, φάρμακον δ' ἀνθρώπῳ καὶ τοῖς μὲν ὄρτυξιν ἑλλέβορος τροφή, τοῖς δ' ἀνθρώποις

in it, it heats and will restore an equality of *krasis*, and will benefit by reason of being among medications, whereas the other, the hot abdomen, it will overheat and greatly harm, and will produce these changes in relation to quality. However, when completely concocted and changed, and has become useful blood, it will increase the innate heat of the animal in relation to substance, and not in relation to quality. For in general, whether it is cold or hot in potentiality, the food, when it is made into blood, will increase the innate heat similarly. However, up to the point of being brought to the form of blood, when it has not yet completely become blood, it cools and heats the body after the manner of a medication. This whole argument is, it would seem, dependent on one principle. On this account, it is always necessary to preserve this principle and to bear it in mind continually—that each of the bodies has acquired a certain specificity of *krasis*, characteristic to its particular nature, but different to some other body, and that, if it changes what is characteristic to its own nature, it will in this way increase the substance of the heat in it. If, on the other hand, it is changed, one of two things will happen to it. It will acquire a certain heat if what changes it is heating, or if it is not heating, it will lose its own [heat]. It is clear, then, from these considerations, that in this all such things are relative. In respect of the specificity of the nature of each of the things administered effecting the change, it will take on the ground of either nutriment or medication, or both. For example, hemlock is a nutriment for a fish but a medication for a person, and hellebore is a

---

[20] αἷμα ⟨γένηται⟩ H; αἱμαχθῇ K
[21] ὡς ἔοικεν add. H

φάρμακον. ἡ μὲν γὰρ τῶν ὀρτύγων κρᾶσις ἐξομοιοῦν ἑαυτῇ δύναται τὸν ἐλλέβορον, ἡ δὲ τῶν ἀνθρώπων οὐ δύναται.

5. Φανερὸν οὖν ἤδη γέγονεν, ὡς ἡ κρίσις τῶν πρὸς ἡμᾶς ὑγρῶν ἢ ξηρῶν ἢ ψυχρῶν ἢ θερμῶν οὐκ ἔξωθέν ποθεν, ἀλλ' ἐξ ὧν ἡμεῖς αὐτοὶ πάσχομεν ἀκριβῶς ἂν γίγνοιτο καὶ ὡς πρῶτον μὲν καὶ μάλιστα ταῦτ' εἴη σκεπτέον, ἐφεξῆς δ', εἴ τι δεήσειε, καὶ τὰ ἔξωθεν. εἰ μὲν γὰρ ἐναργὴς εἴη καὶ σαφὴς αἰσθήσει τοῦ προσαχθέντος ἡ ἐνέργεια φαρμάκου, ταύτῃ πιστευτέον ἀμελοῦντας τῶν ἄλλων ἁπάντων γνωρισμάτων· εἰ δ' ἀμυδρὰ καὶ ἀσαφὴς ἢ ἐπίμικτος ἤ τιν' ὅλως ἀμφισβήτησιν ἔχουσα, τηνικαῦτα καὶ πρὸς τὰ ἐκτὸς ἅπαντα κριτέον αὐτήν, οὔκουν οὐδὲ πρὸς ταῦτα πόρρωθεν, ἀλλ' ἀπ' αὐτῆς τοῦ ζητουμένου τῆς οὐσίας,[22] οἷον εἰ τὸ ἔλαιον θερμόν, οὐχ ὅτι γλίσχρον ἢ ὠχρὸν ἢ κοῦφον, ἀλλ' εἰ ῥᾳδίως ἐκφλογοῦται. τοῦτο γὰρ ἦν αὐτῷ τὸ δυνάμει θερμῷ εἶναι τὸ ταχέως μεταβάλλειν εἰς τὸ ἐνεργείᾳ θερμόν. κατὰ δὲ τὸν αὐτὸν τρόπον κἀπὶ τῶν ἡμετέρων σωμάτων οὐκ εἰ παχυμερὲς ἢ λεπτομερὲς ἢ ὑγρὸν ἢ κοῦφον ἢ γλίσχρον ἢ ὠχρόν, ἀλλ' εἰ θερμαίνει προσαγόμενον· οὕτω δ' οὐδ' εἰ γλυκὺ καὶ κοιλίαν ὑπάγον ἢ εὔρουν ἐργαζόμενον ἐν ταῖς φλεβοτομίαις, εἰ ἐπισταχθείη, τὸ αἷμα. καὶ γὰρ καὶ ταῦτα περιττὰ παρόν γε σκοπεῖν, εἰ θερμαίνει προσαγόμενον. εἰ μὲν οὖν ἐπισήμως τε καὶ ἰσχυρῶς ἐποίει τοῦτο, καθάπερ τὸ πέπερι, πρόδηλον ἂν ἦν πᾶσι καὶ ἀναμφισβήτητον· ἐπεὶ δ' οὐκ ἰσχυρῶς, εὐλόγως εἰς ζήτησιν ἀφι-

nutriment for quails but a medication for people. For the *krasis* of quails is able to assimilate the hellebore to itself but that of people is not able to do this.

5. It has, then, already become clear that the judgment of what to us are wet, or dry, or cold, or hot is not from what is external, but from those things we ourselves are affected by, and would occur entirely from these, and that first and foremost one must consider this. Next in order, if required, one must also consider those things that are external. For if the action of the medication introduced is obvious and clear to perception, one must trust this and set aside all the other signs. If, however, it is indistinct, unclear, or mixed, or altogether has something disputable, under these circumstances one must also determine this in relation to all things external. It is not, therefore, according to those things from afar but from the actual substance of what is being sought. | For example, if olive oil 685K is hot, it is not because it is viscous, pale yellow, or light, but whether it is easily combustible. For this was by it being hot in potentiality that it quickly changes to being hot in actuality. In the same way, in the case of our bodies, it is not whether it is thick or thin-particled, or wet or light or viscous or pale yellow, but whether it heats when administered; likewise, it is not whether it is sweet and laxative, or makes the blood flow freely in the phlebotomies, if it is instilled. For these things too are superfluous to the present consideration of whether something heats when introduced. If it did this distinctly and strongly, as pepper does, it would be clear to everyone and indisputable. Since, however, it does not heat strongly, it is reasonable

---

[22] τῆς οὐσίας *add.* H

κνεῖται, πολὺ δὲ μᾶλλον ἐπὶ ῥοδίνου τε καὶ ὄξους ἠπόρηταί τε καὶ ἠμφισβήτηται τοῖς ἰατροῖς, εἴτε δυνάμει θερμὰ πέφυκεν εἴτε ψυχρά.

Χρὴ τοίνυν ἐξευρεῖν τινας ἐφ᾽ ἅπασι τοῖς δυνάμει λεγομένοις ὑπάρχειν ἢ θερμοῖς ἢ ψυχροῖς ἢ ξηροῖς ἢ ὑγροῖς ἀκριβεῖς καὶ σαφεῖς διορισμούς, ὡς ἔμπροσθεν ἐπὶ τῶν ἐνεργείᾳ | λεγομένων ἐποιησάμεθα. προσήκει δ᾽ οἶμαι τὴν ἀρχὴν ἀπὸ τῶν ἐναργεστάτων ποιήσασθαι· γυμνασάμενος γάρ τις ἐν τούτοις ῥᾷον ἀκολουθήσει τοῖς ἀσαφεστέροις. εὐθὺς οὖν ἐν τῷ προσφέρειν τῷ σώματι τόδε τι τὸ φάρμακον ἢ τὸ σιτίον ἀπηλλάχθω τὸ προσφερόμενον ἁπάσης σφοδρᾶς ἐπικτήτου θερμότητός τε καὶ ψύξεως. ὃν μὲν γὰρ ἐν τοῖς ἔμπροσθεν ἐποιησάμεθα διορισμόν, ἡνίκα τά θ᾽ ὑγρὰ καὶ τὰ ξηρὰ σώματα διαγιγνώσκειν ἐπεχειροῦμεν, οὗτος ἂν οὐδὲν ἧττον εἴη καὶ νῦν χρήσιμος ἐπὶ τῶν δυνάμει θερμῶν καὶ ψυχρῶν. εἴτε γὰρ δυνάμει ψυχρὸν ὂν τὸ προσφερόμενον ἐκθερμήναις σφοδρῶς[23] εἴτε θερμὸν ὂν καταψύξαις, ἡ πρώτη τοῦ σώματος προσβολὴ τὴν αἴσθησιν ἀπὸ τῆς ἐπικτήτου διαθέσεως, οὐκ ἀπὸ τῆς οἰκείας τοῦ προσαχθέντος ἐργάσεται κράσεως. ἵν᾽ οὖν ἀκριβής τε καὶ εἰλικρινὴς ἡ φύσις ἐξετάζηται τοῦ προσαγομένου, χλιαρὸν ὡς οἷόν τε μάλισθ᾽ ὑπαρχέτω μηδεμίαν ἐπίσημον ἔξωθεν ἀλλοίωσιν εἰληφὸς ἤτοι θερμότητος ἢ ψύξεως σφοδρᾶς.

Ἡ μὲν δὴ πρώτη παρασκευὴ τοῦ προσαγομένου φαρμάκου τοιαύτη γιγνέσθω. προσφερέσθω δὲ μὴ πάσῃ διαθέσει | σώματος, ὅταν ἐξετάζῃς αὐτοῦ τὴν

to come to an investigation, while in the cases of rosewater and vinegar, it is much more doubtful and has given rise to contention among doctors as to whether by nature they are hot or cold in potentiality.

Accordingly it is necessary to discover some distinctions, precise and clear, for all the things said to be either hot or cold, or dry or wet in potentiality, as we made previously in the case of the things said to be so in actuality. However, I think it is appropriate to make the start from those things that are most clear, for someone practiced in these will more easily follow those that are more unclear. Thus, immediately on presentation to the body, the particular substance, whether medication or food, that is introduced should be freed from any strong acquired hotness or coldness. This is a distinction we made earlier when we attempted to recognize the wet and dry bodies: it is no less useful now in the case of those things hot and cold potentially. For if you strongly heat something potentially cold or cool something hot [potentially], when it is applied, the first impact on the body will produce the perception from the acquired condition and not from the characteristic *krasis* of what is introduced. Therefore, so that the precise and pure nature of what is presented is examined, let it be as lukewarm as possible, having taken on no notable change from without of either strong hotness or strong coldness.

Certainly, the first preparation of the introduced medication should be such as this. Let it not be administered to every condition of a body when you examine its

---

²³ σφοδρῶς H; σαφῶς K

δύναμιν, ἀλλ' ἁπλουστάταις ὡς ἔνι μάλιστα καὶ ἄκραις. εἰ μὲν οὖν ἐσχάτως θερμῇ διαθέσει προσαχθὲν αἴσθησιν ἐργάζοιτο ψύξεως, εἴη ἂν οὕτω γε ψυχρόν· ὡσαύτως δὲ καὶ εἰ τῇ ψυχρᾷ θερμὸν ἐν τῷ παραυτίκα φαίνοιτο, καὶ τοῦτ' ἂν εἴη θερμόν. εἰ δ' ἤτοι τῇ θερμῇ θερμὸν ἢ τῇ ψυχρᾷ ψυχρὸν φαίνοιτο, μὴ πάντως ἀποφαίνεσθαι τὸ μὲν θερμὸν εἶναι,[24] τὸ δὲ ψυχρόν. ἐνίοτε γὰρ ἄκρως μέν ἐστιν ἡ διάθεσις θερμή, μετρίως δὲ ψυχρὸν ὂν τὸ φάρμακον οὔτ' ἠλλοίωσεν αὐτὴν ἔτι τε πρὸς τούτῳ ψῦξαν καὶ πυκνῶσαν ἅπασαν τὴν ἐκτὸς ἐπιφάνειαν ἀπέκλεισεν εἴσω καὶ διαπνεῖσθαι τὸ θερμὸν ἐκώλυσε κἀκ τούτου μειζόνως ἐξεπύρωσε τὴν διάθεσιν. οὕτω δὲ κἂν εἰ τῇ ψυχρᾷ διαθέσει τὸ προσφερόμενον μηδεμίαν ἐπιφέροι θερμότητα, σκέπτεσθαι, μή τι μετρίως ὑπάρχον θερμὸν οὐδὲν ἔδρασεν εἰς τὴν ἄκρως θερμοῦ δεομένην διάθεσιν. οὔκουν οὔθ' οὕτω χρὴ βασανίζεσθαι τῶν προσφερομένων φαρμάκων τὰς δυνάμεις οὔτ' εἰ κατὰ συμβεβηκὸς ἐργάζοιτό τι καὶ μὴ καθ' αὑτό.

Κρίσις δὲ τοῦ κατὰ συμβεβηκὸς ἥ τε διάθεσις καὶ ὁ χρόνος· ἡ μὲν διάθεσις, εἰ ἁπλῆ καὶ μία· τῷ χρόνῳ δ' ἡ κρίσις διορίζεται κατὰ τάδε. τὸ μὲν ἅμα τῷ προσενεχθῆναι ψύχειν ἢ θερμαίνειν ἐναργῶς φαινόμενον εἴη ἂν δήπου καθ' ἑαυτό τε καὶ δι' ἑαυτὸ τοιοῦτον· τὸ δ' ἐν τῷ χρόνῳ τάχ' ἂν ἔκ τινος συμβεβηκότος εἰς τοῦτ' ἄγοιτο, καθάπερ "ἐπὶ τετάνου θέρεος μέσου νέῳ εὐσάρκῳ ὕδατος ψυχροῦ πολλοῦ

potency, but as far as possible in those that are most absolute and extreme. Therefore, if what is introduced to an extremely hot condition were to create a perception of coldness, it would in this way be cold. And similarly, if to a cold condition, heat would immediately appear in it, this would be hot. If, however, it seems either hot to a hot condition or cold to a cold condition, it could not be said conclusively that it is hot in the first instance and cold in the second. For sometimes, when the condition is exceedingly hot, while the medication is moderately cold, it doesn't customarily change this, and in addition to this, in cooling and condensing the whole external surface, it closes the heat up within and prevents it being transpired, and from this heats up the condition more. In this way, even if what is administered brings no hotness to the cold condition, consider the case of something moderately hot doing nothing to the condition requiring extreme heat. It is not therefore in this way necessary to question the potencies of the administered medications nor whether a medication does something incidentally and not of itself.

Determinants of what is incidental are the condition and the time. It is the condition, if it is simple and single. The determination from the time is distinguished in this way. If the cooling or heating is clearly apparent at the same time as it is administered, such a thing would presumably be in relation to itself and through itself. If, however, it is over time, perhaps it would come to this from something incidental, as in the case of tetanus in midsummer, pouring over a large amount of cold water effects a

---

²⁴ εἶναι *add.* H

κατάχυσις θέρμης ἐπανάκλησιν ποιέεται." ἀλλ' ὅτι γε μὴ καθ' ἑαυτὸ θερμαίνει τὸ ψυχρὸν ὕδωρ, δῆλον ἐκ τῆς πρώτης προσβολῆς· αἴσθησιν γὰρ ἐργάζεται ψύξεως καὶ μὲν δὴ καὶ ψύχει τὸ δέρμα, μέχρις ἂν ἐπιχέηται τούτῳ, καὶ τὴν θερμασίαν οὔτ' ἐπὶ πάντων σωμάτων οὔτ' ἐν τῷ καταχεῖσθαι παρέχεται, ἀλλ' ἐπὶ μόνων εὐσάρκων νέων ἐν θέρει μέσῳ μετὰ τὸ παύσασθαι καταχέοντας. ὥσπερ οὖν οἷς προσπίπτει τὸ ψυχρὸν ὕδωρ, ἐκ τοῦ παραχρῆμα ψύχει ταῦτα, κἂν ἔμψυχα κἂν ἄψυχα τὰ σώμαθ' ὑπάρχῃ κἂν θερμὰ κἂν ψυχρά, κατὰ τὸν αὐτὸν τρόπον, εἴ τις ἦν χρόνος ἢ φύσις σώματος ἢ διάθεσις, ἐφ' ἧς αἴσθησιν ἔφερε τὸ ψυχρὸν ὕδωρ εὐθὺς ἅμα τῷ προσπίπτειν θερμότητος, 689K | ἂν ἐζητεῖτο, πότερα θερμαίνειν ἢ ψύχειν πέφυκε καθ' ἑαυτό. νυνὶ δ' ἐπειδὴ τὰ μὲν ἔμψυχά τε καὶ ἄψυχα πάντα παραχρῆμα καὶ διὰ παντὸς ὁρᾶται ψυχόμενα, οἷς δ' ἐστὶν ἔμφυτος θερμασία καὶ οἷον πηγή τις ἐν τοῖς σπλάγχνοις πυρός, εἰ τούτοις προσενεχθὲν ἐπανάκλησίν τινά ποτε ποιεῖται θερμότητος, εὔλογον οἶμαι κατά τι συμβεβηκός, οὐ καθ' ἑαυτὸ θερμαίνειν αὐτὸ τὰ τοιαῦτα. καὶ δὴ καὶ φαίνεται, κατὰ τί. πυκνώσει γὰρ τῆς ἐκτὸς ἐπιφανείας καὶ κατακλείσει τοῦ θερμοῦ τὴν ἐπάνοδον ποιεῖται τῆς ἐκ τοῦ βάθους θερμασίας, ἅμα μὲν ἀθροισθείσης τῷ μὴ διαπνεῖσθαι, ἅμα δ' εἰς τὸ βάθος ἀποχωρούσης διὰ

---

[20] Hippocrates, *Aphorisms* 5.21, LCL 150 (W. H. S. Jones),

reaction of heat in a well-fleshed youth.[20] But that in fact the cold water does not heat in relation to itself is clear from the first impact. For it produces a perception of cold and certainly also cools the skin for as long as it is being poured onto this. And it produces the heat neither in the case of all bodies, nor in the pouring, but only in the case of well-fleshed youths in the middle of summer after the pouring is stopped. Therefore, just as cold water that falls upon things cools these things immediately from this, whether the bodies are animate or inanimate, and even if they are hot or cold, in the same way, if it were some time, or nature of a body, or a condition in the case of which the cold water immediately at the same time as being poured, brought a perception of heat, it would be reasonable to investigate whether it is naturally heating or cooling in and of itself. Now, however, since all things, both animate and inanimate, are seen to be cooled immediately and throughout, while in those things in which there is an innate heat and, as it were, a source of fire in the internal organs, when administered to these sometimes creates a certain reaction of hotness, it would I think to be reasonable [to assume] such things heat incidentally and not by virtue of themselves. And it is certainly apparent what it is in relation to. For it will condense the external surface and close up the heat within, and create the return of the heat to the depths, while at the same time it is collecting because it is not transpired, but is withdrawing to the depths due to the

162–63. The aphorism in full reads: "Sometimes in a case of tetanus without a wound, the patient being a muscular young man, and the time the middle of summer, a copious affusion of cold water brings a recovery of heat. Heat relieves these symptoms."

τὴν τοῦ περιέχοντος βίαν ψυχροῦ, ἅμα δὲ καὶ τρεφομένης ὑπὸ τῶν ἐνταῦθα χυμῶν. ὅταν γὰρ ἀθροισθεῖσά τε καὶ τραφεῖσα πρὸς τὴν ἐπιφάνειαν ὁρμήσῃ σφοδρότερον, ἐπανάκλησις μὲν γίγνεται τῆς θέρμης, ἔνδειξις δὲ τοῦ μὴ καθ' ἑαυτὸ τὴν θερμασίαν αὐξῆσαι τὸ ψυχρὸν ὕδωρ. καθ' ἑαυτὸ μὲν γὰρ ἔψυξε τὸ δέρμα· τῇ ψύξει δ' αὐτοῦ πύκνωσις θ' ἅμα καὶ εἰς τὸ βάθος ὑπονόστησις ἠκολούθησε τοῦ θερμοῦ, τούτων δ' αὐτῶν τῇ μὲν πυκνώσει κώλυσις τῆς διαπνοῆς, τῇ δ' εἰς τὸ βάθος ὑποχωρήσει κατεργασία τῶν ταύτῃ χυμῶν· ὧν ἡ μὲν κώλυσις τῆς διαπνοῆς τὴν ἄθροισιν τῆς θερμασίας, ἡ δὲ τῶν χυμῶν κατεργασία τὴν γένεσιν αὐτῆς ἐποιήσατο. τούτων δ' ἑκατέρῳ πάλιν ἡ αὔξησις ἕπεται τῆς ἐμφύτου θερμασίας. διὰ μέσων οὖν ἑκατέρων τὸ ψυχρὸν αὔξησίν ποτε τῆς ἐν τῷ ζῴῳ θερμασίας ἐργάζεται, καθ' ἑαυτὸ δ' οὐδέποτε.

Καὶ μὴν καὶ τὸ θερμὸν ἔστιν ὅτε κατὰ συμβεβηκὸς ψύχει διὰ μέσου τοῦ κενοῦν ὡς τὸ κατάπλασμα τὴν φλεγμονήν. ἐπειδὴ γὰρ ὑπὸ θερμοῦ ῥεύματος γίγνεται φλεγμονή, τὸ μὲν ἴδιον αὐτῆς ἴαμα κένωσίς ἐστι τοῦ περιττοῦ, τὸ δὲ τῇ κενώσει πάντως ἑπόμενον ἡ ψῦξις τοῦ διὰ τὴν φλεγμονὴν τεθερμασμένου μορίου. διττῆς οὖν οὔσης ἐν τοῖς φλεγμαίνουσι σώμασι διαθέσεως, ὅσον μὲν ἐπὶ τῇ πλεονεξίᾳ τοῦ περιττοῦ, κατὰ τὸ ποσὸν ἐξισταμένοις τοῦ κατὰ φύσιν, ὅσον δ' ἐπὶ τῇ θερμασίᾳ, κατὰ τὸ ποιόν, ἡ τῆς ἑτέρας τῶν διαθέσεων ἴασις ἑπομένην ἔχει καὶ τὴν ἑτέραν. καὶ γίγνεται κατά τι συμβεβηκὸς τὰ κενωτικὰ τῆς θέρμης

force of the surrounding cold, and at the same time also being nourished by the humors there. For whenever it has accumulated and been nourished, it will be urged more strongly toward the surface and creates a recall of the heat, this being an indication that the cold water does not increase the heat by virtue of itself. For in relation to itself, it cooled the skin and by the cooling of this a condensation and at the same time also a subsiding of the heat to the depths within followed, while these same things, by the condensation, | effected prevention of the transpiration, whereas by the retreat to the depths, a working up of the humors in that place. Of these, the prevention of the transpiration created the accumulation of the heat, while the working up of the humors created the generation of this. The augmentation of the innate heat followed each of these in turn. Through the mediation, then, of each of these two, the cold sometimes produces an increase of the heat in the animal, although not ever of itself.

690K

And further also, the hot sometimes cools incidentally through the mediation of the emptying, like a cataplasm in respect of inflammation. For when inflammation occurs through a hot flux, the specific cure of this is evacuation of the superfluity, while cooling is what always follows the evacuation of the part that has been heated due to the inflammation. Therefore, since the condition is twofold in the inflamed bodies, the extent to which it is due to the predominance of the superfluity relates to the quantity in the deviation from an accord with nature, while the extent to which it is due to the heat relates to the quality, the cure of one of the conditions having as a consequence the cure of the other. And it occurs incidentally that the medica-

ὕλης φάρμακα καὶ τῆς φλογώσεως τῶν μορίων ἐμψυκτικά. ταῦτα τ' οὖν διορίζεσθαι καὶ πειρᾶσθαι κατὰ τὸ ποσὸν τῆς ἁπλῆς διαθέσεως ἐξευρίσκειν τὸ ποσὸν τῆς τοῦ φαρμάκου δυνάμεως, οἷον εἰ ἄκρως ἡ διάθεσις θερμή, καὶ τὸ φάρμακον ἄκρως εἶναι ψυχρόν, εἰ δ' ὀλίγον ἀπολείποιτο τῆς ἀκρότητος ἡ διάθεσις, ὀλίγον χρὴ καὶ τὸ φάρμακον ἀπολείπεσθαι, κἂν εἰ πλέον ἀπέχοι τῆς ἄκρας θερμότητος ἡ διάθεσις, ἀνάλογον ἀπέχειν τῆς ἄκρας ψυχρότητος τὸ φάρμακον. εἰ γὰρ ἀπὸ τοῦ τοιούτου στοχασμοῦ τὴν ἀρχὴν τῆς ἐξετάσεως αὐτῶν ποιοῖο, θᾶττον ἂν ἐξευρίσκοις τὴν οἰκείαν ἑκάστου δύναμιν. ὅλως γὰρ εἰ καθ' ἓν ὁτιοῦν πάθος ἁπλοῦν θερμὸν ὅ τι δήποτε τῶν φαρμάκων προσαχθὲν εὐθὺς ἅμα τῇ πρώτῃ προσφορᾷ ψύξεως αἴσθησιν ἤνεγκε, ψυχρόν ἐστιν ἐκεῖνο τῇ δυνάμει καὶ πολὺ μᾶλλον, εἰ καὶ μετὰ τὴν πρώτην προσφορὰν ἕως παντὸς μένει τοιοῦτον. εἰ δὲ καὶ σαφῶς ἰῷτο τὴν θερμὴν διάθεσιν, ἐξ ἀνάγκης ψυχρόν ἐστι. προσφέρεσθαι δὲ χρὴ πάντως αὐτὸ χλιαρόν, ἐπειδὰν δοκιμάζηται, καθότι καὶ πρόσθεν εἴρηται. γνωρισθὲν δ' ὅτι τοιοῦτόν ἐστιν, εἶτα θεραπείας ἕνεκα παραλαμβανόμενον ἄμεινον ψυχρὸν παραλαμβάνειν, πλὴν εἰ τὸ μὲν φάρμακον ἄκρως εἴη ψυχρόν, οὐκ ἄκρως δὲ θερμὸν εἴη τὸ νόσημα.

Ταυτὶ μὲν οὖν ἐπὶ πλέον ἔν τε τοῖς περὶ φαρμάκων εἰρήσεται κἀν τοῖς τῆς θεραπευτικῆς μεθόδου γράμ-

tions evacuating the hot material are also cooling of the inflammation of the parts. Distinguish these things, then, and attempt to discover in the amount of the simple condition the amount of the potency of the medication. For example, if the condition is extremely hot, the medication should be extremely cold, whereas if the condition is a little short of the extreme, the medication should also be a little short [of the extreme], and if the condition is still further removed from the extreme of the hotness, the medication should be proportionally removed from the extreme of coldness. For if you were to make the beginning of the estimation of these things from such a conjecture, you would more quickly discover the characteristic potency of each. All in all, if any drug taken for any one affection whatsoever that is simply hot, immediately at the time of the first application carried a perception of cold, then that is cold in potentiality, and much more so, if it also remains thus after the first administration. If, however, it were also to clearly cure the hot condition, it is of necessity cold. But one must always administer this lukewarm, when it is being tested, as was also said previously. On the other hand, if it is known to be such a medication, when taken for the purpose of treatment, it is better to take it cold, unless the medication is extremely cold while the disease is not extremely hot.

These things will be spoken of at greater length in the books on medications and in those on the therapeutic

μασιν. ἐν δὲ τῷ παρόντι τό γε τοσοῦτον χρὴ γιγνώσκειν, ὡς, εἴ τι τῇ θερμῇ καὶ ἁπλῇ διαθέσει προσαχθὲν φάρμακον ἔκ τε τοῦ παραχρῆμα κἂν τῷ μετὰ ταῦτα χρόνῳ παντὶ τήν τε τῆς ψύξεως αἴσθησιν ἤνεγκε τῷ κάμνοντι καὶ τὴν τῆς εὐφορίας τε καὶ ὠφελείας, ἐξ ἀνάγκης τοῦτο ψυχρόν ἐστι, κἂν ἐπ᾿ ἄλλων ποτὲ φαίνηται θερμόν. εὑρεθήσεται γὰρ ἐπ᾿ ἐκείνων ἐξεταζόμενον ἀκριβῶς οὐ καθ᾿ ἑαυτὸ θερμὸν ὂν ἀλλὰ κατά τι συμβεβηκός. ὅταν δὲ καθ᾿ ἑαυτὸ λέγωμεν ἢ πρώτως ἢ διὰ μηδενὸς τῶν ἐν τῷ μέσῳ, ταὐτὸν ἐξ ἁπάντων δηλοῦται τῶν ῥημάτων καὶ τὴν γυμνασίαν ἁπάντων τούτων ἅμα τοῖς οἰκείοις παραδείγμασιν ἐν τοῖς περὶ φαρμάκων ὑπομνήμασι ποιησόμεθα.

6. Νυνὶ δὲ πάλιν ἀναμνήσας ὧν ἤδη καὶ πρόσθεν εἶπον ἐπιθεῖναι πειράσομαι τῷ παρόντι λόγῳ τὴν προσήκουσαν τελευτήν. ἐπειδὴ γὰρ τὸ θερμὸν σῶμα πολλαχῶς ἐλέγετο, | καὶ γὰρ καὶ τὸ τὴν ἄκραν ἔχον ποιότητα, τὸ στοιχεῖον αὐτό, καὶ τὸ κατ᾿ ἐπικράτησιν αὐτῆς ὠνομασμένον ἔτι τε τὸ πρὸς ἕτερον λεγόμενον ἤτοι πρὸς τὸ σύμμετρον ὁμογενὲς ἢ πρὸς ὁτιοῦν τὸ τυχόν, οὕτω χρὴ καὶ τὸ δυνάμει μὲν θερμὸν ἐνεργείᾳ δ᾿ οὐδέπω καὶ νοεῖσθαι καὶ δοκιμάζεσθαι πολλαχῶς. ὅθεν οὐκ ὀρθῶς, εἴ τι μὴ ταχέως ἐκπυροῦται, τοῦτ᾿ ἔνιοι νομίζουσιν οὐδὲ πρὸς ἡμᾶς εἶναι δυνάμει θερμόν. εἴτε γὰρ εὔπεπτόν ἐστι καὶ τρέφει ταχέως, εἴη ἂν

---

[21] The three major works on medications are listed in note 13 above. The major work on therapeutics is *De methodo medendi*

method.[21] For the present, this much at least should be known: if some medication, exhibited for a hot and simple condition, immediately from this and even in the whole time afterward, produces the perception of cold in the one suffering, and that of well-being and benefit, of necessity this is cold, even if sometimes in other cases it appears hot. For it will be discovered, in those cases, when investigated accurately, that it is not hot in and of itself, but contingently. Whether we say "in and of itself" or "primarily" or "through no intermediary," it is clear from all the statements and the practice of all these, along with specific examples in the treatises about medication, we will be doing the same thing.

6. Now calling to mind again those things I already said previously, I shall attempt to add the appropriate conclusion to the present discussion. Since "the hot body" was said in many ways—that which has the extreme quality, the element itself, that which is named in relation to the predominance of this, and that which is said relative to something else, or referring to the mean of the same class, or to anything else whatsoever—in this way what is hot in potentiality but not yet in actuality must be conceived of and investigated in many ways. Whence they are not right who think, if something is not quickly flammable, this is not hot potentially toward us. For if it is either well-concocted and nourishes quickly, and would be hot to us,

693K

(*The Method of Medicine*), X.1–1021K; English translation by Johnston and Horsley, LCL 518–520. The short version is *Ad Glauconem de methodo medendi*, XI.1–146; English translation by Johnston, LCL 523. On the matters referred to, see particularly XI.381–400K and X.653K.

ὡς πρὸς ἡμᾶς θερμόν, εἴτε θερμαίνει προσφερόμενον ὡς φάρμακον, εἴη ἂν καὶ τοῦτο δυνάμει θερμὸν ὡς πρὸς ἄνθρωπον. οὕτω δὲ καὶ καθ᾽ ἕκαστον εἶδος ζῴου τὸ δυνάμει θερμὸν εἴθ᾽ ὡς φάρμακον εἴθ᾽ ὡς τροφὴ πρὸς ἐκεῖνο λέγεται μόνον τὸ ζῷον. ἐν γὰρ τῷ πρός τι τὸ δυνάμει πᾶν, ὥστε καὶ ἡ βάσανος ἡ οἰκεία βελτίων τῆς ἔξωθεν. οἰκεία δὲ μία καθ᾽ ἕκαστον, εἰ φαίνοιτο ταχέως γιγνόμενον τοιοῦτον, ὁποῖον ἔφαμεν ὑπάρχειν αὐτὸ δυνάμει. πῦρ μὲν γάρ ἐστι δυνάμει πᾶν ὅ τι ἂν ἐκπυρῶται ταχέως, δυνάμει δ᾽ ὡς πρὸς ἄνθρωπον θερμόν—ὅπερ ἦν ἓν εἶδος τῶν κατ᾽ ἐπικράτησιν θερμῶν—ὅταν ἀνθρώπῳ προσφερόμενον ἢ τὴν ποιότητα τῆς ἐμφύτου θερμασίας ἢ τὴν οὐσίαν αὐξάνῃ. τὰ δ᾽ αὐτὰ κἀπὶ τῶν ἄλλων εἰρῆσθαι χρὴ νομίζειν ὅσα δυνάμει λέγεται ψυχρὰ καὶ ξηρὰ καὶ ὑγρά. καὶ γὰρ καὶ ταῦτα τὰ μὲν ὡς πρὸς αὐτὰ τὰ στοιχεῖα, τὰ δ᾽ ὡς κατ᾽ ἐπικράτησιν ὠνομασμένα καὶ νοεῖσθαι χρὴ καὶ δοκιμάζεσθαι.[25] δῆλον δ᾽ ὡς καὶ τὴν κρίνουσαν ἁφὴν ἁπάσης ἐπικτήτου θερμασίας τε καὶ ψύξεως ἀπηλλάχθαι χρή, καθότι κἀπὶ τῶν φαρμάκων αὐτῶν εἴρηται πρόσθεν.

[25] καὶ διδάσκεσθαι K om.

or if it heats when administered as a medication, this would be hot in potentiality to a person. In this way too, what is hot potentially in relation to each kind of animal, either as a medication or as a nutriment, is said to be so toward that animal alone. For in this everything in potential is relative to something, so that also the test which is specific is better than that from without. However, what is specific is one in relation to each, if it should appear to quickly become of such a kind as we said it to be in potentiality. For fire is everything which in potentiality is quickly ignited, | while it is hot in potentiality as toward a person, which is one kind of heat in relation to prevailing, whenever, if administered to a person, it either increases the quality of the innate heat or the substance. It is necessary to think these same things in the case of the other [qualities] stated—those things said to be cold, dry, and wet potentially. For surely these things, both those named with reference to the elements themselves, and those named in relation to predominance, must be conceived of and tested. However, it is clear that the determining touch must be freed from any acquired heating and cooling, just as in the case of the medications themselves stated previously.

ΠΕΡΙ ΑΝΩΜΑΛΟΥ
ΔΥΣΚΡΑΣΙΑΣ ΒΙΒΛΙΟΝ

ON NON-UNIFORM
DISTEMPERMENT

INTRODUCTION

This short work, a sort of addendum to the main work, *De temperamentis*, is mentioned by Galen in both his *On the Order of My Own Books*[1] and in the list of works at the end of *Art of Medicine*.[2] In the first case, he advises reading *De inaequali intemperie* after reading either the first two books of *De temperamentis* or all three. In the latter, he refers to "another small book which follows the first two on *krasias* (i.e., Books 1 and 2 of *De temperamentis*) and was written on the non-uniform *dyskrasias*." In fact, the short work is an essential addition to the major work on *krasias*, clarifying what was left unclarified in the latter. Indeed, the two works were presented together by Thomas Linacre in particular in his Latin translation of the two, which ran to seven editions between its initial publication in 1521 and the end of the sixteenth century.

Prior to Linacre, there were Latin translations of the *De inaequali intemperie* alone by Pietro d'Abano (ca. 1300) and Niccolò da Reggio (1310–1320) and then by Giorgio Valla in 1498. Durling, in his census of Renaissance translations and editions, lists other sixteenth-

[1] *De ordine librorum propriorum*, XIX.56K
[2] *Ars medica*, I.408K

century Latin translations by N. Leoniceno, L. Fuchs, and J. Lalamantius. Garcia Novo lists in addition ten commentaries on the Greek text from the sixteenth century and four earlier commentaries on the Arabic version from the fourteenth and fifteenth centuries. There is also the late nineteenth-century edition of Linacre's Latin of both works. All in all, this short work received considerable attention from translators and commentators during the second millennium, although it was not included in the *Summaria Alexandrinorum* along with the *De temperamentis*.

As for content, the work begins with a description of what a non-uniform *dyskrasia* is, the essence of the distinction from a uniform *dyskrasia*, and some examples of the former. There is, however, no formal definition. After a brief digression on the necessity of having a sound knowledge of the parts of the body, the focus turns to several important points pertaining to non-uniform *dyskrasias*, exemplified first by inflammation and second by fevers, which numerically are prominent components of this class of diseases. The points are as follows:

1. Non-uniform *dyskrasias* are caused by either focal primary changes in the *krasis* of a part or parts, similar to the systemic changes that cause uniform *dyskrasias*, or by an influx of material into the affected part (predominantly one of the four humors), which alters the *krasis* of the part.
2. The importance of localized pain as evidence of ongoing change in the affected part.
3. The factors determining whether or not a non-

GALEN

uniform *dyskrasia* in a particular part will affect adjacent parts.
4. That fevers, apart from the hectic, are examples of non-uniform *dyskrasia*.
5. That *ēpialos* fevers (agues), which are characterized by the coexistence of chills and fever, are non-uniform *dyskrasias* in which the two components of the hot/cold antithesis are affected simultaneously, signified respectively by the heat of the fever and the coldness leading to shivering and the sensation of coldness.

These points are exemplified by inflammation in a muscle and by the various fevers. There is no systematic consideration of the methods of treatment; of course, general allopathic principles must apply. In the final section Galen gives a list of common focal non-uniform *dyskrasias*.

TEXTS AND TRANSLATIONS

The Greek text in the present work is essentially that of Garcia Novo, which was published in 2012 and is itself based on seven Greek manuscripts that she lists and discusses in her introduction (pp. 19–25). This is the only critical edition, although mention is made of a partially prepared edition by George Helmreich that is preserved in his papers held in the Berlin-Brandenburgische Akademie der Wissenschaften, and which Garcia Novo also examined. In the present text, significant points of difference between Garcia Novo's text and that in the Kühn *Opera omnia* (the only other readily available Greek text) are indicated in footnotes. In certain instances the transla-

tion follows the Kühn Greek or Latin text on grounds of perceived meaning.

There are two modern translations of Galen's *De inaequali intemperie*, both into English. The first is that of Mark Grant in his *Galen on Food and Diet* published in 2000. The second is that of Garcia Novo herself, which closely follows the Greek. Both have been consulted in preparing the present translation.

SYNOPSIS OF SECTIONS

1. Galen begins with a brief introductory statement on the term, non-uniform *dyskrasia*, listing some of the diseases that are regarded as being due to such a state. This is followed by the general classification of the *dyskrasias* (four simple and four compound) and the difference between a uniform and a non-uniform *dyskrasia*. In a final statement of intent, Galen stresses the need to set out all the parts of the body.

2. Galen provides more on the parts, beginning with the subdivision of the major parts (arms, legs, abdomen, chest, and head), and further subdivisions, including that into *homoiomerous* and organic parts, with reference to his work *On Anatomical Procedures*. Specific mention is made of the spaces between the parts and the pores of the skin. Differences in the non-uniform *dyskrasias* in different parts follow the nature of the parts.

3. Galen considers fluxes causing non-uniform *dyskrasias* in various parts—blood vessels, the spaces between various parts, muscle, abdomen, brain, and chest. The outcome depends on whether the flux overcomes the part or vice versa. Pain is a sign of active change. Non-uniform

dyskrasias may be due to heating, cooling, drying, or wetting. Dissolution of continuity is also mentioned with respect to the part(s) involved.

4. On non-uniform *dyskrasia* due to heating, specifically inflammation, and the relation between the heat of the blood in the inflamed part and that contained in the whole body is discussed. The various factors that affect this include both the four humors and *pneuma*. When a humor is involved, relevant factors are its nature and whether it is obstructed, transpired, or putrefied. Also important is the distance of the inflamed part from the blood-containing internal organs.

5. This section continues the discussion of the differential heating of an affected part, as by a flux, and the rest of the body, both specific structures and the blood generally. Consideration is given to the variations in the change of *krasis* in different parts, the left chamber of the heart being particularly susceptible to heating. The limit of the change for each part is the damage to its function. Pain is produced when a quality is undergoing change; when the change ceases and uniformity is reached, pain stops. The latter is the case in the hectic fevers, which are painless and imperceptible to the sufferer. The hectic fever is unique among fevers in not being a non-uniform *dyskrasia*.

6. Non-uniform *dyskrasias* do not cause change in adjacent parts unless the disproportion is severe. Every disproportion is relative to something, and some animals can have humors compatible with each other. In general terms, things are increased and nourished by things similar to themselves but distressed and destroyed by dissimi-

NON-UNIFORM DISTEMPERMENT

lars (opposites). To extend the discussion, Galen regards it as necessary to speak about capacities (faculties).

7. Fevers, apart from the hectic, are diseases compounded non-uniformly. The important point is that the development of non-uniform *dyskrasias* will be variable. Brief consideration is given to fevers being caused by putrefaction as in turn causing non-uniform *dyskrasia*. Heating is the main focus of the section.

8. This section focuses on cooling, in particular severe cooling, and its association with pain. Galen also discusses the coincident occurrence of heating and cooling and the resultant perception of the two together. He returns to the ague fever, in which both fever/heat and shivering/cold occur. He concludes that this fever is combined from two non-uniform *dyskrasias*, as indeed most fevers are apart from the hectic.

9. Galen concludes with a brief account of other diseases due to a non-uniform *dyskrasia*. He lists cancer, erysipelas, anthrax, herpes, edema, inflammation, *phagedaina*, and gangrene. The common feature is a flux of fluids. The nature of the resultant non-uniform *dyskrasia* will depend on what the flux is.

ΠΕΡΙ ΑΝΩΜΑΛΟΥ
ΔΥΣΚΡΑΣΙΑΣ ΒΙΒΛΙΟΝ

1. Ἀνώμαλος δυσκρασία γίγνεται μὲν ἐνίοτε καὶ καθ' ὅλον τοῦ ζώου τὸ σῶμα, καθάπερ ἔν τε τοῖς ἀνασάρκα λεγομένοις ὑδέροις, καὶ τοῖς ἠπιάλοις καλουμένοις πυρετοῖς, καὶ σχεδὸν ἅπασι τοῖς ἄλλοις, πλὴν τῶν ἑκτικῶν ὀνομαζομένων. γίγνεται δ' ἐνίοτε καὶ καθ' ἓν ὁτιοῦν μόριον, οἰδισκόμενον ἢ φλεγμαῖνον ἢ γαγγραινόμενον[1] ἢ τῷ ἐρυσιπέλατι κάμνον ἢ τῷ καρκίνῳ. τούτου δ' ἐστὶ τοῦ γένους καὶ ὁ καλούμενος ἐλέφας καὶ ἡ φαγέδαινα καὶ ὁ ἕρπης. ἀλλὰ ταῦτα μὲν ἅπαντα μετὰ ῥευμάτων· ἄνευ δ' ὕλης ἐπιρρύτου, μόναις ταῖς ποιότησιν ἀλλοιουμένων τῶν μορίων, ἀνώμαλος γίγνεται δυσκρασίαι, | ψυγέντων, ἢ ἐκκαυ-

---

[1] post γαγγραινόμενον: ἢ τῷ ἐρυσιπέλατι κάμνον ἢ τῷ καρκίνῳ. EGN; ἢ ἐρυσιπελατούμενον, ἢ καρκινούμενον. K

---

[1] The term ἀνώμαλος δυσκρασία is rendered "non-uniform *dyskrasia*"; the distinction is between ἀνώμαλος (uneven, non-uniform) and ὁμαλός (even, uniform). It is found in a number of Galen's works, including *Ars medica*, I.408K, and *De methodo medendi*, X.15, 122, 216, and 694K.

ON NON-UNIFORM DISTEMPERMENT

1. A non-uniform *dyskrasia*[1] occurs sometimes in the whole body of the animal, as in the dropsies termed *anasarcas*,[2] in the fevers called agues, in almost all the other fevers, apart from those termed hectic.[3] It also sometimes occurs in relation to any one part whatsoever, when edematous, inflamed, or gangrenous, or affected by erysipelas, or by cancer. However, of this class too are the so-called *elephas*, *phagedaina* and *herpes*.[4] But all these follow fluxes. Non-uniform *dyskrasias* occur without flowing material, when the parts are changed in the qualities alone, that is, when parts are cooled, overheated, or ex-

---

[2] Dropsy may be taken as a general term for edema in Galen's usage. In modern usage, *anasarca* refers to "a generalized infiltration of oedema into subcutaneous connective tissue" (S).

[3] The Greek term ἠπίαλος is defined in LSJ as "ague" with reference to Hippocrates, *De superfoetatione* 34, and Galen, *De febrium differentiis*, VII.347K. The term "ague" came to be applied to a fever in which the sufferer feels both hot and cold, in particular the fever of malaria. On hectic fevers, see, for example, *De methodo medendi*, Book 10, section 6 (X.691–92K), and also section 4 of the present work.

[4] For these diseases, see the General Introduction, section 9, Diseases and Symptoms.

θέντων ἢ γυμνασαμένων ἐπὶ πλέον, ἢ πάντως ἀργησάντων, ἤ τι τοιοῦτον ἕτερον παθόντων. οὐ μὴν ἀλλὰ κἀκ τῶν ἔξωθεν προσπιπτόντων ἀνώμαλοι δυσκρασίαι τοῖς σώμασιν ἡμῶν πλεῖον ἐγγίγνονται, θερμαινομένοις, ἢ ψυχομένοις, ἢ ξηραινομένοις, ἢ ὑγραινομένοις. ἁπλαῖ μὲν γὰρ αὗται τέσσαρές εἰσι δυσκρασίαι,² καθότι κἀν τοῖς περὶ κράσεων ὑπομνήμασιν ἐδείκνυτο· σύνθετοι δὲ ἐξ αὐτῶν εἰσιν ἕτεραι τέσσαρες, ἢ θερμαινομένων τε ἅμα καὶ ξηραινομένων, ἢ ψυχομένων τε ἅμα καὶ ὑγραινομένων, ψυχομένων τε ἅμα καὶ ξηραινομένων, ἢ θερμαινομένων τε ἅμα καὶ ὑγραινομένων. ὅτι δ' αἱ τοιαῦται δυσκρασίαι διαφέρουσι τῶν ὁμαλῶν μόνον τῷ μὴ κατὰ πάντα τὰ μόρια τοῦ δυσκράτως διακειμένου σώματος ὡσαύτως ὑπάρχειν, ἄντικρυς δῆλον.

Ὅστις μὲν οὖν ὁ τρόπος ἐστὶ τῆς γενέσεως ἁπάσαις ταῖς ἀνωμάλοις δυσκρασίαις, ἐν τῷδε τῷ γράμματι πρόκειταί μοι διελθεῖν. ἵνα δὲ σαφὴς ὁ λόγος γένηται, ἀναμνῆσαί σε χρὴ πάντων τῶν τοῦ σώματος μορίων, ἀρξαμένους ἀπὸ τῶν μεγίστων, ἃ δὴ καὶ τοῖς ἰδιώταις γνωρίζεται.³ χεῖρας γάρ τοι καὶ πόδας καὶ γαστέρα καὶ θώρακα καὶ κεφαλὴν οὐκ ἔστιν ὅστις ἀγνοεῖ.

2. Τεμνέσθω δὴ πάλιν ὑφ' ἡμῶν ἓν ὁτιοῦν ἐξ αὐτῶν εἰς τὰ προσεχῆ καλούμενα μόρια, σκέλος μὲν εἰ τύχοι εἰς μηρόν τε καὶ κνήμην καὶ πόδα, χεὶρ δ' αὖ εἰς βραχίονά τε καὶ πῆχυν καὶ ἄκραν χεῖρα. καὶ μὲν δὴ καὶ ἄκρας χειρὸς οἰκεῖα μόρια καρπός τε καὶ μετα-

ercised excessively, are altogether inactive, or are affected by something else of this kind. But further, more non-uniform *dyskrasias* supervene in our bodies when heated, cooled, dried, or wet by things befalling them externally. These are the four simple *dyskrasias*, as shown in the treatises, *On Krasias (Temperaments)*.[5] There are four other *dyskrasias* compounded from these, when bodies are either heated and dried at the same time, or cooled and wet at the same time, or cooled and dried at the same time, or heated and wet at the same time. It is quite clear that such [non-uniform] *dyskrasias* differ from those that are uniform only by not having all the parts of the body similarly disposed *dyskratically*.

Therefore, what lies before me to go over in this book is the manner of genesis of all the non-uniform *dyskrasias*. In order for the discussion to be clear, it is necessary for you to call to mind all the parts of the body, beginning from the largest, which are certainly also known to laymen. For undoubtedly no one is unaware of the arms, legs, abdomen, chest, and head.

2. Let any one of these parts in turn be divided by us into the so-called proximate parts—a leg, for example, into thigh, shank, and foot, or again, an arm into upper arm, forearm, and hand; and further certainly, into specific parts of the hand—wrist, metacarpus, and fingers—while

---

[5] *De temperamentis*, I.509–694K.

---

[2] *post αὐταί;* τέσσαρες εἰσι δυσκρασίαι, *add.* EGN
[3] γνωρίζεται EGN; γινώσκεται K

κάρπιον καὶ δάκτυλοι· δακτύλων δ' αὖ πάλιν, ὀστᾶ καὶ χόνδροι καὶ σύνδεσμοι καὶ νεῦρα καὶ ἀρτηρίαι καὶ φλέβες, ὑμένες τε καὶ σάρκες καὶ τένοντες, ὄνυχές τε καὶ δέρμα καὶ πιμελή. ταῦτα δ' οὐκέτ' ἐγχωρεῖ τέμνειν εἰς ἕτερον εἶδος, ἀλλ' ἔστιν ὁμοιομερῆ τε καὶ πρῶτα, πλὴν ἀρτηριῶν τε καὶ φλεβῶν· αὗται γὰρ ἐξ ἰνῶν σύγκεινταί τε καὶ ὑμένων, καθότι κἂν ταῖς ἀνατομικαῖς ἐγχειρήσεσιν ἐλέγετο. καὶ μὲν δὴ καὶ ὡς χῶραι κεναὶ πολλαί τινες[4] μεταξὺ τῶν εἰρημένων ὁμοιομερῶν τε καὶ πρώτων μορίων ὑπάρχουσι, οὕτω καὶ τούτων ἔτι πλείους τε καὶ μείζους ἐν τῷ μέσῳ τῶν ὀργανικῶν τε καὶ συνθέτων, ἐνίοτε δὲ καὶ καθ' ἓν ὁτιοῦν ὁμοιομερὲς μόριον, ὡς ἐν ὀστῷ καὶ δέρματι. καὶ περὶ τούτων ἁπάντων ἐν ταῖς ἀνατομικαῖς ἐγχειρήσεσιν εἴρηται. τὰ μὲν οὖν μαλακὰ τῶν σωμάτων ἀλλήλοις ἐπιπίπτοντα τὰς μεταξὺ χώρας κενὰς[5] ἀδήλους ἐργάζεται πρὸς τὴν αἴσθησιν· ὅσα δ' ἐστὶ σκληρὰ καὶ ξηρά, πάνυ φωράσαις ἂν αὐτῶν αἰσθήσει τὰ διαλείμματα, καθάπερ ἐν ὀστοῖς τὰς σήραγγας. ἔχουσι δ' αὗται κατὰ φύσιν ὑγρὸν ἐν αὐταῖς παχὺ καὶ λευκὸν εἰς θρέψιν τοῖς ὀστοῖς παρεσκευασμένον. οἱ δ' ἐν τῷ δέρματι πόροι, καθ' ὅντινα γίγνονται τρόπον, ἐν τοῖς περὶ κράσεων εἴρηται. ταυτὶ μὲν οὖν ἀναγκαῖον ἦν ἀναμνῆσαι σαφηνείας ἕνεκα τῶν μελλόντων λεχθήσεσθαι.

[4] χῶραι κεναὶ πολλαί τινες EGN; χῶραι πολλαὶ K
[5] κενὰς add. EGN

NON-UNIFORM DISTEMPERMENT

of the fingers in turn again, there are bones, cartilages, ligaments, sinews (nerves), arteries, veins, membranes, flesh, tendons, nails, skin, and fat. It is no longer possible to divide these into another kind; they are *homoiomerous* and primary [parts], except for arteries and veins; these are compounded from fibers and membranes, as was stated in *On Anatomical Procedures*.[6] And furthermore also, many empty spaces exist between the aforementioned *homoiomeres* and primary parts, and there are more of these and larger in the middle of the organic and compound parts, but sometimes also in relation to any one of the *homoiomerous* parts, as in bone and skin. All these are spoken about in *On Anatomical Procedures*. The soft bodies, then, falling upon one another make the empty spaces between indistinct to perception. However, in the case of those that are hard and dry, you would entirely discover the spaces between by perception, like the medullary cavities in bones. These have in themselves a natural fluid, thick and white, prepared as nutrition for the bones. The pores in the skin arise in a certain way, set out in the treatises *On Krasias (Temperaments)*.[7] It was necessary to make mention of these for the sake of clarity regarding the things that will be said.

736K

---

[6] *De anatomicis administrationibus*, II.215–731K; see particularly VII.5, II.600–603K.

[7] The three books of *De temperamentis* referred to in note 5 above. The reference is presumably to transpiration through the skin and the channels by which this takes place; see Book 2, I.614–18K.

Περὶ δὲ τῆς ἀνωμάλου δυσκρασίας ἤδη ῥητέον, ὁποία τέ τίς ἐστιν ἡ φύσις αὐτῆς καὶ ὁπόσοι τρόποι τῆς γενέσεως. ὅτι μὲν δὴ μία κρᾶσις οὐκ ἔστιν ἐν ἅπασι τοῖς μέρεσι τοῦ πεπονθότος οὕτω σώματος, ἔμπροσθεν εἴρηται. ἀλλὰ τοῦτο μὲν κοινὸν ἁπάσης ἀνωμάλου δυσκρασίας· αἱ διαφοραὶ δὲ ταῖς τῶν πεπονθότων σωμάτων ἕπονται φύσεσιν. ἄλλως μὲν γὰρ ἡ ἁπλῆ σάρξ, ἄλλως δ᾽ ὁ σύμπας μῦς εἰς ἀνώμαλον ἀφικνεῖται δυσκρασίαν.[6]

3. Αὐτίκα γέ τοι ῥεύματος θερμοῦ κατασκήψαντος εἰς μῦν, πρῶται μὲν αἱ μείζους ἀρτηρίαι τε καὶ φλέβες ἐμπίπλανταί τε καὶ διατείνονται, μετὰ ταύτας δ᾽ αἱ μικρότεραι, καὶ τοῦτο γίγνεται μέχρι τῶν σμικροτάτων· ἐν αἷς ὅταν ἰσχυρῶς | σφηνωθῇ καὶ μηκέτι στέγηται τὸ ῥεῦμα, τὸ μέν τι κατὰ τὸ στόμα, τὸ δέ τι καὶ διὰ[7] τῶν χιτώνων αὐτῶν διηθεῖται πρὸς ἐκτός· κἂν τούτῳ πίμπλανται ῥεύματος αἱ μεταξὺ χῶραι κεναὶ τῶν πρώτων σωμάτων, ὥστε καὶ θερμαίνεσθαι καὶ περικλύζεσθαι πανταχόθεν ὑπὸ τῆς ὑγρότητος ἅπαντα. νεῦρα δ᾽ ἐστὶ ταῦτα καὶ σύνδεσμοι καὶ ὑμένες καὶ σάρκες αὐταί. τε πρὸ τούτων αἱ ἀρτηρίαι καὶ φλέβες, αἳ δὴ καὶ πρῶται καὶ μάλιστα ποικίλως ὀδυνῶνται· καὶ γὰρ ἔνδοθεν ὑπὸ τοῦ ῥεύματος θερμαίνονταί τε καὶ διατείνονται καὶ διασπῶνται, κἀκ τῶν ἔξωθεν θερμαίνονταί τε ἅμα καὶ θλίβονται καὶ βαρύνονται· τὰ δ᾽ ἄλλα μόρια, τὰ μὲν τῷ θερμαίνεσθαι

NON-UNIFORM DISTEMPERMENT

I must speak now about the non-uniform *dyskrasia*—of what kind the nature of this is and how many kinds of genesis there are. Certainly, that there is not one *krasis* in all the parts of the body affected in this way was stated previously. But this is common to every non-uniform *dyskrasia*; the differences follow the natures of the affected bodies. For the simple flesh comes to a non-uniform *dyskrasia* in one way, while the whole muscle comes to a non-uniform *dyskrasia* in another way.

3. And in fact, immediately a hot flux falls on a muscle, first the larger arteries and veins are filled and distended,[8] while after these the smaller, and this occurs up to the smallest. When I the flux is strongly wedged in these and can no longer be contained, it is filtered; some in relation to the orifice and some through the walls themselves, toward the outside, and by this, the empty spaces between the primary bodies are filled with a flux, so they are all both heated and washed around on all sides by the wetness. These are the sinews (nerves), ligaments, membranes, and fleshes. Before these, the actual arteries and veins, particularly those which are also first and foremost, are distressed in various ways. For within they are heated, distended, and ruptured by the flux, while from without they are heated, and at the same time compressed and weighed down. Regarding the other parts, some suffer by

[8] There is some textual uncertainty about this clause; see Garcia Novo, *Galen: On the Anomalous Dyskrasia*, 148. KLat is followed.

---

[6] δυσκρασίαν EGN; κρᾶσιν K   [7] *post* διά; τῶν κινούντων αὐτὸ χιτώνων EGN; τῶν χιτώνων αὐτῶν K

μόνον ἢ θλίβεσθαι, τὰ δὲ τῷ συναμφοτέρῳ κάμνει. καὶ καλεῖται μὲν τὸ νόσημα φλεγμονὴ, δυσκρασία δ' ἐστιν ἀνώμαλος τοῦ μυός. τὸ μὲν γὰρ αἷμα τὸ κατ' αὐτὸν ἤδη ζέει· συνεκθερμαίνει δ' αὐτῷ πρώτους μὲν καὶ μάλιστα τοὺς χιτῶνας τῶν ἀρτηριῶν καὶ τῶν φλεβῶν· ἤδη δὲ καὶ τἀκτὸς αὐτῶν, οἷς περικέχυται πάντα.

Καὶ δυοῖν γε θάτερον ἀναγκαῖον ἀπαντῆσαι· νικήσαντος μὲν τοῦ ῥεύματος, φθορὰν γίνεσθαι[8] τῶν νικηθέντων σωμάτων· νικηθέντος δὲ, τὴν εἰς τὸ κατὰ φύσιν ἐπάνοδον τῷ μυΐ. καὶ δὴ καὶ νικάσθω πρότερον τὸ ῥεῦμα. βέλτιον γὰρ ἀπὸ τῶν κρειττόνων ἄρχεσθαι· διττὸς δ' οὖν ἐν τούτῳ ὁ τρόπος ἔσται τῆς ἰάσεως, ἢ διαφορηθέντος ἅπαντος τοῦ κατασκήψαντος ὑγροῦ, ἢ πεφθέντος. ἀλλ' ἡ μὲν διαφόρησις εὐκταιοτάτη τῶν ἰάσεών ἐστιν· τῇ πέψει[9] δὲ δύο ταῦτ' ἐξ ἀνάγκης ἕπεται, πύου τε γένεσις καὶ ἀπόστασις. ἀφίσταται δέ ποτε μέν, εἰς τὴν μεγίστην τε καὶ ἀκυροτάτην τῶν παρακειμένων κοιλιῶν, ἥπερ δὴ καὶ ἡ βελτίστη τῶν ἀποστάσεών ἐστιν· ἐνίοτε δ' εἰς μεγίστην μέν, οὐ μὴν ἄκυρον, ἢ εἰς ἄκυρον μέν οὐ μεγίστην δέ. τοῖς μὲν οὖν κατὰ τὴν γαστέρα, ἡ καλλίστη τῶν ἀποστάσεών ἐστιν, ἡ εἰς τὴν ἐντὸς εὐρυχωρίαν, εἰς ἣν καὶ συρρήγνυνται τὰ πολλά· μοχθηρὰ δ' ἡ ὑπὸ τὸ περιτόναιον.

Οὕτω δὲ καὶ τοῖς κατὰ τὸν ἐγκέφαλον· ἡ μὲν οὖν εἰς τὰς ἐμπροσθίους δύο κοιλίας ἀγαθή, μοχθηρὰ δ' ἥ τε ὑπὸ τὰς μήνιγγας καὶ ἡ εἰς τὴν ὄπισθεν κοιλίαν.

NON-UNIFORM DISTEMPERMENT

being heated alone or compressed, while some are affected by both together. The disease is called inflammation (*phlegmone*) and is a non-uniform *dyskrasia* of the muscle. For the blood on its own account is already seething and heats together with itself first and foremost the walls of the arteries and veins, while when already external to these it drenches them on all sides.

And in fact one or other of the two things is necessarily encountered: either the flux prevails and destruction | of the overcome bodies occurs, or the flux is overcome and there is a return by the muscle to an accord with nature. And further, let the flux be overcome first, for it is better to start from those things that are better. In this, the manner of the cure will be twofold: the dissipation of all the fluid that has rushed down, or its concoction. But dissipation is the most desirable of the cures, while by concoction these two things necessarily follow: the generation of pus and abscess formation. Sometimes, however, it is removed to the largest and most unimportant of the cavities lying adjacent, which assuredly is the best of the abscesses, while sometimes it is removed to the largest and not unimportant, and sometimes to one that is unimportant and not very large. Therefore, for those in the abdomen, the best of the abscesses is that which goes to the open space within, to which the majority run together, whereas that [formed] under the peritoneum is bad.

It is like this too for those in the brain: the flux that goes to the two anterior cavities is good, whereas the one under the meninges and the one that goes to the posterior cavity

---

⁸ γίνεσθαι om. EGN   ⁹ post τῇ πέψει; δὲ δύο ταῦτ' ἐξ ἀνάγκης ἕπεται, EGN; δὲ ἕπεται δύο ταῦτ' ἐξ ἀνάγκης, K

αἱ δὲ κατὰ τὰς πλευρὰς ἀποστάσεις, εἰς τὰς τοῦ θώρακος εὐρυχωρίας ἐκρήγνυνται· καὶ τῶν μὲν μυῶν ὑπὸ τὸ δέρμα, τῶν δὲ σπλάγχνων ἢ εἰς τὰς ἀρτηρίας καὶ τὰς φλέβας τὰς ἐν αὐτοῖς, ἢ ὑπὸ τὸν ὑμένα τὸν περιέχοντα, καθάπερ τι δέρμα καὶ αὐτὸν ὑπάρχοντα τοῖς σπλάγχνοις. εἰ δὲ νικηθείη τὰ σώματα πρὸς τοῦ ῥεύματος, εἰς τοσαύτην μὲν ἀφίξεται δηλονότι δυσκρασίαν, ὡς καὶ τὴν ἐνέργειαν αὐτῶν ἀπολέσθαι καὶ φθαρῆναι τῷ χρόνῳ· παύσεται δ᾽ ὀδυνώμενα τότε πρῶτον, ὅταν ἐξομοιωθῇ τῷ μεταβάλλοντι. οὐ γὰρ ἐν τῷ μεταβεβλῆσθαι τὴν <φύσιν>[κρᾶσιν],[10] ἀλλ᾽ ἐν τῷ μεταβάλλεσθαι πονεῖ τὰ μόρια, καθότι καὶ ὁ θαυμαστὸς Ἱπποκράτης ἔλεγεν· "τοῖσι γὰρ τὴν φύσιν διαλλαττομένοισι καὶ διαφθειρομένοισιν, αἱ ὀδύναι γίγνονται." διαλλάττεται δὲ καὶ διαφθείρεται τὴν φύσιν ἕκαστον, ἢ θερμαινόμενον, ἢ ψυχόμενον, ἢ ξηραινόμενον, ἢ ὑγραινόμενον, ἢ τῆς συνεχείας λυόμενον· ἐπὶ μὲν ταῖς ἀνωμάλοις δυσκρασίαις διὰ τὸ θερμαίνεσθαι ἢ ψύχεσθαι μᾶλλον, δραστικώταται γὰρ αὗται αἱ ποιότητες, ἤδη δὲ καὶ διὰ τὸ ξηραίνεσθαί τε καὶ ὑγραίνεσθαι· ἐν δὲ τῷ πεινῆν ἢ διψῆν, ἐπιλειπούσης ἔνθα μὲν τῆς ξηρᾶς οὐσίας, ἔνθα δὲ τῆς ὑγρᾶς. ἐν δὲ τῷ τιτρώσκεσθαι καὶ ἀναβιβρώσκεσθαι καὶ τείνεσθαι καὶ θλίβεσθαι καὶ διασπᾶσθαι, τῆς συνεχείας λυομένης.

[10] See EGN, note on line 2, p. 152.

are bad. The abscesses in relation to the ribs break out into the open spaces of the thorax. Those of the muscles [break out] under the skin. However, those of the internal organs go to the arteries and veins in them, or under the enveloping membrane which itself exists like some kind of skin for the internal organs. If, on the other hand, the bodies are overcome by the flux, it will clearly come to so great a *dyskrasia* that also the function of these bodies is ruined and destroyed over time. Pain will stop at that time when [the parts] are made similar to what is causing the change. For it is not in the nature (*krasis*) having been changed, but in its being changed that the parts suffer, just as the admirable Hippocrates also said: "The pains arise in those things being changed and destroyed in respect of their nature."[9] Each is changed and destroyed in its nature when it is heated, cooled, dried, or wet, or suffers dissolution of continuity,[10] especially after the non-uniform *dyskrasias* due to heating and cooling—for these are the most active qualities, whereas next are those due to drying and wetting. In being hungry or thirsty, in the first instance they are lacking the dry substance, while in the second the wet substance is lacking. In being injured, eroded, strained, compressed, and torn, there is the dissolution of continuity.

---

[9] See Hippocrates, *Places in Man* (*De locis in homine*) 42, LCL 482 (P. Potter), 84–85. Potter has: "For in each thing that is altered with respect to its nature, pains arise." See also Craik's summary of this work, *The Hippocratic Corpus*, 160–62.

[10] For Galen's concept of dissolution of continuity, see particularly *On the Causes of Diseases*, XI, VIII.37–40K, and *Method of Medicine*, X.117–28K.

4. Εἰ μὲν οὖν ἥ τε θερμότης τοῦ αἵματος ἡ κατὰ τὸ φλεγμαῖνον μόριον ἐπιεικὴς ὑπάρχοι καὶ τὸ περιεχόμενον αἷμα καθ' ὅλον τὸ σῶμα τοῦ ζῴου μετρίως ἔχοι κράσεως, οὐ πάνυ τι ῥᾳδίως συνεκθερμαίνεται τῷ πεπονθότι· εἰ δή τοι ζέοι σφοδρότερον, ἢ εἰ καὶ τὸ καθ' ὅλον[11] τὸ ζῷον αἷμα χολῶδες ὑπάρχοι, παραχρῆμα σύμπαν ἐκθερμαίνεται· πολὺ δὲ μᾶλλον, ἐπειδὰν ἄμφω συνδράμῃ, καὶ τὸ κατὰ τὴν φλεγμονὴν αἷμα θερμὸν ἱκανῶς εἶναι, καὶ τὸ καθ' ὅλον τὸ ζῷον αἷμα[12] χολῶδες. ἐκθερμαίνεται δὲ πρότερον μὲν τὸ κατὰ τὰς ἀρτηρίας, ὅτι καὶ φύσει θερμότερόν ἐστι καὶ πνευματωδέστερον· ἐφεξῆς δὲ καὶ τὸ κατὰ τὰς φλέβας. εἰ δ' ἐγγὺς εἴη τῶν πολυαίμων σπλάγχνων τὸ φλεγμαῖνον μόριον, ἔτι καὶ θᾶττον αὐτῷ συνεκθερμαίνεται τὸ καθ' ὅλον τὸ ζῷον αἷμα. συνελόντι δ' εἰπεῖν, ἐπὶ παντὸς τοῦ θερμαίνοντος ὅσον ἂν εὐαλλοίωτον ᾖ, ἢ φύσει θερμὸν, ἐκθερμαίνεται πρῶτον, ὥσπέρ γε κἀπὶ παντὸς τοῦ ψύχοντος ὅσον ἂν εὐαλλοίωτον ᾖ, ἢ φύσει ψυχρὸν, ἐκεῖνο πρῶτον καταψύχεται.

Εὐαλλοιωτότατον μὲν οὖν τὸ πνεῦμα, διότι καὶ λεπτομερέστατον· θερμοτάτη δὲ φύσει ἡ ξανθὴ χολή, ψυχρότατον δὲ τὸ φλέγμα. τῶν δ' ἄλλων χυμῶν τὸ αἷμα μὲν ἐφεξῆς τῇ ξανθῇ χολῇ θερμόν· ἡ μέλαινα δέ, ψυχρὰ μετὰ τὸ φλέγμα. καὶ δὴ ἀλλοιοῦται μὲν ἡ ξανθὴ χολὴ ῥᾳδίως ὑπὸ παντὸς τοῦ δρῶντος εἰς αὐτήν·

[11] post καθ' ὅλον: τὸ ζῷον αἷμα χολῶδες ὑπάρχοι, EGN; τὸ τοῦ ζῴου σῶμα αἷμα χολωδέστερον ὑπάρχοι, K

NON-UNIFORM DISTEMPERMENT

4. If the heat of the blood in the inflamed part is mild, and the blood contained in the whole body of the animal is moderate in *krasis*, it is not very easy for it to be heated up along with the affected part. If, however, it is either seething more strongly, or if the blood in the whole animal is bilious, immediately all at once, it is heated. But much more [is this so] when both act together and the blood in the inflammation is sufficiently hot while that in the whole animal is bilious. The blood in the arteries is heated up first, because it is both hotter in nature and contains more *pneuma*;[11] next is that in the veins. If, however, the inflamed part is near the blood-filled viscera, the blood in the whole animal is heated with it still more quickly. In short, in the case of everything that is heated, to the extent it is susceptible to change, or is hot in nature, it is first to be heated, just as also in the case of everything cooled, to the extent it is susceptible to change, or is cold in nature, it is first to be cooled.

The *pneuma*, then, is very easily changed because it is also | the most fine-particled. Yellow bile is the hottest in nature, while phlegm is coldest. Of the other humors, the blood is next after yellow bile in hotness, while black bile is next after phlegm in coldness. Furthermore, the yellow bile is easily changed by everything acting on it, whereas

---

[11] The Greek term is πνευμάτωδης, which might be rendered "air-like" (see Aristotle, *Meteorology* 380b17). For a medical meaning, see Hippocrates, *Aphorisms* 5.72, where Jones renders it "flatulence" (LCL 150).

---

[12] αἷμα *add.* EGN

ἡ μέλαινα δέ, δυσκόλως. ἑνὶ δὲ λόγῳ τὸ μὲν λεπτομερὲς ἅπαν, εὐαλλοίωτον· τὸ δὲ παχυμερές, δυσαλλοίωτον. ὥστ' ἀνάγκη πολυειδεῖς γίγνεσθαι τὰς ἐπὶ ταῖς φλεγμοναῖς ἀλλοιώσεις, ὅτι καὶ πολυειδῶς διάκεινται τὰ σώματα. πρώτως μὲν γὰρ ὁ τὴν φλεγμονὴν ἐργαζόμενος χυμὸς ἢ μᾶλλον ἢ ἧττόν ἐστι θερμός· ἐφεξῆς δ' ἡ σῆψις αὐτοῦ κατὰ τὴν οἰκείαν ἁπαντᾷ φύσιν· οὐχ ἥκιστα δὲ καὶ παρὰ τὸ μᾶλλόν τε καὶ ἧττον ἐσφηνῶσθαι τὸν χυμόν.[13] τὰ γὰρ μὴ διαπνεόμενα θᾶττον σήπεται, καθάπερ κἀπὶ τῶν ἐκτὸς ἁπάντων· ὅταν δὲ τὸ αἷμα[14] καὶ θερμὸν καὶ ὑγρὸν τὴν κρᾶσιν ᾖ, τότε δὴ καὶ μάλιστα. καὶ μὲν δὴ καὶ τὸ φλεγμαῖνον μόριον ἢ ἐγγὺς ἢ πόρρω τῶν πολυαίμων ἐστὶ σπλάγχνων, καὶ τὸ πᾶν αἷμα πικρόχολον ἢ μελαγχολικὸν ἢ φλεγματῶδες ἢ πνευματῶδες· καὶ ταῦτα πάντα μᾶλλόν τε καὶ ἧττον. ὥστ' ἀνάγκη πολυειδεῖς γίγνεσθαι τὰς ἀλλοιώσεις, ἑτέρου τε σώματος[15] πρὸς ἕτερον παραβαλλομένου καὶ αὐτοῦ τινος πρὸς ἑαυτό.

5. Αὗται πᾶσαι δυσκρασίαι τοῦ σώματος ἀνώμαλοι γίγνονται, μάλιστα μὲν ἐκπεπυρωμένου[16] τοῦ κατὰ τὴν φλεγμονὴν αἵματος, ἐφεξῆς δὲ τοῦ κατὰ τἆλλα σπλάγχνα καὶ τὴν καρδίαν, καὶ ταύτης μάλιστα τοῦ κατὰ τὴν ἀριστερὰν κοιλίαν, εἰς ἣν εἰ[17] ὑγιαίνοντος ἔτι τοῦ ζῴου καὶ μήπω πυρέττοντος, ἐθελήσαις ἐμβαλεῖν τοὺς δακτύλους, ὡς ἐν ταῖς ἀνατομικαῖς ἐγ-

---

[13] τὸν χυμόν add. EGN
[14] τὸ αἷμα add. EGN
[15] σώματος add. EGN

NON-UNIFORM DISTEMPERMENT

the black bile is changed with difficulty. In a word, everything fine-particled is easily changed, whereas what is thick-particled is difficult to change. As a result, it is necessary for the changes after inflammations to be of many kinds, because the bodily dispositions are also of many kinds. For firstly, the humor producing the inflammation is either more or less hot. Next in order, the putrefaction of this is altogether in relation to its characteristic nature. Not least also, it is according to whether the humor is more or less plugged up. For those things not transpired are putrefied quicker, as in the case of all things external. However, when the blood is both hot and wet in *krasis*, it is particularly so at this time. Furthermore also, the inflamed part is either near to or distant from the blood-filled internal organs, and the whole blood either picrocholic or melancholic, or phlegmatous, or *pneuma*-filled, and all these things are more or less, so that of necessity the changes are of many kinds, of one body compared to another, or the same body compared to itself.

742K

5. All these *dyskrasias* of the body are non-uniform, particularly when the blood is much heated in the inflammation, and next when that in the other internal organs and the heart, and especially in the left chamber of the latter into which, if the animal is still living and not yet febrile, should you wish to place your fingers, as described

---

16 ἐκπεπυρωμένου EGN; ἐκπυρουμένου K

17 *post* εἰ: ὑγιαίνοντος ἔτι τοῦ ζῴου καὶ μήπω EGN; καὶ ζῶντος τοῦ ζῴου καὶ μηδέπω K. KLat has: *si vivo etiam animante, nec adhuc febricitante*. The latter is followed in the translation.

χειρήσεσι γέγραπται, σφοδροτάτης αἰσθήσῃ θερμασίας. οὔκουν ἀπεικός ἐστιν οὐδ' ὁπότε σύμπαν ἐκθερμαίνεται τὸ σῶμα παρὰ φύσιν, εἰς ἄκρον ἥκειν θερμότητος ἐκείνην μάλιστα· καὶ γὰρ λεπτομερέστατον τὸ αἷμα καὶ πνευματωδέστατον ἔχει, καὶ κινεῖται διαπαντός. ἀλλ' ἐν τοῖς τοιούτοις ἅπασι πυρετοῖς ἐκθερμαίνεται μὲν ἐνίοτε σύμπαν ἤδη τὸ αἷμα, ὃ καὶ δέδεκται τὴν παρὰ φύσιν ἐκείνην θερμασίαν τὴν ἐκ τῆς σηπεδόνος τῶν χυμῶν ὁρμηθεῖσαν· οὐ μὴν οὐδ' οἱ χιτῶνες τῶν ἀρτηριῶν ἢ τῶν φλεβῶν, οὐδ' ἄλλό τι τῶν περικειμένων σωμάτων ἤδη πω τελέως ἠλλοίωται τὴν κρᾶσιν, ἀλλ' ἔτι καὶ μεταβάλλεται καὶ ἀλλοιοῦται θερμαινόμενα. εἰ δ' ἐν χρόνῳ πλείονι τοῦτο πάσχοι, κἂν νικηθείη ποτὲ καὶ μεταβληθείη παντάπασιν, ὡς μηκέτι θερμαίνεσθαι μόνον, ἀλλ' ἤδη τεθερμάνθαι παρὰ φύσιν.

Ὅρος δ' ἐστὶ τῆς ἀλλοιώσεως ἑκάστῳ τῶν μορίων ἡ τῆς ἐνεργείας αὐτοῦ βλάβη· τὸ δ' ἄχρι τοῦδε πλάτος ἅπαν, ὁδός[18] ἐστιν εἰς τὸ παρὰ φύσιν, οἷον ἐπίμικτόν τε καὶ κοινὸν καὶ μέσον ἐξ ἀμφοῖν τῶν ἐναντίων γεγονός, αὐτοῦ τε τοῦ κατὰ φύσιν ἀκριβῶς καὶ τοῦ τελέως ἤδη παρὰ φύσιν. ἐν δὴ τούτῳ τῷ χρόνῳ παντὶ τὸ θερμαινόμενον σῶμα τῷ ποσῷ τῆς ἀλλοιώσεως ἀνάλογον ἔχει τὴν ὀδύνην. ὅταν δ' ἐκθερμανθῇ τελέως[19] ⟨αὐτὰ⟩ [ἅπαντα] τὰ στερεὰ μόρια τοῦ

[18] ἅπαν ὁδός EGN; ἐπάνοδός K
[19] post τελέως: ⟨αὐτὰ⟩ [ἅπαντα] τὰ στερεὰ μόρια τοῦ σώματος, EGN; ἅπαντα τοῦ σώματος τὰ στερεὰ μόρια, K

NON-UNIFORM DISTEMPERMENT

in *On Anatomical Procedures*,[12] you would perceive a very strong heat. It is not, therefore, unreasonable, when the whole body is heated contrary to nature, that the left chamber [of the heart] particularly comes to a peak of hotness, for it also has the most fine-particled blood and the most *pneuma*-containing, and is being moved continuously. But in all such fevers, sometimes the whole blood is already heated and has received that heat contrary to nature initiated by the putrefaction of the humors. Neither the walls of the arteries, nor those of the veins, nor any other of the surrounding bodies are, up to this time, already changed completely in *krasis*, but being heated are still being changed | and altered. If, however, over a longer time, this is suffered, and even overcome sometimes and changed altogether, it is not only no longer being heated, but has already been heated contrary to nature.

The limit of the change for each of the parts is the damage to their function. Up to this point, the whole range is a path to a contrariety to nature, like a mixture common and median arising from both of the opposites, one of which is precisely in accord with nature and the other already completely contrary to nature. During this whole time the body being heated has pain in proportion to the amount of the change. When, however, all the actual solid parts of the body are heated completely, such a fever is

---

[12] *De administrationibus anatomicis*, II.205–731K. Vivisection of the heart is described in Book 7, section 12 (II.626–32K); see Singer's English translation, *Galen on Anatomical Procedures*, 190–92.

σώματος, καλεῖται μὲν ὁ τοιοῦτος πυρετὸς ἑκτικός, ὡς ἂν μηκέτι τοῖς ὑγροῖς τε καὶ τῷ πνεύματι μόνον, ἀλλ' ἤδη καὶ τοῖς στερεοῖς[20] ἕξιν ἔχουσι σώμασι περιεχόμενος· ἀνώδυνος δ' ἐστὶ καὶ νομίζουσιν οἱ πυρέττοντες οὕτω μηδὲ πυρέττειν ὅλως. οὐ γὰρ αἰσθάνονται τῆς θερμασίας, ἁπάντων ὁμοίως αὐτοῖς τεθερμασμένων τῶν μορίων. καὶ δὴ καὶ ὡμολογεῖται ταῦτα τοῖς φυσικοῖς ἀνδράσιν ἐν τοῖς περὶ τῶν αἰσθήσεων λογισμοῖς. οὔτε γὰρ χωρὶς ἀλλοιώσεως ἡ αἴσθησις, οὔτ' ἐν τοῖς ἤδη τελέως ἠλλοιωμένοις ἡ ὀδύνη. διὰ ταῦτ' ἄρα καὶ οἱ ἑκτικοὶ πυρετοὶ πάντες, ἄπονοί τε καὶ τελέως εἰσὶν ἀναίσθητοι τοῖς κάμνουσιν· οὐ γὰρ ἔτι τῶν ἐν ἑαυτοῖς μορίων τὸ μὲν ποιεῖ, τὸ δὲ πάσχει, πάντων ὁμοίως ἀλλήλοις ἤδη γεγονότων, καὶ μίαν ἐχόντων ὁμόλογον κρᾶσιν.

6. Εἰ δὲ δὴ καὶ τὸ μὲν αὐτῶν θερμότερον εἴη, τὸ δὲ ψυχρότερον, ἀλλὰ τοσούτῳ γε θερμότερον ἢ ψυχρότερον ὡς μὴ λυπεῖν τὸ πλησιάσαν, ἢ οὕτως ἂν εἴη ἀλλήλοις λυπηρὰ καὶ πάντα τὰ[21] κατὰ φύσιν ἔχοντα μόρια, ὡς διαφέροντά γε καὶ αὐταῖς ταῖς κράσεσι· σὰρξ μὲν γὰρ θερμόν ἐστι μόριον, ὀστοῦν δὲ ψυχρόν· ἀλλὰ καὶ τούτων καὶ τῶν ἄλλων ἁπάντων ἀνώδυνος ἐστιν ἡ ἀνωμαλία τῷ μέτρῳ τῆς ὑπεροχῆς. οὕτω γοῦν καὶ τὸ περιέχον ἡμᾶς οὐκ ἀνιᾷ πρὶν εἰς ἄμετρόν ποτε ψύξιν ἢ θερμασίαν ἐκτραπῆναι. τῶν δ' ἐν τῷ μέσῳ διαφορῶν αὐτοῦ, καίτοι παμπόλλων οὐσῶν καὶ σαφῆ τὴν ὑπεροχὴν ἐχουσῶν, ἀλύπως αἰσθανόμεθα. κινδυνεύει τοιγαροῦν ἐκ τῶνδε κἀκεῖνος ὁ λόγος ἔχειν ἐπι-

NON-UNIFORM DISTEMPERMENT

called hectic, as it would no longer be contained in the fluids or *pneuma* alone, but would already also have an established state (*hexis*) in the solid bodies. However, it is painless and those who are febrile in this way think they are not febrile at all, for they don't perceive the heat as all the parts in them have been heated similarly. And in particular, these things are agreed upon by natural scientists in their conclusions about the perceptions. For there is no perception apart from change and no pain in those things already completely changed. Because of this, then, all the hectic fevers are painless and completely imperceptible to those suffering them. For it is no longer the case that of the parts in them, one is acting while another is being acted upon, since they have all already become similar to each other and have one homologous *krasis*.

6. Certainly if one of the parts were hotter while another colder, but by such an amount hotter or colder as not to cause pain to what is adjacent, or in this way were distressing to each other and to all the parts having an accord with nature, they would differ also in their actual *krasias*. For flesh is a hot part while bone is a cold part, but also of these and of all the others, the non-uniformity is painless due to the moderate degree of the excess. Anyway, in this manner too, what surrounds us does not cause distress before at some point it turns to a disproportion in terms of cooling or heating. Of the differences of this in between, although there are many and they clearly have an excess, we perceive them as painless. Wherefore it is possible from these things for that argument also to

---

[20] στερεοῖς *add.* EGN   [21] πάντα τὰ *add.* EGN

745K εἴκειαν, ὥς που | καὶ Ἱπποκράτης φησὶν ὁ φάσκων· "ἕλκεα πάντα εἶναι τὰ νοσήματα." συνεχείας μὲν γὰρ λύσις τὸ ἕλκος· αἱ δ' ἄμετροι θερμασίαι καὶ ψύξεις πλησίον ἥκουσι τοῦ λύειν τὴν συνέχειαν· ἡ μὲν γὰρ πολλὴ θερμασία τῷ διακρίνειν τε καὶ διατέμνειν τῆς οὐσίας τὸ συνεχές, ἡ δ' ἄκρα ψύξις τῷ πιλεῖν τε καὶ συνωθεῖν εἴσω, τὰ μὲν ἐκπιέζει, τὰ δὲ θλᾷ·

Καὶ τοῦτόν γέ τις ὅρον τιθέμενος τῆς ἀμετρίας τοῦ θερμοῦ καὶ ψυχροῦ, τάχ' ἂν οὐκ ἀπὸ τρόπου γιγνώσκειν δόξειεν.[22]—ἀλλ' εἴθ' δ' οὗτος εἴτ' ἄλλός τις ὅρος ἐστὶ τῆς ἀμετρίας—ἀλλὰ τό γ' ἐν τῷ πρὸς τι πᾶσαν ἀμετρίαν ὑπάρχειν ἤδη που πρόδηλον· οὐ γὰρ ὡσαύτως ὑπὸ τῶν θερμῶν ἢ ψυχρῶν ἅπαν σῶμα διατίθεται. καὶ διὰ τοῦτο τινὰ μὲν οἰκείους ἔχει τοὺς χυμοὺς ἀλλήλοις ζῷα· τινὰ δ'[23] οὐ μόνον οὐκ οἰκείους ἀλλὰ καὶ φθαρτικούς, οἷον ἄνθρωπος καὶ ἔχιδνα. τὸ γοῦν σίελον, ὀλέθριόν ἐστιν ἑκατέρῳ τὸ τοῦ ἑτέρου. οὕτω γ' ἂν καὶ σκορπίον ἀναιρήσειας ἐπιπτύων νῆστις· οὐ μὴν ἂν ἄνθρωπός γ' ἄνθρωπον ἀναιρήσειε δάκνων, οὐδ' ἔχις ἔχιν, οὐδ' ἀσπίς ἀσπίδα. τὸ μὲν

[22] post δόξειεν ante οὐ γὰρ ὡσαύτως K has: εἴτε δὲ οὗτος εἴτε ἄλλός τις ὅρος ἐστὶ τῆς ἀμετρίας, ἀλλὰ τό γ' ἐν τῷ πρὸς τὶ πᾶσαν ἀμετρίαν ὑπάρχειν ἤδη που πρόδηλον.
[23] post τινὰ δ': οὐ μόνον οὐκ οἰκείους EGN; οὐ μόνον οἰκείους ἔχει, K

[13] This is presumably a reference to *On Fractures* 31, which has: "Unless one calls all maladies wounds, for this doctrine also

have an appropriateness, as I Hippocrates also stated somewhere, when he said: "All wounds (ulcers) are diseases."[13] For the wound (ulcer) is a dissolution of continuity, while those things that heat and cool disproportionately come near to dissolving the continuity. Thus, great heat acts by separating and disconnecting the continuity of the substance, while extreme cold, by condensing and compressing inward, squeezes out some things and crushes others.

And someone, if he establishes this as a limit of the disproportion of hot and cold, would perhaps seem to know not without reason whether there is this or some other limit of the disproportion, since it is somehow already clear that every disproportion is relative to something.[14] For not every body is affected similarly by those things that are hot or cold, and because of this, certain animals have humors compatible with each other, while some have those that are not only incompatible but also destructive—for example, a human and a viper. Anyway the saliva in each of the two is deadly to the other. In fact, in this way too, you will kill a scorpion, if you spit on it while fasting, whereas a person will not harm another person by biting, nor a viper another viper, nor an asp another

has reasonableness, since they have affinity one to another in many ways." *Hippocrates* III, LCL 149 (E. T. Withington), 170–71. *Places in Man* is also a possibility.

[14] There is uncertainty about the text for this and the previous sentence; see note 22 to the Greek text. In the translation, the Kühn text (including KLat) is followed. See Garcia Novo's commentary on this passage, *Galen: On the Anomalous Dyskrasia*, 230–31.

GALEN

γὰρ ὅμοιον οἰκεῖον ἐστι καὶ φίλιον, | τὸ δ' ἐναντίον, ἐχθρὸν καὶ ἀνιαρόν.

Αὔξεται γοῦν ἅπαντα καὶ τρέφεται πρὸς τῶν ὁμοίων, ἀναιρεῖται δὲ καὶ φθίνει[24] πρὸς τῶν ἀνομοίων.[25] καὶ διὰ τοῦτο ἡ μὲν τῆς ὑγείας φυλακὴ διὰ τῶν ὁμοίων, ἡ δὲ τῶν νοσημάτων ἀναίρεσις ἔσται[26] διὰ τῶν ἐναντίων. ἀλλὰ περὶ μὲν τούτων ἕτερος ἂν εἴη λόγος. ὁ δ' ἑκτικὸς ἐκεῖνος πυρετὸς ὁ τὴν ἕξιν ἤδη τοῦ ζῴου κατειληφὼς ἀναίσθητός ἐστι τῷ κάμνοντι. τῶν δ' ἄλλων πυρετῶν οὐδεὶς ἀναίσθητος, ἀλλ' οἱ μὲν μᾶλλον, οἱ δ' ἧττον ἀνιαροὶ τοῖς νοσοῦσιν· ἔνιοι δ' αὐτῶν καὶ ῥῖγος ἐπάγουσιν.[27] γίγνεται γὰρ οὖν δὴ καὶ τοῦτο τὸ σύμπτωμα καθάπερ καὶ ἄλλα τινὰ[28] πρὸς τῆς ἀνωμάλου δυσκρασίας. εἰπεῖν δ' οὐκ ἐγχωρεῖ τὸν τρόπον αὐτοῦ τῆς γενέσεως ἐν τῷ νῦν ἐνεστῶτι λόγῳ πρὶν ἀποδεῖξαι περὶ τῶν φυσικῶν δυνάμεων, ὁπόσαι τ' εἰσὶ καὶ ὁποῖαι, καὶ ὅ τι δρᾶν ἑκάστη πέφυκεν. ἀλλ' ἐν ταῖς τῶν συμπτωμάτων αἰτίαις ὑπὲρ ἁπάντων εἰρήσεται.

7. Ἐπάνειμι δὲ πάλιν ἐπὶ τὰς τῆς ἀνωμάλου δυσκρασίας διαφοράς· ὅπως μὲν οὖν ἐπὶ φλεγμονῇ γίγνεται πυρετός,[29] ὅτι τε φλεγμονὴ πᾶσα καὶ πυρετὸς

[24] φθίνει EGN; φθείρεται K
[25] ἀνομοίων EGN; ἐναντίων K   [26] ἔσται add. EGN
[27] ἐπάγουσιν EGN; ἐπιφέρουσι K
[28] τινα EGN; πολλὰ K
[29] post πυρετός: ὅτι τε φλεγμονὴ πᾶσα καὶ πυρετὸς ἅπας, EGN; ἅπας, ὅτι τε καὶ πυρετὸς καὶ φλεγμονὴ πᾶσα, K

NON-UNIFORM DISTEMPERMENT

asp. For what is similar is compatible and friendly, while what is opposite is inimical and distressing.

Anyway, all things are increased and nourished by similar things (similars), while they are distressed and destroyed by dissimilar things (opposites), and because of this, the preservation of health is through similars, while the destruction of the diseases is through opposites.[15] But there is another discussion about these matters. That hectic fever which has already taken hold of the bodily state of the organism is imperceptible to the one suffering. However, none of the other fevers are imperceptible, although there are those that are more and those that are less distressing to those who are sick, while some of the fevers also bring shivering (rigors). For this symptom certainly occurs, just as some others do, in relation to the non-uniform *dyskrasia*. It is not possible, however, to speak about the manner of the actual genesis in the present book prior to a demonstration of the natural capacities (faculties)—how many there are and what kinds, and what each naturally does. There will be a discussion of all these in [the work] *On the Causes of Symptoms*.[16]

7. I am returning once again to the differences of the non-uniform *dyskrasia*. How, then, a fever arises following an inflammation and that every inflammation and every

---

[15] On the basis of this doctrine, see Hippocrates, *Diseases* 4 and *Aphorisms* 2.22, and Galen, *Method of Medicine*, X.634K.

[16] *De causis morborum*, VII.1–46K. See particularly sections III.4, VII.16–18K (Johnston, *Galen: On Diseases and Symptoms*, 166–67).

ἅπας, χωρὶς τῶν ἑκτικῶν ὀνομαζομένων, ἐκ τῶν ἀνωμάλως κεκραμένων ἐστὶ νοσημάτων, ἤδη μοι λέλεκται. γένοιτο δ' ἂν καὶ χωρὶς φλεγμονῆς ἐπὶ σήψει μόνῃ χυμῶν, πυρετός. οὐ γὰρ δὴ τά γε σφηνωμένα τε καὶ μὴ διαπνεόμενα σήπεται μόνον, ἀλλὰ τάχιστα μὲν ταῦτα καὶ μάλιστα. σήπεται δὲ καὶ ἄλλα πάμπολλα τῶν ἐπιτηδείων εἰς σῆψιν. εἰρήσεται δὲ καὶ περὶ τῆς τούτων ἐπιτηδειότητος ἑτέρωθι. καὶ μὲν δὴ καὶ κατ' ἄλλον τρόπον ἀνώμαλος ἔσται δυσκρασία περὶ σύμπαν τὸ σῶμα· ποτὲ μὲν λιγνυώδους διαπνοῆς ἐπισχεθείσης· ποτὲ δ' ἐκ γυμνασίων πλειόνων ἢ πόνων[30] αὐξηθέντος τοῦ θερμοῦ· ποτὲ δ' ἐπὶ θυμῷ ζέσαντος ἀμετρότερον τοῦ αἵματος ἢ δι' ἔκκαυσίν τινα ἔξωθεν ἐκθερμανθέντος.

Ὅτι δὲ κἀν τούτοις ἅπασι τοῖς πυρετοῖς, ὥσπερ κἀπὶ τῆς φλεγμονῆς ἔμπροσθεν ἐλέγετο, παρά τε τὴν ἰσχὺν τοῦ δρῶντος αἰτίου καὶ παρὰ τὴν τοῦ σώματος διάθεσιν, οἱ μὲν μᾶλλον πυρέξουσιν, οἱ δ' ἧττον, οἱ δ' οὐδ' ὅλως, εὔδηλον εἶναι νομίζω. καὶ μὲν δὴ καὶ ὡς ποτὲ μὲν ταύτης τῆς πνευματώδους οὐσίας μόνης, ἐνίοτε δ' ἅπτεται καὶ τῶν χυμῶν ἡ δυσκρασία, πρόδηλον καὶ τοῦτο οὐδὲν ἧττον. καὶ ὡς ἐπὶ πᾶσι τοῖς τοιούτοις πυρετοῖς[31] χρονίζουσιν, ὁ καλούμενος ἑκτικὸς ἀκολουθήσει. καί πως ὁ λόγος ἤδη δείκνυσιν, ὡς ἐνίοτε μὲν οὐσίας θερμῆς ἢ ψυχρᾶς ἐπιρρυείσης μορίῳ τινὶ, γίγνεσθαι συμβαίνει τὴν ἀνώμαλον

NON-UNIFORM DISTEMPERMENT

fever, apart from those termed hectic, are from the diseases compounded non-uniformly, I have already said. However, fever also arises apart from inflammation, following putrefaction of humors alone. For certainly not only does what is plugged up and not transpired putrefy but these things do so very quickly and most of all. And many other of the things inclined to putrefaction do putrefy. Something will also be said about the most inclined of these elsewhere.[17] And certainly there will also be another kind of non-uniform *dyskrasia* involving the whole body. Sometimes [this will be] when a fuliginous transpiration is retained, sometimes from too many exercises or labors when there is augmentation of the heat, and sometimes from anger when the blood seethes more immoderately or is heated up externally by heatstroke.

It is, I think, quite clear that in all these fevers, as was said before in the case of inflammation, because of the strength of the effecting cause and because of the condition of the body, some will be more febrile, some less, and some not at all. And that sometimes the *dyskrasia* attacks the *pneumatic* substance alone and sometimes the humors is no less clear, and that following all such chronic fevers, the so-called hectic fever will follow. And in some way the discussion already shows that sometimes, when a hot or cold substance flows to a certain part, what happens is

---

[17] See, for example, *De methodo medendi*, X.693–97K, and *Art of Medicine*, I.335–39 and 382K.

---

[30] post πόνων: αὐξηθέντος τοῦ θερμοῦ· EGN; αὐξηθέντων, τὴν θερμασίαν ἐπιτεινόντων, K

[31] πυρετοῖς add. EGN

ταύτην δυσκρασίαν, ὥσπερ ἐπὶ τῶν φλεγμαινόντων ἐλέγετο· πολλάκις δ' οὐχ οὕτως, ἀλλ' αὐτῆς τοῦ σώματος τῆς κράσεως ἀλλοιωθείσης κατὰ ποιότητα· καὶ ὡς τῶν ἀλλοιούντων αὐτήν, τὰ μὲν ἐξ αὐτοῦ τοῦ σώματος ὅρμαται, τὰ δ' ἔξωθεν. ὅταν μὲν ἐπὶ σήψεσι μόναις ἤ τισι φλεγμοναῖς ἐγείρηται πυρετός, ἐξ αὐτοῦ τοῦ σώματος. ὅταν δ' ἐπ' ἐκκαύσεσι καὶ γυμνασίοις, ἔξωθεν. εἰρήσεται δὴ καὶ περὶ τούτων ἐπὶ πλέον αὖθις, ἐν ταῖς τῶν νοσημάτων αἰτίαις. ὥσπερ δ' ἐπ' ἐκκαύσεσι γίγνεται πυρετὸς ἀλλοιωθείσης τῆς τοῦ σώματος κράσεως, οὕτως ἐν κρύει πολλάκις ἰσχυρῶς ἐψύγησαν ἔνιοι τὸ σύμπαν σῶμα, καί τινες ἐξ αὐτῶν ἀπέθανον. καὶ μὲν δὴ καὶ ὡς ἀλγοῦσιν οὗτοι πάντες, οὐδὲ τοῦτ' ἄδηλον.

8. Ἀλγοῦσι δὲ καὶ ὅσοι καταψυχόμενοι σφοδρῶς ὑπὸ κρύους καρτεροῦ προὐθυμήθησαν ἐκθερμῆναι ταχέως ἑαυτούς, καὶ πολλοί γε αὐτῶν ἐπενέγκοντες ἀθρόως τῷ πυρὶ τὰς χεῖρας, ἀλγήματος αἰσθάνονται σφοδροτάτου κατὰ τὰς ῥίζας | τῶν ὀνύχων. εἶτα τίς οὕτως ἐναργῶς ὁρῶν ὀδύνης αἰτίαν γιγνομένην τὴν ἀνώμαλον δυσκρασίαν, ἔτ' ἀπιστεῖ περὶ τῶν ἐντὸς ἀλγημάτων, ἢ θαυμάζει πῶς χωρὶς φλεγμονῆς ὀδυνῶνται πολλάκις, ἢ τὸ κόλον,[32] ἢ τοὺς <ὄρχεις> [ὀδόντας],[33] ἢ τῶν ἄλλων τι μορίων; οὔτε γὰρ τῶν τοιούτων διαθέσεων οὐδὲν θαυμαστὸν οὐδὲ πῶς ἅμα καὶ ῥιγῶσι καὶ πυρέττουσιν ἔνιοι τῶν νοσούντων. καὶ γὰρ

[32] τὸ κόλον EGN; τὸ κῶλον K
[33] <ὄρχεις> [ὀδόντας] EGN; ὀδόντας K

NON-UNIFORM DISTEMPERMENT

that this non-uniform *dyskrasia* arises, just as was said in the case of the inflammations. Often, however, it is not like this, but when the actual *krasis* of the body is changed in relation to quality, some of the things changing it start from the body itself and some from without. Whenever a fever is stirred up from putrefactions alone or from certain inflammations, it is from the body itself, whereas when it is from heatstrokes and exercises, it is external. More will be said about these in the treatises *On the Causes of Diseases*.[18] Just as a fever arises after heatstrokes when the *krasis* of the body is changed, in the same way in icy cold often some are cooled strongly in respect of the whole body and some of these people died. And furthermore also, that they all suffer pain in this way is quite clear.

8. And there are those cooled who strongly feel pain due to quite severe cold, and are made eager to heat themselves quickly. And in fact many of these, when they suddenly bring their hands to the fire, perceive a very strong pain in the roots | of the nails. Who then, when he clearly sees in this way the non-uniform *dyskrasia* to be a cause of pain, is still incredulous concerning the internal pains, or wonders how people often suffer pain apart from inflammation in the colon or testicles (teeth)[19] or one of the other parts? For it is not surprising in such conditions, how some of those who are diseased shiver and are feverish at

---

[18] *De causis morborum*, VII.1–41K; see particularly section vi. On heatstroke, see VII.5K.

[19] On the variation between testicles and teeth, see Garcia Novo, *Galen: On the Anomalous Dyskrasia*, 245.

εἰ ὁ φλεγματώδης χυμός, ψυχρὸς ὤν, ὃν ὁ Πραξαγόρας ὑαλώδη προσαγορεύει,[34] καὶ ὁ πικρόχολος καὶ θερμὸς ἅμα πλεονάζοιέν ἐν αὐτοῖς[35] καὶ κινοῖντο διὰ τῶν αἰσθητικῶν σωμάτων, οὐδὲν θαυμαστὸν ἀμφοτέρων ὁμοίως αἰσθάνεσθαι τὸν ἄρρωστον. οὐδὲ γὰρ εἰ στήσας ἄνθρωπον ἐν ἡλίῳ θερμῷ, προσραίνοις[36] ὕδωρ ψυχρόν, ἀδύνατον αὐτῷ τὸ μὴ οὐχ ἅμα καὶ τῆς ἀπὸ τοῦ ἡλίου θέρμης αἰσθάνεσθαι καὶ τῆς ἀπὸ τοῦ ὕδατος ψύξεως.[37] ἀλλ' ἐνταῦθα μέν, ἔξωθέν ἐστιν ἀμφότερα, καὶ κατὰ τὰ μεγάλα προσπίπτει μόρια· κατὰ δὲ τοὺς ἠπιάλους πυρετούς, ἔνδοθέν τε καὶ κατὰ τὰ σμικρά, καὶ διὰ τοῦτο σύμπαν ἀμφοτέρων αἰσθάνεσθαι τὸ σῶμα δοκεῖ. τῷ γὰρ δι' ἐλαχίστου παρεσπάρθαι τό τε ψῦχον καὶ θερμαῖνον[38] οὐδέν ἐστιν αὐτοῦ λαβεῖν μικρὸν μόριον αἰσθητόν, | ἐν ᾧ θάτερον οὐχ ὑπάρχει.

Κατὰ μέντοι τὴν εἰσβολὴν τῶν παροξυσμῶν ἔνιοι τῶν πυρεττόντων καὶ ῥιγῶσι καὶ πυρέττουσιν[39] καὶ ἅμα ἀμφοτέρων αἰσθάνονται, ψύξεως ἀμέτρου καὶ θέρμης ὁμοῦ, ἀλλ' οὐ κατὰ τὸν αὐτὸν τόπον· ἐναργῶς γὰρ ἔχουσι οὗτοι διορίσαι τὰ θερμαινόμενα μόρια τῶν ψυχομένων. ἔνδοθεν μὲν γὰρ καὶ κατ' αὐτὰ τὰ σπλάγχνα τῆς θερμασίας αἰσθάνονται, ἐν δὲ τοῖς

[34] προσαγορεύει EGN; καλεῖ K
[35] ἐν αὐτοῖς add. EGN   [36] προσραίνοις EGN; καὶ προσβρέχοις K   [37] ψύξεως EGN; ψυχρότητος K
[38] ψῦχον καὶ θερμαῖνον EGN; θερμὸν καὶ ψυχρὸν K
[39] πυρέττουσιν EGN; διψῶσι K

NON-UNIFORM DISTEMPERMENT

the same time. Surely, if the phlegmatic humor is cold—that which Praxagoras calls *hyaloid* (glassy)[20]—and the picrocholic and hot at the same time is in excess in them, and these move through the perceiving bodies, it is no wonder that the sick person similarly perceives both. For if you were to stand a person in hot sun and were to drench him with cold water in addition, it would be impossible for him not to feel at the same time both the heat from the sun and the cold from the water. But in this case both are external and fall on large parts, whereas in the ague (*ēpialos*) fevers,[21] they are internal and [fall on] small [parts], and because of this the whole body seems to perceive both at the same time. For certainly, when the hot and the cold are dispersed through the smallest parts, there is no small perceiving part of this to be found in which either one of the two doesn't exist.

Nevertheless, in relation to the attack of the paroxysms, some of those with fevers shiver and are feverish and perceive both at the same time—that is, disproportionate cold and heat at the same time—but not in the same place. For clearly they are able to distinguish the heated parts from the cooled. They perceive the heat internally and in relation to the actual viscera, whereas they perceive the cold

---

[20] On Praxagoras' elaboration of the list of humors and particularly the *hyaloid* mentioned here and below, see F. Steckerl, *The Fragments of Praxagoras of Cos and His School* (Leiden, 1957), frag. 53, p. 73; and O. Lewis, *Praxagoras of Cos on Arteries, Pulse and Pneuma* (Leiden, 2017), 202.

[21] See note 3 above.

ἔξωθεν μορίοις ἅπασι, τῆς ψύξεως αἰσθάνονται. εἴσι δὲ καὶ οἱ λειπυρίαι καλούμενοι πυρετοὶ διἀπαντὸς τοιοῦτοι, καί τι γένος ὀλεθρίων καύσων. ὅπερ οὖν ἐν τούτοις κατὰ τὰ μεγάλα μόρια, τοῦτ' ἐν τοῖς ἠπιάλοις κατὰ τὰ σμικρὰ συμπέπτωκεν. ἀνώμαλος μὲν γὰρ ἐστι δυσκρασία καὶ ἡ τῶν συνθέτων[40] πυρετῶν·

Ἀνώμαλος δὲ καὶ ἡ τῶν ἄλλων ἁπάντων, πλὴν τῶν ἑκτικῶν ὀνομαζομένων· ἀνώμαλος δὲ καὶ τοῖς ῥιγῶσι μέν, οὐκ ἐπιπυρέττουσι δέ. σπάνιον μὲν γὰρ τὸ σύμπτωμα· γίγνεται μὴν ὅμως[41] ‹κατὰ τὴν διάθεσιν› καὶ ταῖς γυναιξὶ καί τισιν ἀνδράσιν ἐνίοτε. χρὴ δὲ πάντως ἀργὸν προηγήσασθαι βίον, ἤ τι πλῆθος ἐδεσμάτων ἐν χρόνῳ πλείονι προσενηνέχθαι τὸν ἄνθρωπον, ἐξ οὗ χυμὸς ἀργὸς καὶ ψυχρὸς καὶ ὠμὸς καὶ φλεγματώδης γεννᾶται, ὁποῖόν τινα καὶ ὁ Πραξαγόρας εἰσηγήσατο τὸν | ὑαλώδη.

Πάλαι δέ, ὡς ἔοικεν, οὐδεὶς οὕτως ἔπασχεν ὅτι μηδεὶς τοσοῦτον ἀργῶς καὶ πλησμονωδῆς διῃτᾶτο, καὶ διὰ τοῦτο γέγραπται παρὰ τοῖς παλαιοῖς ἰατροῖς ἐξ ἀνάγκης ἕπεσθαι ῥίγει πυρετόν. ἀλλὰ καὶ ἡμῖν αὐτοῖς καὶ ἄλλοις πολλοῖς τῶν νεωτέρων ἰατρῶν, ὦπται πολλάκις ῥῖγος ᾧ πυρετὸς οὐκ ἐπηκολούθησεν. σύνθετος δ' οὖν ἐστιν ἐκ ταύτης τῆς δυσκρασίας, καὶ προσέτι τῆς τῶν πυρεττόντων, ὁ ἠπίαλος. οὕτω δ' ὀνομάζω τὸν πυρετὸν ἐκεῖνον, ᾧ διὰ παντὸς ἄμφω συμ-

[40] συνθέτων EGN; τούτων K   [41] post ὅμως/ὅλως: ‹κατὰ τὴν διάθεσιν› καὶ ταῖς γυναιξὶ EGN; καὶ γυναιξὶ K

NON-UNIFORM DISTEMPERMENT

externally in all the parts. There are also the so-called malignant intermittent fevers such as this throughout and a class of the fatal bilious remittent fevers.[22] Therefore, what in these has fallen on the large parts, has fallen on the small parts in the ague fevers, for this is a non-uniform *dyskrasia* and of the compound fevers.

It is also non-uniform in all the other [fevers] apart from those termed hectic. And it is non-uniform in those who have rigors but do not have a subsequent fever. The symptom is rare. Nevertheless, it arises in relation to the condition, in women, and in certain men sometimes. It is necessarily preceded by a completely idle life or for the person to have been given an abundance of food over a long time, from which a humor that is inactive, cold, raw, and phlegmatic is generated—a kind which Praxagoras deemed | *hyaloid* (glassy).

751K

In ancient times, however, it would seem that nobody suffered in this way because no one lived his life so idly and replete (surfeited),[23] and because of this it has been written by the doctors of ancient times that a fever follows a rigor (shivering) of necessity. But also a rigor which a fever doesn't follow has been observed by ourselves and by many others of the more recent doctors. Thus the ague fever is compounded from this *dyskrasia* and besides from the *dyskrasia* of those who are febrile. This is my term for that fever in which both (i.e., fever and rigors) happen

---

[22] The two terms are λιπυρία and καῦσος (see Hippocrates, *On Crises and the Days of Crisis* 11, *On Diseases* 2.51; and Galen, XVII(2).728K); on the latter, the endemic fever of the Levant, see Hippocrates, *Aphorisms* 3.21.   [23] See Hippocrates, *Aphorisms* 2.4 and *On Regimen in Acute Diseases* 56.

βέβηκεν. ᾧ δ' ἡγεῖται μὲν τὸ ῥῖγος, ἕπεται δ' ὁ πυρετός, ὡς ἐν τριταίοις καὶ τεταρταίοις, οὐ καλῶ τοῦτον ἠπίαλον. ὥστ' ἐκ δύο τῶν ἀνωμάλων δυσκρασιῶν ὁ ἠπίαλος συμπέπλεκται, καὶ οἱ λοιποὶ δὲ πυρετοὶ σχεδὸν ἅπαντες, πλὴν τῶν ἑκτικῶν ὀνομαζομένων.

9. Ὡσαύτως δὲ καὶ ὅσα μορίου τινὸς ἑνός ἐστι νοσήματα μετ' ὄγκου, καὶ ταῦτα σύμπαντα παραπλησίως τῇ φλεγμονῇ, κατὰ δυσκρασίαν ἀνώμαλον ἀποτελεῖται· καρκῖνος, ἐρυσίπελας, ἄνθραξ, ἕρπης, οἴδημα, φλεγμονή,[42] φαγέδαινα, γάγγραινα. κοινὸν μὲν γὰρ αὐτοῖς ἅπασι τὸ ἐξ ἐπιρροῆς ὑγρῶν γεγονέναι· διαφέρει δέ, τῷ τὰ μέν, ὑπὸ φλεγματικοῦ χυμοῦ, τὰ δέ, ὑπὸ χολώδους, ἢ μελαγχολικοῦ χυμοῦ, τὰ δέ, ὑφ' αἵματος ἤτοι θερμοῦ καὶ λεπτοῦ καὶ ζέοντος, ἢ ψυχροῦ καὶ παχέος, ἤ πως ἄλλως διακειμένου, γίγνεσθαι. δηλωθήσεται μὲν γὰρ ἀκριβῶς ὑπὲρ τῆς κατ' εἶδος ἐν τούτοις διαφορᾶς ἑτέρωθι· πρὸς δὲ τὸν ἐνεστῶτα λόγον, ἀρκεῖ καὶ τοῦτ' εἰρῆσθαι μόνον, ὡς ὁποῖον ἂν ᾖ τὸ ῥεῦμα, κατὰ τὸν αὐτὸν λόγον ἕκαστον τῶν εἰρημένων ἐργάσεται παθῶν, καθ' οἷον ἔμπροσθεν ὑπὸ τοῦ θερμοῦ καὶ αἱματώδους ἐδείκνυτο γίγνεσθαι φλεγμονήν· καὶ ὡς τῶν ὁμοιομερῶν τε καὶ ἁπλῶν καὶ πρώτων[43] σωμάτων ἕκαστον, ὑπὸ τοῦ

[42] φλεγμονὴ add. K
[43] καὶ πρώτων add. EGN

NON-UNIFORM DISTEMPERMENT

throughout. The fever in which the rigor precedes and the fever follows, as in tertian and quartan fevers, I do not, however, call an ague. Consequently, the ague fever is combined from two of the non-uniform *dyskrasias*, as are almost all the remaining fevers, apart from those termed hectic.

9. In like manner also, those diseases that are of one particular part with swelling, and all those similar to inflammation are produced in relation to a *dyskrasia* that is non-uniform—cancer, erysipelas, anthrax, herpes, edema, inflammation, *phagedaina*, and gangrene.[24] Common to all these is that they arise from a flux of fluids.[25] On the other hand, they differ in that there are those due to a phlegmatic humor, | those due to a bilious or melancholic humor, those due to blood, either hot, thin and seething, or cold and thick, or somehow otherwise disposed. It will be shown in detail elsewhere that the difference in these pertains to kind.[26] However, in regard to the present discussion, it is enough to say this alone—that of whatever kind the flux might be, it will produce each of the aforementioned affections according to the same reason. For example, it was shown before that inflammation arises due to a hot and bloody [flux] and that each of the *homoiomeres* and simple and primary bodies, when they have

752K

---

[24] On the several diseases listed here, see the Introduction, section 9, Diseases and Symptoms. Inflammation, absent in Garcia Novo, *Galen: On the Anomalous Dyskrasia*, is included following Kühn.   [25] A distinction is made between ὑγρός (fluid) and χυμός (humor).

[26] See, for example, *On the Differentiae of Diseases*, VI.874–76K, and *On the Causes of Diseases*, VII.21–23K.

ῥεύματος τοῦδε διατεθειμένον, εἰς ἀνώμαλον ἀφίξεται καὶ αὐτὸ δυσκρασίαν· ἔξωθεν μέντοι θερμαινόμενον ἢ ψυχόμενον ἢ ξηραινόμενον ἢ ὑγραινόμενον, ὁποῖον ἂν ᾖ καὶ τὸ ῥεῦμα· μέχρι δὲ τοῦ βάθους μήπω διακείμενον ὁμοίως. εἰ δὲ πᾶν ὅλον δι᾽ ὅλου μεταβάλλοι καὶ ἀλλοιωθείη, γενήσεται μὲν εὐθὺς ἀνώδυνον· ἐν χαλεπωτέρᾳ δ᾽ ἂν οὕτω γε διαθέσει κατασταίη. ταῦτ᾽ ἀρκεῖν μοι δοκεῖ προεγνῶσθαι τοῖς μέλλουσι τῇ τε περὶ φαρμάκων ἀκολουθήσειν πραγματείᾳ καὶ μετ᾽ αὐτὴν τῇ τῆς θεραπευτικῆς μεθόδου.

Τέλος Γαληνοῦ περὶ τῆς ἀνωμάλου δυσκρασίας.[44]

[44] *On the addition of this concluding sentence see* EGN, *note 10, p. 172.*

NON-UNIFORM DISTEMPERMENT

been so disposed by this flux, will itself also come to a non-uniform *dyskrasia*. Externally, however, it is heated, cooled, dried, or wetted, but whatever kind of flux there may be, to this point is not yet similarly disposed in the depths. If, on the other hand, the whole changes completely and is altered, it will immediately become painless, although in this way, it might in fact be established in a more difficult condition. It seems to me these things are enough to know beforehand for those who are going to follow the treatise on medications and after this, the one on the therapeutic method.[27]

(The end of Galen's *On the Non-Uniform Dyskrasias*.)

---

[27] *De simplicium medicamentorum temperamentis ac facultatibus*, XI.369–802 and XII.1–377K; *De methodo medendi*, X.1–1021K; English translation by Johnston and Horsley, *Galen's Method of Medicine*.

ΟΤΙ ΤΑΙΣ ΤΟΥ ΣΩΜΑΤΟΣ ΚΡΑΣΕΣΙΝ ΑΙ ΤΗΣ ΨΥΧΗΣ ΔΥΝΑΜΕΙΣ ΕΠΟΝΤΑΙ

THE SOUL'S TRAITS DEPEND ON BODILY TEMPERAMENT

INTRODUCTION

This work is one of a triad of relatively short treatises on the soul (*psyche*) collected together at the end of volume 4 and the beginning of volume 5 of Kühn's *Galeni Opera Omnia*.[1] It is included here with the other two works on *krasis/eukrasia/dyskrasia* because it extends these concepts to the structure and function of the rational soul, which Galen, like Plato, localizes in the brain. The essence of Galen's argument is, as the title makes clear, that the characteristics and capacities of the rational soul are dependent on the *krasis* of the body, and specifically of the brain and meninges, if this is truly the location of the rational soul, as Galen believes. Indeed, he says in the work itself that "it will also be necessary, then, for those who postulate the soul to be a specific substance, to concede that it is itself a slave to the *krasias* of the body." Having stated his main point, in the opening two sections he also articulates three other points deemed important, as follows:

---

[1] The three are: *Quod animi mores corporis temperamenta sequantur*, IV.767–822K; *De propriorum animi cuiuslibet affectuum dignotione et curatione*, V.1–57K; *De animi cuiuslibet peccatorum dignotione et curatione*, V.58–103K.

THE SOUL'S TRAITS

1. From observations on young children it is apparent there are from a very early age individual differences in the capacities of the rational soul.
2. The soul has a substance, which is responsible for its functions.
3. The soul has a number of capacities, the number being equal to the number of its functions, this being in keeping with his concept of capacity (*dunamis*) as applied to other organs.

The next eight sections consist mainly of long quotations from Galen's three most revered authorities—Plato (3–6, 10), Aristotle (7), and Hippocrates (8–9)—interspersed with his own comments and observations. A summary of the views of the three men, accepted or rejected, follows:

Plato: (a) Galen accepts his tripartite division of the soul into rational (in the brain), spirited (in the heart), and appetitive (in the liver); (b) he opposes Plato's claim that the rational soul is incorporeal and immortal, and can survive without a body; (c) he claims Plato's recognition that *kakochymia* and drinking wine can adversely affect the rational soul commits him to the belief that it is affected by bodily states (*krasis*); (d) he considers Plato's view that the wetness of the receiving body at birth can adversely affect intelligence.

Aristotle: Galen claims that the quotations given support two suppositions: (a) the capacities of the soul at birth follow the *krasis* of the maternal blood; (b) the capacities of the soul follow the *krasis* (nature) generally of the body.

Hippocrates: Galen claims that the quotations given show: (a) Hippocrates subscribed to the four elements/

four qualities theory; (b) Hippocrates held that both bodily states and mental dispositions are affected by physical and climatic conditions, mediated through bodily *krasis*.

In the final section, Galen ventures into more philosophical matters, in particular whether the human soul or nature is innately good or innately evil. Indeed, while the Stoics and their contemporaries were grappling with the issue, the two leading Confucians who followed the sage, Meng Ke and Xun Qing, were likewise engaged. As far as I am aware, the matter remains unresolved. Another less pressing issue on which there is some uncertainty is the date of composition of this work of Galen's and, related to this, the concordance of the views expressed with those in his other works on the same subject, most notably his major work, *On the Opinions of Plato and Hippocrates*. On present evidence it seems reasonable to accept a late date of composition for *The Soul's Traits Depend on Bodily Temperament*, probably the final decade of the century.

The Greek text is essentially that of Bazou (B). Notable differences from Müller (*Scripta Minora*, SM) or Kühn (K) are indicated by a footnote. The title literally translated is *That the Capacities of the Soul Follow the Krasias of the Body*. The alternative title is used to avoid the assumption of prior knowledge of two Galenic technical terms.

SYNOPSIS OF SECTIONS

1. A brief statement is given of the basic claim of the work: that the state and functioning of the soul is dependent on the *krasis* of the body as a whole and can be influenced by various factors that influence this, including foods, drinks, and regimen generally.

THE SOUL'S TRAITS

2. Galen makes two initial points: (a) even in small children there are apparent individual differences in the capacities of the soul; (b) the soul has a substance that is responsible for its functions. Next, he enlarges on these capacities and the numerical correspondence between the number of capacities and the number of functions. He lists the capacities of the soul and introduces Plato's subdivision of these, giving a summarized account of his concept of a tripartite soul.

3. Galen accepts Plato's idea of a tripartite soul and the location of the rational soul in the brain, the spirited soul in the heart, and the appetitive soul in the liver. He refers to the basic division into matter and form and elaborates on the Aristotelian position. The capacities of the soul follow its substance—a central statement in Galen's position. He questions Plato's claim that the rational soul is incorporeal and immortal. He cites a number of examples of the rational soul being affected by a demonstrable physical factor, which must support the view that it is corporeal rather than incorporeal. Relatively detailed consideration is given to the good and bad effects of wine on the rational soul, with several quotations from Homer. This, Galen avers, is through the effect of wine on *krasias* and humors. He concludes: "It will also be necessary, then, for those who postulate the soul to be a specific substance, to concede that it is itself a slave to the *krasias* of the body."

4. Galen considers Plato's theory that the wetness of the embryonic body into which the immortal rational soul is received has a deleterious effect, since, generally, wetness is associated with lack of understanding (foolishness) and dryness with quick understanding (sagacity). He briefly considers the views of Andronicus the Peripatetic

and of Aristotle, and then more fully the Stoic position, which includes the role of *pneuma*.

5. Galen states his adherence to the four elements/four qualities (continuum) theory of matter as opposed to the particle/void theory (atomism, corpuscularism). He affirms his position that the soul is a corporeal substance, as is the rest of the body, and has a *krasis* subject to the same changes like any other part.

6. This is a short section largely consisting of quotations from Plato's *Timaeus* to show that he recognized the adverse effects that *kakochymia* can have on the functioning of the soul.

7. A long section comprising mainly quotations from two of Aristotle's works, *Parts of Animals* and *History of Animals*, interspersed with comments by Galen. The Aristotelian passages are taken as supporting two suppositions: (a) the capacities of the soul at birth follow the *krasis* of the maternal blood; (b) The capacities of the soul follow the *krasis* (nature) generally of the body. The pseudo-Aristotelian works *Physiognomica* and *Problems* are also referenced.

8. This section comprises a series of quotations from the Hippocratic works, *Airs, Waters, Places* and *Epidemics II* to support the view that Hippocrates himself adhered to the four elements/four qualities concept of basic structure (Galen also makes mention of *On the Nature of Man* in this regard), and both bodily state and mental disposition (i.e., the functions of the rational soul) are influenced, indeed determined, by the physical and climatic conditions through their effects on the *krasis* of the body.

9. Galen praises Hippocrates for his views but adds that the effects of place and climate on the soul are obvious to

everyone. He then returns to Plato, or rather his later followers, quoting from the *Timaeus* and *Laws* to show he too accepted that external factors, including climate, soil, and nutriments, dependent on the soil, affected all three components of the soul.

10. This section centers on three quotations from Plato, two from *Laws* on the drinking of wine, which clearly show that Plato believed wine could influence, for better or worse, the functioning of the soul, the effects depending in part on the stage of life and the underlying *krasis*. In the quotation from *Timaeus*, the focus is on pursuits, studies, and nurture, the last involving not only nutriments but also regimen as a whole. Galen refers to his own treatises, and particularly that on *euchymia* and *kakochymia* as determined by foods.

11. Should people be held responsible for their good and bad qualities given that these are dependent on their *krasias*? Regardless of the answer, people respond to others as if they were responsible for their nature. Galen dismisses the Stoic view that all are possessed of goodness but are perverted by those with whom they associate. On the age-old question of nature or nurture, Galen's view, as stated in this final section, is this: bad habits arise naturally in the nonrational parts of the soul, and false opinions in the rational part. By proper nurture, good habits and true opinions may develop. Sagacity and stupidity in the rational soul, however, largely follow the *krasias*, while these in turn follow the first genesis and *euchymous* regimens. *Dyskrasias*, arising for whatever reason, may have various adverse effects according to the predominant quality or qualities.

ΟΤΙ ΤΑΙΣ ΤΟΥ ΣΩΜΑΤΟΣ ΚΡΑΣΕΣΙΝ ΑΙ ΤΗΣ ΨΥΧΗΣ ΔΥΝΑΜΕΙΣ ΕΠΟΝΤΑΙ

1. Ταῖς τοῦ σώματος κράσεσιν ἔπεσθαι τὰς δυνάμεις τῆς ψυχῆς, οὐχ ἅπαξ ἢ δὶς ἀλλὰ πάνυ πολλάκις, οὐδ' ἐπ' ἐμαυτοῦ μόνου βασανίσας τε καὶ πολυειδῶς ἐρευνήσας ἀλλ' ἀπ' ἀρχῆς μὲν ἅμα τοῖς διδασκάλοις, ὕστερον δὲ σὺν τοῖς ἀρίστοις φιλοσόφοις ἀληθῆ τε διὰ παντὸς εὗρον τὸν λόγον ὠφέλιμόν τε τοῖς κοσμῆσαι τὰς ἑαυτῶν ἐθέλουσι ψυχάς, ἐπειδήπερ [γάρ], ὡς διῆλθον ἐν τῇ Περὶ τῶν ἐθῶν πραγματείᾳ, [καὶ] διὰ τῶν ἐδεσμάτων τε καὶ πομάτων ἔτι τε τῶν ὁσημέραι πραττομένων εὐκρασίαν[1] ἐργαζόμεθα τῷ σώματι κἀκ ταύτης εἰς ἀρετὴν τῇ ψυχῇ συντελέσομεν, ὡς οἱ περὶ Πυθαγόραν τε καὶ Πλάτωνα καί τινες ἄλλοι τῶν παλαιῶν ἱστοροῦνται πράξαντες.

2. Ἀρχὴ δὲ παντὸς τοῦ μέλλοντος εἰρήσεσθαι λό-

---

[1] post εὐκρασίαν: ἐργαζόμενοι τῷ σώματι, K; ἐργαζόμεθα SM; ἐργαζόμεθα τῷ σώματι, B

THE SOUL'S TRAITS DEPEND ON BODILY TEMPERAMENT

1. That the capacities of the soul follow the *krasias* of the body, I have not only put to the test myself and inquired into in many ways, and not just once or twice, but very often, from the beginning, along with the teachers and later with the best philosophers, I have found the argument true throughout and beneficial to those who wish to bring order to their own souls. After that, as I went over in the | work *On Customs*,[1] it is through both foods and drinks, and besides these the daily activities, we create *eukrasia* in the body and from this we shall bring completion to excellence of the soul, as the followers of Pythagoras and Plato and certain others of the ancients are recorded as doing.[2]

2. The beginning of the whole argument that is going

---

[1] On this work, not included in Kühn, see Müller, Preface xxxv–xxxvi, and also J. H. Mattock, "A Translation of the Arabic Epitome of Galen's Book *Peri ethon*," in S. M. Stern et al., *Islamic Philosophy and the Islamic Tradition* (Oxford, 1972), 236–60.

[2] The reference is presumably to dietary and other practices associated with the two schools. For another mention of the two men together, see also Galen's *De propriorum animi cuiuslibet affectuum dignotione et curatione*, V.37–38K.

γου γνῶσις τῆς διαφορᾶς τῶν ἐν τοῖς μικροῖς παιδίοις φαινομένων ἔργων τε καὶ παθῶν τῆς ψυχῆς, ἐξ ὧν αἱ δυνάμεις αὐτῆς κατάδηλοι γίγνονται. τινὰ μὲν γὰρ αὐτῶν φαίνονται δειλότατα, τινὰ δὲ καταπληκτικώτατα καὶ τινὰ μὲν ἄπληστα καὶ λίχνα,[2] τινὰ δ' ἐναντίως διακείμενα, καὶ τινὰ μὲν ἀναίσχυντα, τινὰ δ' αἰσχυντηρά . . . καὶ πολλὰς ἑτέρας ἔχοντα τοιαύτας διαφοράς, ἃς ἁπάσας ἐν ἑτέροις διῆλθον. ἐνταῦθα δ' ἀρκεῖ παραδείγματος ἕνεκα τῶν τριῶν αὐτῆς εἰδῶν τε καὶ μερῶν ἐνδεδεῖχθαι τὰς δυνάμεις ἐναντίας ὑπαρχούσας φύσει τοῖς βρέφεσιν. ἐκ τούτου γὰρ ἐνέσται συλλογίζεσθαι μὴ τὴν αὐτὴν ἅπασιν εἶναι φύσιν τῆς

769K ψυχῆς | [εὔδηλον δ' ὅτι τὸ τῆς φύσεως ὄνομα κατὰ τοὺς τοιούτους λόγους ταὐτὸν σημαίνει τῷ τῆς οὐσίας]· εἴπερ γὰρ ἦν ἀπαράλλακτος αὐτῶν ἡ οὐσία τῆς ψυχῆς, ἐνήργουν τ' ἂν τὰς αὐτὰς ἐνεργείας ἔπασχόν τ' ἂν ἀπὸ τῶν αὐτῶν αἰτίων ταὐτὰ πάθη.

Δῆλον οὖν, ὅτι διαφέρουσιν ἀλλήλων οἱ παῖδες εἰς τοσοῦτον ταῖς τῶν ψυχῶν οὐσίαις, εἰς ὅσον καὶ ταῖς ἐνεργείαις τε καὶ τοῖς παθήμασιν αὐτῶν· εἰ δὲ τοῦτο, καὶ ταῖς δυνάμεσι. συγκεχυμένοι δ' εἰσὶν εὐθὺς ἐν τούτῳ πολλοὶ τῶν φιλοσόφων[3] ἀδιάρθρωτον ἔννοιαν ἔχοντες τῆς δυνάμεως· ὡς γὰρ ἐνοικοῦντός τινος πράγματος ταῖς οὐσίαις, ὡς ἡμεῖς ταῖς οἰκίαις, οὕτω

---

[2] See SM, praef. p. xxxvi on the opening clauses of this sentence.

[3] φιλοσόφων SM; σαφῶν K

THE SOUL'S TRAITS

to be stated is a knowledge of the difference of the apparent actions and affections of the soul in small children; from these the capacities of the soul become manifest. For some of them appear very timid, some very quick of apprehension, and some insatiable and gluttonous, while others are oppositely disposed. And some are shameless whereas others have a sense of shame, and they have many other such differences, all of which I went over in other places.[3] Here it suffices for the sake of an example to have pointed out the capacities of the three kinds and parts of the soul which are opposite in nature in infants. From this it will be possible to infer that the nature of the soul is not the same in all. | It is quite clear, however, that the term "the nature" signifies the same as that of "the substance" in such arguments, for if the substance of the soul in them was indistinguishable,[4] they would customarily carry out the same functions and would suffer the same affections from the same causes.

It is clear, therefore, that children differ from each other to such a degree in the substances of their souls as they differ also in the actions and the affections of these, and if this is so, also in the capacities. If this is so, many of the philosophers are openly confused, having an incorrect concept of "the capacity." For just as if some matter were dwelling in the substances, as we dwell in

---

[3] See Galen, *De propriorum animi cuiuslibet affectuum dignotione et curatione*, V.1–57K.

[4] On the term ἀπαράλλακτος meaning the "same" or "indistinguishable," see H. Von Arnim, ed., *Stoicorum Veterum Fragmenta* (Leipzig, 1903), 2.26.

GALEN

μοι δοκοῦσι περὶ τῶν δυνάμεων φαντάζεσθαι μὴ γιγνώσκοντες, ὅτι τῶν γιγνομένων ἑκάστου ποιητική ⟨τίς⟩ ἐστιν αἰτία νοουμένη κατὰ τὸ πρός τι καὶ ταύτης τῆς αἰτίας ὡς μὲν πράγματος τοιοῦδέ τινος ἰδίᾳ καὶ καθ' ἑαυτὸ κατηγορία τίς ἐστιν, ἐν δὲ τῇ πρὸς τὸ γιγνόμενον ἀφ' ἑαυτῆς σχέσει δύναμίς ἐστι τοῦ γιγνομένου καὶ διὰ τοῦτο τοσαύτας δυνάμεις ἔχειν τὴν οὐσίαν φαμέν, ὅσας ἐνεργείας, οἷον τὴν ἀλόην καθ-
770K αρτικήν τε δύναμιν ἔχειν | καὶ τονωτικὴν στομάχου καὶ τραυμάτων ἐναίμων κολλητικὴν ⟨καὶ⟩ ἰσοπέδων ἑλκῶν ἐπουλωτικὴν ⟨καὶ⟩ ὑγρότητος βλεφάρων ξηραντικήν, οὐ δήπου τῶν εἰρημένων ἔργων ἕκαστον ἄλλου τινὸς ποιοῦντος παρ' αὐτὴν τὴν ἀλόην. αὕτη γάρ ἐστιν ἡ ταῦτα δρῶσα καὶ διὰ τὸ δύνασθαι ποιεῖν αὐτὰ τοσαύτας ἐλέχθη δυνάμεις ἔχειν, ὅσα τὰ ἔργα. λέγομεν οὖν τὴν ἀλόην καθαίρειν δύνασθαι καὶ ῥωννύναι τὸν στόμαχον καὶ κολλᾶν τραύματα καὶ ἕλκη συνουλοῦν καὶ ὀφθαλμοὺς ὑγροὺς ξηραίνειν, ὡς οὐδὲν διαφέρον ἢ καθαίρειν δύνασθαι φάναι τὴν ἀλόην ἢ δύναμιν ἔχειν καθαρτικήν· [οὕτω δὲ καὶ τὸ ξηραίνειν ὑγροὺς ὀφθαλμοὺς δύνασθαι ταὐτὸν σημαίνει τῷ δύναμιν ἔχειν ὀφθαλμῶν ξηραντικήν].[4]

---

[4] *On the section in brackets, see* SM, *praef. p.* xxxvii.

[5] A distinction is made between a temporary state (*schesis*) and a permanent state (*hexis*).

[6] This is a complex sentence. For a clearer account of capacities/faculties (δυνάμεις), see the General Introduction, section 7

THE SOUL'S TRAITS

houses, they seem to me to have a mental picture of the capacities, not realizing that for each of the things that come about, there is some effecting cause conceived of in relation to something, and of this cause, as of any such matter, there is a specific name pertaining to this itself, while in the state (*schesis*),[5] in relation to what comes about from itself, there is a capacity of what comes about, and because of this, we say the substance has as many capacities as it has functions.[6] For example, we say aloe has a cathartic capacity, | a strengthening capacity of the *stomachus*,[7] a coagulating capacity for wounds that are bleeding, a healing capacity for superficial ulcers, and a drying capacity for wetness of the eyelids, there clearly not being anything else that performs each of these actions mentioned apart from the aloe itself. For it is the aloe itself that has done these things, and because of its being able to do them, it is said to have as many capacities as there are actions. Therefore, we say the aloe is able to purify and strengthen the *stomachus*, conglutinate wounds, cicatrize ulcers and dry wet eyelids, and that there is no difference in saying either the aloe is able to purify or the aloe has a capacity that is purifying, and in this way also the ability to dry wet eyes signifies the same as having a capacity that is drying for eyes.

770K

on terminology, and the opening sections of Galen's *De naturalibus facultatibus*, particularly section IV, and A. J. Brock's introduction (LCL 71, 1963 [1916]).

[7] The term *stomachus* refers variably to the esophagus, the cardiac orifice of the stomach, and the stomach itself, among other meanings. For "stomachical affections," see Galen, *De alimentis facultatibus*, VI.577K, and Soranus 2.48.

Κατὰ δὲ τὸν αὐτὸν τρόπον, ὅταν εἴπωμεν· "ἡ ἐν ἐγκεφάλῳ καθιδρυμένη λογιστικὴ ψυχὴ δύναται μὲν αἰσθάνεσθαι διὰ τῶν αἰσθητηρίων, δύναται δὲ καὶ μεμνῆσθαι [διὰ]⁵ τῶν αἰσθητῶν αὐτὴ καθ᾿ ἑαυτὴν ἀκολουθίαν τε καὶ μάχην ἐν τοῖς πράγμασιν ὁρᾶν ἀνάλυσίν τε καὶ σύνθεσιν," οὐκ ἄλλο τι δηλοῦμεν ἢ εἰ περιλαβόντες εἴποιμεν· "ἡ λογιστικὴ ψυχὴ δυνάμεις ἔχει πλείους, αἴσθησιν καὶ μνήμην καὶ σύνεσιν ἑκάστην ⟨τε⟩ τῶν ἄλλων." ἐπεὶ δ᾿ οὐ μόνον αἰσθάνεσθαι δύνασθαί φαμεν αὐτὴν ἀλλὰ καὶ κατ᾿ εἶδος ὁρᾶν ἀκούειν ὀσμᾶσθαι γεύεσθαι ἅπτεσθαι, πάλιν αὖ δυνάμεις αὐτὴν ἔχειν λέγομεν ὀπτικήν, ἀκουστικήν, ὀσφρητικήν, γευστικήν, ἁπτικήν. οὕτω δὲ καὶ τὴν ἐπιθυμητικὴν αὐτῇ δύναμιν ὁ Πλάτων ὑπάρχειν ἔλεγεν, ἥν γε δὴ κοινῶς ἐπιθυμητικήν, οὐκ ἰδίως ὀνομάζειν ἔθος αὐτῷ. πλείους μὲν γὰρ εἶναι ταύτης τῆς ψυχῆς ἐπιθυμίας φησί, πλείους δὲ καὶ τῆς θυμοειδοῦς, πολὺ δὲ πλείους καὶ ποικιλωτέρας τῆς τρίτης, ἣν δι᾿ αὐτὸ τοῦτο κατ᾿ ἐξοχὴν ὠνόμασεν ἐπιθυμητικὴν εἰωθότων οὕτως τῶν ἀνθρώπων ἐνίοτε τὰ πρωτεύοντα τῶν ἐν τῷ γένει τῷ τοῦ γένους ὅλου προσαγορεύειν ὀνόματι, καθάπερ ὅταν εἴπωσιν ὑπὸ μὲν τοῦ ποιητοῦ λελέχθαι τόδε τὸ ἔπος, ὑπὸ δὲ τῆς ποιητρίας τόδε· πάντες γὰρ ἀκούομεν Ὅμηρον ⟨μὲν⟩ λέγεσθαι ποιητήν, Σαπφὼ δὲ ποιήτριαν [οὕτω δὲ καὶ θῆρα λέγουσιν ἐξαιρέτως τὸν λέοντα καὶ ἄλλα τοιαῦτα κατ᾿ ἐξοχὴν ὀνομάζουσιν].⁶

THE SOUL'S TRAITS

In the same manner, when we say, "the rational soul, situated in the brain, is able to sense through the organs of sense, is able to remember the things sensed through itself, is able to see coherence and contradiction in matters and to analyze and synthesize," we are not indicating anything else, if we say inclusively, "the rational soul | has a number of capacities—perceiving, memorizing, comprehending and each of the others." However, since we say it is not only able to perceive but also according to kind to see, hear, smell, taste, and touch, in turn again we say it has capacities—visual, auditory, olfactory, gustatory, and tactile. In this way too, Plato said there is the desiring capacity, which in fact is desiring generally, it being customary for him not to name specifically.[8] For he says there are more desires of this soul, while there are also more of the spirited (passionate), and much more and more varied of the third, which on this account he termed desiderative (appetitive) above all, since people are accustomed sometimes to call in this way what is preeminent in the class by the name of the class as a whole, as when they say this particular verse has been said by the poet or this by the poetess, for we all understand Homer when poet is said and Sappho when poetess is said. In this way too, they say beast in respect of the lion which is exceptional and they name other such things in relation to prominence. |

771K

---

[8] Plato, *Republic* 580D.

---

[5] SM *has in part*: Alterum διὰ *quod etiam* Nicolaus Rheg. *legit vertens 'memorari per obiecta', ex priore* διὰ *tractum uncis inclusimus. See* SM, *praef. p. xxxviii.*

[6] *On the section in brackets, see* SM, *praef. p. xxxvii.*

GALEN

Ἐπιθυμητικὸν οὖν ἐστι κατὰ τὸ κοινὸν τῆς ἐπιθυμίας σημαινόμενον ἀληθείας μὲν καὶ ἐπιστήμης καὶ μαθημάτων καὶ συνέσεως καὶ μνήμης καὶ συλλήβδην εἰπεῖν ἁπάντων τῶν καλῶν ἐκεῖνο τὸ μέρος τῆς ψυχῆς, ὃ καλεῖν εἰθίσμεθα λογιστικόν· ἐλευθερίας δὲ καὶ νίκης καὶ τοῦ κρατεῖν καὶ ἄρχειν καὶ τοῦ δοξάζεσθαι καὶ τοῦ τιμᾶσθαι τὸ θυμοειδές· ἀφροδισίων δὲ καὶ τῆς ἐξ ἑκάστου τῶν ἐσθιομένων τε καὶ πινομένων ἀπολαύσεως τὸ κατ᾽ ἐξοχὴν ὀνομαζόμενον ὑπὸ Πλάτωνος ἐπιθυμητικόν, οὔτε τῆς ἐπιθυμητικῆς[7] ψυχῆς ὄρεξιν τοῦ καλοῦ ἔχειν δυναμένης οὔτε τῆς λογιστικῆς ἀφροδισίων ἢ βρωμάτων ἢ πομάτων ὥσπερ οὐδὲ νίκης[8] ἢ ἀρχῆς ἢ δόξης ἢ τιμῆς, κατὰ δὲ τὸν αὐτὸν λόγον οὐδὲ τῆς θυμοειδοῦς ὧν ὀρέξεις εἶχεν ἥ τε λογιστικὴ καὶ ἡ ἐπιθυμητική.

3. Ὅτι μὲν οὖν τρία τῆς ψυχῆς ἐστιν εἴδη καὶ ὅτι ὁ Πλάτων βούλεται ταῦτα, δι᾽ ἑτέρων ἐπιδέδεικται, καθάπερ γε καὶ ὅτι τὸ μὲν ἐν ἥπατι, τὸ δ᾽ ἐν καρδίᾳ, τὸ δ᾽ ἐν ἐγκεφάλῳ καθίδρυται· ὅτι δ᾽ ἐκ τούτων τῶν εἰδῶν τε καὶ μερῶν τῆς ὅλης ψυχῆς τὸ λογιστικὸν ἀθάνατόν ἐστι, Πλάτων μὲν φαίνεται πεπεισμένος, ἐγὼ δ᾽ οὔθ᾽ ὡς ἔστιν οὔθ᾽ ὡς οὐκ ἔστιν ἔχω διατείνεσθαι πρὸς αὐτόν. πρῶτον μὲν οὖν ἐπισκεψώμεθα περὶ τῶν ἐν καρδίᾳ καὶ ἥπατι τῆς ψυχῆς εἰδῶν, ἃ κἀκείνῳ κἀμοὶ συνωμολόγηται φθείρεσθαι κατὰ τὸν θάνατον. ἔχοντος δ᾽ ἰδίαν οὐσίαν ἑκατέρου τῶν σπλάγχνων, ἥτις μέν ἐστιν ἀκριβῶς αὕτη, μηδέπω ζητῶμεν, ἀνα-

THE SOUL'S TRAITS

Therefore, desiring in the general sense is that part of the soul which signifies the desire for truth, scientific knowledge, learning, understanding, and memory, and in a word, for of all the good things we are accustomed to call rational, while the spirited, on the other hand, is desirous of freedom, conquest, overcoming, ruling, of being held in high regard, and of being honored, and while that of desiring sexual activity and enjoyment of each the things eaten and drunk was named by Plato desiderative (appetitive) above all, for of the appetitive soul there cannot be desire of the good, nor of the rational soul of sexual activity, drink and food, just as it is not able to have desires for victory, power, reputation, and honor. And by the same argument, the spirited does not have the desires which either the rational or the desiderative (appetitive) souls have.

772K

3. Therefore, that there are three kinds of the soul, and that Plato favors these, has been shown through other writings,[9] just as, in fact, that he favors the view they are situated in liver, heart, and brain. That of these kinds and | parts of the whole soul, the rational is immortal, Plato has obviously been persuaded. I, however, am unable to oppose him on whether this is so or not so. First, then, let us consider the kinds of soul in heart and liver, which it is agreed by Plato and myself are destroyed at the time of death. Each of the internal organs has a characteristic substance; what exactly this is, let us not yet inquire into.

773K

---

[9] See particularly Galen's *De placitis Hippocratis et Platonis*, V.181–805K; English translation by de Lacy *Galen on the Doctrines*.

---

[7] ἐπιθυμητικῆς SM; αὐτῆς K   [8] νίκης SM; εἰκὸς K

μνησθῶμεν δὲ περὶ τῆς κοινῆς οὐσίας⁹ ἁπάντων σωμάτων συστάσεως, ὡς ἐκ δυοῖν ἀρχῶν ἡμῖν ἐδείχθη σύνθετος ὑπάρχειν, ὕλης τε καὶ εἴδους,¹⁰ ὕλης ⟨μὲν⟩ ἀποίου κατ' ἐπίνοιαν, ἐχούσης δ' ἐν ἑαυτῇ ποιοτήτων τεττάρων κρᾶσιν, θερμότητος, ψυχρότητος, ξηρότητος, ὑγρότητος. ἐκ τούτων καὶ χαλκὸς καὶ σίδηρος καὶ χρυσὸς ἥ τε σὰρξ νεῦρόν τε καὶ χόνδρος καὶ πιμελὴ καὶ πάνθ' ἁπλῶς τὰ πρωτόγονα μὲν ὑπὸ Πλάτωνος, ὁμοιομερῆ δ' ὑπ' Ἀριστοτέλους ὀνομαζόμενα γέγονεν.

Ὥσθ' ὅταν αὐτὸς οὗτος Ἀριστοτέλης εἶδος εἶναι τοῦ σώματος εἴπῃ τὴν ψυχήν, ἐρωτητέον αὐτὸν ἢ τούς γ' ἀπ' αὐτοῦ, πότερον τὴν μορφὴν εἶδος εἰρῆσθαι πρὸς | αὐτοῦ νοήσωμεν, ὥσπερ ἐν τοῖς ὀργανικοῖς σώμασιν, ἢ τὴν ἑτέραν ἀρχὴν τῶν φυσικῶν σωμάτων σῶμα δημιουργοῦσαν, ὅπερ ὁμοιομερές τ' ἐστὶ καὶ ἁπλοῦν, ὡς πρὸς αἴσθησιν οὐκ ἔχον ὀργανικὴν σύνθεσιν. ἀποκρινοῦνται γὰρ ἐξ ἀνάγκης τὴν ἑτέραν ἀρχὴν τῶν φυσικῶν σωμάτων, εἴ γε δὴ τούτων εἰσὶ πρώτως ἐνέργειαι· δέδεικται γὰρ τοῦθ' ἡμῖν ἑτέρωθι καὶ νῦν, ἂν δεήσῃ, πάλιν εἰρήσεται. καὶ μὴν εἴπερ ἐξ ὕλης τε καὶ εἴδους ἅπαντα συνέστηκε τὰ τοιαῦτα σώματα, δοκεῖ δ' αὐτῷ τῷ Ἀριστοτέλει τῶν τεττάρων ποιοτήτων ἐγγιγνομένων τῇ ὕλῃ τὸ φυσικὸν γίγνεσθαι σῶμα, τὴν ἐκ τούτων κρᾶσιν ἀναγκαῖον αὐτοῦ τίθεσθαι τὸ εἶδος, ὥστε πως καὶ ἡ τῆς ψυχῆς οὐσία

⁹ οὐσίας add. SM
¹⁰ ὕλης τε καὶ εἴδους, add. SM (see note, p. 36)

THE SOUL'S TRAITS

Rather, let us call to mind the common substance of all bodies as was shown by us to be compounded from two principles—matter and form. Matter is without quality conceptually but has in it a *krasis* (mixture) of four qualities, which are hotness, coldness, dryness, and wetness. From these also, copper, iron, and gold or flesh, sinews (nerves), cartilage and fat, and in short all the things termed *prōtogona* by Plato[10] and *homoiomeres* by Aristotle have come into being.

Consequently, when this same Aristotle says the soul is form of the body, one must ask him or those who follow on from him, whether we should think form is said by him in respect of the shape,[11] as in the organic bodies, or in respect of the other principle of the physical bodies which fabricates a body which is *homoiomerous* and simple as regards perception, not having an organic composition. They must of necessity answer that it is the other principle of physical bodies, if in fact functions are primarily of these, for this has been shown by us elsewhere, and will be stated again now should you require it to be. And further, if all such bodies are compounded from matter and form, it seems to Aristotle himself that if the physical body arises from the material of the four qualities inherent in the material, the form of this (body) necessarily inheres in the *krasis* of these (qualities). Consequently, in some way the substance of the soul will also be a *krasis*

774K

---

[10] On πρωτόγονος as a Platonic term meaning "firstborn" or "first-created," LSJ refers to this passage.   [11] Aristotle defines matter (ὕλη) as τὸ ὑποκείμενον γενέσεως καὶ φθορᾶς δεκτικόν (*On Coming-to-be and Passing Away*, 320a2), while form σχῆμά is defined and exemplified at *Categories* 10a11.

κρᾶσίς τις ἔσται τῶν τεττάρων εἴτε ποιοτήτων ἐθέλεις λέγειν, θερμότητός τε καὶ ψυχρότητος, ξηρότητός τε καὶ ὑγρότητος, εἴτε σωμάτων, θερμοῦ καὶ ψυχροῦ, ξηροῦ τε καὶ ὑγροῦ.

[Τῇ δὲ τῆς ψυχῆς οὐσίᾳ τὰς δυνάμεις αὐτῆς δεικνύναι ἑπομένας, εἴ γε καὶ τὰς ἐνεργείας].[11] εἰ μὲν οὖν τὸ λογιζόμενον εἶδος τῆς ψυχῆς ἐστι, θνητὸν ἔσται· καὶ γὰρ καὶ αὐτὸ κρᾶσίς τις ἐγκεφάλου καὶ πάνθ' οὕτως τὰ τῆς ψυχῆς εἴδη τε καὶ μέρη τὰς δυνάμεις ἑπομένας ἕξει τῇ κράσει·[12] τουτέστιν αὐτὴ οὖν ἡ τῆς ψυχῆς οὐσία· εἰ δ' ἀθάνατον ἔσται, ὡς ὁ Πλάτων βούλεται, διὰ τί χωρίζεται ψυχθέντος σφοδρῶς ἢ ὑπερθερμανθέντος ἢ ὑπερξηρανθέντος ἢ ὑπερυγρανθέντος τοῦ ἐγκεφάλου, καλῶς ἂν ἐπεποιήκει γράψας αὐτὸς ὥσπερ καὶ τἆλλα τὰ κατ' αὐτὴν[13] ἔγραψε. γίγνεται γὰρ ὁ θάνατος κατὰ Πλάτωνα χωριζομένης τῆς ψυχῆς ἀπὸ τοῦ σώματος. διὰ τί δ' αὐτὴν αἵματος κένωσις χωρίζει πολλὴ καὶ κώνειον ποθὲν καὶ πυρετὸς διακαής, εἰ μὲν ὁ Πλάτων αὐτὸς ἔζη, παρ' ἐκείνου πάντως ἂν ἠξίωσα μαθεῖν. ἐπεὶ δ' οὔτ' ἐκεῖνος ἔστιν ἔτι καὶ τῶν Πλατωνικῶν διδασκάλων οὐδεὶς οὐδεμίαν αἰτίαν ἐδίδαξέ με, δι' ἣν ὑφ' ὧν εἶπον ἡ ψυχὴ τοῦ σώματος ἀναγκάζεται χωρίζεσθαι, τολμῶ λέγειν αὐτός, ὡς οὐ πᾶν εἶδος σώματος ἐπιτήδειόν ἐστιν ὑποδέξασθαι τὴν λογιστικὴν ψυχήν. ἀκόλουθον γὰρ ὁρῶ τοῦτο τῷ περὶ ψυχῆς δόγματι τοῦ Πλάτωνος, ἀπόδειξιν δ' οὐδεμίαν ἔχω λέγειν αὐτοῦ διὰ τὸ μὴ γιγνώσκειν με τὴν οὐσίαν τῆς ψυχῆς ὁποία τίς ἐστιν,

THE SOUL'S TRAITS

of these four—if you wish to speak either of qualities, hotness, coldness, dryness and wetness, or of bodies, hot, cold, dry and wet.

And it is shown that if the capacities of the soul itself follow its substance, then in fact the functions do too. If, then, there is the rational form of the soul, it will be mortal, for surely also, the *krasis* in a particular brain, and in like manner all the forms and parts will have the capacities following the *krasis,* and such will be the substance of the soul. If, however, the rational soul is immortal, as Plato would have it, why would it be separated when the brain is cooled strongly, or overheated, or excessively dried or moistened, he would have done well to write himself, just as he wrote the other things in relation to this. For death occurs, according to Plato, when the soul is separated from the body. And why a large evacuation of blood would separate it, or drinking hemlock, or a very high fever, I would altogether expect to learn from Plato, if that man were alive [today]. However, since he is no more, and none of the Platonist teachers taught me any cause due to which the soul is compelled to separate from the body by those things I mentioned, I am myself emboldened to say that not every form of body is suitable to receive the rational soul. For I see this as a consequence of Plato's doctrine about the soul, but in fact I am unable to articulate any demonstration of this because I don't know of what kind

---

[11] *On the text in brackets, see* SM, *praef. p. xxxviii.*

[12] *post* τῇ κράσει: τουτέστιν αὐτὴ οὖν SM; καὶ ἔσται τοιαύτη K

[13] κατὰ ταὐτὸν SM; κατ' αὐτὴν B, K

ἐκ τοῦ γένους τῶν ἀσωμάτων ὑποθεμένων ἡμῶν ὑπάρχειν αὐτήν.

Ἐν μὲν γὰρ σώμασί γε τὰς κράσεις ὁρῶ πάμπολύ τε διαφερούσας ἀλλήλων καὶ παμπόλλας οὔσας· ἀσωμάτου δ' οὐσίας αὐτῆς καθ' ἑαυτὴν εἶναι δυναμένης, οὐκ οὔσης δὲ ποιότητος ἢ εἴδους σώματος, οὐδεμίαν ἐπινοῶ διαφοράν, καίτοι πολλάκις ἐπισκεψάμενός τε καὶ ζητήσας ἐπιμελῶς, ἀλλ' οὐδὲ πῶς οὐδὲν οὖσα τοῦ σώματος εἰς ὅλον αὐτὸ δύναιτ' ἂν ἐκτείνεσθαι. τούτων μὲν οὐδὲν οὐδ' ἄχρι φαντασίας ἐννοῆσαι δεδύνημαι καίτοι προθυμηθεὶς χρόνῳ παμπόλλῳ· γιγνώσκω δ' ἐκεῖνο σαφῶς ⟨καὶ⟩ ἐναργῶς φαινόμενον, ὡς ἡ μὲν τοῦ αἵματος κένωσις καὶ ἡ τοῦ κωνείου πόσις καταψύχουσι τὸ σῶμα, πυρετὸς δὲ σφοδρὸς ὑπερθερμαίνει. καὶ πάλιν ἐρῶ ταὐτόν· διὰ τί ψυχόμενον σφοδρῶς[14] ἢ ὑπερθερμαινόμενον τὸ σῶμα καταλείπει τελέως ἡ ψυχή; πολλὰ ζητήσας οὐχ εὗρον ὥσπερ γ' οὐδὲ διὰ τί χολῆς μὲν ξανθῆς | ἐν ἐγκεφάλῳ πλεοναζούσης εἰς παραφροσύνην ἑλκόμεθα, διὰ τί δὲ τῆς μελαίνης εἰς μελαγχολίαν, διὰ τί δὲ τὸ φλέγμα καὶ ὅλως τὰ ψυκτικὰ παραίτια ληθάργων, ἐξ ὧν καὶ μνήμης καὶ συνέσεως βλάβαις ἁλισκόμεθα, καὶ μέντοι καὶ διὰ[15] τί μωρίαν αὐτῇ ἐργάζεται κώνειον ποθέν, ᾧ καὶ τοὔνομα ἔνθεν παρώνυμον ἀπὸ τοῦ πάθους, ὃ πάσχον ὁρῶμεν ὑπ' αὐτοῦ τὸ σῶμα.

[14] On the addition of σφοδρῶς, see SM, praef. p. xxxix.
[15] post καὶ διὰ: τί μωρίαν αὐτῇ B; τί μωρίαν [αὐτὴν] SM; τίμωρίαν αὐτὴν K

THE SOUL'S TRAITS

the substance of the soul is, since I am assuming it is from the class of incorporeal things.

What I do, in fact, see in bodies is that the *krasias* differ very much from each other and they are very numerous. But since the actual substance of the incorporeal is able to exist in itself not being a quality or form of a body, I observe no difference, although I have considered this matter often and looked into it carefully, but I do not observe how what is not of a body would itself be able to extend through the whole of the body. I have not, up until now, been able to form an impression of any of these things, although I have been ardently desirous [of doing so] over a very long time. I do, however, know that it is clearly apparent that the evacuation of blood and a drink of hemlock cool the body, while a strong fever excessively heats it. And again I shall say this: why does the soul completely leave the body which is severely cooled or overheated? I have not found out after much investigation, just as I have not found out why, when yellow bile abounds in the brain, we are drawn to delirium (*paraphrosune*), or why, when black bile abounds, to melancholia, or why phlegm and cooling agents in general are partial causes of the lethargies, from which also we are seized by harms to memory and comprehension. And further also, why a drink of hemlock creates folly in it, by which the name too is derived from the affection which we see the body affected by.[12]

---

[12] There is some uncertainty about this sentence. Singer, *Galen: Psychological Writings*, favors the Arabic version rather than the Greek; see his note 4.17, p. 414, which is persuasive, the essence being that κώνειον is wrong.

GALEN

Λύπης δ' ἀπάσης καὶ δυσθυμίας κουφίζει σαφῶς οἶνος πινόμενος, ἑκάστης γὰρ ἡμέρας[16] τούτου πειρώμεθα· καὶ Ζήνων, ὥς φασιν, ἔλεγεν, ὅτι καθάπερ οἱ πικροὶ θέρμοι βρεχόμενοι τῷ ὕδατι γίγνονται γλυκεῖς, οὕτω καὶ αὐτὸς ὑπ' οἴνου διατίθεσθαι. φασὶ δὲ καὶ τὴν οἰνοπίαν ῥίζαν ἔτι καὶ μᾶλλον ἐργάζεσθαι τοῦτο καὶ ταύτην εἶναι τὸ τῆς Αἰγυπτίας ξένης φάρμακον, ὅ φησιν ὁ ποιητής·

αὐτίκ' ἄρ' εἰς οἶνον βάλε φάρμακον, ἔνθεν ἔπινον,
νηπενθές τ' ἄχολόν τε κακῶν ἐπίληθες ἁπάντων.

ἡ μὲν οὖν οἰνοπία ῥίζα χαιρέτω· δεόμεθα γὰρ αὐτῆς οὐδὲν | εἰς τὸν λόγον ὁρῶντες ὁσημέραι τὸν οἶνον ἐργαζόμενον ὅσαπερ οἱ ποιηταὶ λέγουσιν·

οἶνός σε τείρει μελιηδής, ὅστε καὶ ἄλλους
βλάπτει, ὃς ἄν μιν χανδὸν ἕλῃ μηδ' αἴσιμα πίνῃ·
οἶνος καὶ Κένταυρον, ἀγακλυτὸν Εὐρυτίωνα
ἆσ' ἐνὶ μεγάρῳ μεγαθύμου Πειριθόοιο
ἐς Λαπίθας ἐλθόνθ'. ὃ δ' ἐπεὶ φρένας ἄασεν οἴνῳ,
μαινόμενος κάκ' ἔρεξε δόμον κάτα Πειριθόοιο.'

καὶ ἀλλαχόθι περὶ αὐτοῦ φησιν·

---

[16] post ἡμέρας: τούτου πειρώμεθα· SM; τούτῳ χρώμεθα. K

THE SOUL'S TRAITS

Drinking wine clearly relieves all distress and *dysthymia* (despondency)—we have proof of this every day. And Zeno, so they say, said that just as bitter lupines become sweet when soaked in water,[13] so too was he himself settled by wine. However, they also say the "the wine-like root" does this even more, and this was the drug of the Egyptian stranger, of whom the poet says:

> At once she cast into the wine which they were drinking a drug to quiet all pain and strife, and bring forgetfulness of every ill.[14]

Let the wine-like root be done with then, for we have no need of this | for the argument, seeing each day those things the poets say wine does:

> It is wine that wounds you, honey-sweet wine, which works harm to others also, whoever takes it in great gulps, and drinks beyond measure. It was wine that made foolish the centaur too, glorious Eurytrion, in the hall of great-hearted Peirithous, when he went to the Lapithae, and when he had made his heart foolish with wine, in his madness he did evil in the house of Peirithous.[15]

And elsewhere he said about this:

---

[13] For this statement, see Diogenes Laertius, *The Lives of Eminent Philosophers*, LCL 185 (R. D. Hicks), 136–37

[14] Homer, *Odyssey* 4.220–21, LCL 104 (A. T. Murray), 135. See also, Theophrastus, *History of Plants* 9.15.1.

[15] Homer, *Odyssey* 21.293–98, LCL 105 (Murray), 331–33.

ἠλεός, ὅς τ' ἐφέηκε πολύφρονά περ μάλ' ἀεῖσαι
καί θ' ἁπαλὸν γελάσαι καί τ' ὀρχήσασθαι ἀνῆκεν
καί τι ἔπος προέηκεν, ὅπερ τ' ἄρρητον ἄμεινον.

οὕτω δὲ καὶ Θέογνις ἔλεγεν·

οἶνος πινόμενος πολὺς κακόν· εἰ δέ τις αὐτὸν
πίνει ἐπισταμένως, οὐ κακὸν ἀλλ' ἀγαθόν.

ὄντως γάρ, εἰ συμμέτρως ποθείη, καὶ πέψει καὶ ἀναδόσει καὶ αἱματώσει καὶ θρέψει. μεγάλα δὲ συντελεῖ μετὰ τοῦ καὶ τὴν ψυχὴν ἡμῶν ἡμερωτέραν θ' ἅμα καὶ θαρσαλεωτέραν ἐργάζεσθαι διὰ μέσης δηλονότι τῆς κατὰ τὸ σῶμα κράσεως, ἥντινα πάλιν αὐτὴν ἐργάζεται διὰ μέσων τῶν χυμῶν. οὐ μόνον δ' ὡς ἔφην ἡ κρᾶσις τοῦ σώματος ὑπαλλάττει καὶ τὰς ἐνεργείας τῆς ψυχῆς ἀλλὰ καὶ χωρίζειν αὐτὴν ἀπὸ τοῦ σώματος δύναται. τί γὰρ ἂν ἄλλο τις εἴποι θεώμενος τὰ ψύχοντά τε καὶ ὑπερθερμαίνοντα φάρμακα παραχρῆμα τὸν προσενεγκάμενον ἀναιροῦντα; τοῦ γένους δ' εἰσὶ τούτου καὶ οἱ τῶν θηρίων ἰοί. τοὺς δηχθέντας γοῦν ὑπὸ τῆς ἀσπίδος ὁρῶμεν ἀποθνήσκοντας αὐτίκα παραπλήσιον τοῖς ἀποθνήσκουσιν[17] ὑπὸ τῆς κωνείου πόσεως, ὡς καὶ τοῦ ταύτης ἰοῦ ψύχοντος. ἀναγκαῖον οὖν ἔσται καὶ τοῖς ἰδίαν οὐσίαν ἔχειν ὑποθεμένοις τὴν

[17] ἀποθνήσκουσιν add. SM

THE SOUL'S TRAITS

> Hear me now Eumaeus, and all the rest of you, his men, while I tell a boasting tale, for the wine bids me, befooling wine, which sets one, even though he be very wise, to singing and soft laughter, and makes him stand up and dance, and sometimes brings forth a word that were better left unspoken.[16]

In the same way too, Theognis said:

> Drinking wine in excess is bad, whereas if someone were to drink it wisely, it is not bad but good.[17]

For truly, if wine is drunk in moderation, it will concoct, distribute, form blood, and nourish. It accomplishes great things along with this, making our soul more gentle and at the same time also more courageous, clearly through the mediation of the *krasis* in the body, which in turn it brings about through the mediation of the humors. However, it not only changes the *krasis* of the body, as I said, and the functions of the soul, but is also able to separate this from the body. For what else would someone say, seeing the cooling and excessively heating medications immediately cause destruction of the one to whom they are administered? Of this class also are the poisons of wild animals. Thus we see those bitten by the asp killed immediately in a similar manner to the drink from hemlock, since the poison of this is cooling. It will also be necessary, then, for those who postulate the soul to be a specific

779K

---

[16] Homer, *Odyssey* 14.462–66, LCL 105 (Murray), 71. More of the original is included in the translation than is given in Galen's text.

[17] Theognis, frag. 8, Douglas Young (Teubner, 1998).

ψυχὴν ὁμολογῆσαι δουλεύειν αὐτὴν ταῖς τοῦ σώματος κράσεσιν, εἴ γε καὶ χωρίζειν ἐξουσίαν ἔχουσι καὶ παραφρονεῖν ἀναγκάζουσι καὶ μνήμην καὶ σύνεσιν ἀφαιροῦνται καὶ λυπηροτέραν καὶ ἀτολμοτέραν καὶ ἀθυμοτέραν ἐργάζονται, καθάπερ ἐν ταῖς μελαγχολίαις φαίνεται, καὶ τούτων ἔχειν τἀναντία τὸν πίνοντα τὸν οἶνον συμμέτρως.

4. Ἆρ᾽ οὖν ὑπὸ μὲν τῆς κατὰ τὸ θερμόν τε καὶ ψυχρὸν κράσεως ὑπαλλάττεσθαι πεφύκασιν αἱ δυνάμεις τῆς ψυχῆς, ὑπὸ δὲ τῆς κατὰ τὸ ξηρόν τε καὶ ὑγρὸν οὐδὲν πάσχειν; καὶ μὴν καὶ τούτου πολλὰ τεκμήρια κατά τε τὰ φάρμακα καὶ τὴν ὁσημέραι δίαιταν ἔχομεν, ἃ τάχ᾽ ἂν ἐφεξῆς εἴποιμι ξύμπαντα πρότερον ἀναμνήσας ὃν ὁ Πλάτων ἔγραψε λόγον, ὑπὸ τῆς τοῦ σώματος ὑγρότητος εἰς λήθην ἔρχεσθαι τὴν ψυχὴν ὧν πρότερον ἠπίστατο πρὶν ἐνδεθῆναι τῷ σώματι. λέγει γὰρ ὧδέ πως αὐτοῖς ῥήμασιν ἐν Τιμαίῳ κατ᾽ ἐκεῖνο τὸ χωρίον τοῦ συγγράμματος, ἔνθα φησὶ τοὺς θεοὺς δημιουργῆσαι τὸν ἄνθρωπον ἐνδοῦντας τὴν ἀθάνατον ψυχὴν "εἰς ἐπίρρυτον σῶμα καὶ ἀπόρρυτον," εὔδηλον ὅτι τὴν ὑγρότητα τῆς τῶν βρεφῶν οὐσίας αἰνιττόμενος. ἐφεξῆς γοῦν τούτοις τάδε φησίν· "αἱ δ᾽ εἰς ποταμὸν ἐνδεθεῖσαι πολὺν οὔτ᾽ ἐκράτουν οὔτ᾽ ἐκρατοῦντο, βίᾳ δ᾽ ἐφέροντό τε καὶ ἔφερον"· καὶ μετ᾽ ὀλίγα πάλιν· "πολλοῦ γὰρ ὄντος τοῦ κατακλύζοντος καὶ ἀπορρέοντος κύματος, ὃ τὴν τροφὴν παρεῖχεν, ἔτι μείζονα θόρυβον ἀπειργάζετο τὰ τῶν προσπιπτόντων παθήματα ἑκάστῳ."

THE SOUL'S TRAITS

substance, to concede that it is itself a slave to the *krasias* of the body. If in fact they have a power to separate, they compel it to be deranged, take away memory and understanding, and make it more distressed, less courageous and more spiritless, as is seen in the melancholias, while the one drinking wine in moderation has the opposite effects to these.

4. Are, then, the capacities of the soul naturally changed by the *krasis* relating to the hot and cold, whereas they are not affected at all by the *krasis* relating to the dry and wet? Surely we have much evidence on this in relation to medications and the daily regimen. Perhaps I should speak about all these next, calling to mind beforehand the argument Plato set out—that due to wetness of the body, the soul comes to forgetfulness of those things it knew before it was bound to the body. For he says something like this in the statements in the *Timaeus*,[18] in that part of the work where he says the gods craft the person when they give the immortal soul "into a body subject to influx and efflux." It is clear that he is intimating the wetness of the substance of infants. Anyway, subsequent to these things, he says this: "The souls, then, being thus bound within a mighty river, neither mastered it nor were mastered, but with violence they rolled along and were rolled along themselves . . ." And again, a little later: "For while the flood which foamed in and streamed out, as it supplied the food, was immense, still greater was the tumult produced within each creature . . ."

[18] The following three quotations from Plato's *Timaeus* are to be found closely adjacent at 43A–B. The translations follow R. G. Bury (1929), LCL 234, 94–95.

GALEN

Καὶ μέντοι καὶ διελθὼν αὐτὰ πάλιν ἐφεξῆς φησι·

διὰ δὴ ταῦτα πάντα τὰ πάθη κατ' ἀρχὰς ἄνους ἡ ψυχὴ γίγνεται τὸ πρῶτον, ὅταν εἰς σῶμα ἐνδεθῇ θνητόν· ὅταν δὲ τὸ τῆς αὐξήσεως καὶ τροφῆς[18] ἔλαττον ἐπίῃ ῥεῦμα, πάλιν ⟨δ' αἱ⟩ περίοδοι λαμβανόμεναι γαλήνης τὴν ἑαυτῶν ὁδὸν ἴωσι καὶ καθιστῶνται μᾶλλον ἐπιόντος τοῦ χρόνου, τότε ἤδη πρὸς τὸ φύσει ἰόντων σχῆμα ἑκάστων τῶν κύκλων αἱ περιφοραὶ κατευθυνόμεναι τό τε θάτερον καὶ τὸ ταὐτὸν προσαγορεύουσαι[19] κατ' ὀρθὸν ἔμφρονα τὸν ἔχοντα αὐτὰς γιγνόμενον ἀποτελοῦσιν.

ὅταν φησίν "τὸ τῆς αὐξήσεως καὶ τροφῆς ἔλαττον ἐπίῃ ῥεῦμα," τὴν ὑγρότητα δηλονότι λέγων τὴν ἔμπροσθεν ὑπ' αὐτοῦ εἰρημένην τῆς κατὰ ψυχὴν ἀνοίας αἰτίαν γιγνομένην, ὡς τῆς μὲν ξηρότητος εἰς σύνεσιν, τῆς δ' ὑγρότητος εἰς ἄνοιαν ἀγούσης τὴν ψυχήν. ἀλλ' εἴπερ ὑγρότης μὲν ἄνοιαν ἐργάζεται, ξηρότης δὲ σύνεσιν, ἡ μὲν ἄκρα ξηρότης ἄκραν ἐργάζεται σύνεσιν, ἡ δ' ἐπίμικτος ὑγρότητι τοσοῦτον ἀφαιρέσει τῆς τελείας συνέσεως, ὅσον ἐκοινώνησεν ὑγρότητος. τίνος οὖν[20] θνητοῦ ζῴου τοιοῦτον σῶμα τίνος ἄμοιρον

[18] τροφῆς SM; πρὸς K
[19] προσαγορεύουσαι SM: προσαγόμεναι K
[20] post τίνος οὖν: θνητοῦ ζῴου τοιοῦτον σῶμα, ὥστ' ἄμοιρον ⟨ὑπάρχειν⟩ ὑγρότητος, SM; ἦν τοῦ ζῴου τοιουτόνδε σῶμα, τίνος ἄμοιρον ὑγρότητος, K

THE SOUL'S TRAITS

And indeed, also going over these in turn, he says next:

> Hence it comes about that, because of all these affections, now as in the beginning, so often as the soul is bound within a mortal body it becomes at the first irrational. But as soon as the stream of increase and nutriment enters in less volume, and the revolutions calm down and pursue their own path, becoming more stable as time proceeds, then at length, as the several circles move, each according to its natural track, their revolutions are straightened out and they announce the Same and the Other aright, and thereby render their possessor intelligent.[19]

When he says "the stream of increase and nutriment enters in less volume," he is clearly speaking of the wetness mentioned by him previously as being a cause of the folly (lack of understanding) in relation to the soul, for as the dryness leads the soul to understanding so the wetness leads it to want of understanding. But if wetness produces lack of understanding while dryness produces quick understanding,[20] the peak of dryness produces maximum understanding while the admixture of wetness will detract from complete understanding to the extent that the wetness is shared. Therefore, of what mortal animal is there such a body that there is no share of wetness, like those

---

[19] Bury, LCL 234, 44A–B, pp. 96–99.

[20] The two terms used here are ἄνοια, defined by LSJ as "want of understanding, folly" (*Timaeus* 86B), and σύνεσι, defined as "the faculty of quick understanding, sagacity" (*Cratylus* 412A).

ὑγρότητος, ὥσπερ τὰ τῶν ἄστρων; οὐδενὸς οὐδ' ἐγγύς. ὥστ' οὐδὲ συνέσεως ἄκρας ἐγγύς ἐστί τι σῶμα θνητοῦ ζῴου, πάντα δ' ὥσπερ ὑγρότητος οὕτω καὶ ἀνοίας μετέχει. ὁπότ' οὖν τὸ λογιστικὸν τῆς ψυχῆς μονοειδῆ τὴν οὐσίαν ἔχον τῇ τοῦ σώματος κράσει συμμεταβάλλεται, τί χρὴ νομίσαι πάσχειν τὸ θνητὸν εἶδος αὐτῆς; ἢ δῆλον ὅτι πάντη δουλεύει τῷ σώματι; ἄμεινον δὲ φάναι μὴ δουλεύειν ἀλλ' αὐτὸ δὴ τοῦτ' εἶναι τὸ θνητὸν τῆς ψυχῆς, τὴν κρᾶσιν τοῦ σώματος. ἐδείχθη γὰρ ἔμπροσθεν ἡ θνητὴ ψυχὴ κρᾶσις οὖσα τοῦ σώματος. ἡ μὲν οὖν τῆς καρδίας κρᾶσις τὸ θυμοειδὲς εἶδός ἐστι τῆς ψυχῆς, ἡ δὲ τοῦ ἥπατος τὸ καλούμενον ὑπὸ Πλάτωνος μὲν ἐπιθυμητικόν, θρεπτικὸν δὲ καὶ φυτικὸν ὑπ' Ἀριστοτέλους. Ἀνδρόνικον δὲ τὸν Περιπατητικόν, ὅτι μὲν ὅλως ἐτόλμησεν ἀποφήνασθαι τὴν οὐσίαν τῆς ψυχῆς ⟨κρᾶσιν ἢ δύναμιν εἶναι τοῦ σώματος⟩[21] ὡς ἐλεύθερος ἀνὴρ ἄνευ τοῦ περιπλέκειν ἀσαφῶς, ἐπαινῶ τε πάνυ καὶ ἀποδέχομαι τὴν προαίρεσιν τἀνδρός (εὑρίσκω γὰρ αὐτὸν καὶ κατ' ἄλλα πολλὰ τοιοῦτον)·[22] | ὅτι δ' ἤτοι κρᾶσιν εἶναί φησιν ἢ δύναμιν ἑπομένην τῇ κράσει, μέμφομαι τῇ προσθήκῃ τῆς δυνάμεως. εἰ γὰρ ἡ ψυχὴ πολλὰς ἔχει δυνάμεις οὐσία τις οὖσα καὶ τοῦτ' ὀρθῶς Ἀριστοτέλει λέλεκται καὶ τούτῳ διώρισται καλῶς ἡ ὁμωνυμία—λεγομένης γὰρ οὐσίας καὶ τῆς ὕλης καὶ τοῦ εἴδους καὶ τοῦ συναμφοτέρου τὴν κατὰ τὸ εἶδος οὐσίαν ἀπε-

[21] add. SM, *but see Singer, note 42, p. 386*
[22] *On the material in parentheses, see* SM, *note, p. 44.*

THE SOUL'S TRAITS

of the stars? There is none that is near this. As a result, there is no body of a mortal animal which is near to peak understanding. Just as all partake of wetness, so too do they also partake of want of understanding. Therefore, whenever the rational part of the soul, having a substance single in form,[21] is changed along with the *krasis* of the body, what must we think the mortal kind of soul suffers? Or is it clear that in every way it is a slave to the body? Rather, it is better to say, not that it is a slave to, but that in fact it is itself a mortal part of the soul which is the *krasis* of the body. For it was shown previously that the mortal soul is the *krasis* of the body.[22] Therefore, the *krasis* of the heart is the spirited form of the soul, while the *krasis* of the liver is what is called by Plato the desiderative (appetitive) form, while by Aristotle it is called the nutritive and vegetative part. However, Andronicus the Peripatetic, because he altogether dared to declare the substance of the soul to be a *krasis* or capacity (*dunamis*) of the body, as a free man and without being obscurely complicated, I praise him greatly and accept his way of speaking, for I find him to also be like this in many other matters. However, when he says it is either a *krasis* or a capacity following the krasis, I find fault with the addition of "the capacity." For the soul has many capacities, itself being a substance, and this has been correctly stated by Aristotle, and by this the homonymy is properly distinguished. Thus, when he says substance applies to the matter and the form both together, he declares the soul to be substance in rela-

783K

---

[21] Plato, *Timaeus* 59B.

[22] On the textual issues involving the two preceding sentences, see Singer, note 4.19, p. 415. The text given and translated is that given in Kühn.

φήνατο ψυχὴν ὑπάρχειν—οὐκ ἐγχωρεῖ λέγειν ἄλλο τι παρὰ τὴν κρᾶσιν, ὡς ὀλίγον ἔμπροσθεν ἐδείκνυτο.

Ἐν ταὐτῷ δὲ γένει τῆς οὐσίας καὶ ἡ τῶν Στωϊκῶν περιέχεται δόξα. πνεῦμα μὲν γάρ τι τὴν ψυχὴν εἶναι βούλονται καθάπερ καὶ τὴν φύσιν, ἀλλ' ὑγρότερον μὲν καὶ ψυχρότερον τὸ τῆς φύσεως, ξηρότερον δὲ καὶ θερμότερον τὸ τῆς ψυχῆς. ὥστε καὶ τοῦθ' ὕλη μέν τις οἰκεία τῆς ψυχῆς ἐστι τὸ πνεῦμα, τὸ δὲ τῆς ὕλης εἶδος ἡ ποιὰ κρᾶσις[23] ἐν συμμετρίᾳ γιγνομένη τῆς ἀερώδους τε καὶ πυρώδους οὐσίας· οὔτε γὰρ ἀέρα μόνον οἷόν τε φάναι τὴν ψυχὴν οὔτε πῦρ, ὅτι μήτε ψυχρὸν ἄκρως ἐγχωρεῖ[24] γίγνεσθαι ζῴου σῶμα μήτ' ἄκρως θερμὸν ἀλλὰ μηδ' ἐπικρατούμενον ὑπὸ θατέρου κατὰ μεγάλην ὑπεροχήν, ὅπου γε, κἂν βραχεῖ πλεῖον γένηται τοῦ συμμέτρου, πυρέττει μὲν τὸ ζῷον ἐν ταῖς τοῦ πυρὸς ἀμέτροις ὑπεροχαῖς, καταψύχεται δὲ καὶ πελιδνοῦται καὶ δυσαίσθητον ἢ παντελῶς ἀναίσθητον γίγνεται κατὰ τὰς τοῦ ἀέρος ἐπικρατήσεις· οὗτος γὰρ αὐτός, ὅσον μὲν ἐφ' ἑαυτῷ, ψυχρός ἐστιν, ἐκ δὲ τῆς πρὸς τὸ πυρῶδες στοιχεῖον ἐπιμιξίας εὔκρατος γίγνεται. δῆλον οὖν ἤδη σοι γέγονεν, ὡς ἡ τῆς ψυχῆς οὐσία κατὰ ποιὰν κρᾶσιν ἀέρος τε καὶ πυρὸς γίγνεται κατὰ τοὺς Στωϊκούς· καὶ συνετὸς μὲν ὁ Χρύσιππος ἀπείργασται διὰ τὴν τούτων εὔκρατον μίξιν, οἱ δ' Ἱπποκράτους υἱεῖς ὑώδεις,[25] οὓς ἐπὶ μωρίᾳ σκώπτουσιν οἱ κωμικοί, διὰ τὴν ἄμετρον θέρμην. ἴσως οὖν τις ἐρεῖ

[23] ἡ ποιὰ κρᾶσις SM: ἤτοι κράσεως K [24] ἐγχωρεῖ SM: ἐμφανῇ K [25] On ὑώδεις see SM, note, line 26, p. 45.

THE SOUL'S TRAITS

tion to the form, it not being possible to say anything else besides the *krasis*, as was shown a little earlier.

In this class of the substance the notion of the Stoics is also contained, for really they understand the soul to be *pneuma*, just as they also do the nature, but that of the nature is wetter and colder, while that of the soul is dryer and hotter. As a further consequence of this, the *pneuma* is a material characteristic of the soul, whereas the form of the material, or the kind of the *krasis* arises in a balance of the airy and fiery substance. For it cannot be said that the soul is air alone, or fire, because the body of an animal is not completely cold, | nor completely hot, but neither 784K does one prevail over the other by a great excess. In fact, in one case, even if one [of the elemental qualities] becomes a little more than the moderate, the animal is febrile in the immoderate excesses of fire, whereas it is cooled, made livid, and becomes dysaesthetic or completely anesthetic in the preponderances of air. For this itself (i.e., air) to the extent it is cold in itself, is cold, while from the admixture of the fiery element it becomes *eukratic*. It has, then, already become clear to you that the substance of the soul arises in relation to what kind of *krasis* of air and fire there is, according to the Stoics, and Chrysippus was rendered intelligent due to the *eukratic* mixture of these, while the sons of Hippocrates, whom the comic poets mocked for their folly, became swinish due to the immoderate heat.[23] Perhaps, then, someone might say

---

[23] See Aristophanes, *Clouds* 1000ff. Henderson (LCL 488) has the following note: "Hippocrates, nephew of Pericles, was killed at Delium in 424; his three sons (Demophon, Pericles and Telesippus) are ridiculed elsewhere in comedy as being swinish and uneducated."

μήτε Χρύσιππον ἐπαινεῖσθαι δεῖν ἐπὶ συνέσει μήτ᾽ ἐκείνους ἐπὶ μωρίᾳ ψέγεσθαι (μήτ᾽ αὖ πάλιν ἐπὶ τῷ τῆς ψυχῆς ἐπιθυμητικῷ τοὺς ἐγκρατεῖς μὲν ἐπαινεῖσθαι, τοὺς δ᾽ ἀκολάστους ψέγεσθαι,)[26] παραπλησίως δὲ καὶ εἰς τὰ τοῦ θυμοειδοῦς ἔργα καὶ πάθη μήτε τοὺς εὐτόλμους ἐπαινεῖσθαι [χρὴ] μήτε τοὺς ἀτόλμους ψέγεσθαι.

5. Περὶ μὲν οὖν τούτων ὀλίγον ὕστερον ἐπισκεψόμεθα.[27] | νυνὶ δ᾽ οἷς ἐξ ἀρχῆς προὐθέμην, προσθήσω τὰ λείποντα τοσοῦτον ἔτι πάλιν ἐπειπών, ὡς οὐχ οἷόν τ᾽ ἐστὶ πάντα δεικνύειν ἐν ἅπασι καὶ ὡς δυοῖν οὐσῶν αἱρέσεων ἐν φιλοσοφίᾳ [κατὰ τὴν πρώτην τομήν][28] ἔνιοι μὲν γὰρ ἡνῶσθαι τὴν κατὰ τὸν κόσμον οὐσίαν ἅπασαν, ἔνιοι δὲ διῃρῆσθαί φασι κενοῦ παραπλοκῇ—τὴν δευτέραν αἵρεσιν ἐφωράσαμεν οὐκ ἀληθῆ δι᾽ ἐκείνων τῶν ἐλέγχων, οὓς ἐν τῷ Περὶ τῶν καθ᾽ Ἱπποκράτην στοιχείων ἐγράψαμεν. πρὸς δὲ τὸν παρόντα λόγον ὑπόθεσιν λαβόντες τὸ ἀλλοιοῦσθαί τὴν οὐσίαν ἡμῶν καὶ τὴν κρᾶσιν αὐτῆς ἐργάζεσθαι τὸ φυσικὸν σῶμα ἐν ὁμοιομερεῖ, τὴν τῆς ψυχῆς οὐσίαν ἐδείξαμεν κατὰ τὴν κρᾶσιν συνισταμένην, ἐάν γε μή τις αὐτὴν ὑπόθηται, καθάπερ ὁ Πλάτων, ἀσώματον ὑπάρχειν καὶ ἄνευ τοῦ σώματος εἶναι δυναμένην. ὑποθεμένοις δὲ τοῦτο τὸ ὑπὸ τῆς τοῦ σώματος κράσεως

---

[26] The text in parentheses, present here in SM, is placed at the end of the paragraph in B and omitted altogether in K.

[27] This is the final sentence of section 4 in K.

[28] On [κατὰ τὴν πρώτην τομήν] see SM, praef. p. xlii.

THE SOUL'S TRAITS

we need not praise Chrysippus for his intelligence nor censure those sons for their folly. Nor in turn should those disciplined in the desiderative component of the soul be praised, or the undisciplined censured. Similarly also, toward the actions and affections of the spirited soul, we must neither praise the courageous nor censure the cowardly.

5. We shall, then, consider these matters a little later. Now, however, let me add what remains to those things I proposed at the beginning, once more saying this much—that it is not possible to demonstrate everything in every work, and that there are two sects in philosophy, according to the first division. Some make the whole substance in the universe (Cosmos) one, while others say it is divided by an interweaving of emptiness.[24] We found the second sect not to be true through those refutations we wrote in *On the Elements according to Hippocrates*.[25] Regarding the present argument, we have taken as an hypothesis that our substance is changed and that the *krasis* (composition) of this in a *homoiomere* creates the physical body, and we have shown the substance of the soul to be put together according to the *krasis,* if in fact someone does not suppose, as Plato did, that it is incorporeal and able to exist without the body. However, to those who do suppose this, it has already been adequately shown it is prevented by

[24] The distinction is between continuum theories and atomic theories. On these two basic positions in a medical context, see *De morborum differentiis* (VI.836–880K), sections II and V, Johnston, *Galen: On Diseases and Symptoms*, 134–36 and 140–44.

[25] Galen, *De elementis secundum Hippocratem*, I.415–508K; English translation by de Lacy, *Galen on the Elements*.

ἐνεργεῖν κωλύεσθαι τὰς οἰκείας ἐνεργείας ἱκανῶς μὲν ἤδη δέδεικται, προστεθήσονται δὲ καὶ ἄλλαι τινὲς ἀποδείξεις.

Ἀλλὰ νῦν γε καὶ τούτου ἐστὶ τοῦ τρόπου εἰρημένου τὸν περὶ τῶν κράσεων λόγον προσθεῖναι δοκεῖ μοι βέλτιον εἶναι. δυνήσονται γὰρ λέγειν οἱ τὴν ψυχὴν εἶδος εἶναι τοῦ σώματος ἡγούμενοι τὴν συμμετρίαν τῆς κράσεως, οὐ τὴν ξηρότητα, συνετωτέραν αὐτὴν ἐργάζεσθαι· καὶ ταύτῃ διαφωνήσουσι τοῖς ἡγουμένοις, ὅσῳπερ ἂν ἡ κρᾶσις γίγνηται ξηροτέρα, τοσούτῳ καὶ τὴν ψυχὴν ἀποτελεῖσθαι συνετωτέραν. ἀλλ' οὐ καὶ ξηρότητα συγχωρήσομεν αἰτίαν εἶναι συνέσεως ὥσπερ οἵ γ' μὴν ἀμφ' Ἡράκλειτον; καὶ γὰρ καὶ οὗτος εἶπεν "αὐγὴ ξηρή, ψυχὴ σοφωτάτη" τὴν ξηρότητα πάλιν ἀξιῶν συνέσεως εἶναι αἰτίαν· τὸ γὰρ τῆς αὐγῆς ὄνομα τοῦτ' ἐνδείκνυται· καὶ βελτίονά γε δόξαν ταύτην νομιστέον ἐννοήσαντας τοὺς ἀστέρας αὐγοειδεῖς θ' ἅμα καὶ ξηροὺς ὄντας ἄκραν σύνεσιν ἔχειν· εἰ γὰρ μή τις αὐτοῖς ὑπάρχειν τοῦτο φαίη, δόξει τῆς τῶν θεῶν ὑπεροχῆς ἀναίσθητος εἶναι.

Διὰ τί τοίνυν εἰς ἔσχατον γῆρας ἀφικνούμενοι παρελήρησαν οὐκ ὀλίγοι τῆς τοῦ γήρως ἡλικίας ἀποδεδειγμένης εἶναι ξηρᾶς; οὐ διὰ τὴν ξηρότητα φήσομεν ἀλλὰ διὰ τὴν ψυχρότητα· φανερῶς γὰρ αὕτη πᾶσι τοῖς ἔργοις τῆς ψυχῆς λυμαίνεται. ἀλλὰ ταῦτα μέν, εἰ καὶ πάρεργά ἐστιν, ἀλλ' ἐναργῶς γε τὸ τῆς προκειμένης νῦν ἡμῖν πραγματείας ἐνδείκνυται, τὰ τῆς ψυχῆς ἔργα καὶ πάθη ταῖς τοῦ σώματος ἕπεσθαι

THE SOUL'S TRAITS

the *krasis* of the body from carrying out the specific functions, while certain other demonstrations will be added also.

But now in fact | it seems to me to be better, having spoken in this way, to add the discussion about the *krasias*.[26] For those who think the soul is a form of the body will be able to say it is the balance of the *krasis* and not the dryness that makes it more intelligent, and in this way they will disagree with those who think by as much as the *krasis* is drier, so the soul is made more intelligent. But should we not also agree with the followers of Heraclitus that dryness is a cause of intelligence? For he himself said: "Dry ray, wisest soul,"[27] thinking again the dryness to be a cause of intelligence, for the term "ray" indicates this. And one must in fact deem this opinion to be better, considering the stars, which are ray-like and at the same time also dry, to have peak intelligence. If someone were to say this is not in them, he will seem to be indifferent to the preeminence of the gods.

For what reason, therefore, were those coming to the extreme of age in not a few instances made crazy when the stage of life of old age has been shown to be dry? We shall say not because of the dryness, but because of the coldness. Clearly | this is damaging to all the actions of the soul. But these things, even if they are secondary to the main argument, are clearly part of the matter now before us, and show the actions and affections of the soul follow the

786K

787K

---

[26] On this statement, see Singer's textual notes 4.24 and 4.25, p. 416.

[27] Heraclitus, frag. 118, DK. On this apparent quotation and its original source, see SM Preface, xlii–xliii.

κράσεσιν. εἰ μὲν γὰρ εἶδός ἐστιν ὁμοιομεροῦς σώματος ἡ ψυχή, τὴν ἀπόδειξιν ἐξ αὐτῆς τῆς οὐσίας ἕξομεν ἐπιστημονικωτάτην· εἰ δ' ὑποθοίμεθα ταύτην ἀσώματον εἶναι φύσιν ἰδίαν ἔχουσαν, ὡς ὁ Πλάτων ἔλεγεν, ἀλλὰ τό γε δεσπόζεσθαι καὶ δουλεύειν τῷ σώματι καὶ κατ' αὐτὸν ἐκεῖνον ὁμολογεῖται διά τε τὴν τῶν βρεφῶν ἄνοιαν καὶ τὴν τῶν ἐν γήρᾳ ληρούντων ἔτι τε τῶν εἰς παραφροσύνην ἢ μανίαν ἢ ἐπιλησμοσύνην ἢ ἄνοιαν ἀφικνουμένων ἐπὶ φαρμάκων δόσεσιν ἢ τισιν ἐν τῷ σώματι γεννηθεῖσι μοχθηροῖς χυμοῖς. ἄχρι μὲν γὰρ τοῦ λήθην ἢ ἄνοιαν ἢ ἀκινησίαν ἢ ἀναισθησίαν ἕπεσθαι τοῖς εἰρημένοις, ἐμποδίζεσθαι φαίη τις ἂν αὐτὴν ἐνεργεῖν αἷς ἔχει φύσει δυνάμεσιν· ὅταν δέ τις οἴηται βλέπειν τὰ μὴ βλεπόμενα καὶ ἀκούειν ἃ μηδεὶς ἐφθέγξατο, καὶ φθέγγηται τι τῶν αἰσχρῶν ἢ ἀπορρήτων ἢ ὅλως ἀδιανοήτων, οὐ μόνον ἀπωλείας ἐστὶ τεκμήριον ὧν εἶχε δυνάμεων ἡ ψυχὴ συμφύτων ἀλλὰ καὶ τῆς τῶν ἐναντίων ἐπεισόδου.

Τοῦτο μὲν οὖν ἤδη καὶ ὑποψίαν τινὰ φέρει μεγάλην ὅλῃ τῇ τῆς ψυχῆς οὐσίᾳ, μὴ οὐκ ἀσώματος ᾖ. πῶς γὰρ ἂν ὑπὸ τῆς τοῦ σώματος κοινωνίας εἰς τὴν ἐναντίαν ἑαυτῆς φύσιν ἀχθείη μήτε ποιότης τις οὖσα τοῦ σώματος μήτ' εἶδος μήτε πάθος μήτε δύναμις; ἀλλὰ τοῦτο μὲν ἐάσωμεν, ἵνα μὴ τὸ πάρεργον ἡμῖν γένηται αὖ πολὺ μεῖζον ἔργου οὗ προυθέμεθα. τὸ δ' ὑπὸ[29] τῶν τοῦ σώματος κακῶν δυναστεύεσθαι τὴν ψυχὴν ἐναρ-

---

[29] τὸ δ' ὑπὸ SM: οὐδ' ὑπὸ K

THE SOUL'S TRAITS

krasias of the body. If the soul is a form of *homoiomerous* body, we shall have the most scientific demonstration from the actual substance. If, however, we are to assume the soul is incorporeal (immortal),[28] having a specific nature, as Plato said, but is in fact dominated and enslaved by the body, which is also agreed by that same Plato, through the mindlessness of infants and that of those talking nonsense in old age, as well as those coming to delirium or mania or loss of memory or mindlessness following administrations of drugs, or when some of the bad humors are generated in the body. For as far as the forgetfulness, mindlessness, immobility, or anesthesia follow the things mentioned, someone might say the soul is being impeded in functioning with the capacities it has naturally. However, when someone thinks he sees things that are not visible or hears things no one has uttered, or says something that is shameful, or unspeakable, or completely unthinkable, not only is this evidence of loss of the innate capacities which the soul has, but also of the entrance of the opposites.

788K

This, then, already carries a certain strong suspicion regarding the whole substance of the soul, that it is not incorporeal. For how would it be led to the nature opposite to itself by the association with the body, if it were not some quality of the body, or form, or affection, or capacity? But let us disregard this so it does not become for us a secondary matter much greater than the task which we set ourselves. It is clearly apparent that the soul is over-

---

[28] On the issue of incorporeal (SM, B) versus immortal (K), see Singer, note 4.27, p. 417.

γῶς ἐν μελαγχολίαις καὶ φρενίτισι καὶ μανίαις φαίνεται. τὸ μὲν γὰρ ἀγνοῆσαι διὰ νόσημα σφᾶς τ' αὐτοὺς καὶ τοὺς ἐπιτηδείους, ὅπερ ὅ τε Θουκυδίδης συμβῆναι[30] πολλοῖς φησιν ἔν τε τῇ λοιμώδει νόσῳ τῇ νῦν γενομένῳ ἔτεσιν οὐ πολλοῖς ἦν καὶ ἡμεῖς ἐθεασάμεθα, παραπλήσιον εἶναι δόξει τῷ μὴ βλέπειν διὰ λήμην ἢ ὑπόχυσιν οὐδὲν αὐτῆς τῆς ὀπτικῆς δυνάμεως πεπονθυίας· τὸ δ' ἀνθ' ἑνὸς τρία βλέπειν αὐτῆς τῆς ὀπτικῆς δυνάμεώς ἐστι μέγιστον πάθος, ᾧ τὸ φρενιτίζειν ἔοικεν.

6. Ὅτι δὲ καὶ ὁ Πλάτων αὐτὸς οἶδε βλαπτομένην τὴν ψυχὴν ἐπὶ τῇ κακοχυμίᾳ τοῦ σώματος, ἡ ἑξῆς ῥῆσις ἤδη δηλώσει·

> ὅπου γὰρ ἂν οἱ τῶν ὀξέων καὶ τῶν ἁλυκῶν φλεγμάτων ἢ καὶ ὅσοι πικροὶ καὶ χολώδεις χυμοὶ κατὰ τὸ σῶμα πλανηθέντες ἔξω μὲν μὴ λάβωσιν ἀναπνοήν, ἐντὸς δ' ἑλκόμενοι[31] τὴν ἀφ' ἑαυτῶν ἀτμίδα[32] τῇ τῆς ψυχῆς διαθέσει φορᾷ συμμίξαντες ἀνακερασθῶσι, παντοδαπὰ νοσήματα ψυχῆς ἐμποιοῦσι μᾶλλον καὶ ἧττον καὶ ἐλάττω καὶ πλείω πρός τε τοὺς τρεῖς τόπους[33] ἐνεχθέντα τῆς ψυχῆς, πρὸς ὃν ἂν ἕκαστ' αὐτῶν προσπίπτῃ, ποικίλα μὲν εἴδη δυσκολίας καὶ δυσθυμίας παντοπαδάς, πολλάκις δὲ θρασύτητός τε καὶ δειλίας, ἔτι δὲ λήθης ἅμα καὶ δυσμαθείας.

[30] συμβῆναι SM: ἐκβῆναι K
[31] εἰλλόμενοι SM: ἑλκόμενοι K

THE SOUL'S TRAITS

come by the evils of the body in melancholias, *phrenitides*, and manias. For the failure through disease to recognize themselves and their families that Thucydides says befell many people during the plague which occurred not many years ago now,[29] and which we are also accustomed to see, will seem to be similar to the failure of vision due to rheum or a cataract, when the visual capacity itself has not suffered anything. However, the greatest affection of the visual capacity itself is to see three [images] instead of one, which is like that in the *phrenitides*.

6. That Plato himself also knew the soul is harmed by the *kakochymia* of the body, the following statement will immediately show:

> For wherever the humors arise from acid and salty phlegms, and all humors that are bitter and bilious wander through the body and find no external vent but are confined within, and mingle their vapor with the movement of the soul and are blended therewith, they implant diseases of the soul of all kinds, varying in intensity and extent; and as these humors penetrate to the three regions of the soul, according to the region they severally attack, they give rise to all kinds of discontent and despondency, but often of rashness and cowardice, and of forgetfulness also, as well as stupidity.[30]

---

[29] Thucydides 2.47–54.
[30] Plato, *Timaeus* 86E–87A, LCL 234 (Bury), 234–37 (translation after Bury with minor modifications).

---

[32] *post* ἀτμίδα: τῇ τῆς ψυχῆς φορᾷ SM: τὸ τῆς ψυχῆς διαθέσει σφόδρα K   [33] τόπους SM: τρόπους K; *sedes* KLat

ἐν ταύτῃ τῇ ῥήσει σαφῶς ὁ Πλάτων ὡμολόγησε τὴν ψυχὴν ἐν κακίᾳ τινὶ γίγνεσθαι διὰ τὴν ἐν τῷ σώματι κακοχυμίαν, ὥσπερ δὲ πάλιν ἐν νόσῳ καθίσταται διὰ τὴν τοῦ σώματος ἕξιν κατὰ τήνδε τὴν ῥῆσιν·

τὸ δὲ σπέρμα ὅτῳ[34] πολὺ καὶ γλοιῶδες[35] περὶ τὸν μυελὸν γίγνεται καθαπερεὶ δένδρον πολυκαρπότερον τοῦ συμμέτρου πεφυκὸς ᾖ, πολλὰς μὲν καθ' ἕκαστον ὠδῖνας, πολλὰς δ' ἡδονὰς κτώμενος ἐν ταῖς ἐπιθυμίαις καὶ τοῖς περὶ τὰ τοιαῦτα τόκοις ἐμμανὴς τὸ πλεῖστον τοῦ βίου γιγνόμενος διὰ τὰς μεγίστας ἡδονὰς καὶ λύπας νοσοῦσαν καὶ ἄφρονα ἴσχων ὑπὸ τοῦ σώματος τὴν ψυχὴν οὐχ ὡς νοσῶν ἀλλ' ὡς ἑκὼν κακὸς <κακῶς>[36] δοξάζεται· τὸ δ' ἀληθὲς ἡ περὶ τὰ ἀφροδίσια ἀκολασία κατὰ τὸ πολὺ μέρος διὰ τὴν ἑνὸς γένους ἕξιν ὑπὸ μανότητος[37] ὀστῶν ἐν τῷ σώματι ῥυώδη καὶ ὑγραίνουσαν νόσος ψυχῆς γέγονε.

ἱκανῶς μὲν οὖν κἂν ταύτῃ τῇ ῥήσει τὴν ψυχὴν νοσεῖν ἀπεφήνατο διὰ τὴν μοχθηρὰν ἕξιν τοῦ σώματος. ἀλλ' οὐδὲν ἧττον ἔτι καὶ διὰ τῶν ἐφεξῆς ὑπ' αὐτοῦ γεγραμμένων ἡ γνώμη κατάδηλος γίγνεται τοῦ φιλοσόφου. τί γάρ φησι;

[34] ὅτῳ add. SM   [35] post γλοιῶδες: περὶ τὸν μυελὸν γίγνεται καθαπερεὶ δένδρον πολυκαρπότερον SM, B: ἔχων, καθάπερ εἰ δένδρον πολύκαρπον K

THE SOUL'S TRAITS

In this same statement Plato clearly agreed that the soul comes into a certain bad state through the *kakochymia* in the body, just as in turn it is established in disease due to the state of the body, according to the following statement.

> And whenever a man's sperm grows to abundant volume and is glutinous in his marrow, as it were a tree that is overladen beyond measure with fruit, he brings on himself time after time many pangs and many pleasures owing to his desires and the issue thereof, and comes to be in a condition of madness for the most part of his life because of those greatest of pleasures and pains, and keeps his soul diseased and senseless by reason of the action of the body. Yet such a man is reputed to be voluntarily wicked and not diseased, although in truth, this sexual incontinence, which is due in the most part to the abundance and fluidity of one substance because of the porosity of the bones, constitutes a disease of the soul.[31]

It is sufficient, then, that in this statement he declared the soul to be diseased due to a bad (abnormal) state of the body. But no less is the opinion of the philosopher quite clear through those things written next by him. For what does he say?

---

[31] Plato, *Timaeus* 86C–D, LCL 234 (Bury), 234–35 (translation after Bury with minor modifications).

---

[36] κακῶς Plat. *om. codd.* (SM, *p. 50*)
[37] *post* μανότητος: ὀστῶν ἐν τῷ σώματι SM, B; ὡς τὴν ἐν σώματι K

καὶ σχεδὸν ἅπανθ᾽ ὁπόσα ἡδονῶν ἀκρασία καὶ
ὄνειδος ὡς ἑκόντων λέγεται τῶν κακῶν, οὐκ ὀρθῶς ὀνειδίζεται· κακὸς μὲν γὰρ ἑκὼν οὐδείς, διὰ
δὲ πονηρὰν ἕξιν τοῦ σώματος καὶ ἀπαιδεύτους
τροφὰς κακὸς γίγνεται, παντὶ δὲ ταῦτα ἐχθρὰ
καὶ ἄκοντι προσγίγνεται.

791K ὅτι μὲν οὖν ὁ Πλάτων αὐτὸς | ὁμολογεῖ τὰ προαποδε-
δειγμένα ὑπ᾽ ἐμοῦ, ἔκ τε τούτων αὐτῶν τῶν ῥήσεών
ἐστι δῆλον ἐξ ἄλλων τε πολλῶν, ὧν τινὰς μὲν ἐν τῷ
Τιμαίῳ, καθάπερ καὶ τάσδε τὰς νῦν εἰρημένας, τινὰς
δ᾽ ἐν ἄλλοις αὐτοῦ βιβλίοις ἔστιν εὑρεῖν.

7. Ὅτι δὲ καὶ Ἀριστοτέλης τῇ κράσει τοῦ τῆς μητρὸς αἵματος, ἐξ οὗ τὴν γένεσιν ἔχειν ἡμῶν φησι τὸ
αἷμα, τὰς τῆς ψυχῆς δυνάμεις ἀκολουθεῖν οἴεται,
δῆλόν ἐστιν ἐκ τῶν ῥήσεων αὐτοῦ. κατὰ μέν γε τὸ
δεύτερον Περὶ ζῴων μορίων οὕτως ἔγραψεν·

ἔστι δ᾽ ἰσχύος μὲν ποιητικώτερον τὸ παχύτερον
αἷμα καὶ θερμότερον, αἰσθητικώτερον δὲ καὶ
νοερώτερον[38] τὸ λεπτότερον καὶ ψυχρότερον.
τὴν αὐτὴν δ᾽ ἔχει διαφορὰν καὶ τῶν ἀνάλογον[39]
ὑπαρχόντων πρὸς τὸ αἷμα· διὸ καὶ μέλιτται καὶ
ἄλλα τοιαῦτα ζῷα φρονιμώτερα τὴν φύσιν
ἐστὶν ἐναίμων πολλῶν, καὶ τῶν ἐναίμων τὰ ψυχρὸν ἔχοντα καὶ λεπτὸν αἷμα φρονιμώτερα τῶν

[38] νοερώτερον add. SM
[39] ἀνάλογον SM: ἀνάπαλιν K

THE SOUL'S TRAITS

And indeed almost all [cases of] incontinence of pleasures, however many there are, as though the wicked acted voluntarily, are wrongly so reproached; for no one is voluntarily wicked, but the wicked man becomes wicked by reason of some evil condition of the body and unskilled nurture. And these are experiences which are hateful to everyone and are involuntary.[32]

Therefore, that Plato himself accepts those things previously demonstrated by me is clear from these same statements and from many others, some of which are found in the *Timaeus* like those I just now quoted, and some in other books of his.

7. That Aristotle too thinks the capacities of the soul follow the *krasis* of the mother's blood, from which he says our blood has its genesis, is clear from his statements. In fact, in the second book of *On the Parts of Animals*, he wrote thus:

The thicker and hotter the blood is, the more it makes for strength; if it tends to be thinner and colder, it is more conducive to sensation and intelligence. The same difference holds good with the counterpart of blood in other creatures; and thus we can explain why bees and other such creatures are of a more intelligent nature than many sanguineous [creatures], and among those that are sanguineous, those having cold and thin blood are more intelli-

[32] Plato, *Timaeus* 86D–E, LCL 234 (Bury), 234–35.

ἐναντίων ⟨ἐστίν⟩. ἄριστα δὲ τὰ τὸ θερμὸν ἔχοντα καὶ λεπτὸν καὶ καθαρόν⁴⁰ αἷμα· πρός τε γὰρ ἀνδρείαν τὰ τοιαῦτα καὶ πρὸς φρόνησιν ἔχει καλῶς· διὸ | καὶ τὰ ἄνω μόρια πρὸς τὰ κάτω ταύτην ἔχει τὴν διαφορὰν καὶ πρὸς τὸ θῆλυ τὸ ἄρρεν καὶ τὰ δεξιὰ πρὸς τὰ ἀριστερὰ τοῦ σώματος.

εὔδηλος μὲν οὖν ἐστιν ὡς ὁ Ἀριστοτέλης κἀκ ταύτης τῆς ῥήσεως ἕπεσθαι βουλόμενος τὰς τῆς ψυχῆς δυνάμεις τῇ φύσει⁴¹ τοῦ αἵματος οὐδὲν δ' ἧττον καὶ κατωτέρω τοῦ συγγράμματος ἀποφαινόμενος τὴν αὐτὴν δόξαν ἔγραψεν ὧδε·

τὰς δὲ καλουμένας ἶνας τὸ μὲν ἔχει αἷμα, τὸ δ' οὐκ ἔχει, οἷον τὸ τῶν ἐλάφων καὶ δορκάδων, διόπερ οὐ πήγνυται τὸ τοιοῦτον αἷμα· τοῦ γὰρ αἵματος τὸ μὲν ὑδατῶδες μᾶλλον ψυχρόν ἐστι, διὸ καὶ οὐ πήγνυται, τὸ δὲ γεῶδες πήγνυται ἐξατμιζομένου τοῦ ὑγροῦ ἐν ταῖς πήξεσι· αἱ δ' ἶνες γῆς εἰσιν. συμβαίνει δ' ἔνια καὶ γλαφυρωτέραν ἔχειν τὴν διάνοιαν τῶν τοιούτων, οὐ διὰ τὴν ψυχρότητα τοῦ αἵματος ἀλλὰ διὰ τὴν λεπτότητα μᾶλλον καὶ διὰ τὸ καθαρὸν εἶναι· τὸ γὰρ γεῶδες οὐδέτερον ἔχει τούτων. εὐκινητο-

---

⁴⁰ post καθαρόν: ἅμα γὰρ πρός τ' SM; αἷμα· πρός τε γὰρ K, B   ⁴¹ post τῇ φύσει: τοῦ σώματος SM; τοῦ αἵματος ἀπεφήνατο ἕπεσθαι K—see B, p. 39

THE SOUL'S TRAITS

gent than the opposites. Best of all are those animals whose blood is hot and also thin and clear. For such are well disposed regarding strength and practical wisdom. Consequently too, the upper parts of the body have the same preeminence over the lower parts, the male over the female, and the right side of the body over the left.[33]

It is, therefore, quite clear from this statement that Aristotle understood the capacities of the soul to follow the nature of the blood. No less is this also so, further on in the book, when expressing the same opinion, he wrote as follows:

The blood of some animals contains what are called fibers; the blood of others, for example, the blood of deer and gazelles, does not. Blood which lacks fibers does not congeal (coagulate), for the following reason. Part of the blood is of a more watery nature and therefore does not congeal, while the other part, which is earthy, congeals when the watery part evaporates away. The fibers are this earthy part. Now some of the animals whose blood is watery have a more subtle intellect, not due to the coldness, but due to the greater thinness and purity; the earthy has neither of these. Those [creatures]

---

[33] Aristotle, *Parts of Animals* 648a3–13, LCL 323 (A. L. Peck), 118–21 (translation after Peck with minor modifications).

τέραν γὰρ ἔχουσι τὴν αἴσθησιν τὰ λεπτοτέραν ἔχοντα τὴν ὑγρότητα καὶ καθαρωτέραν· διὰ γὰρ τοῦτο καὶ τῶν ἀναίμων ἔνια συνετωτέραν ἔχει τὴν ψυχὴν ἐνίων ἐναίμων, καθάπερ εἴρηται καὶ πρότερον, οἷον ἡ μέλιττα καὶ τὸ γένος τὸ τῶν μυρμήκων κἂν εἴ τι ἕτερον τοιοῦτόν ἐστι. δειλότερα δὲ τὰ λίαν ὑδατώδη· ὁ γὰρ φόβος καταψύχει. προωδοποίηται τοίνυν τῷ πάθει τὰ τοιαύτην ἔχοντα τὴν ἐν τῇ καρδίᾳ κρᾶσιν· τὸ γὰρ ὕδωρ τῷ ψυχρῷ πηκτόν[42] ἐστι· διὸ καὶ τἆλλα τὰ ἄναιμα δειλότερα τῶν ἐναίμων ἐστίν, ὡς ἁπλῶς εἰπεῖν, καὶ ἀκινητίζει τε φοβούμενα καὶ προίεται περιττώματα καὶ μεταβάλλει ἔνια τὰς χρόας αὐτῶν.

Τὰ δὲ πολλὰς ἔχοντα λίαν ἶνας καὶ παχείας γεωδέστερα τὴν φύσιν ἐστὶ καὶ θυμώδη τὸ ἦθος καὶ ἐκστατικὰ διὰ τὸν θυμόν· θερμότητος γὰρ παρεκτικὸς[43] ὁ θυμός. τὰ δὲ στερεὰ θερμανθέντα μᾶλλον[44] θερμαίνει τῶν ὑγρῶν, αἱ δ' ἶνες στερεὸν καὶ γεῶδες, ὥστε γίγνονται οἷον πυρίαι ἐν τῷ αἵματι καὶ ζέσιν ποιοῦσιν ἐν τοῖς θυμοῖς· διὸ οἱ ταῦροι καὶ οἱ κάπροι θυμώδεις καὶ ἐκστατικοί· τὸ γὰρ αἷμα τούτων ἰνωδέστατον καὶ τό γε

[42] πηκτόν SM: ποιητόν K
[43] ποιητικὸν SM: παρεκτικὸς K
[44] post μᾶλλον: θερμαίνει τῶν ὑγρῶν, αἱ δ' ἶνες στερεὸν καὶ γεῶδες, ὥστε γίγνονται οἷον πυρίαι ἐν τῷ αἵματι SM; εἰσὶν ἢ προσήκει θερμά, ὡς ἡ τῶν ἰνῶν φύσις ἐν τῷ ὑγρῷ.

THE SOUL'S TRAITS

having a thinner and more pure watery part have a perceptive faculty that is more easily moved. Due to this also some of the non-sanguineous [creatures] have a more intelligent soul than some of the sanguineous creatures, just as I also said earlier. For example, the bee and the class of the ants and even if there is another of such a class. Those that have excessively watery blood are more timid. This is because water is congealed by cold, and coldness also accompanies fear. Therefore, in those creatures whose hearts contain a predominantly watery *krasis*, the way is already prepared for a timorous disposition. This, too, is why, generally speaking, the nonsanguineous creatures are more timid than sanguineous ones, and why they become motionless when afraid, let loose their excretions, and why some change their colors.

On the other hand, those having very many and thick fibers are more earthy in nature and are spirited in character, and more excitable due to the heat, for the passion is able to cause heat. Solid bodies, when heated, give off more heat than those that are watery, so the fibers, which are solid and earthy, become as it were embers in the blood and cause it to seethe when the fits of passion come on. That is why bulls and boars are spirited and excitable, for in these animals the blood has the most fibers. Indeed, the blood of the bull congeals (co-

---

αἱ γὰρ ἶνες στερεώτεραι καὶ γεωδέστεραι γίνονται, οἷον αἱ πυρίαι ἐν τῷ σώματι K

GALEN

τοῦ ταύρου τάχιστα πήγνυται πάντων. ἐξαιρουμένων δὲ τούτων τῶν ἰνῶν οὐ | πήγνυται τὸ αἷμα, [αἱ γὰρ ἶνες γεώδεις] καθάπερ γὰρ ἐκ πηλοῦ εἴ τις ἐξέλοι τὸ γεῶδες, οὐ πήγνυται τὸ ὕδωρ, οὕτω καὶ τὸ αἷμα· αἱ γὰρ ἶνες γῆς· μὴ ἐξαιρουμένων δὲ πήγνυται, οἷον ἂν πάθῃ ἡ ὑγρὰ γῆ ὑπὸ ψύχους.[45] τοῦ γὰρ θερμοῦ ὑπὸ τοῦ ψυχροῦ ἐκθλιβομένου συνεξατμίζεται τὸ ὑγρόν, καθάπερ εἴρηται πρότερον, καὶ πήγνυται οὐχ ὑπὸ θερμοῦ ἀλλ᾽ ὑπὸ ψυχροῦ ξηραινόμενον· ἐν δὲ τοῖς σώμασιν ὑγρόν ἐστι διὰ τὴν θερμότητα τὴν ἐν τοῖς ζῴοις.

ταῦτα προειπὼν ὁ Ἀριστοτέλης ἐφεξῆς αὐτοῖς συνάπτει ταῦτα·

πολλῶν δ᾽ ἐστὶν αἰτία ἡ τοῦ αἵματος φύσις καὶ κατὰ τὸ ἦθος τοῖς ζῴοις καὶ κατὰ τὴν αἴσθησιν, εὐλόγως· ὕλη γάρ ἐστι παντὸς τοῦ σώματος· ἡ γὰρ τροφὴ ὕλη, τὸ δ᾽ αἷμα ἡ ἐσχάτη τροφή. πολλὴν οὖν ποιεῖ διαφορὰν θερμὸν ὂν καὶ ψυχρὸν καὶ λεπτὸν καὶ παχὺ καὶ καθαρὸν καὶ θολερόν.

οὐσῶν δὲ καὶ ἄλλων αὐτοῦ ῥήσεων ἐν ταῖς περὶ τῶν ζῴων αὐτοῦ πραγματείαις καὶ τοῖς τῶν Προβλημάτων

[45] post ψύχους: ἂν πάθοι (K) om.

THE SOUL'S TRAITS

agulates) quickest of all. However, just as when the fibers are removed, | the blood doesn't congeal for the fibers are earthy. But just as when the earthy component is taken out of mud, the watery component that remains does not congeal, so when the fibers, which are earthy, are taken out of the blood, it no longer congeals. When they remain, however, it does congeal, like wet earth would when affected by cold. For when the heat is forced out by the cold, the fluid is caused to evaporate, as was said before, so it is due to the solidifying effect of the cold and not of the hot, that what remains becomes congealed. And while it is in the body the blood is fluid on account of the heat which is there.[34]

794K

Aristotle, having previously said these things, next adds to them the following:

It is reasonable, however, that the nature of the blood is a cause of many things, both in regard to character in the animals and to sensation, for blood is a material for the whole body, material is nutriment, and blood is the ultimate nutriment. Therefore, it makes a considerable difference whether it is hot or cold, thin or thick, pure or muddy.[35]

And there are his other statements in his treatises about animals and in the books of his *Problems*; it seemed to me

---

[34] Aristotle, *Parts of Animals* 650b14 to 651a14, LCL 323 (Peck), 136–41 (translation after Peck with minor modifications).

[35] Aristotle, *Parts of Animals* 651a14–17, LCL 323 (Peck), 140–41.

αὐτοῦ βιβλίοις ἔδοξέ μοι περιττὸν εἶναι παραγράφειν ἁπάσας· ἀρκεῖ γὰρ μίαν [τὴν] Ἀριστοτέλους ἐνδείξασθαι γνώμην, ἣν ἔχει περί τε τῶν κράσεων τοῦ σώματος καὶ τῶν τῆς ψυχῆς δυνάμεων· ὅμως δὲ προσθήσω καὶ τὰ κατὰ τὸ πρῶτον εἰρημένα τῆς τῶν ζῴων ἱστορίας, ὧν τινὰ μὲν ἄντικρυς εἰς τὴν κρᾶσιν ἀνάγεται, τινὰ δὲ διὰ μέσων τῶν φυσιογνωμονικῶν σημείων καὶ μάλιστα κατὰ τὸν Ἀριστοτέλην. βούλεται γὰρ οὗτος τοῖς τῆς ψυχῆς ἤθεσί τε καὶ δυνάμεσιν οἰκείαν γίγνεσθαι τὴν διάπλασιν ὅλου τοῦ σώματος ἑκάστῳ γένει τῶν ζῴων. οἷον αὐτίκα τῶν ἐναίμων ἡ γένεσις μέν ἐστιν ἐκ τοῦ τῆς μητρὸς αἵματος, ἀκολουθεῖ δὲ τῇ κράσει τούτου τὰ τῆς ψυχῆς ἤθη, καθότι καὶ διὰ τῶν προγεγραμμένων ῥήσεων ἀπεφήνατο· τῶν δ' ὀργανικῶν μορίων ἡ διάπλασις οἰκεία τοῖς τῆς ψυχῆς ἤθεσι γίγνεται κατ' αὐτὸν τὸν Ἀριστοτέλην· καὶ κατὰ τοῦτο δὴ λοιπὸν οὐκ ὀλίγα πέφυκε γνωρίσματα[46] περί τε τῶν τῆς ψυχῆς ἠθῶν καὶ τῆς τοῦ σώματος κράσεως. ἔνια δὲ τῶν φυσιογνωμονικῶν ἄντικρύς τε καὶ δι' οὐδενὸς μέσου τὴν κρᾶσιν ἐνδείκνυται. τοιαῦτα δ' ἐστὶ κατὰ τὰς χρόας καὶ τρίχας, ἔτι δὲ τὰς φωνὰς καὶ τὰς ἐνεργείας τῶν μορίων. ἀκούσωμεν οὖν ἤδη τῶν ὑπ' Ἀριστοτέλους γεγραμμένων ἐν τῷ πρώτῳ Περὶ τῶν ζῴων ἱστορίας·

[46] γνωρίσματα add. SM; ⟨σημεῖα⟩ B

THE SOUL'S TRAITS

superfluous to subjoin them all. It is enough for me to have pointed out one opinion Aristotle holds regarding the *krasias* of the body and the capacities of the soul. Nevertheless, I shall add also those things mentioned in the first book of the *History of Animals*, some of which refer directly to the *krasis,* and some through the medium of the physiognomical signs, and this particularly applies to Aristotle. For this man wishes [to attribute] the proper conformation of the whole body in each class of animals to the characteristics and capacities of the soul. For example, the genesis of sanguineous animals is directly from the blood of the mother, while the characteristics of the soul follow the *krasis* of this, just as he also declared in the previously quoted statements. However, the conformation of the organic parts is specific to the characteristics of the soul, according to this same Aristotle. And in relation to what remains, not a few signs naturally concern the characteristics of the soul and the *krasis* of the body. However some of the physiognomical signs indicate the *krasis* directly and not through any intermediary. Such things pertain to complexions and hair, and further, to voices and the functions of the parts. Therefore, let us now hear what is written by Aristotle in the first book of *History of Animals*.[36]

---

[36] Aristotle, *History of Animals*, Book 1, LCL 437 (Peck). The six statements run from 491b11 to 492a3, pp. 38–45, with interspersed passages (translation after Peck with minor modifications).

προσώπου δὲ τὸ μὲν ὑπὸ τὸ βρέγμα μεταξὺ τῶν
ὀμμάτων μέτωπον· τοῦτο δ' οἷς μὲν μέγα, βρα-
δύτεροι, οἷς δὲ μικρόν, εὐκίνητοι, καὶ οἷς μὲν
πλατύ, ἐκστατικοί.

μία μὲν αὕτη ῥῆσις· ἑτέρα δ' οὐ μετὰ πολὺ τῆσδε
τόνδε τὸν τρόπον ἔχουσα·

ὑπὸ δὲ τῷ μετώπῳ ὀφρύες διφυεῖς,[47] ὧν αἱ μὲν
εὐθεῖαι φύουσαι μαλακοῦ ἤθους σημεῖον, αἱ δὲ
πρὸς τὴν ῥῖνα τὴν καμπυλότητα ἔχουσαι στρυ-
φνοῦ, αἱ δὲ πρὸς τοὺς κροτάφους μωκοῦ καὶ
εἴρωνος, αἱ δὲ κατεσπασμέναι φθόνου.

εἶτα πάλιν μετ' οὐ πολύ

κοινὸν δὲ τῆς βλεφαρίδος μέρος τῆς ἄνω καὶ
κάτω κανθοὶ δύο, ὁ μὲν πρὸς τῇ ῥινί, ὁ δὲ πρὸς
τοῖς κροτάφοις, οἳ ἂν μὲν ὦσι μακροί, κακοη-
θείας σημεῖον,[48] ἂν δὲ βραχεῖς, ἤθους βελτίονος
ἐὰν δ' οἷον οἱ κτένες κρεώδες ἔχωσι πρὸς τῷ
μυκτῆρι, πονηρίας.

[εἶτ' ἐφεξῆς πάλιν "αἱ ὀφρύες ἂν ὦσι κατεσπασμέναι,
φθόνου."][49] καὶ μετὰ τοῦτο πάλιν.

[47] διφυεῖς (K, SM) om. B
[48] post σημεῖον,: ἂν/οἷς δὲ βραχεῖς, ἤθους βελτίονος. K,
B; om. SM
[49] εἶτ' ἐφεξῆς πάλιν "αἱ ὀφρύες ἂν ὦσι κατεσπασμέναι,
φθόνου" add. SM

THE SOUL'S TRAITS

> That part of the face which is below the bregma and between the eyes is the *metopon* (brow, forehead). Those in whom this is large are rather slow; those in whom it is small are easily moved (changeable); those in whom it is flat are excitable; [those in whom it is rounded are irascible].[37]

This is one statement; another, not much further on, has the following form:

> Below the forehead are two eyebrows. Eyebrows that grow straight are a sign of a soft character; those that bend toward the nose are a sign of a harsh character; those that grow toward the temples are a sign of a mocking and dissimulating character; those that are drawn down, of envy.[38]

Then in turn, not much after:

> Common to the upper and lower parts of the eyelid are two canthi—one toward the nose and one toward the temples. Those that are long are a sign of *kakoethia* (a malicious disposition); if they are short, of a better disposition. If the *carunculi lachrymalis* are fleshy toward the nostrils, it is a sign of wickedness.[39]

And again after this:

---

[37] This last statement is not included by Galen.
[38] See Peck's note 2 on this final clause.
[39] There are some textual issues with this sentence; see Peck, note 1, p. 40.

ὀφθαλμῶν δὲ τὸ μὲν λευκὸν ὅμοιον ὡς ἐπὶ τὸ πολὺ πᾶσι, τὸ δὲ καλούμενον μέλαν διαφέρει· τοῖς μὲν γάρ ἐστι μέλαν, | τοῖς δὲ σφόδρα γλαυκόν, τοῖς δὲ χαροπόν, [ἐνίοις δ' αἰγωπόν, ὃ ἤθους][50] βελτίστου σημεῖον [ἐστί].[51]

καὶ δὴ καὶ τούτων ἐφεξῆς τάδε γράφει·

τῶν δ' ὀφθαλμῶν οἱ μὲν μεγάλοι, οἱ δὲ μικροί, οἱ δὲ μέσοι· οἱ μέσοι βέλτιστοι. καὶ ἡ ἐκτὸς σφόδρα ἢ ἐντὸς ἢ μέσως· τούτων οἱ ἐντὸς μάλιστα ὀξυωπέστατοι ἐπὶ παντὸς ζῴου, τὸ δὲ μέσον ἤθους βελτίστου σημεῖον. καὶ ἢ σκαρδαμυκτικοὶ ἢ ἀτενεῖς ἢ μέσοι· βελτίστου δ' ἤθους οἱ μέσοι, ἐκείνων δ' οἱ μὲν ἀναιδεῖς, οἱ δ' ἀβέβαιοι.

καὶ πάλιν οὐ μετὰ πολὺ κατὰ τὸν περὶ τῶν ὤτων λόγον ὡδί πως ἔγραψε περὶ τοῦ μεγέθους αὐτῶν·

καὶ ἢ μεγάλα ἢ μικρὰ ἢ μέσα ἢ ἐπανεστηκότα σφόδρα ἢ οὐδὲν ἢ μέσον· τὰ δὲ μέσα βελτίστου ἤθους σημεῖον, τὰ δὲ μεγάλα καὶ ἐπανεστηκότα μωρολογίας καὶ ἀδολεσχίας.

ταῦτα μὲν ἐν τῷ πρώτῳ Περὶ ζῴων ἱστορίας ὁ Ἀριστοτέλης ἔγραψεν, οὐκ ὀλίγων δὲ μέμνηται καὶ κατ'

[50] ἐνίοις δ' αἰγωπόν, ὃ ἤθους add. SM
[51] post ἐστί: καὶ πρὸς ὀξύτητα ὄψεως κράτιστον. (K) om.

THE SOUL'S TRAITS

> The white of the eyes is similar in almost everyone but the so-called black shows differences. For in some it is black, | in some it is strongly gray, in some grayish-blue, and in some goat-like, which is a sign of the best [character] and the best for sharpness of sight.

797K

And further on after these, he writes as follows:

> Eyes may be large, small, or intermediate. The intermediate are the best. And there are those that are strongly protruding, those that are deep-set and those that are intermediate. Of these, those that are deep-set are particularly the most keen-sighted in the case of every animal, while those that are intermediate are a sign of the best character. And there are those that tend to blink, those that are unblinking, and those that are intermediate. The intermediate are a sign of the best character. Of the others, the first indicates instability and the second impudence.

And again, not much after, in the discussion about the ears, he wrote about the size of these as follows:

> They may be large, or small, or of medium size. They may be strongly prominent, or not at all, or intermediate. The last are a sign of the best character, whereas the large and prominent are a sign of garrulousness and foolish talk.

Aristotle wrote these things in the first book of his *History of Animals*. Quite a few are also mentioned in another

GALEN

ἄλλο σύγγραμμα φυσιογνωμονικῶν θεωρημάτων, ὧν καὶ παρεθέμην ἄν τινας ῥήσεις, εἰ μήτε μακρολογίας ἔμελλον ἀποίσεσθαι δόξαν ἀναλίσκειν τε τὸν χρόνον μάτην ἐξὸν ἐπὶ τὸν πάντων ἰατρῶν τε καὶ φιλοσόφων πρῶτον εὑρόντα τὴν θεωρίαν ταύτην ἀφικέσθαι μάρτυρα, τὸν θεῖον Ἱπποκράτην.

8. Γράφει τοίνυν αὐτὸς ἐν τῷ βιβλίῳ, ἐν ᾧ περὶ ἀέρων καὶ τόπων καὶ ὑδάτων διδάσκει, πρῶτον μὲν ἐπ' ἐκείνων τῶν πόλεων, ἃς ἐστράφθαι φησὶ πρὸς ἄρκτους, ὡδί πως αὐτοῖς ὀνόμασι· "τά τε ἤθεα ἀγριώτερα ἢ ἡμερώτερα" καὶ μετὰ ταῦτα πάλιν ἐπὶ τῶν πρὸς ἀνατολὴν ἐστραμμένων ὡδί· "λαμπρόφωνοί τε οἱ ἄνθρωποι ὀργήν τε καὶ ξύνεσιν βελτίους τῶν πρὸς βορέην."

Ἔπειτα προελθὼν ἐπὶ πλέον περὶ τῶν αὐτῶν ὡδί πως διεξέρχεται·

> τὴν Ἀσίην πλεῖστον διαφέρειν φημὶ τῆς Εὐρώπης ἐς τὰς φύσιας τῶν ξυμπάντων τῶν τ' ἐκ τῆς γῆς φυομένων καὶ τῶν ἀνθρώπων· πολὺ γὰρ καλλίονα καὶ μείζονα πάντα γίγνεται ἐν τῇ Ἀσίῃ· ἥ τε χώρη τῆς χώρης ἡμερωτέρη καὶ τὰ

---

[40] This work, *Physiognomics*, is attributed to an unknown author writing around 300 BC and is included in the Aristotelian corpus; see, for example, LCL 307 (W. S. Hett), 83–137. The opening sentence reads: "Dispositions (*dianoiai*) follow bodies and are not in and of themselves unaffected by the changes of the body."

THE SOUL'S TRAITS

book about physiognomical speculations,[40] from which I would add some statements except this would bring with it a reputation for long-windedness and wasting time in vain, when we have, to come to as a witness, the first of all doctors and philosophers to discover this theory—the divine Hippocrates.

8. Accordingly, he writes in this way in the book in which he teaches about airs, places, and waters,[41] first in the case of those cities which he says are turned toward north, with these very words as follows: "The characteristics are more savage rather than more civilized." (1) And after these again, in the case of those turned toward east: "The people are clear-voiced and better in respect of spirit and intelligence than those toward north." (2)

Then, going on still further about these same matters, he continues as follows:

> I hold that Asia differs very greatly from Europe in the nature of all its inhabitants, and of all those things grown from the earth, and of the people. For everything in Asia grows to far greater beauty and size; the one region is less wild than the other, the

---

[41] Hippocrates, *Airs, Waters, Places*, LCL 147 (W. H. S. Jones). The fourteen separated quotations are as follows (numbered in parentheses): (1) 4.32, pp. 78–79; (2) 5.19–20, pp. 80–81; (3) 12.7–14, pp. 104–7; (4) 12.24–26, pp. 106–7; (5) 12.40–44, pp. 106–9; (6) 16.3–8, pp. 114–15; (7) 16.39–43, pp. 116–17; (8) 23.19–21, pp. 132–33; (9) 24.7–22, pp. 132–35; (10) 24.28–31, pp. 134–35; (11) 24.32–36, pp. 134–35; (12) 24.43–45, pp. 136–37; (13) 24.45–53, pp. 136–37; (14) 24.53–63, pp. 136–37.

ἤθεα τῶν ἀνθρώπων[52] ἠπιώτερα καὶ εὐοργητό-
τερα. τὸ δὲ αἴτιον τούτων ἡ κρῆσις τῶν ὡρέων.

τὴν κρᾶσιν εἶναι αἰτίαν φησὶν οὐ μόνον τῶν ἄλλων
ὧν διῆλθεν ἀλλὰ καὶ τῶν ἠθῶν. ὅτι δὲ τὴν τῶν ὡρῶν
κρᾶσιν ἐν τῇ θερμότητι καὶ ψυχρότητι, ξηρότητί τε
καὶ ὑγρότητι διαλλάττειν ἑτέραν τῆς ἑτέρας φησί,
παμπόλλας ῥήσεις παρεθέμην ἐν τῇ πραγματείᾳ,
καθ' ἣν ἐπιδείκνυμι τὴν αὐτὴν φυλάττοντα δόξαν
αὐτὸν περί τε τῶν στοιχείων ἔν τε τῇ Περὶ φύσιος
ἀνθρώπου βίβλῳ καὶ κατ' ἄλλα πάντα συγγράμματα.
ἀλλὰ καὶ κατὰ τὰς ἑπομένας ῥήσεις τῇ προκειμένῃ
ταὐτὸν διδάσκων ὡδὶ γράφει περὶ τῆς εὐκράτου
χώρας, ἣν καὶ τὰ τῶν ἀνθρώπων ἤθη ποιεῖν φησιν
εὔκρατα· "οὔτε γὰρ ὑπὸ τοῦ θερμοῦ ἐκκέκαυται λίην[53]
οὔτε ὑπὸ αὐχμῶν καὶ ἀνυδρίης ἀνεξήρανται οὔτε ὑπὸ
ψύξιος πέττεται." διὰ τοῦτ' οὖν ἐφεξῆς φησι

τὸ δὲ ἀνδρεῖον καὶ τὸ ταλαίπωρον καὶ τὸ εὔτονον[54] καὶ τὸ θυμοειδὲς οὐκ ἂν δύναιτο ἐν τῇ
τοιαύτῃ φύσει ἐγγίγνεσθαι οὔτε ὁμοφύλου οὔτε
ἀλλοφύλου ἀλλὰ τὴν ἡδονὴν ἀνάγκη κρατέειν.

καὶ μέντοι κατωτέρω πάλιν ἐν ταὐτῷ γράμματι τάδε
γράφει·

[52] post ἀνθρώπων: ἠπιώ‹τερα καὶ εὐοργητό›τερα. SM; ἡμερώτερα Κ   [53] λίην SM, B; λέγει Κ
[54] καὶ τὸ εὔτονον add. SM, B

THE SOUL'S TRAITS

characters of the people milder and more gentle. The *krasis* of the seasons is the cause of these things. (3)

He says the *krasis* is not only a cause of the other things he went through, but also of the characteristics. He says that in respect of the *krasis*, the seasons differ from one another in terms of hot and cold, dry and wet; I set down very many statements in the treatise in which I show him to maintain the same opinion regarding the elements in the book, *On the Nature of Man*,[42] and all his other works. But also in the statements following what was previously said before, he teaches the same things when he writes in this way about the *eukratic* place which he says also makes the characteristics of the people *eukratic*: "For it has not been burned by the heat overmuch, nor dried up by the effects of drought and lack of water, nor by cold." (4) Due to these things, then, he says next:

Bravery, endurance, vigor and spiritedness could not come into being in such a nature, neither of the same nor of another race, but of necessity pleasure prevails. (5)

And indeed, again lower down in this work, he writes thus:

---

[42] Hippocrates, *On the Nature of Man*, LCL 150 (W. H. S. Jones), 1–42; Craik, *The Hippocratic Corpus* 36, pp. 207–13.

περὶ δὲ τῆς ἀθυμίης τῶν ἀνθρώπων καὶ τῆς
ἀνανδρείης ὅτι ἀπολεμώτεροί εἰσι τῶν Εὐρω-
παίων οἱ Ἀσιηνοὶ καὶ ἡμερώτεροι τά τε ἤθεα,
αἴ τε ὧραι αἴτιαι μάλιστα οὐ μεγάλας τὰς
μεταβολὰς ποιεόμεναι οὔτε ἐπὶ τὸ θερμὸν οὔτε
ἐπὶ τὸ ψυχρὸν ἀλλὰ παραπλήσιαι.

καὶ μέντοι καὶ μετ᾽ ὀλίγα πάλιν οὕτως εἶπεν·

εὑρήσεις δὲ καὶ τοὺς Ἀσιηνοὺς διαφέροντας
αὐτοὺς ἑωυτῶν καὶ τοὺς μὲν βελτίονας, τοὺς δὲ
φαυλοτέρους ἐόντας· τούτων δὲ αἱ μεταβολαὶ
αἴτιαι τῶν ὡρέων, ὥσπερ μοι εἴρηται ἐν τοῖσι
προτέροισι

καὶ κατωτέρω τοῦ συγγράμματος, ἡνίκα περὶ τῶν τὴν
Εὐρώπην οἰκούντων ὁ λόγος αὐτῷ γίγνεται, τάδε γρά-
φει· "τό τε ἄγριον καὶ τὸ ἀμείλικτον[55] καὶ τὸ θυμοειδὲς
ἐν τῇ τοιαύτῃ φύσει ἐγγίγνεται." καὶ μετὰ ταῦτα
πάλιν ἐν ἑτέρᾳ ῥήσει γράφει ταυτί·

ὅσοι μὲν ὀρεινὴν χώρην οἰκέουσι καὶ τρηχεῖαν
καὶ ὑψηλὴν καὶ εὔυδρον καὶ αἱ μεταβολαὶ αὐ-
τέοισι γίγνονται τῶν ὡρέων μεγάλαι,[56] ἐνταῦθα
εἰκὸς εἴδεα μεγάλα εἶναι καὶ πρὸς τὸ ταλαίπωρον καὶ τὸ ἀνδρεῖον εὖ πεφυκότα, καὶ τό τε
ἄγριον καὶ τὸ θηριῶδες αἱ τοιαῦται φύσιες οὐχ
ἥκιστα ἔχουσιν· ὅσοι δὲ κοῖλα χωρία καὶ λειμακώδεα[57] καὶ πνιγηρὰ καὶ τῶν θερμῶν πνευ-
μάτων πλέον μέρος μετέχουσιν ἢ τῶν ψυχρῶν

THE SOUL'S TRAITS

> Concerning the lack of spirit of the people and their want of bravery, the main reason | Asian men are less warlike and more civilized in their characters than the Europeans is particularly the uniformity in the seasons which show no major changes, either toward hot or cold, but are nearly the same. (6)

800K

And in fact, a little further on, he again speaks in this way:

> You will also find differences among the Asians themselves, some being better and some worse. The reason for this, as I have said above, is the changes of the seasons. (7)

And further down in the book, when his discussion comes to be about those living in Europe, he writes as follows: "Wildness, antisocial and spirited behavior arise in such a nature." (8) And after this again, in another statement, he writes thus:

> Inhabitants of a region that is mountainous, rugged, high and well-watered, where the changes of the seasons show sharp contrasts, are likely to be tall in stature and well developed toward endurance and bravery, and in fact such natures have no little wildness and brutality. However, those living in hollow places which are pestilential and stifling and partake of hot winds more than cold, and where

---

55 ἀμείλικτον SM; ἄμικτον K
56 μεγάλαι SM; μέγα δὲ τὸ διάφορον K
57 λειμακώδεα SM; λοιμώδεα K

ὕδασί τε χρέωνται θερμοῖσιν, οὗτοι δὲ μεγάλοι
μὲν οὐκ ἂν εἴησαν οὐδὲ κανονίαι,[58] ἐς εὖρος δὲ
πεφυκότες καὶ σαρκώδεες καὶ μελανότριχες· καὶ
αὐτοὶ μέλανες μᾶλλον ἢ λευκότεροι φλεγματίαι
τε ἧσσον ἢ χολώδεες, τὸ δὲ ἀνδρεῖον καὶ τὸ
ταλαίπωρον ἐν τῇ ψυχῇ φύσει μὲν οὐκ ἂν
ὁμοίως ἐνείη, νόμος δὲ προσγενόμενος ἀπεργά-
σοιτ᾽ ἄν.

νόμον εἴρηκε δὲ δηλονότι τὴν νόμιμον ἐν ἑκάστῃ
χώρᾳ τοῦ βίου διαγωγήν, ἣν δὴ καὶ τροφὴν καὶ παι-
δείαν καὶ συνήθειαν ἐπιχώριον ὀνομάζομεν, οὗ καὶ
αὐτοῦ μεμνήσομαι πρὸς τὸν ὀλίγον ὕστερον εἰρησό-
μενον λόγον. ἐν γὰρ τῷ παρόντι προσθεῖναι ταύτας
ἔτι βούλομαι τὰς ῥήσεις αὐτοῦ·

ὅσοι δὲ ὑψηλὴν οἰκέουσι χώρην καὶ λείην καὶ
ἀνεμώδεα καὶ εὔυδρον, εἶεν ἂν εἴδεα μεγάλοι
καὶ ἑωυτοῖσι παραπλήσιοι, καὶ ἀνανδρότεραι
καὶ ἡμερώτεραι τούτων αἱ γνῶμαι·

καὶ συνάπτων γε τὰ ἐφεξῆς περὶ τὸ χωρίον ἔτι τάδε
γράφει·

ὅσοι δὲ λεπτὰ καὶ ἄνυδρα καὶ ψιλὰ καὶ τῇσι
μεταβολῇσι τῶν ὡρέων οὐκ εὔκρητα, ἐν ταύτῃ
τῇ χώρῃ τὰ εἴδεα εἰκὸς[59] σκληρὰ καὶ εὔτονα[60] ἢ
καὶ ξανθότερα εἶναι ἢ μελάντερα καὶ τὰ ἤθεα
καὶ τὰς ὀργὰς αὐθάδεάς τε καὶ ἰδιογνώμονας.

THE SOUL'S TRAITS

hot waters are used are large, | not tall but broad in stature and well-fleshed, dark-haired, more dark than pale, and less phlegmatic than bilious. Similarly, courage and endurance are not in the soul by nature, but will be brought about when custom (*nomon*) is added. (9)

801K

I have said *nomon* obviously in respect of the customary life followed in each country, which certainly we also call nurture, upbringing, and the customary practice of the country—this is also something that will be mentioned when the discussion is continued a little later. For the present, I would still like to add these statements of his.

Those who live in a high country that is flat, windy and well-watered will be tall in stature and like each other in their dispositions, but will be rather unmanly and more gentle. (10)

And he subjoins these things next about the place, when he writes further as follows:

In those who dwell in places that are thin, dry and bare and where the changes of the seasons are not *eukratic* in this place, the bodily forms are likely to be hard and strong, they will be fairer rather than darker, | and in their characteristics and impulses, stubborn and independent. (11)

802K

---

[58] κανονίαι SM; εὐμήκεες K
[59] εἰκὸς SM; ἀλλά K
[60] post εὔτονα: [ἢ] καὶ ξανθότερα εἶναι ἢ μελάντερα καὶ τὰ ἤθεα SM, B; καὶ ξανθότερα, οἷς μελανώτερα εἶναι τὰ εἴδεα K

GALEN

καὶ τί γὰρ ἵνα μὴ πολλῶν αὐτοῦ μνημονεύσω ῥήσεων, ἐφεξῆς ἐρεῖ

εὑρήσεις γὰρ ἐπὶ τὸ πολὺ τῆς χώρης τῇ φύσει ἀκολουθέοντα καὶ τὰ εἴδεα τῶν ἀνθρώπων καὶ τοὺς τρόπους.

αὐτὴν δὲ δηλονότι τὴν χώραν κατὰ τὸ θερμόν τε καὶ ψυχρόν, ὑγρόν τε καὶ ξηρὸν ἑτέρας χώρας διαφέρειν εἴρηκε πολλάκις ἐν τῷ συγγράμματι. διὰ τοῦτ' οὖν καὶ πάλιν ἐφεξῆς ἐρεῖ

ὅκου μὲν γὰρ ἡ γῆ πίειρα καὶ μαλθακὴ καὶ εὔυδρος καὶ τὰ ὕδατα κάρτα μετέωρα, ὥστε θερμὰ εἶναι τοῦ θέρεος καὶ τοῦ χειμῶνος ψυχρὰ καὶ τῶν ὡρέων καλῶς κεῖται, ἐνταῦθα καὶ οἱ ἄνθρωποι σαρκώδεές εἰσι καὶ ἄναρθροι καὶ ὑγροὶ καὶ ἀταλαίπωροι καὶ τὴν ψυχὴν κακοὶ ὥστε ἐπὶ τὸ πολύ· τό τε ῥάθυμον καὶ τὸ ὑπνηλὸν ἔνεστιν ἐν αὐτοῖσιν,[61] ἔς τε τὰς τέχνας παχέες καὶ οὐ λεπτοὶ οὐδὲ ὀξέες.

ἐν τούτοις πάλιν ἐδήλωσε σαφέστατα μὴ μόνον τὰ ἤθη ταῖς τῶν ὡρῶν κράσεσιν ἀλλὰ καὶ τὴν ἀμβλύτητα τῆς διανοίας ὥσπερ οὖν καὶ τὴν σύνεσιν ἑπομένην. ὅμοια δὲ τούτοις καὶ τὰ κατὰ τὴν ἐχομένην ῥῆσιν ὑπ' αὐτοῦ· γέγραπται γὰρ αὐτοῖς ὀνόμασιν

ὅκου δέ ἐστιν ἡ γῆ ψιλή τε καὶ ἄνυδρος καὶ τρηχεῖα καὶ ὑπὸ τοῦ χειμῶνος πιεζομένη καὶ

THE SOUL'S TRAITS

What, so that I will not call to mind too many of his statements, does he say next?

> For in general you will find assimilated to the nature of the place both the physique and characteristics of the inhabitants. (12)

He has said often in the work that the country itself clearly differs in relation to hot, cold, dry and wet from other countries. Because of this then, he again says next:

> For where the land is rich, soft, and well-watered and the waters are very near the surface, so as to be hot in the summer and cold in winter, and if the situation is favorable as regards the seasons, there the people are well-fleshed and without visible joints, wet and miserable, and bad in respect of the soul. So for the most part it is possible to see laziness and sleepiness in them, and in the arts they are thick-witted and not subtle or sharp-witted. (13)

In these again, he showed very clearly that not only the characteristics follow the *krasias* of the seasons but also the dullness of the intellect, just as also the understanding. Similar to these and pertaining to his adjacent statement, he has written these very words:

> Where, | however, the land is bare, waterless and 803K
> rough, and is weighed down by the winter and

---

61 αὐτοῖσιν SM; αὐτοῖς ἰδεῖν K

ὑπὸ τοῦ ἡλίου κεκαυμένη, ἐνταῦθα δὲ σκληρούς
τε καὶ ἰσχνοὺς καὶ διηρθρωμένους καὶ εὐτόνους
καὶ δασέας τό τε ἐργαστικὸν ὀξὺ ἐνεὸν ἐν τῇ
φύσει τῇ τοιαύτῃ καὶ τὸ ἄγρυπνον τά τε ἤθεα
καὶ τὰς ὀργὰς αὐθάδεας καὶ ἰδιογνώμονας τοῦ
τε ἀγρίου μᾶλλον μετέχοντας ἢ τοῦ ἡμέρου ἔς
τε τὰς τέχνας ὀξυτέρους τε καὶ ξυνετωτέρους
καὶ τὰ πολέμια ἀμείνονας εὑρήσεις.

ἐν τούτῳ πάλιν τῷ λόγῳ σαφῶς οὐ μόνον τὰ ἤθη ταῖς
τῆς χώρας κράσεσιν ἀκόλουθά φησιν ὑπάρχειν ἀλλὰ
καὶ πρὸς τὰς τέχνας τοὺς μὲν ὀξυτέρους, τοὺς δ' ἀσυνετωτέρους[62] γίγνεσθαι, τουτέστιν τοὺς μὲν συνετω-
τέρους, τοὺς δ' ἀμβλυτέρους τε καὶ παχέας τὴν διάνοιαν. οὐκέτ᾽ οὖν δέομαι τῶν κατὰ τὸ δεύτερον καὶ
ἕκτον τῶν Ἐπιδημιῶν γεγραμμένων φυσιογνωμονικῶν γνωρισμάτων ἁπάντων μνημονεύειν ἀλλ᾽ ἀρκέσει
παραγράψαι παραδείγματος ἕνεκα τήνδε τὴν λέξιν
αὐτοῦ·

ᾧ ἂν ἡ φλὲψ ἡ ἐν ἀγκῶνι σφύζῃ, μανικὸς καὶ
ὀξύθυμος, ᾧ δὲ ἂν ἀτρεμέῃ, τυφώδης.

κατὰ δὲ τὴν ῥῆσιν ὁ λόγος ἐστὶ τοιοῦτος· ὧν ἀνθρώπων ἡ κατὰ τὸν ἀγκῶνα ἀρτηρία σφοδροτάτην ποι-
εῖται κίνησιν αὐτῆς, οὗτοι μανικοί εἰσι. φλέβας μὲν

[62] post ἀσυνετωτέρους: γίγνεσθαι, τουτέστιν τοὺς μὲν συ-
νετωτέρους, add. SM

THE SOUL'S TRAITS

burned up by the sun, there you will see those who are hard, lean, with prominent joints, vigorous and hirsute, while there is keen industriousness in such a nature, and sleeplessness, and with respect to characteristics, also the impulses are stubborn and opinionated; they partake more of fierceness than of gentleness, while when it comes to the arts, they are sharper and more intelligent and you will find them better in matters of war. (14)

In this description, he again says clearly that not only are the characteristics consequent upon the *krasias* of the country, but also that some are sharper in regard to the arts, while some are more unintelligent. That is, some are more astute while some are more obtuse and thick-witted in respect of the intellect. I do not, therefore, still need to call to mind all the physiognomonic signs of the second and sixth books of the *Epidemics* he has written. It is sufficient to write in addition for the sake of an example, the following statement of his:

One in whom the blood vessel in the elbow (antecubital fossa) pulsates is manic and quick to anger; one in whom it is still is stuporous.[43]

The argument in this statement is this: Those people in whom the artery in the antecubital fossa (elbow) | makes a very strong movement are manic. For the ancients also

---

[43] Hippocrates, *Epidemics* 2.16, LCL 477 (W. D. Smith), 78–79. "Stuporous" is Smith's rendering of τυφώδης. LSJ gives the meaning as "delirious," citing *Epidemics* 4.2 and other sources.

γὰρ καὶ τὰς ἀρτηρίας ἐκάλουν οἱ παλαιοί, ὡς δέδεικται πολλάκις, οὐδέπω δὲ πᾶσαν ἀρτηριῶν κίνησιν ὠνόμαζον σφυγμὸν ἀλλὰ μόνον τὴν αἰσθητὴν αὐτὴν ἐν τῷ ἀνθρώπῳ πάντως οὖσαν σφοδράν. Ἱπποκράτης δ' ὡς καὶ πρῶτος ἄρξας τοῦ μετὰ ταῦτα κρατήσαντος ἔθους εἴρηκέ γὰρ που σφυγμὸν ἅπασαν τῶν ἀρτηριῶν τὴν κίνησιν, ὁποία τις ἂν ᾖ· κατὰ μέντοι τήνδε τὴν ῥῆσιν ἔτι τῷ παλαιῷ τρόπῳ τῆς ἑρμηνείας χρώμενος ἐκ σφοδρᾶς κινήσεως ἀρτηρίας ἐτεκμήρατο μανικὸν καὶ ὀξύθυμον ἄνθρωπον, ἐπεὶ διὰ τὸ πλῆθος τῆς ἐν τῇ καρδίᾳ θερμασίας οὕτω σφύζουσιν αἱ ἀρτηρίαι· μανικοὺς μὲν γὰρ καὶ ὀξυθύμους τὸ πλῆθος τῆς θερμασίας ἐργάζεται, νωθροὺς δὲ καὶ βαρεῖς καὶ δυσκινήτους ἡ τῆς κράσεως ψυχρότης.

9. Ἱπποκράτης μὲν οὖν ἐπιδείξας ἐν ὅλῳ τῷ περὶ ὑδάτων καὶ ὡρῶν κράσεως[63] συγγράμματι ταῖς [περὶ] τοῦ σώματος κράσεσιν ἕπεσθαι [καὶ] τὰς τῆς ψυχῆς δυνάμεις, οὐ μόνον ὅσαι κατὰ τὸ θυμοειδὲς ἢ ἐπιθυμητικὸν αὐτῆς εἰσιν, ἀλλὰ καὶ τὰς κατὰ τὸ λογιστικόν, ἁπάντων ἀξιοπιστότατός ἐστι μάρτυς, εἴ τις ἐπιμαρτύροιτο, καθάπερ ἐνίοις ἔθος ἐστί, τὴν τῶν δογμάτων ἀλήθειαν. ἐγὼ δ' οὐχ ὡς μάρτυρι πιστεύω τἀνδρὶ τοῖς πολλοῖς ὡσαύτως ἀλλ' ὅτι τὰς ἀποδείξεις αὐτοῦ βεβαίας ὁρῶ, διὰ τοῦτο γοῦν καὶ αὐτὸς ἐπαινῶ τὸν Ἱπποκράτην. τίς γὰρ οὐχ

[63] post κράσεως: ἑπομένας ταῖς τοῦ σώματος κράσεσι SM; συγγράμματι, σώματος κράσει ἕπεσθαι K

called arteries veins, as has been shown often, having not yet named every movement of arteries a pulse (pulsation), but only that which is perceptible and which in the person is altogether strong. Hippocrates, however, as he was first originator of the prevailing customs after these, said somewhere that every movement of the arteries is a pulse, whatever kind it may be.[44] Nevertheless, in this particular statement, he was still using the ancient method of expression, taking his proof from a strong movement of an artery in respect of a manic or choleric person, since it is due to the abundance of heat in the heart that the arteries pulsate like this. For the abundance of heat also renders them manic and quick to anger, whereas the coldness of the *krasis* renders them sluggish, torpid, and hard to move.

9. Hippocrates, who showed in the whole book about the *krasis* of waters and seasons that the capacities of the soul follow the *krasias* of the body—and not only those that are in relation to the spirited and desiring [parts] of this, but also those in relation to the rational [part]—is a witness deserving of trust most of all, if in fact someone were to make the truth of the opinions depend on a witness, as is the custom for some. I, however, do not believe the man as a witness in a similar way to the majority, but because I see his demonstrations as secure. Anyway, because of this, I too praise this Hippocrates, for who does not see the

[44] It is generally accepted that the distinction between arteries and veins was due to Praxagoras. On this particular statement by Galen, see C. R. S. Harris, *The Heart and Vascular System in Ancient Greek Medicine* (Oxford, 1973), 185ff. See also Hippocrates I, LCL 150 (Jones), 339 (Introduction to *Nutriment*) and 358 (*Nutriment* 48).

ὁρᾷ τὸ σῶμα καὶ τὴν ψυχὴν ἁπάντων τῶν ὑπὸ τοῖς ἄρκτοις ἀνθρώπων ἐναντιώτατα διακείμενα τοῖς ἐγγὺς τῆς διακεκαυμένης ζώνης; ἢ τίς οὐκ οἶδε τοὺς ἐν τῷ μέσῳ τούτων, ὅσοι τὴν εὔκρατον οἰκοῦσι χώραν, ἀμείνους τά τε σώματα καὶ τὰ τῆς ψυχῆς ἤθη καὶ σύνεσιν καὶ φρόνησιν ἐκείνων τῶν ἀνθρώπων;

Ἀλλὰ διά τινας τῶν Πλατωνικοὺς μὲν ἑαυτοὺς ὀνομαζόντων, ἡγουμένους δ' ἐμποδίζεσθαι μὲν ἐν ταῖς νόσοις τὴν ψυχὴν ὑπὸ τοῦ σώματος, ὑγιαίνοντος δὲ[64] τὰς ἰδίας ἐνεργείας ἐνεργεῖν οὔτ' ὠφελουμένην οὔτε βλαπτομένην ὑπ' αὐτοῦ, παραγράψω τινὰς ῥήσεις τοῦ Πλάτωνος, ἐν αἷς ἀποφαίνεται διὰ τὴν τῶν τόπων κρᾶσιν εἰς φρόνησιν ὠφελουμένους τε καὶ βλαπτομένους τοὺς ἀνθρώπους[65] ἄνευ τοῦ νοσεῖν τὸ σῶμα. ἐν μέν γε τῷ Τιμαίῳ κατὰ τὰ πρῶτα τῶν λόγων ἔγραψε·

> ταύτην δὴ ξύμπασαν τὴν διακόσμησιν καὶ σύνταξιν ἡ θεὸς προτέρους ὑμᾶς[66] διακοσμήσασα κατῴκισεν ἐκλεξαμένη τὸν τόπον, ἐν ᾧ γεγένησθε, τὴν εὐκρασίαν τῶν ὡρῶν ἐν αὐτῷ κατιδοῦσα, ὅτι ἄνδρας οἴσοι φρονιμωτάτους.

ἀλλὰ καὶ συνάπτων ἐφεξῆς

---

[64] post ὑγιαίνοντος δὲ: τὰς ἰδίας ἐνεργείας ἐνεργεῖν SM, B; ἰδίᾳ ἐνεργεῖν K
[65] τοὺς ἀνθρώπους add. SM
[66] ὑμᾶς SM; ἡμᾶς K

THE SOUL'S TRAITS

body and soul of all people under the Bears (Ursa Major and Minor—i.e., the northern region) most oppositely disposed to those near the torrid zone? Or who doesn't know that those in between these [two zones], those who dwell in an *eukratic* place, are better in respect of the bodies and the characteristics of the soul, and intelligence and practical wisdom than those people?

But because certain of those who call themselves Platonists think that, although the soul is hindered by the body in the diseases, when healthy it carries out its own functions, being neither benefitted nor harmed by the body, I shall put on record some statements of Plato's in which he declares that people are benefitted and harmed in practical wisdom due to the *krasis* of the places without the body being diseased. In fact, in the first of the books in the *Timaeus*, | he wrote:[45]

> So when, at that time, the Goddess had furnished you, before all others, with all this orderly and regular system, she established your state, choosing the place in which you were born, seeing that therein there was *eukrasia* of the seasons so that it would bring forth the most prudent people.

But he also subjoins the following:

[45] The quotations are consecutive passages from Plato's *Timaeus* 24C–D, LCL 234 (Bury), 38–41 (translation after Bury with minor modifications).

ἅτε οὖν φιλοπόλεμός τε καὶ φιλόσοφος ἡ θεὸς
οὖσα τὸν προσφερεστάτους αὐτῇ μέλλοντα οἴ-
σειν τόπον ἄνδρας τοῦτον ἐκλεξαμένη πρῶτον
κατῴκισεν.

ὅτι μὲν οὖν πολὺ δίδωσι τοῖς τόποις, τουτέστι ταῖς
ἐπὶ γῆς οἰκήσεσιν, εἴς τε τὰ τῆς ψυχῆς ἤθη καὶ
σύνεσιν καὶ φρόνησιν, ἤδη μὲν κἀκ τούτων ἐστὶ
δῆλον, ἀλλὰ κἂν τῷ πέμπτῳ τῶν Νόμων ὡδί πως
ἔγραψε·[67]

καὶ γάρ, ὦ Μέγιλλέ τε καὶ Κλεινία, μηδὲ τοῦθ᾽
ἡμᾶς λανθανέτω περὶ τόπων,[68] ὡς εἰσὶν ἄλλοι
τινὲς διαφέροντες ἄλλων τόπων πρὸς τὸ γεννᾶν
ἀνθρώπους ἀμείνους καὶ χείρους.

ἐναργῶς πάλιν ἐνταῦθα γεννᾶν τοὺς τόπους φησὶν
ἀμείνους τε καὶ χείρους ἀνθρώπους. ἐφεξῆς δὲ πάλιν
ἐπιφέρων τοῖσδέ φησιν

οἱ μέν γέ που διὰ πνεύματα παντοῖα καὶ εἰλήσεις ἀλλόκοτοί τ᾽ εἰσὶ καὶ ἀναίδεοι αὐτῶν, οἱ δὲ
δι᾽ ὕδατα, οἱ δὲ δι᾽ αὐτὴν τὴν ἐκ τῆς γῆς τροφὴν
ἀναδιδοῦσαν οὐ μόνον τοῖς σώμασιν ἀμείνω
καὶ χείρω, ταῖς ψυχαῖς δ᾽ οὐχ ἧττον δυναμένην
πάντα τὰ τοιαῦτα | ἐμποιεῖν.

[67] post ἔγραψε·: καὶ γάρ, ὦ Μέγιλλέ τε καὶ Κλεινία, add.
SM   [68] post περὶ τόπων: ὡς εἰσὶν ἄλλοι τινὲς διαφέροντες ἄλλων τόπων add. B

THE SOUL'S TRAITS

> So it was that the Goddess, being herself both a lover of war and a lover of wisdom, chose the place which was likely to bring forth men most like herself, and this she established first.

It is then already clear from these things that he gives great importance to the places—that is to say, to the dwelling places of the earth—toward the characteristics of the soul, and intelligence and practical wisdom. And even in the fifth book of the *Laws*, he wrote as follows:[46]

> For that, too, is a point, O Megillus and Clinias, which we must not fail to notice—that some places are naturally superior to others for the generation of better and worse people.

Clearly again here he says the places generate better and worse people, Next in turn, continuing [the topic], he says this:

> Some places are ill-conditioned and some well-conditioned owing to a variety of winds or to sunshine, others owing to their waters, others owing simply to the produce of the soil, which offers produce either good or bad for their bodies, and is equally able to effect similar results in their souls as well.

---

[46] The two quotations are almost continuous; see Plato, *Laws* 747D–E, LCL 187 (Bury), 388–89 (translation after Bury with minor modifications).

ἐν τούτῳ τῷ λόγῳ "πνεύματα" φησὶ σαφῶς,[69] καὶ [τὰς] εἰλήσεις [τουτέστι τὰς ἐξ ἡλίου θερμότητας] εἰς τὰς τῆς ψυχῆς δυνάμεις προσισταμένας, εἰ μή τι ἄρα νομίζουσι διὰ μὲν τὰ πνεύματα καὶ τὴν τοῦ περιέχοντος ἀέρος θερμότητα καὶ ψυχρότητα καὶ τὴν τῶν ὑδάτων τε καὶ τὴν τῆς τροφῆς φύσιν ἀμείνους τε καὶ χείρους τὴν ψυχὴν ἀνθρώπους δύνασθαι γενέσθαι, ταῦτα δ' αὐτὰ μὴ διὰ μέσων τῶν κράσεων ἐργάζεσθαί τε κατὰ τὴν ψυχὴν ἀγαθά τε καὶ φαῦλα· καὶ γὰρ καὶ ταῦτ' ἂν εἴη τῇ συνέσει καὶ παιδείᾳ[70] τῶν ἀνδρῶν ἀκόλουθα.

Ἀλλ' ἡμεῖς γε σαφῶς ἴσμεν, ὡς ἕκαστον ἔδεσμα καταπίνεται μὲν πρῶτον εἰς τὴν γαστέρα, προκατείργασται δ' ἐν αὐτῇ καὶ μετὰ ταῦτα διὰ τῶν ἐξ ἥπατος εἰς αὐτὴν καθηκουσῶν φλεβῶν ἀναληφθὲν ἐργάζεται τοὺς ἐν τῷ σώματι χυμούς, ἐξ ὧν τρέφεται τἆλλα μόρια πάντα καὶ σὺν αὐτοῖς ἐγκέφαλός τε καὶ καρδία καὶ ἧπαρ, ἐν δὲ δὴ τῷ τρέφεσθαι θερμότερα σφῶν αὐτῶν γίγνεται ψυχρότερά τε καὶ ὑγρότερα συνεξομοιούμενα τῇ δυνάμει τῶν ἐπικρατούντων χυμῶν. ὥστε σωφρονήσαντες καὶ νῦν γοῦν οἱ δυσχεραίνοντες, ὅτι τροφὴ δύναται τοὺς μὲν σωφρονεστέρους, τοὺς δ' ἀκολαστοτέρους ἐργάζεσθαι καὶ τοὺς μὲν ἐγκρατεστέρους, τοὺς δ' ἀκρατεστέρους καὶ θαρ-

[69] post σαφῶς,: καὶ [τὰς] εἰλήσεις [τουτέστι τὰς ἐξ ἡλίου θερμότητας] εἰς τὰς τῆς ψυχῆς δυνάμεις προσισταμένας, B; τουτέστι τοὺς ἀνέμους, καὶ "εἰλήσεις," τουτέστι τὰς ἐξ ἡλίου θερμότητας, δύνασθαί φησιν εἰς τὰς τῆς ψυχῆς δυνάμεις,

THE SOUL'S TRAITS

In this discussion, he clearly says *pneumata* (that is to say the winds) and *eilēseis* (that is to say the heat from the sun)[47] are able to be applied to the capacities of the soul, unless perhaps they think it is due to the winds, and the heat and cold of the ambient air, and of the waters, and the nature of the nutriments that are able to bring about better and worse people in the respect of the soul, whereas these same things do not bring about good and bad in the soul through the mediation of the *krasias*, and furthermore would also have consequences for the understanding and education of the men.

But we, in fact, know clearly that each food is swallowed down first to the stomach, and is worked upon in this, after which it is taken up through the veins which come down to it from the liver and creates the humors in the body, from which all the other parts are nourished, and with these the brain, heart, and liver. Certainly, in the nourishing, they become hotter than they were, and colder and wetter, being assimilated by the capacity of the prevailing humors. As a consequence, let those men who are now at least displeased [with the idea] that nutriment | is 808K able to make some people more sensible, some more undisciplined, some more self-controlled and some more intemperate, and some more courageous, cowardly, gentle

---

[47] On this term, defined as "sun-heat" in LSJ, see Plato, *Republic* 380E, 404B, and Aristotle, *Physics* 197a23.

SM; φησι καὶ τὰς ἡλιάσεις, τουτέστι τὰς ἐξ ἡλίου θερμότητας εἰς τὰς τῆς ψυχῆς δυνάμεις προσθησομενας K

[70] καὶ παιδείᾳ add. SM

σαλέους καὶ δειλοὺς ἡμέρους τε καὶ πρᾴους ἐριστικούς τε καὶ φιλονείκους, ἡκέτωσαν πρός με μαθησόμενοι, τίνα μὲν ἐσθίειν αὐτοὺς χρή, τίνα δὲ πίνειν. εἴς τε γὰρ τὴν ἠθικὴν φιλοσοφίαν ὀνήσονται μέγιστα καὶ πρὸς ταύτῃ κατὰ τὰς τοῦ λογιστικοῦ δυνάμεις ἐπιδώσουσιν εἰς ἀρετὴν συνετώτεροι καὶ μνημονικώτεροι γενόμενοι καὶ φιλομαθέστεροι καὶ φρινιμώτεροι.[71] πρὸς γὰρ ταῖς τροφαῖς καὶ τοῖς πόμασι καὶ τοὺς ἀνέμους αὐτοὺς διδάξω καὶ τὰς τοῦ περιέχοντος κράσεις ἔτι τε τὰς χώρας, ὁποίας μὲν αἱρεῖσθαι προσήκει, ὁποίας δὲ φεύγειν.

10. Ἀναμνήσω δὲ πάλιν αὐτούς, κἂν μὴ θέλωσιν, ὡς ὁ Πλάτων αὐτός, ἀφ᾽ οὗ παρονομάζουσιν ἑαυτούς, οὐχ ἅπαξ ἢ δὶς ἀλλὰ πολλάκις ἔγραψε περὶ τούτων. ὡσαύτως δ᾽ οὖν ἀρκέσει μοι προσθεῖναι τρεῖς ῥήσεις τινὰς[72] κατὰ τὸν ἐνεστῶτα λόγον,[73] ὧν αἱ μὲν δύο περὶ οἴνου πόσεώς εἰσιν ἐκ τοῦ δευτέρου τῶν Νόμων, ἄλλη δὲ περὶ τροφῆς ἐκ Τιμαίου. ἡ δ᾽ οὖν ῥῆσις ἐκ τοῦ δευτέρου τῶν Νόμων ἥδε·

ἆρ᾽ οὐ νομοθετήσομεν πρῶτον μὲν τοὺς παῖδας μέχρις ἐτῶν ὀκτωκαίδεκα τὸ παράπαν οἴνου μὴ γεύεσθαι, διδάσκοντες, ὡς οὐ χρὴ πῦρ ἐπὶ πῦρ ὀχετεύειν εἴς τε τὸ σῶμα καὶ τὴν ψυχήν, πρὶν ἐπὶ τοὺς πόνους ἐγχειρεῖν πορεύεσθαι, τὴν ἐμ-

[71] καὶ φιλομαθέστεροι καὶ φρινιμώτεροι add. K, B
[72] ῥήσεις τινὰς add. SM
[73] post λόγον: ὧν αἱ μὲν δύο περὶ οἴνου πόσεώς εἰσιν ἐκ

and mild, captious and contentious, come to me and they will learn what they must eat and what they must drink, for they will benefit very greatly regarding the ethical philosophy and in addition to this they will advance toward excellence in relation to the capacities of the rational [soul], becoming more intelligent with respect to virtue and more capable of memorizing, more fond of learning and more prudent. For in addition to the foods and drinks, I shall teach them about winds, and about the *krasias* of the ambient air, and further, the countries—which kinds it is appropriate to choose and which to avoid.

10. I shall, however, remind them again, even if they don't wish it, that Plato himself, from whom they derive their own name, not once only or twice but often wrote about these things. In like manner, then, it will be sufficient for me to add three statements in relation to the present discussion, of which two about drinking wine are from the second book of *Laws*, while the other about nutrition is from the *Timaeus*. The statement from the second book of *Laws* is this:

> [How then shall we encourage them to take readily to singing?] Shall we not pass a law that, in the first place, no children under eighteen shall taste wine at all, teaching that it is wrong to pour fire upon fire either in body or in soul, before they progress to physical labors, and thus guarding against the excit-

809K

τοῦ δευτέρου τῶν Νόμων, ⟨ἄλλη δὲ περὶ τροφῆς ἐκ Τιμαίου. Ἡ δ' οὖν ῥῆσις ἐκ τοῦ δευτέρου τῶν Νόμων⟩ ἥδε· SM; πρὸς τῷ περὶ τροφῆς ἐκ Τιμαίου καὶ ἐκ τοῦ δευτέρου τῶν Νόμων, ὧν οἱ μὲν δύο περὶ οἴνου πόσεώς εἰσιν. K

μανῆ εὐλαβούμενοι ἕξιν τῶν νέων· μετὰ δὲ
τοῦτο οἴνου μὲν δεῖ γεύεσθαι τοῦ μετρίου μέχρι
τριάκοντα ἐτῶν, μέθης δὲ καὶ πολυοινίας τὸ
παράπαν τὸν νέον ἀπέχεσθαι· τετταράκοντα
δὲ ἐπιβαίνοντα ἐτῶν ἐν τοῖς ξυσσιτίοις εὐωχη-
θέντα καλεῖν τούς τε ἄλλους θεοὺς καὶ δὴ καὶ
Διόνυσον παρακαλεῖν εἰς τὴν τῶν πρεσβυτῶν
τελετὴν ἅμα καὶ παιδιάν, ἣν τοῖς ἀνθρώποις
ἐπίκουρον τῆς τοῦ γήρως αὐστηρότητος ἐδωρή-
σατο τὸν οἶνον φάρμακον, ὥστε ἀνίας καὶ δυσθυμίας λήθην γίγνεσθαι μαλακώτερόν τ' ἐκ
σκληροτέρου τὸ τῆς ψυχῆς ἦθος, καθάπερ εἰς
πῦρ σίδηρον ἐντεθέντα, γιγνόμενον καὶ οὕτως
εὐπλαστότερον εἶναι;

ἐκ ταύτης τῆς ῥήσεως μεμνῆσθαι[74] παρακαλῶ τοὺς
γενναίους Πλατωνικοὺς οὐ μόνον ἃ λέλεκται περὶ πόσεως οἴνου κατ' αὐτὴν ἀλλὰ καὶ περὶ τῆς τῶν ἡλικιῶν
διαφορᾶς. ἐμμανῆ μὲν γὰρ εἶναί φησι τὴν τῶν μειρακίων φύσιν, αὐστηρὰν δὲ καὶ δύσθυμον καὶ σκληρὰν τὴν τῶν γερόντων, οὐ δήπου διὰ τὸν ἀριθμὸν τῶν
ἐτῶν ἀλλὰ κατὰ τὴν τοῦ σώματος ἔχουσαν κρᾶσιν
τὴν οὖσαν ἑκάστῃ τῶν ἡλικιῶν. ἡ μὲν γὰρ τῶν μειρακίων θερμὴ καὶ πολύαιμος, ἡ δὲ τῶν γερόντων ὀλί-
γαιμός τε καὶ ψυχρὰ καὶ διὰ τοῦτό γ' αὖ τοῖς μὲν
γέρουσιν ὠφέλιμος οἴνου πόσις εἰς συμμετρίαν θερμασίας ἐπανάγουσα τὴν ἐκ τῆς ἡλικίας ψυχρότητα,
τοῖς δ' αὐξανομένοις ἐναντιωτάτη· ζέουσαν γὰρ αὐτῶν

THE SOUL'S TRAITS

> able disposition of the young? And next, we shall rule that young men under thirty may take wine in moderation but that they must entirely abstain from intoxication and heavy drinking. But when a man has reached the age of forty, he may join in the convivial gatherings and invoke Dionysus above all other gods, inviting his presence at the rite (which is also the recreation) of the elders, which he bestowed on mankind as a medicine potent against the crabbedness of old age, that thereby we men may renew our youth, and that, through forgetfulness of care, the temper of our souls may lose its hardness and become softer and more ductile, just as iron does when forged in the fire?[48]

From this statement, I call upon the noble Platonists to remember, not only what has been said in it about drinking wine, but also about the difference of the stages of life. For he says the nature of young lads is frantic while that of old men is harsh, despondent, and hard, and this is not, of course, due to the number of years, but relates to the *krasis* of the body they have, which exists in each of the stages of life, for that of young lads is hot and full of blood while that of old people is lacking in blood and cold. And in fact because of this in turn, the drinking of wine is beneficial to old men, returning to a due proportion of heat the coldness due to the stage of life, while it is most inimical to those who are growing, as it overheats their

---

[48] Plato, *Laws* 666, A3–C2, LCL 187 (Bury), 132–35 (translation after Bury with minor modifications).

---

[74] *post* μεμνῆσθαι: παρακαλῶ τοὺς γενναίους SM, B; ποιεῖ καλοὺς καὶ γενναίους τοὺς K

τὴν φύσιν καὶ σφοδρῶς κινουμένην ὑπερθερμαίνει τε καὶ εἰς ἀμέτρους καὶ σφοδρὰς ἐκβαίνει κινήσεις. ἀλλὰ Πλάτωνι μὲν καὶ ταῦτα καὶ ἄλλα πολλὰ κατὰ τὸ δεύτερον τῶν Νόμων εἴρηται περὶ πόσεως οἴνου τοῖς βουλομένοις ἀναγιγνώσκειν ὠφέλιμα· μόνης δ' ἔτι μιᾶς ἐγὼ ῥήσεως αὐτοῦ μνημονεύσω κατὰ τὸ τέλος εἰρημένης ἁπάντων τῶν περὶ πόσεως οἴνου[75] λόγων,[76] ἔνθα προῄρηται τὸν τῶν Καρχηδονίων νόμον. ἔχει δ' ἡ ῥῆσις οὕτως·

ἀλλ' ἔτι μᾶλλον τῆς Κρητῶν καὶ Λακεδαιμονίων χρείας προσθείμην ἂν τῷ τῶν Καρχηδονίων | νόμῳ, μηδέποτε μηδένα ἐπὶ στρατοπέδου γενέσθαι τούτου τοῦ πόματος ἀλλ' ὑδροποσίαις ξυγγίγνεσθαι τοῦτον τὸν χρόνον ἅπαντα καὶ κατὰ πόλιν μήτε δοῦλον[77] μήτε δούλην γενέσθαι μηδέποτε μηδὲ ἄρχοντας τοῦτον τὸν ἐνιαυτὸν ὃν ἂν ἄρχωσι, μηδ' αὖ κυβερνήτας μηδὲ δικαστὰς ἐνεργοὺς ὄντας οἴνου μὴ[78] γεύεσθαι τὸ παράπαν μηδὲ ὅστις βουλευσόμενος εἰς βουλὴν ἀξίαν τινὰ λόγου ξυνέρχεται μηδέ γε μεθ' ἡμέραν τὸ παράπαν μηδένα, εἰ μὴ σωμασκίας ἢ νόσων ἕνεκα, μηδ' αὖ νύκτωρ, ὅταν ἐπινοῇ τις παῖδας ποιεῖσθαι ἀνὴρ ἢ γυνή. καὶ ἄλλα δὲ πολλὰ ἄν τις λέγοι, ἐν οἷς τοῖς νοῦν τε καὶ νόμον ἔχουσιν ὀρθὸν οὐ ποτέος οἶνος.

[75] οἴνου add. SM
[76] post λόγων,: ἔνθα προῄρηται τὸν τῶν Καρχηδονίων νόμον. SM, B; ἔνθαπερ εἴρηκε τὸν τῶν Καρχηδονίων λόγον K

THE SOUL'S TRAITS

seething and strongly moving nature and drives it to immoderate and violent movements. But there are these and many other things said by Plato about drinking wine in the second book of *Laws* which are beneficial to those who wish to read them. I shall call to mind only one more of his statements said at the end of all discussions about drinking wine. Here I choose in addition the law of the Carthaginians. The statement is as follows:

> I would go even beyond the practice of the Cretans and Lacedaemonians; and to the | Carthaginian law, which ordains that no soldier on the march should ever taste of this potion, but confine himself for the whole of the time to drinking water only, I would add this—that in the city no male or female slave should ever drink wine, nor should magistrates during their year of office, nor should pilots or judges while on duty taste any wine at all. Nor should any councilor, while attending a council of discussion of any importance. Nor should anyone whatever taste of it at all, except for purposes of bodily training or health, in the daytime; nor should anyone do so at night—be they man or woman—when proposing to procreate children. Many other occasions also might be mentioned when wine should not be drunk by men who are swayed by right reason and law.[49]

811K

---

[49] Plato, *Laws* 674A5–B9, LCL 187 (Bury), 160–63 (translation after Bury with minor modifications).

---

[77] μήτε δοῦλον add. SM
[78] μὴ add. SM

ταῦτα τοῦ Πλάτωνος εἰρηκότος οὐκ ἐπὶ τῶν νοσούντων σωμάτων ἀλλ' ἐπὶ τῶν ἀμέμπτως ὑγιαινόντων, εἴ γ' ὑμῖν δοκοῦσιν, ὦ γενναιότατοι Πλατωνικοί, στρατεύειν καὶ ἄρχειν καὶ δικάζειν[79] καὶ κυβερνᾶν ναῦς ὑγιαίνοντες ἄνδρες, ἀποκρίνασθέ μοι τοὐντεῦθεν ἐρωτῶντι, πότερον οὐχ ὥσπερ τις τύραννος ὁ ποθεὶς οἶνος κελεύει τὴν ψυχὴν μήτε νοεῖν ἀκριβῶς, ἃ πρόσθεν ἐνόει, μήτε πράττειν ὀρθῶς, ἃ πρόσθεν ἔπραττε, καὶ διὰ τοῦτο φυλάττεσθαί φησιν ὁ Πλάτων ὡς πολέμιον; εἰ γὰρ ἅπαξ εἴσω τοῦ σώματος ἀφίκοιτο, καὶ τὸν κυβερνήτην κωλύει, ὡς προσήκει, μεταχειρίζεσθαι τοὺς οἴακας τῆς νεὼς καὶ τοὺς στρατευομένους μὴ[80] σωφρονεῖν ἐν ταῖς παρατάξεσι καὶ τοὺς δικαστάς, ὁπότε οὖν δικαίους εἶναι χρή, ποιεῖ σφάλλεσθαι καὶ πάντας τοὺς ἄρχοντας ἄρχειν κακῶς καὶ προστάττειν μὲν οὐδὲν ὑγιές. ἡγεῖται γὰρ τὸν οἶνον ἀτμῶν θερμῶν ὅλον τε τὸ σῶμα καὶ μάλιστα τὴν κεφαλὴν πληροῦντα κινήσεως μὲν ἀμετροτέρας αἴτιον γίγνεσθαι τῷ τ' ἐπιθυμητικῷ μέρει τῆς ψυχῆς καὶ τῷ θυμοειδεῖ, βουλῆς δὲ προπετεστέρας τῷ λογιστικῷ. καὶ μήν, εἴπερ οὕτως ἔχει ταῦτα, διὰ μέσης τῆς κράσεως αἱ εἰρημέναι τῆς ψυχῆς ἐνέργειαι φαίνονται βλαπτόμεναι πινόντων ἡμῶν τὸν οἶνον, ὥσπερ γε πάλιν ὠφελούμεναι τινές.

Ἀλλὰ γὰρ τοῦθ' ὑμᾶς, ἐὰν βούλησθε, διδάξω καθ' ἕτερον καιρόν, ἐκ τοῦ θερμαίνειν ὅσα βλάπτειν τε καὶ ὠφελεῖν ἡμᾶς οἶνος πέφυκεν· ἐν δὲ τῷ παρόντι τὴν ἐν

THE SOUL'S TRAITS

Since Plato has said these things not in the case of diseased bodies but in the case of those that are faultlessly healthy, if at least, most notable Platonists, those men seem to you healthy who fight wars, govern, judge, and pilot ships, answer for me the question I ask now: Does the drinking of wine not command the soul like some tyrant, so it doesn't think precisely those things it thought before, nor does it correctly do those things it did before, and because of this Plato says to guard against it as an enemy? For if, all at once, it comes within the body, it prevents the helmsman from handling the rudders of the ship as he should, and the soldiers of exercising self-control in the ranks, and it makes judges err when they must make judgments, and all the magistrates adjudicate badly, and to order nothing sound. For he was accustomed to think the wine, when it fills the whole body and particularly the head with hot vapors, becomes a cause of movements that are more immoderate in the desiderative and spirited parts of the soul, while deliberation in the rational part is more rash. For surely, if these things are so through the mediation of the *krasis*, the stated functions of the soul are manifestly damaged when we drink wine just as some conversely are in fact benefitted.

But on this point, if you wish, I shall instruct you at another time which things wine harms and benefits us naturally from the heating. For the present, I shall write

---

⁷⁹ δικάζειν SM; δοκιμάζειν K
⁸⁰ μὴ *add.* SM

τῷ Τιμαίῳ ῥῆσιν παραγράψω, καθ' ἣν προειπὼν ὁ Πλάτων οὕτως·

> ταύτῃ κακοὶ πάντες οἱ κακοὶ[81] διὰ δύο ἀκουσιώτατα γιγνόμεθα· ὧν αἰτιατέον μὲν ἀεὶ τοὺς φυτεύοντας τῶν φυτευομένων μᾶλλον καὶ τοὺς τρέφοντας τῶν τρεφομένων

ἐφεξῆς φησι

> πειρατέον μήν, ὅπῃ τις δύναται, καὶ διὰ τροφῆς καὶ δι' ἐπιτηδευμάτων μαθημάτων τε φυγεῖν μὲν κακίαν, τοὐναντίον δ' ἑλεῖν.[82]

ὥσπερ γὰρ "ἐπιτηδεύματα καὶ μαθήματα" κακίας μὲν ἀναιρετικά, γεννητικὰ δ' ἀρετῆς ἐστιν, οὕτω καὶ ἡ 'τροφή'· λεγομένης τ' ἐνίοτε τροφῆς [ὑπ' αὐτῶν] οὐ μόνον τῆς ἐπὶ σιτίοις ἀλλὰ καὶ πάσης τῶν παίδων τε διαίτης οὐχ οἷόν τε φάναι κατὰ τὸ δεύτερον σημαινόμενον εἰρῆσθαι μὲν νῦν ὑπ' αὐτοῦ τὴν τροφήν· οὐ γὰρ τοῖς παισὶν ἀλλὰ τοῖς τελείοις παρακελευόμενος ἔφη "πειρατέον μήν,[83] ὅπῃ τις δύναται, καὶ διὰ τροφῆς καὶ δι' ἐπιτηδευμάτων μαθημάτων τε φυγεῖν μὲν κακίαν, τοὐναντίον δ' ἑλεῖν." ἐπιτηδεύματ' οὖν λέγει τὰ [γε] κατὰ γυμναστικήν τε καὶ μουσικήν, μαθήματα ⟨δὲ⟩ τά τε κατὰ γεωμετρίαν καὶ ἀριθμητικήν· τροφὴν δ' οὐκ ἄλλην τινὰ νοεῖν οἷόν τε παρὰ τὴν ἐκ τῶν σιτίων

---

[81] οἱ κακοὶ SM; ἢ καλοὶ K. SM (p. 71) has: οἱ κακοὶ Corn. ἢ καλοὶ codd. vel mali omnes vel boni N

THE SOUL'S TRAITS

besides the statement as in the *Timaeus* in which Plato spoke thus:[50]

> Thus it comes to pass that all of us who are wicked become wicked owing to two quite involuntary causes. And for these we must always blame the begetters rather than the begotten, and the nurturers rather than the nurtured.

And next he says:

> And yet each man must endeavor, in whatever way is ever possible, by means of nurture and by his pursuits and studies, to flee the evil and conversely to choose the good.

For just as "pursuits and studies" are destructive of badness but productive of goodness, so too is "nurture" (*trophē*). Sometimes, when he says nurture, not only is this through foods but also through the whole regimen of children, and it is not possible to say that nurture is now said by him according to the second significations. To make a direct command not for children but for adults, he said: "One must try in whatever way possible, through nurture, pursuits, and studies to avoid the bad and conversely choose the good." Therefore, he says pursuits are those things related to gymnastic and musical activity, while studies are those things related to geometry and arithmetic. It is not possible to conceive of nurture in any other

[50] Plato, *Timaeus* 87B3–6 and 6–9, LCL 234 (Bury), 236–37 (translation after Bury with minor modifications).

---

[82] *post* ἐλεεῖν: ἀρετήν (K) *om.*   [83] μήν SM; μὴ K

GALEN

καὶ ῥοφημάτων καὶ πομάτων, ἐξ ὧν ἐστι καὶ οἶνος, ὑπὲρ οὗ πολλὰ διῆλθεν ὁ Πλάτων | ἐν τῷ δευτέρῳ τῶν Νόμων.

Ὅστις δὲ βούλεται καὶ χωρὶς τούτου γνῶναί τι περὶ πάσης τῆς ἐν ταῖς τροφαῖς δυνάμεως, ἔνεστιν ἀναγιγνώσκειν αὐτῷ[84] τρία περὶ τοῦδε ἡμέτερα ὑπομνήματα. καὶ τέταρτον ἐπ' αὐτοῖς τὸ Περὶ εὐχυμίας τε καὶ κακοχυμίας, οὗ μάλιστα χρῄζομεν εἰς τὸ παρόν. πολλὰ μὲν οὖν ἡ κακοχυμία ταῖς τῆς ψυχῆς ἐνεργείαις λυμαίνεται, διασῴζει δ' ἀβλαβεῖς ἡ εὐχυμία.

11. Οὐκ οὖν ἀναιρετικὸς ὅδ' ὁ λόγος ἐστὶ τῶν ἐκ φιλοσοφίας καλῶν ἀλλ' ὑφηγητικός τε καὶ διδασκαλικὸς καίτοι γ'[85] ἀγνοούμενος[86] ἐνίοις τῶν φιλοσόφων· οἵ τε γὰρ ἡγούμενοι πάντας ἀνθρώπους ἐπιδεκτικοὺς ἀρετῆς ὑπάρχειν[87] ὅπερ ἴσον ἐστὶ τῷ μηδένα γενέσθαι φύσει κακόν, οἵ τε μηδένα τὴν δικαιοσύνην αἱρεῖσθαι[88] αὐτήν, ἐξ ἡμίσεος ἑκάτεροι τὴν ἀνθρωπίνην ἑωράκασι φύσιν. οὔτε γὰρ ἅπαντες [φίλοι] ἐχθροὶ φύονται δικαιοσύνης οὔθ' ἅπαντες φίλοι, διὰ τὰς κράσεις τῶν σωμάτων ἑκάτεροι τοιοῦτοι γενόμενοι. πῶς οὖν, φασί, δικαίως ἄν τις ἐπαινοῖτο καὶ ψέγοιτο καὶ μισοῖτο καὶ φιλοῖτο γενόμενος πονηρὸς ἢ ἀγαθὸς οὐ δι' ἑαυτὸν | ἀλλὰ διὰ τὴν κρᾶσιν, ἣν ἐξ ἄλλων αἰτίων φαίνεται λαμβάνων;

---

[84] post αὐτῷ: τρία ⟨περὶ τοῦδε βιβλία τὰ ἡμέτερα καὶ τέταρτον ἐπ' αὐτοῖς τὸ⟩ SM; τοὺς τρεῖς περὶ τοῦδε τῶν ἡμετέρων ὑπομνήσεις. καὶ τὸ τέταρτον ἐπ' αὐτοῖς K   [85] καίτοι γ' add. SM   [86] post ἀγνοούμενος: ⟨μέχρι⟩ τινὸς ἐνίοις SM; τινὸς ἐν αὐτοῖς K

way besides that from foods, gruel, and drinks, of which there is also wine, many aspects of which Plato has gone through in detail | in the second book of *Laws*.

It is possible for someone who wishes to know about every capacity in the nutriments apart from this to read my three treatises about these and the fourth in addition to these *On Euchymia and Kakochymia* which we use particularly in the present context.[51] Thus *kakochymia* impairs the functions of the soul in many ways, whereas *euchymia* preserves them unharmed.

11. This argument is not therefore destructive of the good things from philosophy but is expository and didactic of something unknown to some of the philosophers, for there are those who think all people are receptive to good, which is equivalent to there being nothing bad in nature, and those who think no-one chooses justice for itself, each group seeing one half of human nature. For not all are born inimical to justice, nor are all lovers of justice—it is through the *krasias* of the bodies each becomes like this. How then, they say, is someone reasonably praised or blamed, hated or loved, when he has become worthless or good not through himself | but through the *krasis* which has obviously been taken from other causes?

---

[51] *De alimentis facultatibus libri tres*, VI.453–748K, and *De bonis et malis alimentorum succis*, VI.749–815K.

---

[87] *post* ὑπάρχειν: ὅπερ ἴσον ἐστὶ τῷ μηδένα γενέσθαι φυσικόν SM; οἵ τε μηδένα τὴν δικαιοσύνην K (*see* SM, *note, p. 73, and* B, *note, p. 77*)

[88] *post* αἱρεῖσθαι: αὐτήν, SM, B; ὅπερ ἴσον ἐστὶ τῷ μηδένα γίγνεσθαι φυσικὸν ὅρον, K

Ὅτι, φήσομεν, ὑπάρχει τοῦτο πᾶσιν ἡμῖν, ἀσπάζεσθαι μὲν τὸ ἀγαθὸν καὶ προσίεσθαι καὶ φιλεῖν, ἀποστρέφεσθαι δὲ καὶ μισεῖν καὶ φεύγειν τὸ κακόν, οὐκέτι προσκεπτομένοις οὔτ᾽ εἰ γενητόν ἐστιν οὔτ᾽ εἰ μὴ γενητόν[89] οὔτ᾽ εἴ τι ἕτερον αὐτὸ τοιοῦτον ἐποίησεν οὔτ᾽ εἴ τι κατεσκεύασεν αὐτὸ τοιοῦτον. τούς τε οὖν σκορπίους καὶ τὰ φαλάγγια καὶ τὰς ἐχίδνας ἀναιροῦμεν ὑπὸ τῆς φύσεως γεγονότα τοιαῦτα καὶ οὐχ ὑφ᾽ ἑαυτῶν. ἀγένητόν τε καὶ τὸν πρῶτον καὶ μέγιστον θεὸν ὁ Πλάτων λέγων ὁμοίως ἀγαθὸν ὀνομάζει, καὶ ἡμεῖς δὲ πάντες φύσει φιλοῦμεν αὐτὸν ὄντα τοιοῦτον ἐξ αἰῶνος, οὐχ ὑφ᾽ ἑαυτοῦ γενόμενον ἀγαθόν· ὅλως γὰρ οὐδ᾽ ἐγένετό ποτε διὰ παντὸς ἀγένητος ὢν καὶ ἀίδιος. εἰκότως οὖν καὶ τῶν ἀνθρώπων τοὺς πονηροὺς μισοῦμεν οὐ προσλογιζόμενοι τὸ ποιῆσαν αἴτιον αὐτοὺς τοιούτους, ἔμπαλιν δὲ προσιέμεθα καὶ φιλοῦμεν τοὺς ἀγαθούς, εἴτ᾽ ἐκ φύσεως εἴτ᾽ ἐκ παιδείας καὶ διδασκαλίας εἴτ᾽ ἐκ προαιρέσεως καὶ ἀσκήσεως ἐγένοντο τοιοῦτοι. καὶ μέν γε καὶ ἀποκτείνομεν τοὺς ἀνιάτως πονηροὺς διὰ τρεῖς αἰτίας εὐλόγως· ἵνα μήθ᾽ ἡμᾶς ἀδικήσωσι ζῶντες εἰς φόβον τε τοὺς ὁμοίους αὐτοῖς ἐνάγωσιν ὡς κολασθησομένους ἐφ᾽ οἷς ἂν ἀδικήσωσι, καὶ τρίτον ὅτι καὶ αὐτοῖς ἐκείνοις ἄμεινόν ἐστι τεθνάναι διεφθαρμένοις οὕτω τὴν ψυχὴν ὡς ἀνίατον ἔχειν τὴν κακίαν, ὡς μηδ᾽ ὑπὸ τῶν Μουσῶν αὐτῶν παιδεύεσθαι, μή τί γ᾽ ὑπὸ Σωκράτους ἢ καὶ Πυθαγόρου βελτιοῦσθαι δύνασθαι.[90]

THE SOUL'S TRAITS

We shall say that this exists in all of us—to embrace, accept, and love the good, but to turn away from, hate, and avoid the bad, not yet having considered beforehand whether it is innate or not innate, or whether or not one of these made it so, or it has contrived to be so by itself. Anyway, we destroy scorpions, venomous spiders, and vipers, which have become as they are by nature and not by themselves. Plato, when he calls the first and greatest god ungenerated, similarly terms him good, and that we all naturally love him, being as he is from eternity and not having become good through himself. For he did not at any time become at all, being perpetually ungenerated and eternal. Naturally then, we also hate those who are wicked without considering beforehand what cause made them so, while conversely, we accept and love those who are good, whether they became so from nature, or from upbringing and teaching, or from choice and training. And indeed also, we put to death the incorrigibly | wicked reasonably on three grounds: so that while they live they do not harm us; to promote in those similar to them a fear of being punished for those things they might do wrong; and third, because it is better for those people themselves to die, since the soul is destroyed that is incurably evil, so that not even by the Muses themselves are they educable, nor indeed can they be made better by a Socrates or Pythagoras.

816K

---

[89] *post* γενητόν: οὐκέτι ἕτερον αὐτῷ τοιοῦτον ἐποίησεν οὔτε κατεσκεύασεν αὐτῷ τοιοῦτον]. τοὺς γοῦν SM; οὐ γὰρ τὸ ἕτερον αὐτῶν τοιοῦτον ἐποίησεν, οὔτε κατεσκεύασεν αὐτὸ τοιοῦτον. τοὺς τ' οὖν K

[90] βελτιοῦσθαι δύνασθαι SM; βελτίους γενέσθαι K

GALEN

Θαυμάζω δ' ἐν τῷδε τῶν Στωϊκῶν ἅπαντας μὲν ἀνθρώπους εἰς ἀρετῆς κτῆσιν ἐπιτηδείως ἔχειν οἰομένων, διαστρέφεσθαι δ' ὑπὸ τῶν συζώντων.[91] ἵνα γὰρ ἐάσω τἆλλα πάντα τὰ καταβάλλοντα τὸν λόγον αὐτῶν, ἓν δὲ μόνον ἐρωτήσω περὶ τῶν πρώτων γενομένων ἀνθρώπων, οἳ μηδένα πρὸ ἑαυτῶν ἔσχον ἕτερον· ἡ διαστροφὴ πόθεν ἢ ὑπὸ τίνων αὐτοῖς ἐγένετο; οὐχ ἕξουσι λέγειν, ὥσπερ γε κἂν τῷ νῦν χρόνῳ μικρὰ παιδία θεώμενοι πονηρότατα, τίς ἐδίδαξεν αὐτὰ τὴν πονηρίαν, ἀδυνατοῖεν ἂν λέγειν, καὶ μάλισθ' ὅταν ᾖ πολλὰ ἅμα μὲν τρεφόμενα τὴν αὐτὴν τροφὴν ὑπὸ τοῖς αὐτοῖς γονεῦσιν ἢ διδασκάλοις ἢ παιδαγωγοῖς, ἐναντιώτατα δὲ ταῖς φύσεσιν. τί γὰρ ἂν ἐναντιώτερον εἴη τοῦ κοινωνικοῦ παιδίου τῷ φθονερῷ, καθάπερ καὶ τοῦ ἐλεήμονος τῷ ἐπιχαιρεκάκῳ, τοῦ δὲ δειλοῦ πρὸς ἅπαντα τῷ θαρσαλέῳ καὶ τοῦ μωροτάτου τῷ συνετωτάτῳ καὶ τοῦ φιλαλήθους τῷ φιλοψευδεῖ; καὶ φαίνεταί γε τὰ παιδία, κἂν ὑπὸ τοῖς αὐτοῖς γονεῦσι καὶ διδασκάλοις καὶ παιδαγωγοῖς τρέφηται, κατὰ τὰς εἰρημένας ἐναντιώσεις ἀλλήλων διαφέροντα.

Παραφυλάττειν οὖν δεῖ εἰπεῖν τὰ τοιαῦτα κατὰ τῶν νῦν φιλοσόφων· ἄμεινον δ' ἐστὶν ἴσως εἰπεῖν οὐ φιλοσόφων ἀλλ' ἐπαγγελλομένων φιλοσοφεῖν· ὡς, εἴ γ' ὄντως ἐφιλοσόφουν, ἐφύλαττον ἂν αὐτὸ τοῦτο πρῶτον, τὸ ἀπὸ τῶν ἐναργῶς φαινομένων τὰς ἀρχὰς τῶν ἀποδείξεων ποιεῖσθαι. καὶ τοῦτ' ἐοίκασι μάλιστα πάντων

---

[91] συζώντων SM; οὐ καλῶς ζώντων K

THE SOUL'S TRAITS

What amazes me in this is when the Stoics think all people are suitably possessed of goodness but are perverted by those they live with. Consequently, I shall leave aside all the other things overthrowing their argument and will ask this one thing alone, about the first people to exist, who had no one else before themselves. From where and by whom did the perverting of these same people occur? They will not be able to say, as, in fact, even in the present time, when we say small children, whom we see are very bad, were we to ask who taught them the wickedness, they would be unable to say, and particularly when there are many | brought up at the same time and nurtured in the same way by the same parents, teachers, or tutors, who are completely opposite in their natures. For what would be more opposite than the sharing child to the one who is envious, as also the merciful child to one who rejoices in others' misfortune, the timid child to the child courageous in every respect, and the very stupid child to one who is very intelligent, and the one who loves truth to the one fond of lying? And it is in fact apparent that children, even when brought up by the same parents, teachers, and pedagogues, differ from each in regard to the opposites mentioned.

817K

Therefore, one needs to guard carefully against those of the present philosophers who say such things. However, it would perhaps be better not to say philosophers but those who claim to be philosophizing. If, in fact, they really philosophized in this way, they would maintain this very thing first of all—to make the beginning of their demonstrations from those things that are clearly apparent.

οἱ παλαιότατοι[92] πρᾶξαί τε καὶ κληθῆναι σοφοὶ παρὰ τοῖς ἀνθρώποις οὔτε συγγράμματα γράφοντες οὔτε διαλεκτικὴν ἢ φυσικὴν ἐπιδεικνύμενοι θεωρίαν ἀλλ' ἐξ αὐτῶν μὲν τῶν[93] ἐναργῶς φαινομένων τῆς θεωρίας ἀρξάμενοι τῶν ἀρετῶν, ἀσκήσαντες δ' αὐτὰς ἔργοις, οὐ λόγοις. οὗτοι γοῦν οἱ φιλόσοφοι βλέποντες εὐθὺς ἐξ ἀρχῆς τὰ παιδία, κἂν ἄριστα παιδεύηται καὶ μηδὲν ἔχῃ θεάσασθαι παράδειγμα κακίας, ὅμως ἁμαρτάνοντα τινὲς μὲν αὐτῶν φύσει κακοὺς ἅπαντας ἀνθρώπους ἀπεφήναντο, τινὲς δ' ὀλίγου δεῖν ἅπαντας—σπάνιον γὰρ ὄντως ἔστι θεάσασθαι παιδίον ἄμεμπτον—οἱ μὲν οὐδενὸς ὄντος τοιούτου πάντας ἀνθρώπους ἀπεφήναντο φύσει κακοὺς ὑπάρχειν, οἱ δ' ἕνα ἢ δύο που κατὰ τὸ σπάνιον ἰδόντες οὐ πάντας ἀλλὰ τοὺς πλείστους ἔφασαν εἶναι κακούς. εἰ γάρ τις, οὐκ ἂν ἐπιτρίπτων τε καὶ φιλονείκων, ἐθελήσειεν ἂν ἐλευθερίᾳ γνώμῃ καθάπερ οἱ παλαιοὶ φιλόσοφοι τὰ πράγματα θεάσασθαι, παντάπασιν ὀλίγους παῖδας εὑρήσει πρὸς ἀρετὴν εὖ πεφυκότας καὶ παύσεται πάντας μὲν ἡμᾶς ἡγούμενος εὖ πεφυκέναι, διαστρέφεσθαι δ' ὑπὸ τῶν ἐξεπιτιμώντων[94] γονέων τε καὶ παιδαγωγῶν καὶ διδασκάλων· οὐ γὰρ δὴ ἄλλοις γέ τισιν ἐντυγχάνει τὰ παιδία.

Πάνυ δ' εὐήθεις[95] εἰσὶ καὶ οἱ διαστρέφεσθαι λέγοντες ἡμᾶς ὑπό τε τῆς ἡδονῆς[96] καὶ τοῦ ἄλγους, τῆς

[92] παλαιότατοι SM; πάλαι θειότατοι K [93] post τῶν: ἐναργῶς φαινομένων τῆς θεωρίας ἀρξάμενοι τῶν add. SM

THE SOUL'S TRAITS

And this, it would seem, most ancients did most of all—men who were called wise among men, despite not writing treatises or setting out a dialectical or natural theory, but | beginning from these same clearly apparent things of the 818K theory of the virtues, practicing them by actions and not by words. Anyway, these philosophers now, when they observe children right from the start, even when they receive the best education and have no example of badness to see, nevertheless fall into error, some of them declaring all are bad by nature, while some almost all, for it is truly rare to see a child without fault. There are those who declare that, since none are such (i.e., without fault), all people are bad in nature, whereas there are those who, having seen rarely one or two (without fault), say that not all but the majority are bad. But if someone who is not an irritating rascal and fond of contention were to wish, with a free mind, (like the philosophers of ancient times) to observe the matters, he will find altogether few children who naturally tend toward virtue and will cease to think that we are all naturally good but are turned aside from this by honored parents, pedagogues, and teachers, for children do not meet with any others.

However, those who say we are diverted by | pleasure 819K and pain are very foolish, the former drawing toward itself

---

94 B: ἐξεπιτιμό[ώ]ντων WMB Lat.: ἐξεπιστημόντων P: ἐξ ἐπιστημόνων Morellus Chartier: ἐξεστηκότων Nauck: om. Ar.: secl. Müller

95 εὐήθεις SM: ἀληθεῖς K

96 post τῆς ἡδονῆς: καὶ τοῦ ἄλγους, τῆς μὲν ἑλκούσης ⟨πρὸς ἑαυτήν⟩,SM; καίτοι γε αὐτῆς μὲν ἐχούσης πολὺ K

GALEN

μὲν ἑλκούσης πρὸς ἑαυτήν, τοῦ δ' ἀποτρεπτικοῦ τε καὶ τραχέος ὄντος. εἰ μὲν γὰρ πάντες ᾠκειώμεθα πρὸς τὴν ἡδονήν, οὐκ οὖσαν ἀγαθὸν ἀλλ', ὡς ὁ Πλάτων ἔφη, "δέλεαρ μέγιστον κακοῦ," φύσει κακοὶ πάντες ἐσμέν, εἰ δ' οὐ πάντες ἀλλά τινες, ἐκεῖνοι μόνοι φύσει μοχθηροὶ τυγχάνουσιν ὄντες. καίτοι γ' εἰ μὲν μηδεμίαν ἔχομεν ἑτέραν ἐν ἡμῖν δύναμιν[97] ᾠκειωμένην μᾶλλον ἡδονῆς, ἰσχυροτέρα πως ἔσται ἡδονὴ τῆς ἀρετῆς, ἥτις κρείττων ἐστὶ τῆς πρὸς τὴν ἡδονὴν ἀγούσης ἡμᾶς ἐστι δυνάμεως κἂν οὕτως εἴημεν ἅπαντες κακοὶ τὴν μὲν κρείττονα δύναμιν καὶ ἀσθενεστέραν, ἰσχυροτέραν δὲ τὴν μοχθηρὰν ἔχοντες· εἰ δ' ἡ κρείττων ἐστὶν ἰσχυροτέρα, τίς τοὺς πρώτους ἀνθρώπους ἀνέπεισεν ὑπὸ τῆς ἀσθενεστέρας νικηθῆναι;

Ταῦτ' οὖν αὐτὰ τῶν Στωϊκῶν ἐμέμψατο καὶ ὁ πάντων ἐπιστημονικώτατος Ποσειδώνιος, ὃς ἐν οἷς ἐπαινῶν ἐστι μεγίστων ἄξιος, ἐν τούτοις αὐτοῖς μὲν ὑπὸ τῶν ἄλλων οὐκ ἐπαινεῖται Στωϊκῶν· ἐκεῖνοι μὲν γὰρ ἔπεισαν ἑαυτοὺς τὴν πατρίδα μᾶλλον ἢ δόγματα προδοῦναι, Ποσειδώνιος δὲ τὴν τῶν Στωϊκῶν αἵρεσιν μᾶλλον ἢ τὴν ἀλήθειαν. διὰ τοῦτο κατά τε τὴν Περὶ τῶν παθῶν πραγματείαν ἐναντιώτατα φρονεῖ Χρυσίππῳ κἂν τοῖς Περὶ τῆς διαφορᾶς τῶν ἀρετῶν, πολλὰ

[97] post δύναμιν ante ἅπαντες; ᾠκειωμένην [ἡδονὴν μᾶλλον] ἀρετῇ μᾶλλον ἡδονῆς ἤ τινα ἰσχυροτέραν τῆς πρὸς τὴν ἡδονὴν ἀγούσης ἡμᾶς φύσει, οὕτως ἂν εἴημεν B; οἰκειωμένην ἡδονὴν ἢ μᾶλλον ἀρετήν, μᾶλλον ἡδονῆς ἤ τις ἰσχυροτέρα τῆς πρὸς τὴν ἡδονὴν ἀγούσας ἡμᾶς φύσει, K (see B, p. 85)

THE SOUL'S TRAITS

while the latter repels and is harsh. For if we are all associated with pleasure, which is not a good, but is as Plato said, "the greatest lure of bad,"[52] we will all be bad in nature. If, however, we are not all bad, but only some are, then only those will be naturally bad. And indeed, if we have in us no other capacity than associating more with pleasure, then somehow pleasure will be stronger than virtue, which itself is stronger than the capacity that leads us toward pleasure. And in this way we would all be bad, the greater capacity being weaker while the abnormal (bad) capacity is stronger. If, however, the greater is stronger, who persuaded the first people to be overcome by the weaker?

Therefore, Posidonius,[53] the most scientifically knowledgeable of all the Stoics, and a man worthy of the highest praise among them, censured these very views, and is not praised by the other Stoics on these same points. For those men persuaded themselves to betray their fatherland rather than the doctrines, whereas Posidonius was persuaded to betray the sect of the Stoics rather than the truth. Because of this, in the treatise *On the Affections* he thinks the most opposite things to Chrysippus, and even in the treatise *On the Difference of the Virtues*, he finds

820K

---

[52] *Timaeus* 69D.

[53] Posidonius (ca.135–ca. 51 BC) was a major figure in later Stoic philosophy. His copious writings, including his *On Emotions* referred to in Galen's *On the Opinions of Hippocrates and Plato*, V.181–895K, survive only as fragments; see the brief summary in the Introduction and the very informative entry by I. G. Kidd in the *OCD* (pp. 1231–33).

μὲν ὧν εἶπε Χρύσιππος ἐν τοῖς λογικῶς ζητουμένοις περὶ τῶν παθῶν τῆς ψυχῆς μεμψάμενος, ἔτι δὲ πλείω τῶν ἐν τοῖς Περὶ τῆς διαφορᾶς τῶν ἀρετῶν. οὐ τοίνυν οὐδὲ Ποσειδωνίῳ δοκεῖ τὴν κακίαν ἔξωθεν ἐπεισιέναι τοῖς ἀνθρώποις οὐδεμίαν ἔχουσαν ἰδίαν ῥίζαν ἐν ταῖς ψυχαῖς ἡμῶν, ὅθεν ὁρμωμένη βλαστάνει τε καὶ αὐξάνεται, ἀλλ' αὐτὸ τοὐναντίον εἶναι. καὶ γὰρ καὶ τῆς κακίας ἐν ἡμῖν αὐτοῖς σπέρμα· καὶ δεόμεθα πάντες οὐχ οὕτω τοῦ φεύγειν τοὺς πονηροὺς ὡς τοῦ διώκειν τοὺς καθαρίσοντάς τε καὶ κωλύσοντας ἡμῶν τὴν αὔξησιν τῆς κακίας. οὐ γάρ, ὡς οἱ Στωϊκοί φασιν, ἔξωθεν ἐπεισέρχεται ταῖς ψυχαῖς ἡμῶν τὸ σύμπαν τῆς κακίας, ἀλλὰ τὸ πλέον ἐξ ἑαυτῶν ἔχουσιν οἱ πονηροὶ τῶν ἀνθρώπων, ἔξωθεν δ' ἔλαττον τούτων πολλῷ τὸ ἐπεισερχόμενόν ἐστιν.

Ἐκ τούτου μὲν οὖν ἐθισμοί τε γίγνονται μοχθηροὶ τῷ τῆς ψυχῆς ἀλόγῳ μέρει καὶ δόξαι ψευδεῖς τῷ λογιστικῷ, καθάπερ | ὅταν ὑπὸ καλοῖς καὶ ἀγαθοῖς ἀνδράσι παιδευώμεθα, δόξαι[98] μὲν ἀληθεῖς, ἐθισμοὶ δὲ χρηστοί. ταῖς κράσεσι δ' ἕπεται κατὰ μὲν τὸ λογιστικὸν ἀγχίνοιά τε καὶ μωρία κατὰ τὸ μᾶλλόν τε καὶ ἧττον· αἱ κράσεις δ' αὐταὶ τῇ τε πρώτῃ γενέσει καὶ ταῖς εὐχύμοις διαίταις ἀκολουθοῦσιν, ὥστε συναυξάνειν ἄλληλα ταῦτα. διὰ γοῦν τὴν θερμὴν κρᾶσιν οἱ ὀξύθυμοι γιγνόμενοι ταύταις πάλιν ταῖς ὀξυθυμίαις ἐκπυροῦσι τὴν ἔμφυτον θερμασίαν· ἔμπαλιν δ' οἱ σύμμετροι ταῖς κράσεσι συμμέτρους τὰς τῆς ψυχῆς

THE SOUL'S TRAITS

fault with many things Chrysippus said in his *Logical Investigations* about the affections of the soul, and still more in the writings *On the Difference of the Virtues*. Accordingly, it does not seem to Posidonius that badness enters people from without, having no specific root in our souls starting from which it sprouts and grows; rather, it is the very opposite. For surely there is a seed of badness in us ourselves, and we all need not so much to avoid those who are wicked as to pursue those who will purify us and prevent the growth of the badness. For the whole of the badness does not, as the Stoics say, enter our souls from without, but the people who are wicked have the greater part from themselves, and what enters from without is less in them by far.

From this then, bad habits arise in the nonrational part of the soul, and false opinions in the rational part, just as when we are educated by good and virtuous men, true 821K opinions and good habits arise. However, to a greater or lesser degree sagacity and stupidity in the rational part follow the *krasias*, while the *krasias* themselves follow the first genesis and the *euchymous* regimens, so that these jointly increase each other. Anyway, those who are quick to anger become so due to the hot *krasis* and with their bursts of anger fire up the innate heat. Contrariwise, those moderate in the *krasias* have moderate movements of the soul which are beneficial to tran-

---

98 B: ante δόξαι add. die vernünftige Seele Ar. δόξαι: Ansichten und Meinungen Ar. ante ἐθισμοὶ add. die unvernünftige Seele Ar. δὲ KLat: τε MBP Chartier: καὶ W. post χρηστοί lacunam susp. Müller

κινήσεις ἔχοντες εἰς εὐθυμίαν ὠφελοῦνται. ὥσθ᾽ ὁ μὲν ἡμέτερος λόγος ὁμολογεῖ τοῖς ἐναργῶς φαινομένοις[99] ἐξηγούμενος τὰς αἰτίας ὧν μὲν ὑπ᾽ οἴνου καὶ φαρμάκων τινῶν πάσχομεν, ὧν[100] δ᾽ ὑπὸ διαίτης ἀγαθῆς καὶ κακῆς,[101] ὥσπερ γε καὶ ὠφελούμεθά τε καὶ βλαπτόμεθα[102] δι᾽ ἐπιτηδευμάτων τε καὶ μαθημάτων, οὐχ ἥκιστα δὲ καὶ τῆς φυσικῆς διαφορᾶς τῶν παίδων ἀποδιδοὺς τὴν αἰτίαν. οἱ δ᾽ οὐκ ἐκ τῆς τοῦ σώματος κράσεως ἡγούμενοι τὴν ψυχὴν ὠφελεῖσθαί τε καὶ βλάπτεσθαι περί τε τῆς τῶν παίδων διαφορᾶς οὐδὲν ἔχουσι λέγειν ὧν τ᾽ ἐκ τῆς διαίτης ὠφελούμεθα, οὐδενὸς ἔχουσιν αἰτίαν ἀποδοῦναι, καθάπερ οὐδὲ τῆς ἐν τοῖς ἤθεσι διαφορᾶς, καθ᾽ ἣν τὰ μὲν θυμικά, τὰ δ᾽ ἄθυμα καὶ τὰ μὲν συνετά, τὰ δ᾽ οὐ φαίνεται. ἐν Σκύθαις μὲν γὰρ εἷς ἀνὴρ ἐγένετο φιλόσοφος, Ἀθήνησι δὲ πολλοὶ τοιοῦτοι· πάλιν δ᾽ ἐν Ἀβδήροις ἀσύνετοι πολλοί, τοιοῦτοι δ᾽ Ἀθήνησιν ὀλίγοι.

[99] post φαινομένοις: ἐξηγούμενος τὰς αἰτίας ὧν μὲν SM; κατὰ τὰς αἰτίας, ὥσθ᾽ K
[100] πάσχομεν, ὧν SM; φασκομένων K
[101] κακῆς SM; καλῆς K
[102] βλαπτόμεθα add. SM

THE SOUL'S TRAITS

quility.[54] So then, our argument agrees with the obvious appearances, explaining the causes from which we are affected by wine and certain medications, and by a good and bad regimen, just as we are also benefitted and harmed by both practices and studies. Not least does this also yield the cause of the natural difference of children. However, those who do not think the soul is benefitted and harmed by the *krasis* of the body have nothing to say about the difference in children, and none | [of them] has a cause to give of those benefits we derive from regimen, just as they do not on the difference in the characteristics which appears in relation to the high-spirited and spiritless, and the intelligent and unintelligent. Among the Scythians, there has been one man who became a philosopher,[55] while among the Athenians there are many such men. Conversely, among Abderians many are unintelligent but among Athenians such men are few.

822K

[54] The two contrasting terms are *oxythymos*, "quick to anger" (see Aristotle, *Rhetoric* 1368b20), and *euthymos*, "kind, generous, cheerful" (*Odyssey* 14.63).

[55] The "single" Scythian philosopher referred to is presumably Anacharsis, "a largely legendary Scythian prince who came to exemplify the wise barbarian" (*OCD*).

APPENDIX
TWO SHORT TREATISES

INTRODUCTION

In these two short treatises, which Galen states were prepared at the request of friends,[1] he offers some thoughts on what constitutes the best state or condition of the human body, based primarily on his concept of *krasis* and its variants, and his division of the components of the body into *homoiomeres* and organic structures. They clearly supplement his major treatise on *krasias* (temperaments).

In the first work, *On the Best Constitution of our Body*, Galen poses two fundamental questions:

> What the best composition of the human body is.
> What the best composition of the human body does.

His concise answer to the first question is given in the final sentence of the work: ". . . we must lodge the best constitution in two things—in *eukrasia* of the *homoiomeres* and in balance of the organic parts." The short answer to the second question is that the best constitution ensures excellence of all functions, facilitates *euchymia*, and makes the body most resistant to factors, both external and internal, that threaten to adversely affect this otherwise stable state.

Three other issues considered are as follows:

---

[1] Galen refers to the treatises having been written at the request of friends in his *On the Order of My Own Books*, XIX.56K. He refers to them more generally and in relation to his other works in his *Art of Medicine*, I.407–8K.

APPENDIX

1. The harms that can befall the body, which he divides into "external" (to include heating, cooling, wetting, or drying beyond what is appropriate, and other factors, among which he lists fatigue, insomnia, grief, and anxiety) and "internal," which are due to the superfluities or residues of nutriment, divided into quantitative and qualitative abnormalities.

2. The numerical distribution of the types of constitution that may be encountered in medical practice: (a) the best constitution, as defined above, which is encountered rarely; (b) the constitution that falls somewhat short of the best, but not by enough to fall into the third category. Such bodies are the most commonly encountered and presumably include both healthy and diseased bodies, as defined by him; (c) the constitution that he describes as having "... already acquired a manifest and large defect."

3. The recognition of abnormal states (i.e., those that fall short of the best constitution strictly defined), which basically depends on visual and tactile evaluation centered on certain qualities or actions that have a certain range: he lists hardness and softness, rarefaction and condensation, hairiness and hairlessness, veins that are dilated or constricted, and pulses that are large or small. Not mentioned here, but certainly relevant are hot/cold and wet/dry. On this point, Galen notes that small deviations from the mean may at times be beneficial. He gives the example of a body somewhat harder than normal being more resistant to adverse external factors.

The second treatise, *On Good Bodily State* (*Euexia*), is even briefer than the first. Galen's chief purpose is to clarify the distinction between the term *euexia* when used in an absolute sense and when used with some qualification (addition—$προσθήκη$). The former is characterized

INTRODUCTION

by *eukrasia* of all parts of the body, excellence of functions, and *euchymia*. There is also due proportion in the amount of blood and in body mass. In the latter, exemplified by the *euexia* of athletes, due to "overfilling," there is a disproportionate amount of blood and a disproportionately large body mass. Galen briefly considers the consequences of these disproportions and cites Hippocrates' statement on the *euexia* of athletes being a danger to health.

In both treatises, there is consideration of a number of terms used to describe the state of the body—specifically, "constitution" (κατασκευή), "condition" (διάθεσις), "state," divided into that which is stable and relatively permanent (ἕξις) and that which is unstable and relatively impermanent (σχέσις), and "nature" (φύσις). Galen stresses, as he does in a number of other places, that the actual term is not what is important; it is the matter itself. Both these short treatises complement the major work on *krasias*, while the work on the best constitution is especially relevant to Galen's *Hygiene*.

The Greek text for both these works is that of G. Helmreich's 1901 edition.[2] There are minor differences from the more readily available Kühn Greek text, the more substantial of which are indicated by footnotes. None of these affect meaning in any important way. There is an English translation of *On the Best Constitution of Our Body* by R. J. Penella and T. S. Hall (1973)[3] and of both works by Singer (1997). All three were consulted.

---

[2] *Galeni De optima nostri corporis constitutione, De bonu habitu* (Hof, 1901).

[3] R. J.Penella, and T. S. Hall, "Galen's *On the Best Constitution of Our Body*: Introduction, Translation and Notes," *Bulletin of the History of Medicine* 47 (1973): 282–96. Singer, *Galen: Selected Works*.

ΓΑΛΗΝΟΥ ΠΕΡΙ ΑΡΙΣΤΗΣ ΚΑΤΑΣΚΕΥΗΣ ΤΟΥ ΣΩΜΑΤΟΣ

1. Τίς ἡ ἀρίστη κατασκευὴ τοῦ σώματος ἡμῶν; ἆρά γ᾽ ἡ εὐκρατοτάτη, καθάπερ ἔδοξε πολλοῖς τῶν παλαιῶν ἰατρῶν τε καὶ φιλοσόφων; ἢ τὴν μὲν ἀρίστην ἀναγκαῖον εὐκρατοτάτην ὑπάρχειν, οὐ μὴν τήν γ᾽ εὐκρατοτάτην ἐξ ἀνάγκης ἀρίστην; ἡ μὲν γὰρ ἐκ θερμοῦ καὶ ψυχροῦ καὶ ξηροῦ καὶ ὑγροῦ σύμμετρος κρᾶσις ὑγίεια τῶν ὁμοιομερῶν[1] ἐστι τοῦ σώματος ἡμῶν· ἡ δ᾽ ἐκ τούτων ἁπάντων | τοῦ ζῴου διάπλασις ἐν θέσει καὶ μεγέθει καὶ σχήματι καὶ ἀριθμῷ τῶν συνθέτων ὑπάρχει καὶ δόξειεν ἂν ἐγχωρεῖν[2] ἐξ ἁπάντων ἢ ἐκ τῶν πλείστων γε μορίων εὐκράτων συγκείμενόν τι σῶμα περὶ τὸ μέγεθος αὐτῶν ἢ τὸν ἀριθμὸν ἢ τὴν διάπλασιν ἢ τὴν πρὸς ἄλληλα σύνταξιν ἡμαρτῆσθαι.

Πειρατέον οὖν ὑπὲρ ἁπάντων τούτων ἐφεξῆς διελθεῖν ἀρξαμένους ἀπὸ τῶν ὀνομάτων, οἷς ἐξ ἀνάγκης χρώμεθα κατὰ τόνδε τὸν λόγον, ἐπειδὴ καὶ περὶ τούτων ἐρίζουσί τινες, οἱ μὲν κατασκευὴν ἀρίστην, οἱ δὲ διάθεσιν, οἱ δ᾽ ἕξιν ἢ σχέσιν ἢ φύσιν σώματος ἢ ὅπως ἂν

[1] post ὁμοιομερῶν add. μελῶν K
[2] post ἐγχωρεῖν: ἐξ ἁπάντων ἢ H; τὸ K

1. ON THE BEST CONSTITUTION OF OUR BODY

1. What is the best constitution of our body? Is it, in fact, the most *eukratic*, as it seemed to many of the doctors and philosophers of ancient times? Or is it that the best constitution is necessarily the most *eukratic*, but in fact the most *eukratic* is not of necessity the best? Health is a balanced *krasis* of hot, cold, wet, and dry of the *homoiomerous* parts of our body, whereas | the conformation of the animal is in the combination of all these [*homoiomerous* parts] in terms or position, magnitude, form, and number. And it would seem possible that even a body compounded from all or a very large number of *eukratic* parts could be at fault regarding their magnitude, number, conformation, or arrangement with each other.

We must attempt, then, to go over all these matters in order, beginning from the terms we use of necessity in this discussion, since there are some who contend about these. Some think fit to say best "constitution," some "condition," some "state" (*hexis*), some "state" (*schesis*),[1] or "nature" of

---

[1] Galen consistently makes a distinction between *hexis* as a stable and relatively permanent state and *schesis* as an unstable and temporary state.

ἑκάστῳ δόξῃ λέγειν ἀξιοῦντες. ἐγὼ δὲ τὸ μὲν ὡς ἄν τῳ παραστῇ ποιεῖσθαι τὴν ἑρμηνείαν οὐ μέμφομαι, τὸ δ' ἐγκαλεῖν τοῖς ἑτέρως ὀνομάζουσιν οὐκ ἐπαινῶ, τὴν πλείστην καὶ μεγίστην φροντίδα τῶν πραγμάτων αὐτῶν, ὑπὲρ ὧν ὁ λόγος ἐστίν, ἡγούμενος χρῆναι ποιεῖσθαι μᾶλλον ἢ τῶν ὀνομάτων. εἴτ' οὖν ἀρίστην κατασκευὴν σώματος εἴτε διάθεσιν εἴθ' ἕξιν εἴτε σχέσιν εἴτε φύσιν εἴθ' ὁπωσοῦν ἑτέρως ὀνομάζειν ἐθέλοι τις, οὕτω θέμενος, ἐὰν ἀπό τε τῆς ὁμολογουμένης ἐννοίας | ἄρξηται καὶ προϊὼν ἐπὶ τὴν τῆς οὐσίας εὕρεσιν ἐν τάξει τινὶ καὶ μεθόδῳ ποιῆται τὴν ζήτησιν, ἐπαινέσομαι τοῦτον ἐγὼ πολὺ μᾶλλον ἢ εἰ περὶ τῶν ὀνομάτων πάνδεινός τίς μοι φαίνοιτο. καὶ τοίνυν καὶ ἡμεῖς οὕτω ποιῶμεν ἀπὸ τῆς κοινῆς ἐννοίας ἀρξάμενοι καὶ ταύτην διορισάμενοι μεθόδῳ προΐωμεν ἐπὶ τὸ συνεχὲς τῆς σκέψεως.

739K

2. Τίς οὖν ἔννοια κοινὴ πᾶσιν ἀνθρώποις ἐστὶν ἀρίστης κατασκευῆς σώματος; ἀκοῦσαι μὲν γὰρ ἔστι λεγόντων αὐτῶν οὐχ ὁμοίως τῇ λέξει, νοούντων δ' ἁπάντων ἓν καὶ ταὐτὸν πρᾶγμα. τὸ γοῦν ὑγιεινότατον σῶμα πάντες μὲν ἑξῆς ἐπαινοῦσιν, ὥσπερ οὖν καὶ τὸ εὐεκτικώτατον, εἰς ἕν μέν τι πρᾶγμα βλέποντες ἀμφότεροι καὶ τούτῳ τὴν διάνοιαν ἐπιβάλλοντες, οὐ μὴν οὔτε διηρθρωμένως αὐτὸ νοοῦντες οὔθ' ἑρμηνεῦσαι σαφῶς ἐπιστάμενοι. καὶ γὰρ καὶ τὰς ἐνεργείας ἁπάντων τῶν τοῦ σώματος μορίων εὐρώστους ἔχειν ἀξιοῦσι καὶ μὴ ῥᾳδίως ὑπὸ τῶν νοσωδῶν αἰτίων νικᾶσθαι. τούτων δὲ τὸ μὲν ἐν ταῖς ἐνεργείαις ταῖς κατὰ φύσιν ὑγίεια, τὸ δὲ

TWO SHORT TREATISES

a body, or whatever it might seem to each person to be best to say. I would not find fault with someone standing by the explanation he makes, but I would not approve of his objecting to those who name otherwise. What I think it is necessary to do is to give the most and greatest thought to the matters themselves, about which the argument is, rather than to the names. Therefore, if someone should wish to use the term best "constitution" of a body (*kataskeuē*), or "condition" (*diathesis*), or "state" (*hexis*), or "state" (*schesis*), or "nature" (*physis*), or whatever else, if after establishing it in this way, | he begins from the agreed concept and proceeds to the discovery of the essence, carrying out the investigation with a certain order and method, I would commend him much more than I would someone who seems to me clever with names. And accordingly, let us also proceed in this way, beginning from the common concept, and having defined this, go forward by method to the continuation of our inquiry. 739K

2. What, then, is the concept of the best constitution of a body common to all men? It is possible to hear the same things said dissimilarly in speech, although all are thinking about one and the same matter. At any rate, all, one after another, praise the most healthy body, just as they also do the most well-conditioned body, both looking at this one matter, directing their thought to it, although neither conceiving of it distinctly nor knowing how to explain it clearly. For truly, they hold the view that the functions of all the parts of the body should be strong and not easily overcome by the causes of diseases. For them, health is in the functions being in accordance with nature, while this,

APPENDIX

740K μετὰ ῥώμης τινὸς εὐεξία. κοινὸν δ' ἀμφοῖν τὸ μὴ ῥᾳδίως ἁλίσκεσθαι νόσοις, ὥστ' εὐεκτικὴ μὲν πάντως[3] ἐστὶν ἡ ὑγιεινοτάτη κατάστασις, ἧς ἅπαντες ἄνθρωποι γλίχονται· συμβέβηκε δ' αὐτῇ τό τ' ἐν ταῖς ἐνεργείαις κατωρθωμένον καὶ τὸ δύσλυτον. ταῦτά τοι καὶ δεόντως εὐεξία κέκληται, ἐμφαίνοντος μὲν ἤδη καὶ αὐτοῦ τοῦ τῆς ἕξεως ὀνόματος τὸ μόνιμόν τε καὶ δύσλυτον, ἀλλ'[4] ἐπὶ μᾶλλον ἔτι τοῦ τῆς εὐεξίας, ὡς ἂν ἀρίστης τινὸς ἕξεως ὑπαρχούσης. ὥστε καὶ κατασκευὴν ἀρίστην σώματος εἴτε τὴν ὑγιεινοτάτην εἴτε τὴν εὐεκτικωτάτην εἴποιμεν, οὐχ ἁμαρτησόμεθα καὶ κρινοῦμεν αὐτὴν τῷ δυσλύτῳ τῶν κατωρθωμένων ἐνεργειῶν.

Ἐπεὶ δὲ τοῦτ' ἤδη διώρισται, σκεπτέον ἐφεξῆς, ἥτις ἐστὶν ἡ οὐσία τῆς τοιαύτης τοῦ σώματος ἕξεως. ἀρχὴ δὲ κἀνταῦθα τῆς εὑρέσεως, εἰ ζητήσαιμεν, ὅπως διακειμένου τοῦ σώματος ἐνεργοῦμεν ἄριστα. χρὴ τοίνυν εἰς τοῦτο τῶν ἤδη δεδειγμένων ἐν ἑτέροις ὑπομνήμασιν ἀναμνησθῆναι, πρῶτον μὲν ὡς ἐκ θερμοῦ καὶ ψυχροῦ 741K καὶ ξηροῦ καὶ ὑγροῦ τὰ σώμαθ' ἡμῶν κέκραται· δέδεικται δὲ περὶ τούτων ἐν τῷ Περὶ τῶν καθ' Ἱπποκράτην στοιχείων γράμματι· δεύτερον δὲ τοῦ διορίσασθαι τὰς κράσεις τῶν μορίων· εἴρηται δὲ καὶ περὶ τούτων ἐν τοῖς Περὶ κράσεων ὑπομνήμασιν· ἐφεξῆς δὲ τούτων, ὡς ἕκαστον μὲν τῶν ὀργανικῶν τοῦ σώματος μελῶν ἓν ἔχει

[3] πάντως H; πάντων K
[4] ante ἀλλ' add. οὐ μὴν K

accompanied by a certain strength, is *euexia*. Common to both | is not easily being overcome by diseases, so that a good bodily condition (*euektikos*) is altogether the healthiest condition, which all men long for, and in which contingently there is proper accomplishment and difficulty of dissolution in the functions. Certainly [the term] *euexia* has been properly applied to this, the term itself already reflecting stability and difficulty of dissolution of the state (*hexis*). But still more is this so of the term *euexia* as being an excellent state. As a result, if we also call a best constitution of a body either the most healthy or the most well-conditioned (*euektic*), we shall not be mistaken, and we shall judge this by the difficulty of dissolution of the properly working functions.

740K

Since this is already determined, what is considered next is what the essence of such a state of the body (*hexis*) is. And here too, the beginning of the discovery is, if we inquire how the body is disposed when we function best. Accordingly, to this end it is necessary for these things we have already shown in other treatises to be recalled. First is that | our bodies are compounded from hot, cold, dry, and wet. These have been shown in the work, *On the Elements according to Hippocrates*.[2] Second, there is the distinguishing of the *krasias* of the parts. There was discussion about these in the treatises, *On the Krasias* (*Temperaments*).[3] Next after these, we must recall that each of the organic parts of the body has in itself one of the

741K

---

[2] *De elementis secundum Hippocratem*, I.413–508K; English translation by de Lacy, *Galen on the Elements*.

[3] *De temperamentis*, II.509–694K; English translation in this volume.

APPENDIX

τῶν ἐν ἑαυτῷ μορίων αἴτιον τῆς ἐνεργείας, τὰ δ' ἄλλα σύμπαντα τὰ συμπληροῦντα τὸ πᾶν ὄργανον ἐκείνου χάριν ἐγένετο. δέδεικται δὲ καὶ περὶ τούτων αὐτάρκως ἐν τῇ Περὶ χρείας μορίων πραγματείᾳ.

Εἴη ἂν οὖν ἀρίστη κατασκευὴ τοῦ σώματος, ἐν ᾗ τὰ μὲν ὁμοιομερῆ πάντα—καλεῖται δ' οὕτω δηλονότι τὰ πρὸς αἴσθησιν ἁπλᾶ—τὴν οἰκείαν ἔχει κρᾶσιν, ἡ δ' ἐκ τούτων ἑκάστου τῶν ὀργανικῶν σύνθεσις ἔν τε τοῖς μεγέθεσιν αὐτῶν καὶ τοῖς πλήθεσι καὶ ταῖς διαπλάσεσι καὶ ταῖς πρὸς ἄλληλα συντάξεσιν εὐμετρότατα κατεσκεύασται. ὅ τι γὰρ ἁπάσαις ταῖς ἐνεργείαις ἄριστα διάκειται, τοῦτο καὶ δυσπαθέστατον εἶναι τῶν ἄλλων σωμάτων ἁπάντων οὐ χαλεπῶς ἄν τις ἐξεύροι. ὃ γὰρ ἂν ἐνεργῇ μόριον ἄριστα, τοῦτο τῆς μὲν τῶν ὁμοιομερῶν | εὐκρασίας καὶ τῆς τῶν ὀργανικῶν συμμέτρου κατασκευῆς ἔκγονον ὑπάρχει. τοιοῦτον δ' ἐστὶ τὸ προειρημένον σῶμα. δῆλον οὖν, ὡς ἐνεργήσει πάντων ἄριστα. ὅτι δὲ καὶ δυσπαθέστατόν ἐστιν, ὧδ' ἂν μάλιστα μάθοις.

3. Αἱ βλάβαι τοῖς σώμασιν ἡμῶν αἱ μὲν ἀπὸ τῶν ἔξωθεν αἰτίων, αἱ δ' ἀπὸ τῶν τῆς τροφῆς ὁρμῶνται περιττωμάτων. αἱ μὲν ἀπ' αὐτῶν τῶν ἔξωθεν αἰτίων ἐγκαυθεῖσί τε καὶ ψυχθεῖσιν ὑγρανθεῖσί τε καὶ ξηρανθεῖσι πέρα τοῦ προσήκοντος. ἐν τούτῳ δὲ τῷ γένει καὶ κόπους καὶ ἀγρυπνίας καὶ λύπας καὶ φροντίδας ὅσα τ' ἄλλα τοιαῦτα θετέον. ἀπὸ δὲ τῶν τῆς τροφῆς περιττωμάτων διτταὶ μὲν κατὰ γένος, ὅτι καὶ ταῦτα διττά, τὰ

TWO SHORT TREATISES

parts which is the cause of the function, while all the other parts combine to fill up the whole organ for the sake of that part. There has been sufficient elucidation of these matters in the treatise, *On the Use of the Parts*.[4]

Therefore, the best constitution of the body would be that in which all the *homoiomeres* (for obviously the parts simple to perception are termed thus) have their own proper *krasis* The compoundings of each of these organs is in the magnitudes of these, and the amounts, conformations, and arrangements with each other, prepared in the most balanced way. For such is best disposed in all the functions, and someone should not have any trouble finding this as being the most difficult to affect of all the other bodies. For a part would function best if there is | *eukrasia* of the *homoiomerous* parts, and is a product of the duly proportioned constitution of the organs. The previously described body is such a thing. It is clear, therefore, that it will function best of all. And that it is also most difficult to affect, you would learn particularly in what follows.

3. The harms to our bodies start in some cases from external causes and in some cases from the superfluities of the nutriment. The harms from actual external causes are by heatings, coolings, wettings, and dryings beyond what is appropriate. In this class too, we must place fatigues, insomnias, griefs, anxieties, and other such things. Harms from the superfluities of nutriment are twofold in class, because these are also two in number—those that

---

[4] *De usu partium*, III.1–933K and IV.1–366K; English translation by May, *Galen on the Usefulness*.

μὲν τῷ ποσῷ, τὰ δὲ τῷ ποιῷ διοχλοῦντα, πολυειδεῖς δὲ κατ' εἶδος.

Ὅτι μὲν οὖν τοῖς ἔξωθεν αἰτίοις τὸ συμμέτρως διακείμενον σῶμα δυσάλωτόν ἐστι, πρόδηλον μὲν κἀξ αὐτῆς αὐτοῦ τῆς εὐκρασίας, χαλεπῶς καὶ μόγις εἰς ἀμετρίαν κράσεως ἀφικνουμένης τῷ πάντη τῶν ἄκρων ἀφεστάναι πλεῖστον. | οὐ μὴν ἀλλὰ κἀκ τοῦ καλῶς ἐνεργεῖν ἄριστ' ἂν εἴη παρεσκευασμένον εἰς δυσπάθειαν, ἥκιστα καμάτοις ἁλισκόμενον. ὑπάρχει δ' εὐθὺς τῷ τοιούτῳ σώματι καὶ εὐχυμοτάτῳ τῶν ἄλλων ἁπάντων εἶναι, ὥστε καὶ λύπης καὶ θυμοῦ καὶ ἀγρυπνίας καὶ φροντίδος, ὄμβρων τε καὶ αὐχμῶν καὶ λοιμῶν καὶ πάντων ἁπλῶς εἰπεῖν τῶν νοσερῶν αἰτίων ῥᾷον τῶν ἄλλων ἀνέξεται σωμάτων. μάλιστα γὰρ δὴ τὰ κακόχυμα πρὸς τῶν τοιούτων αἰτίων ἐξελέγχεται ῥᾳδίως, ὡς ἂν ἤδη καὶ καθ' ἑαυτὰ πλησίον ἥκοντα⁵ νόσων. οὕτω μὲν ὑπὸ τῶν ἔξωθεν ἡμῖν προσπιπτόντων καὶ λυπούντων τὸ σῶμα δυσάλωτός⁶ ἐστιν ἡ εἰρημένη διάθεσις. ὅτι δ' οὐδ' ὑπὸ τῶν τῆς τροφῆς περιττωμάτων εὐάλωτος ὑπάρχει νόσοις, ὧδ' ἂν καὶ τόδε μάλιστα μάθοις, εἰ λογίσαιο μήτε πλῆθος ἐν τῇ τοιαύτῃ φύσει μήτε κακοχυμίαν [οὐδεμίαν] ἢ ἀθροίζεσθαι ῥᾳδίως ἢ ἀθροισθεῖσαν λυμαίνεσθαί τι τοῖς ζῴοις. ἥ τε γὰρ συμμετρία τῶν φυσικῶν ἐνεργειῶν πρὸς ἀλλήλας καὶ ἡ καθ' ἑκάστην αὐτῶν ἀρετὴ καὶ γίνεσθαι | κωλύει τὰ περιττώματα καὶ γενόμενα ῥᾳδίως ἐκκρίνει, κἂν εἰ μείνειε δέ ποτε μέχρι πλείονος, ἥκιστα νικᾶται πρὸς αὐτῶν. τὸ μὲν γὰρ ὑπὸ τῶν νοσερῶν αἰτίων ῥᾳδίως νικᾶσθαι ταῖς ἀσθενέσι τε

cause trouble due to quantity and those that do so due to quality; in terms of kind, there are many forms.

That the body disposed in a well-balanced way is difficult to overcome by external causes is clear from the actual *eukrasia* of this—it is with difficulty and unusually that it comes to an imbalance of *krasis* because it is to the greatest extent removed from the extremes in every way. Not otherwise than from functioning well would there be the best preparation for resisting affection and the least likelihood of being seized by troubles. Immediately also, for such a body, there is the greatest *euchymia* compared to all others, so that it will hold up more easily than other bodies against distress, anger, insomnia, anxiety, heavy rains and droughts, plagues, and in short all morbid causes. For certainly, the *kakochymous* bodies in particular are readily put to the test by such causes, since already in themselves they come near to being diseased. In this way, the stated condition is very hard to overcome by those things befalling us externally and things that distress the body. That it is not easily overcome by diseases due to the superfluities of nutriment is something you would learn particularly, if you were to consider that neither an abundance in such a nature, nor any *kakochymia*, either collects easily or if it does collect, causes harm to the animals. In fact, a balance of the natural functions in relation to each other and the excellence of each of them prevents the occurrence of the superfluities and easily excretes those that do occur, while even if sometimes they do remain to excess, it is least overpowered by them. It is easy

---

⁵ καθ' ἑαυτὰ πλησίον ἥκοντα H; καθ' ἑαυτῶν πλησίον ἡκουσῶν K   ⁶ δυσάλωτός H; δυσάλωτότατός K

APPENDIX

καὶ δυσκράτοις ὑπάρχει φύσεσι, τὸ δ' ἀντέχειν ἐπὶ πλεῖστον ταῖς εὐκράτοις τε καὶ ἰσχυραῖς, ὁποίαν εἶναι τὴν ἀρίστην ἔφαμεν. ἔχεις δ' αὐτῆς ἐν μὲν τοῖς Περὶ κράσεων ὑπομνήμασιν ὡς εὐκράτου τὰ γνωρίσματα, κατὰ δὲ τὸ ἑπτακαιδέκατον τῶν Περὶ χρείας μορίων ὡς συμμέτρως ἐχούσης τοῖς ὀργανικοῖς μορίοις, καὶ νῦν δ' ἂν οὐδὲν ἧττον ἐκ τῶν εἰρησομένων ἀναμνησθείης αὐτῶν.

Ἐπειδὴ γὰρ οὐκ ἐστενωμένον οὐδ' ἁπλοῦν ἀκριβῶς οὐδ' ἄτμητον πρᾶγμα τὴν ὑγίειαν ἐδείξαμεν ἐν τοῖς περὶ αὐτῆς λόγοις, ἀλλ' εἰς ἱκανὸν πλάτος ἐκτείνεσθαι δυναμένην, δοκεῖ μοι καλῶς ἔχειν, εἰ μέλλει χρήσιμος ὁ λόγος ἔσεσθαι τοῖς ἐργαζομένοις τὴν τέχνην, μὴ μόνον τὸ σπάνιον ἐν αὐτῇ σῶμα καὶ οἷον παράδειγμά τι τοῦ Πολυκλείτου κανόνος ἐν τῷ λόγῳ πλάττειν ἀλλὰ καὶ τῶν ἀπολειπομένων μὲν αὐτῆς κατά τι, μὴ μέντοι κατάφωρον ἤδη | καὶ μέγα τὸ σφάλμα κεκτημένων ἀναμνησθῆναι. τό τε γὰρ ἑτοίμως γνωρίζειν τὴν ἀρίστην κατασκευὴν τοῦ σώματος, εἰ καὶ σπάνιος ἡ γένεσις αὐτῆς, καὶ τὸ τὰς ἄλλας ἁπάσας, αἷς ὁμιλοῦμεν ὁσημέραι, διαγινώσκειν ῥᾳδίως οὕτως ἂν μάλισθ' ἡμῖν ὑπάρξαι. τὸ μὲν γὰρ ἄκρως ἐν ἅπασι κατωρθωμένον, ὡς μήτε τῶν ὁμοιομερῶν μήτε τῶν ὀργανικῶν μορίων μηδὲν ἀμέτρως ἔχειν διακείμενον, οὐ πάνυ τι συνεχῶς, ἀλλ' ἐν μακροτέραις χρόνων περιόδοις εἴωθε γίνεσθαι,

---

[5] Polyclitus was an Argive sculptor active during the middle of the third century BC. His most famous work was the Dory-

for weak and *dyskratic* natures to be overcome by disease-producing causes, whereas it is easy for *eukratic* and strong natures to withstand these to the greatest extent— the kind of nature we said to be the best. You have this in the treatises *On Krasias* (*Temperaments*) as the signs of *eukrasia*, while in the seventeenth book of *On the Use of the Parts*, you have what is a balance in the organic parts, and now no less would you be reminded of these from the things I am going to say.

Since, in the discussions about health, we have shown that this is not a matter to be contained within a narrow compass, nor is it entirely simple, or indivisible, but is able to extend over a considerable latitude, it seems to me right, if the intention is for the discussion to be useful for those practicing the [medical] art, to form a notion not only of the body seldom seen in this practice, and the example as it were of the Canon of Polyclitus[5] in the discussion, but also of those bodies that fall short of this relative to it, without however | those bodies which have already acquired a manifest and large defect being called to mind. In this way particularly we shall readily begin to know the best constitution of the body, even if the creation of this is rare, and to easily diagnose all the others with which we come into contact day to day. For what is perfectly functioning in all aspects, so that neither the *homoiomeres* nor the organic parts have any imbalance in terms of disposition is not very frequent, but over longer periods of time

745K

ophorus (Spearbearer). He is said to have written a book (his Canon) setting out the principles of his art, to which Galen refers here and in a number of other places. The key feature was the proportion of the parts of the sculpted parts.

τὸ δ' ἀπολειπόμενον βραχὺ τοῦδε κἂν συνεχῶς θεάσαιο.

4. Τὸ μὲν οὖν ἀκριβῶς εὔκρατον μέσον ἐστὶν ἁπαλοῦ τε καὶ σκληροῦ καὶ δασέος καὶ ψιλοῦ τριχῶν καὶ φλέβας εὐρείας ἔχοντος ἢ στενὰς καὶ σφυγμοὺς μεγάλους ἢ μικρούς. τὸ δ' ἀκριβῶς σύμμετρον τοῖς ὀργανικοῖς μορίοις ἑνὶ κεφαλαίῳ περιληφθὲν οἷόσπερ ὁ Πολυκλείτου κανὼν ὑπάρχειν ἐλέγετο. ὅσα δ' ἤδη θερμότερα τοῦ προσήκοντός ἐστιν, οὐ μὴν πολλῷ γε, καὶ ψυχρότερα δὴ καὶ ξηρότερα καὶ ὑγρότερα | μετρίως ἤ τι μέρος[7] ἓν οὐκ ὀρθῶς διαπεπλασμένον ἔχοντα, ταῦτα σύμπαντα καὶ πλεονεκτεῖν ἐνίοτε δόξειεν τοῦ συμμέτρου. αὐτίκα τὸ μὲν σκληρότερον αὐτοῦ σῶμα δυσπαθέστερόν ἐστιν ἅπασι τοῖς ἔξωθεν αἰτίοις, τὸ δ' ἁπαλώτερον τοῖς ἔνδοθεν. οὕτω δὲ καὶ τὸ μὲν πυκνότερον τοῖς ἔξωθεν, τὸ δ' ἀραιότερον τοῖς ἔνδοθεν.

Τὸ γοῦν ὑφ' Ἱπποκράτους εἰρημένον ἐν τῷ Περὶ τροφῆς· "Ἀραιότης σώματος εἰς διαπνοήν, οἷς πλεῖον ἀφαιρέεται, ὑγιεινότερον, οἷς δ' ἔλασσον, νοσερώτερον" περὶ τῶν ἐκ τῆς τροφῆς περιττωμάτων εἰς ὑγίειαν καὶ νόσον συντελούντων εἴρηται. οὐ γὰρ δὴ περί γε τῶν ὑγιεινῶν ἁπλῶς ἢ νοσερῶν σωμάτων ἐν ἐκείνῳ τῷ βιβλίῳ προὔκειτο λέγειν αὐτῷ, ἀλλὰ περὶ πάντων τῶν ἐκ τῆς τροφῆς γινομένων ἀγαθῶν τε καὶ κακῶν τὸν λόγον ποιούμενος εὐλόγως ἐμνημόνευσε καὶ τῶν ὅσον ἐπὶ τοῖς ἐξ αὐτῆς περιττώμασιν ὑγιεινῶν καὶ νοσερῶν

---

[7] μετρίως ἤ τι μέρος H; μετρίως, καίτοι μέρος K

is accustomed to occur, whereas what falls short of this by a little, you will see frequently.

4. The precisely *eukratic* [body], then, is midway between soft and hard, hairy and hairless, having dilated or narrow veins, and large or small pulses. The body precisely median in the organic parts, encompassed under one heading, was said to be like the Canon of Polyclitus. Now those bodies that are already hotter than is fitting, but not in fact by much, and moderately colder, and drier, and wetter, or have some one part not properly formed, may all seem at some time to have an advantage over the body of due proportionality. To begin with, the body that is harder than normal is more difficult to affect by all the external causes, while the softer body is more difficult to affect by internal causes. In this way too, the more dense body is more difficult to affect by causes from without and the more rarefied by those from within.

Anyway, this was said by Hippocrates in *On Nutriment*: "Rarefaction of a body is healthier for transpiration in those from whom more is taken away, but more morbid in those from whom less [is taken away]"[6] This was said about the superfluities from nutriment which contribute to health and disease. For it was certainly not proposed by him in that book to discuss bodies that are absolutely healthy or diseased, but in making this discussion about all the good and bad effects arising from the nutriment, he reasonably made mention of the healthy and diseased bodies inasmuch as they follow the superfluities from this.

---

[6] Hippocrates, *On Nutriment* 28, LCL 147 (W. H. S. Jones), 352–53.

APPENDIX

σωμάτων. τὸ μὲν γὰρ ἀραιότερον ὑγιεινότερον, ὅσον ἐπὶ τοῖσδε, τὸ δὲ πυκνότερον νοσερώτερον. ἔμπαλιν δὲ τοῖς ἔξωθεν αἰτίοις ἅπασι τὸ μὲν ἀραιότερον εὐαλωτό-
747K τερον, | τὸ δὲ πυκνότερον δυσπαθέστερον. ὥστε τὸ σύμμετρον σῶμα πρὸς τοῖς ἄλλοις ἀγαθοῖς οὐδ' ἀραιὸν ἢ πυκνὸν ἔχομεν εἰπεῖν ἀλλ' ὥσπερ τῶν ἄλλων ὑπερβολῶν μέσον, οὕτω καὶ τῶνδε· πλεονεκτεῖ δὲ κατά τι τῶν ὑπερβαλλόντων ἑκάτερον. τὸ μὲν γὰρ πυκνότερον ἧττον τοῖς ἔξωθεν αἰτίοις εὐάλωτον, τὸ δ' ἀραιότερον τοῖς ἔνδοθεν. ἀμφοτέροις δὲ δυσάλωτον ἀκριβῶς μὲν οὐκ ἂν εὕροις οὐδέν, μετρίως δέ πως τὸ μέσον τῶν ὑπερβολῶν ἁπασῶν, ὃ δὴ καὶ πάντων ὑγιεινότατον τῶν σωμάτων εἶναί φαμεν. οὕτω δὲ καὶ τὸ μὲν ξηρότερον τοῦ συμμέτρου σῶμα τοῖς ὑγραίνουσιν αἰτίοις ἅπασίν ἐστι δυσαλωτότερον, τὸ δ' ὑγρότερον τοῖς ξηραίνουσιν. ὥσθ' ὅπερ ἐλέχθη μικρῷ πρόσθεν, οὐ πάντη πάντων ἐστὶ δυσπαθέστατον τὸ μέσον, ἀλλ' ἑκάστου μὲν καθ' ἕν τι χεῖρον, ἁπάντων δ' αἱρετώτατον.

Ὅτι δ' οὐκ ἀναγκαῖον ἢ μέγα τὸ τοιοῦτον ἢ μικρὸν ἀλλὰ μέσον ὑπάρχειν, εἴρηται μὲν κἀν τοῖς Περὶ κράσεων, οὐδὲν δ' ἧττον εἰρήσεται καὶ νῦν. ὡς τὸ μὲν μέγα καὶ διὰ πλῆθος ὕλης, τὸ δὲ σμικρὸν καὶ διὰ βραχύτητα
748K γίνεται τοιοῦτον, | ὥσπερ ἀνδριὰς μέγας μὲν ἐκ χαλκοῦ πολλοῦ, σμικρὸς δ' ἐξ ὀλίγου, σύμμετρον δ' εἶναι τοῖς μορίοις ἑκάτερον οὐδὲν κωλύει. καὶ δὴ καὶ σῶμα τὸ μὲν μήτε πυκνὸν ἐπιδήλως μήτ' ἀραιὸν μήτε σκληρὸν μήτε μαλακὸν μήτε λάσιον μήτε ψιλὸν τριχῶν εὐκρατότατόν ἐστιν, ὁπηλίκον ἂν ᾖ μεγέθει. εἰ δὲ καὶ τὰς συμμετρίας

For the more rarefied [body] is the more healthy to the extent it follows these, while the more dense is the more diseased. Conversely, the more rarefied body is the more susceptible to all the external causes, | whereas the more dense body is the more difficult to affect. As a consequence, we can say the well-balanced body, in addition to the other good features, is neither rarefied nor dense, but just as it is midway between the other excesses, so too it has the advantage relatively over each of the excesses. For the more dense body is less susceptible to the external causes, whereas the more rarefied is less susceptible to those that are internal. You would not find any body that is entirely resistant to both, although moderation is somehow the median of all excesses, which, what is more, we say is the most healthy of all bodies. In this way too, the body that is drier than the median is more resistant to all the wetting causes, while the wetter body is more resistant to the drying causes. As a result, we said a little earlier that the median, although it is not in every way the most resistant of all, but is worse than each in one particular way, is still the most desirable of all.

747K

That it is not necessary for such a body to be either large or small, but median, was said in the treatises, *On Krasias (Temperaments)*, and will be said no less again now. A large body becomes such due to an abundance of material, while a small body becomes such due to a small amount, | just as a large statue arises from a large amount of bronze, while a small statue arises from a small amount, although nothing prevents each from being duly proportioned in the parts. For the body that is neither clearly dense nor rarefied, neither hard nor soft, neither hairy nor hairless, is the most *eukratic*, however large it may be. And if it

748K

APPENDIX

τῶν ὀργανικῶν μορίων τὰς πρὸς ἄλληλα σώζοι, κάλλιστόν τ' ἂν οὕτως ἰδέσθαι καὶ κατωρθωμένον ἐν τῇ κατασκευῇ τελέως ὑπάρχοι.

Τὸ δὲ μεῖζον τοῦ δέοντος ἢ σμικρότερον κατὰ διττὰς αἰτίας γίνεται, τὸ μὲν μεῖζον ἢ διὰ πλεονεξίαν ὑγρότητος ἢ ὕλης, τὸ δ' ἔλαττον ἢ διὰ ξηρότητα κρατοῦσαν ἢ ὕλης ἔνδειαν. αὔξησις μὲν γάρ, ἔστ' ἂν τὰ ὀστᾶ κρατυνθῇ, κρατύνεται δ' ἐπὶ σμικρῷ μεγέθει δι' ὕλης ἔνδειαν ἢ ξηρότητα. καὶ τοίνυν καὶ παύεται τῆς αὐξήσεως ἢ πρωϊαίτερον ἢ ὀψιαίτερον ἕκαστον ἐπὶ διττῇ προφάσει·[8] ὥστ' οὐχ ἁπλῶς οὔτε τὸ μέγεθος ὑγρότητος σημεῖον οὔθ' ἡ σμικρότης ξηρότητος· ἀλλ' εἰ τὸ μὲν μέγα καὶ μαλακὸν εὐθὺς ὑπάρχει, τὸ δὲ σμικρὸν σκληρόν, εἴη ἂν οὕτω τὸ μὲν ὑγρὸν, τὸ δὲ ξηρόν. ἀλλ' εἰ ταῦτ' ἀχώριστα, περιττὸν εἰς μέγεθος ἢ σμικρότητα βλέπειν. ἀρκεῖ γὰρ τὰ τῶν κράσεων ἴδια γνωρίσματα. διῄρηται δ' ἐπὶ πλέον ὑπὲρ αὐτῶν ἐν τοῖς Περὶ κράσεων. εἴπερ οὖν ταῦθ' οὕτως ἔχει, τὴν ἀρίστην κατασκευὴν ἐν δυοῖν τούτοιν θετέον, εὐκρασίᾳ μὲν τῶν ὁμοιομερῶν, συμμετρίᾳ δὲ τῶν ὀργανικῶν.

[8] In K, the sentence καὶ τοίνυν ... προφάσει·, with καὶ τοίνυν καὶ omitted, follows ἡ σμικρότης ξηρότητος· above.

preserves the due proportions of the organic parts to each other, it would be in this way the most beautiful to behold and completely successful in constitution.

The body that is larger or smaller than it ought to be arises in relation to two causes. It is larger due to a preponderance of wetness or material, while it is smaller due to a prevailing dryness or a lack of material. For there is growth while the bones are being hardened, but they are hardened in a small size due to lack of material or dryness. And accordingly, when growth ceases either sooner or later, each is due to a twofold cause, so that largeness is not absolutely a sign of wetness, nor smallness of dryness. If the large body is also frankly soft, or the small body hard, in this way the former would be wet while the latter would be dry. But if these are inseparable, it is superfluous to look at the largeness or smallness, for the specific signs of the *krasias* are sufficient. These were distinguished to a greater extent in the treatises, *On Krasias* (*Temperaments*). If, then, this is the case, we must lodge the best constitution in two things—in *eukrasia* of the *homoiomeres* and in balance of the organic parts.

ΓΑΛΗΝΟΥ ΠΕΡΙ ΕΥΕΞΙΑΣ

Τὸ τῆς ἕξεως ὄνομα κατὰ παντὸς ἐπιφέρειν εἰθίσμεθα τοῦ μονίμου τε καὶ δυσλύτου καὶ οὐδὲν μᾶλλον ἐπαινοῦντες ἢ ψέγοντες. ἀλλ' ὅταν εὐεξίαν ἢ καχεξίαν εἴπωμεν, ἤδη τηνικαῦτα διοριζόμεθ',[1] ὁποίαν τινὰ τὴν ἕξιν εἶναί φαμεν. ἀγαθὴ μὲν οὖν ἁπλῶς ἕξις ἐν ἀρίστῃ κατασκευῇ γίνεται σώματος, οὐχ ἁπλῶς δὲ καθ' ἑκάστην φύσιν σώματος. ἡ μέντοι καχεξία περὶ πᾶσαν συνίσταται κατασκευὴν σώματος, εἴθ' ἁπλῶς εἴτ' ἐν τῷ πρός τι λέγοιτο. χρὴ τοίνυν ἀναμνησθῆναι τῶν περὶ τῆς ἀρίστης κατασκευῆς εἰρημένων ἰδίᾳ τὸν βουλόμενον ἀκριβῶς ἐπιγνῶναι, τί ποτ' ἐστὶν ἁπλῶς εἰπεῖν εὐεξία.

Πλάτος γὰρ ἱκανὸν ἐχούσης τῆς ὑγιείας, ὡς πολλάκις ἐν ἑτέροις ἐπιδέδεικται, τὴν μὲν ἐπίτασιν αὐτῆς εὐεξίαν ὀνομάζουσιν οἱ παλαιοὶ φιλόσοφοί τε καὶ ἰατροί, τὴν δ' ἔκλυσιν ἰδίῳ μὲν οὐκέτι προσαγορεύουσιν ὀνόματι, τῷ δὲ τοῦ παντὸς γένους ὡσαύτως ὑγίειαν καλοῦσιν, ὥστ' ἀρίστη τις ὑγίεια ἡ εὐεξία, καὶ διὰ

[1] διοριζόμεθ' H; διορίζομεν K

2. ON GOOD BODILY STATE (*EUEXIA*)

We are accustomed to apply the term *hexis* (bodily state)[1] to any state which is stable and difficult to dissipate, neither praising nor blaming more. But whenever we say "good bodily state" (*euexia*) or "bad bodily state" (*cachexia*), we are, under these circumstances, already making a distinction as to what kind we say the bodily state to be. Thus an excellent state occurs absolutely in an excellent constitution of a body, but not absolutely in each nature of a body. Of course, a bad bodily state (*cachexia*) may exist involving every constitution of a body, and may be said either absolutely or in relation to something. Accordingly, for one who wishes to know precisely, what must be borne in mind are those things said specifically about the best constitution, what at any time *euexia* is when said absolutely.

Since health has a considerable range, as has been demonstrated often in other places, the doctors and philosophers of ancient times called the intensification of this *euexia* (a good bodily state) while they did not yet call the dissipation by a specific term, calling the whole class similarly health. Consequently, *euexia* is optimum health, and

[1] The contrast is with *schesis*—see p. 435.

APPENDIX

τοῦτ᾽ ἐν τοῖς ἄριστα κατεσκευασμένοις γίνεται σώμασιν. εἰ γάρ τι μὴ τοιοῦτον, οὐκ ἂν δέξαιτο τὴν ἀρίστην ὑγίειαν, ὥστ᾽ οὐδὲ τὴν εὐεξίαν. ἡ² δ᾽ ἐν τῷ πρός τι κατὰ τὴν ἑκάστου φύσιν γίνεται καὶ διὰ τοῦτο μετὰ προσθήκης λέγεται Δίωνος, εἰ οὕτως ἔτυχεν, ἢ Μίλωνος εὐεξία, οὐχ ἁπλῶς εὐεξία. ἡ μέν γε τοῦ Μίλωνος καὶ ἡ τοῦ Ἀχιλλέως καὶ ἡ τοῦ Ἡρακλέους ἁπλῶς τ᾽ εἰσὶν εὐεξίαι καὶ χωρὶς προσθήκης ὀνομάζονται, καθάπερ καὶ καλὸς μὲν ὁ Ἀχιλλεὺς ἁπλῶς, ὁ δὲ πίθηκος οὐχ ἁπλῶς, ἀλλ᾽ ὡς πίθηκος καλός. ἐκ τῶν μετὰ προσθήκης ἐστὶ λεγομένων καὶ ἡ τῶν ἀθλητῶν εὐεξία καὶ δεόντως ὑπὲρ αὐτῆς ὁ Ἱπποκράτης ἔλεγεν· "Ἐν τοῖσι γυμναστικοῖσιν αἱ ἐπ᾽ ἄκρον εὐεξίαι σφαλεραί." οὐ γὰρ δὴ τήν γ᾽ ἁπλῶς ὀνομαζομένην εὐεξίαν, ἐπειδὰν εἰς ἄκρον ἥκῃ, σφαλερὰν εἶναί φησιν. αὐτὸ γὰρ δὴ τοῦτ᾽ ἔστιν αὐτῇ τὸ εἰς ἄκρον ἥκειν, τὸ πασῶν τοῦ σώματος τῶν διαθέσεων ὑπάρχειν ἀσφαλεστάτην. ἀλλ᾽ ἡ τῶν ἀθλητῶν ἢ γυμναστικῶν ἢ ὅπως ἂν ἐθέλῃ τις ὀνομάζειν εὐεξία, διότι μὴ ἁπλῶς ἐστιν εὐεξία ἡ ἀρίστη διάθεσις σώματος, εὐλόγως εἰς ἄκρον ἰοῦσα σφαλερωτάτη γίνεται. "Διάθεσις γάρ,"

² ἡ H; εἰ K

---

[2] Milo of Croton was an athlete and a wrestler famous for his strength, particularly in his hands. Galen gives a disparaging and somewhat amusing account of some of his most famous feats in his *Protrepticus*, I.34–35K. Hercules (Heracles), son of Zeus and

because of this occurs in the bodies constituted best. If it were not such a thing, it would not receive the best health and so would not receive *euexia*. If, however, it is relative to something, it occurs in relation to the nature of each person, and because of this, *euexia* is said to be added to Dion, if this happens to be the case, or to Milo, and not *euexia* absolutely. In fact, the *euexias* of Milo, Achilles and Hercules[2] are *euexias* said absolutely, and are named apart from any addition, just as also Achilles is beautiful absolutely whereas a monkey is not beautiful absolutely but as a monkey that is beautiful, and is among those things said with an addition, as also is the *euexia* of athletes. On this matter, Hippocrates rightly said: "In those who exercise, the *euexias* that reach an extreme are dangerous."[3] For he is certainly not saying *euexia* is a term applied absolutely, which, when it comes to a peak is dangerous, for *euexia* itself, when it comes to a peak, is the safest of all the conditions of the body. But when it is of athletes, or of those who are devoted to gymnastics, or however someone might wish to apply the term, *euexia* is not absolutely the best condition of a body, and with good reason, because when it comes to an extreme it is very dangerous. For Hippocrates says: "The athletic condition

752K

the mortal woman Alcmene, was a legendary hero noted for his great strength. Achilles, son of Peleus and Thetis, was the greatest of the Greek heroes in the Trojan War and the central character of Homer's *Iliad*.

[3] Hippocrates, *Aphorisms* 1.3. W. H. S. Jones has: "In athletes, a perfect condition that is at its highest pitch is treacherous" (*Hippocrates* IV, LCL 150, 98–99).

APPENDIX

φησίν, "ἀθλητικὴ οὐ φύσει, ἕξις ὑγιεινὴ κρείσσων." τῆς μὲν οὖν ὑγιεινῆς ἕξεως ἡ τελειότης εὐεξία ἐστί. τῆς δὲ τῶν ἀθλητῶν διαθέσεως οὐχ ἁπλῶς, ἀλλὰ μετὰ προσθήκης, ὡς εὔμορφος πίθηκος καὶ πῆχυς μέγας καὶ ἄδικος χοῖνιξ καὶ ἀδόκιμος δραχμή. εἴτε γὰρ ὁ πῆχυς μέγας, οὐκέθ᾽ | ἁπλῶς πῆχυς, ἀλλ᾽ ὅλον τοῦτο μέγας πῆχυς, εἴθ᾽ ἡ χοῖνιξ ἄδικος, οὐκέθ᾽ ἁπλῶς χοῖνιξ, ἀλλ᾽ ὅλον τοῦτο χοῖνιξ ἄδικος.

Ὡσαύτως δὲ κἀπὶ τῶν ἄλλων ἁπάντων τὸ [χωρὶς τῆς προσθήκης] ἁπλῶς ὀνομαζόμενον οὐ τῆς αὐτῆς ἐστι φύσεως τῷ μετὰ προσθήκης λεγομένῳ, ἀλλ᾽ ἐνίοτε τὸ μὲν ἄκρως ἐπαινετόν ἐστι, τὸ δ᾽, εἰ οὕτως ἔτυχε, ψεκτόν, ὥσπερ γε καὶ ἡ τῶν ἀθλητῶν εὐεξία. τοσοῦτον γὰρ ἐνδεῖ³ τὸ ἐπαινετὸν ἔχειν, ὥστε καὶ ψέγεσθαι δεόντως, οὐχ ὑφ᾽ Ἱπποκράτους μόνον ἢ τῶν ἄλλων ἰατρῶν τῶν παλαιῶν, ἀλλὰ καὶ πρὸς τῶν ἀρίστων φιλοσόφων, ὥσπερ καὶ Πλάτωνος ἐν τῷ τρίτῳ τῆς Πολιτείας τήν τ᾽ ἀχρηστίαν αὐτῆς ἅπασαν εἰς τὰς κατὰ φύσιν ἐνεργείας ἐπιδεικνύντος καὶ ὡς σφαλερὰ πρὸς ὑγίειάν ἐστι διεξιόντος· οὐ γὰρ εὐκρασίαν ἁπλῶς τοῦ σώματος, ἀλλὰ καὶ μέγεθος ὄγκου μεταδιώκοντες, ὃ χωρὶς ἀμέτρου πληρώσεως οὐκ ἂν γένοιτο, οὕτω καὶ σφαλερὰν αὐτὴν ἀπεργάζονται καὶ πρὸς τὰς πολιτικὰς ἐνεργείας ἄχρηστον. ἵνα τοίνυν τῆς ὄντως εὐεξίας εἰς ἀκριβῆ | γνῶσιν ἀφικώμεθα, παραβάλλειν αὐτῇ χρὴ τὴν ὁμώνυμον εὐεξίαν τὴν

³ γὰρ ἐνδεῖ H; δὴ K

is not natural; a healthy state is better."[4] Thus, *euexia* is the perfection of the healthy bodily state; not, however, absolutely of the condition of athletes, but with an addition, like a well-formed monkey, and a big cubit, and a false *choenix*, and a counterfeit *drachma*. For if it is a big cubit, it is no longer | absolutely a cubit, but wholly this big cubit, or the false *choenix* is no longer absolutely a *choenix* but wholly this false *choenix*.

In a similar fashion, in the case of all other things, that which is named absolutely apart from the addition is not of the same nature as that named with the addition. And sometimes the one is praised exceedingly, while the other, as may happen, is blamed. In fact, the *euexia* of athletes is lacking in praise to such an extent that it is rightly censured—and not only by Hippocrates or other doctors of old, but also by the best philosophers. For example, Plato too, in the third book of the *Republic*, showed its uselessness in respect of every natural function, going through the danger it presents to health. What it pursues is not simply *eukrasia* of the body, but also a large body mass, which would not occur without immoderate filling.[5] In this way it is also dangerous, and as regards civic functions, useless. Accordingly, so that | we may come to a precise knowledge of what is truly *euexia*, we must compare to it

---

[4] Hippocrates, *Nutriment* 34. The full section reads: "Nutriment sometimes pertains to growth and being, sometimes to being only, as is the case with old men, and sometimes in addition it pertains to strength. The condition (*diathesis*) of the athlete is not natural; a healthy state (*hexis*) is superior to all."

[5] For Plato's consideration of this term, see *Republic* 3.13.

APPENDIX

ἀθλητικὴν καὶ σκέψασθαι, τί ταὐτὸν ἑκατέραις ὑπάρχει τί τ' ἐναντίον.

Ἡ μὲν δὴ τῶν μορίων ἁπάντων τοῦ σώματος εὐκρασία κοινὸν ἀμφοῖν. οὕτω δὲ καὶ ἡ τῶν ἐνεργειῶν ἀρετὴ καὶ εἴπερ ταῦτα, καὶ ἡ εὐχυμία. ταυτὶ μὲν τὰ κοινά. τὰ δ' ἐναντία συμμετρία μὲν αἵματός τε καὶ τοῦ τῶν στερεῶν σωμάτων ὄγκου παντὸς ἐν ταῖς ὄντως εὐεξίαις, ἀμετρία δὲ τῶν αὐτῶν τούτων καὶ μάλιστα τοῦ σαρκώδους γένους ἐν ταῖς ἀθλητικαῖς, αἷς ἐξ ἀνάγκης ἕπεται τὸ σφαλερόν, ἐπειδὰν εἰς ἄκρον ἀφίκηται. ὅταν γὰρ ἐσθίωσι μὲν πρὸς ἀνάγκην, πέττῃ δ' ἡ γαστὴρ ἐρρωμένως καὶ ἡ ἀνάδοσις ἐπὶ τῇ πέψει γίνηται ῥᾳδίως αἱμάτωσίς τε καὶ πρόσθεσις καὶ πρόσφυσις καὶ θρέψις ἕπηται τοῖσδε, κίνδυνος ὑπερπληρωθῆναι τὴν ἕξιν, ὡς μηκέτ' εἶναι τῇ φύσει χώραν προσθέσεως κἂν τῷδε πληροῦνται μὲν αἵματος αἱ φλέβες ἀμέτρως, καταπνίγεται δὲ καὶ σβέννυται τὸ ἔμφυτον θερμὸν ἀποροῦν τῆς | διαπνοῆς. εἴπερ δ' ἔτ' ἀντέχει τοῦτο, ῥήγνυταί τι τῶν ἐπικαίρων ἀγγείων, ἃ δὴ καθ' ἧπάρ τε καὶ πνεύμονα καὶ θώρακα τέτακται. καὶ γὰρ δὴ καὶ μαλακώτερα τοῖς χιτῶσι τῶν ἐν τοῖς κώλοις ὑπάρχει ταῦτα καὶ πρότερα τὴν τροφὴν δέχεται καὶ διὰ τὸ πλῆθος τῆς ἐν αὐτοῖς φυσικῆς θερμασίας ἔτι τε τῶν ἐνεργειῶν τὸ διηνεκὲς ὅμοιόν τι τῇ ζέσει πάσχον τὸ αἷμα τοὺς χιτῶνας αὐτῶν ἀναρρήγνυσιν, ὥσπερ ὁ γλευκίνης οἶνος τοὺς πίθους.

---

[6] On the second and third of these four terms, I have used

the similarly named athletic *euexia* and consider what is the same in each and what is opposite.

Certainly the *eukrasia* of all the parts of the body is common to both. So too is the excellence of the functions, and if there are these excellent functions, there is also *euchymia*. These are what is common. On the other hand, those that are opposite are a due proportion of blood and of the whole mass of the solid bodies in the true *euexias*, in contrast to an imbalance of these same things, and in particular of the fleshy class, in the athletic *euexias*, which the danger follows of necessity when it comes to a peak. When people eat according to need, while the stomach concocts strongly, and distribution occurs readily after concoction, blood formation, apposition, attachment, and nutrition follow these,[6] there is a danger of the bodily state (*hexis*) being overfilled, so there is no longer a natural place of apposition. In this the veins are disproportionately filled with blood, while also the innate heat is quenched and choked up, being without means of | transpiration. If, however, it still withstands this, some one of the important vessels which are situated in relation to liver, lungs, and chest is ruptured, for truly these are softer in the walls than those in the limbs and receive the nutriment earlier. And through the amount of natural heat in them, and in addition through the continuous nature of their functions, the blood suffers something similar to seething, and they rupture their walls, just as partly fermented wine ruptures wine casks.

755K

A. J. Brock's translation and the associated notes 5 and 6; see Galen's *De naturalis facultatibus*, I.11 (II.24–26K). The notes are on page 39 (LCL 71, 1916).

APPENDIX

Ταῦτά τ' οὖν οὕτω γίνεται πάντα ταῖς ἀμέτροις πληρώσεσιν ἐξ ἀνάγκης ἑπόμενα καὶ αἱ περὶ αὐτῶν ἀποδείξεις τοῖς φυσικοῖς ἕπονται λόγοις. ὅτι δὲ σβέννυται τὸ ἔμφυτον θερμὸν ὑπερπληρωθεισῶν αἵματος τῶν φλεβῶν, ἐν τοῖς Περὶ χρείας ἀναπνοῆς εἴρηται, ὅτι δ' αἱ φλέβες ῥήγνυνται ἐν τοῖς Ἀνατομικοῖς. οὕτω δ' ἂν καὶ ὁ Ἱπποκράτης φανείη γινώσκων, οὐ μόνον ἐπειδὰν φῇ τὴν ἐν τοῖς γυμναστικοῖς ἐπ' ἄκρον εὐεξίαν εἶναι σφαλεράν, ἀλλὰ κἀπειδὰν ἑτέρωθι γράφῃ· "Τὸ δὲ ἐξαίφνης ἄφωνον γενέσθαι, φλεβῶν ἀπολήψιες[4] | λυπέουσι." τὰς γὰρ αἰφνιδίους παραλύσεις τῶν ἐνεργειῶν ἁπασῶν διὰ μιᾶς ἐπικαιροτάτης ἐδήλωσεν. ἀπολήψεις δὲ φλεβῶν τὰς ὑπερπληρώσεις εἶπεν, ἐπειδὰν ἀπορῶσιν εἰς ἀνάψυξιν διαπνοῆς.

[4] post ἀπολήψιες add. τὸ σῶμα K

All these things, then, occur in this way necessarily following the disproportionate fillings, and the demonstrations of these follow the arguments of the natural philosophies That the innate heat is quenched when the veins are overfilled with blood is spoken about in the writings *On the Use of Respiration*,[7] while that the veins are ruptured is spoken about in *On Anatomical Procedures*.[8] In this way too, Hippocrates would seem to have known, not only when he says *euexia* in those who exercise, when it comes to an extreme, is dangerous, but also from what he writes elsewhere: "When there is sudden aphonia, stoppage of veins | distresses the body."[9] He showed the sudden paralyses of all the functions due to one that is very important. He said there are stoppages of veins that are overfilled when they lack transpiration for cooling.

[7] *De respiratione usu*, IV.470–511K.
[8] *De anatomicis administrationibus*, II.215–731K.
[9] On the first statement, see note 3 above. I have been unable to locate the second.

INDEXES

The index is divided into four sections, as follows: (a) Personal and Place Names, (b) Books and Treatises, (c) Foods and Medications, and (d) General. Entries relating to the General Introduction and the specific introductions to the separate treatises are given first according to the actual book page. The entries relating to the translations are then given according to the Kühn page, divided into those for *On Temperaments*, marked T; *On Non-Uniform Distemperment*, marked N; *The Soul's Traits Depend on Bodily Temperament*, marked S; and the Appendix, marked A.

a
PERSONAL AND PLACE NAMES

Achilles, A 751
Aetius of Amida, lxxi
Alcmaeon, xv, xxix
Anaxagoras, lxxxix; T 589
Andronicus the Peripatetic, lxxxix, 329; S 782
Antiochus the doctor, liii
Arabs/Arabian, lxii, lxxii–lxxiii; T 618, 628
Archimedes, lxxxix, 11; T 657
Aristotle, xvi, xviii–xx, xxv, xxviii, xli–xlii, li, lviii, lxiv, lxv, lxxx, lxxxiv, xc, 11, 327, 329; T 523, 535, 554, 566, 581, 624, 628, 636, 666, 672; S 773–74, 782–83, 791–97
Asia, T 657; S 798–800
Athenaeus, xxi–xxii, xc; T 522–23
Athens/Athenians, S 822
Ayyub al-Rahawi, lxxiii

Burgundio of Pisa, lxxv

Carthage/Carthaginians, S 811
Celts, T 627

463

INDEXES

Chrysippus, xxix, xli, xc; S 784, 820
Crete/Cretans, S 810

Da Reggio, N., lxxv, 282
Dalmations, T 618
Democritus, xvi

Egypt/Egyptians, T 618
Empedocles, xv–xvi, xx–xxi
Erasistratus/Erasistratean, xl, xlii, xci; T 599
Ethiopians, T 616, 618, 628
Eudemus, xci; T 632
Europe, xiii; S 800

Gerard of Cremona, lxxv
Germans, T 618, 627

Heraclitus, xci; S 786
Hercules, A 751
Hippocrates, xvi–xviii, xxix, xxxvi, xli, lxv, lxviii, lxx, lxxvii, xcii, 11, 328–29, 433; T 509, 527, 530–32, 554, 603, 605, 640, 660, 673; N 739, 745; S 784, 798–805; A 746, 752–53, 755–56
Hippocrates (nephew of Pericles), xci; S 784
Homer, xcii; T 513; S 771, 777–78
Hunayn ibn Ishaq, lxxiii

Illyrians, T 618
Indians, T 618

Kühn, C-G, lxxxix, 4, 284, 326, 433

Kuhn, Thomas, lxix

Leucippus, xvi
Linacre, Thomas, lxxv–lxxvi, lxxxi, 4, 282–83

Medea, xcii, 11; T 658
Milo, A 751
Mnesitheus, lxxxiv
Mysia/Mysians, T 657

Nutton, Vivian, lxvi

Oribasius, lxxi–lxxii

Paul of Aegina, lxviii, lxxii–lxxiii
Phrygians, xli
Plato, xvii–xviii, xxviii, xli–xlii, lxiii, lxxxiv, xcii, 326–27, 329, 331; T 544; S 768, 771–73, 775, 780–91, 805–6, 808, 811, 815; A 753
Polyclitus, lxxxii; T 566; A 744
Posidonius, xx–xxi, xxix, xciii; S 819–20
Praxagoras, xciii; N 749–51
Premigenes the Peripatetic, lv
Pyrrho/Pyrrhonists, T 589
Pythagoras, xciii; S 768, 816

Richet, Charles, lxvi
Rome, T 630

Sappho, xciii; S 771
Sauromations, T 618
Scythians, T 618, 627; S 822
Socrates, S 816
Sophist(s), T 549

INDEXES

Stoics, xx–xxi, xxix, xciii, 328; T 523; S 784, 816, 819–20

Telephus the grammarian, liii
Theognis, xciv; S 778
Theophrastus, xli, xciv; T 523, 535, 544

Thrace/Thracians, T 627
Thucydides, xciv; S 788

Zeno of Citium, xli, xciv; S 777

b
BOOKS AND TREATISES

Aristotle
  History of Animals, xviii, lxxxii, 330; S 795–97
  Metaphysics, xx, lxxx, lxxxii
  Meteorologica, lxxxviii
  Parts of Animals, xxviii, lxxxiii, 330; T 566; S 791
  Physics, lxxxix
  Physiognomica, 10, 330
  Problems, 330; S 784, 794

Galen
  Ars medica, xiv, xxxvi–xl, lxxii, 3, 282
  De anatomicis administrationibus, xci, 285; N 735, 742; A 755
  De bono habitu, xiii, 3, 432
  De causis morborum, xxxii–xxxiv; N 746, 748
  De demonstratione (lost work), T 593
  De differentiis morborum, xxix–xxxiii, xliii, li; T 610
  De dignoscendis pulsibus, 7; T 538, 540
  De elementis secundum Hippocratem, xiv, xvi, xliii, xlvi, li, 3; S 785; A 741
  De inaequali intemperie, xii, xxvi–xxvii, xliii, xlv, lxx, lxxiii, 4, 282
  De methodo medendi, xiv, xxvi, xl–lii, lxiv, lxxxiv, 2, 5, 13; T 692; N 752
  De naturalibus facultatibus, xxv, lxxxii, 11; T 654
  De optima corporis nostri constitutione, xiii, 3, 431
  De ordine librorum suorum ad Eugenianum, 3, 282
  De partium homeoemerium differentia, lxxxix
  De placitis Hippocratis et Platonis, lxxxviii, 328; S 785; A 741
  De sanitate tuenda, xiv, xxv, liii–lx, lxviii, 2, 433
  De simplicium medicamentorum temperamentis et facultatibus, 3; T 650; N 752
  De symptomatum causis, xxxv–xxxvi; N 746
  De symptomatum differentiis, xxxv–xliii, liii
  De temperamentis, xii, xxii–xxvi, xliii, li, lxiii, lxviii, lxx, lxxiii, lxxv, 2–4, 282; N 734–36; A 741, 744, 747, 749

INDEXES

De usu partium, xliii; T 619; A 741, 744
De usu respiratione, A 755
De voce, lix
Definitiones medicae, lxxxiv, lxxxvi
Quod animi mores corporis temperamenta sequantur, xii, xviii, xx, xxvii–xxix, lxiii, lxxv, 328

Hippocrates
Airs Waters Places, xxix, 330; S 798–803
Aphorisms, lxxxii; T 527
Epidemics (II and III), xxix; T 530 (II), 531–32 (III); S 803
Nature of Man, xiv, xvii, lxx, 238, 330; T 603; S 799
On Nutriment, A 746

Plato
Laws, xxviii, 331; S 806–10
Republic, S 771; A 753
Timaeus, xxviii, 331; S 780–91, 805–6, 808, 812–13

c
FOODS AND MEDICATIONS

Aloes (ἀλόη), c; its various capacities, S 769–70
Asphalt (ἄσφαλτος), c; T 649; potentially hot, T 669; readily flammable, T 658

Barley cakes (μάζα), T 656
Beef, T 655
Beet, T 656
Bread, T 633, 656, 673

Cantharides (κανθαρίδες), c, 12; use in dropsy, T 667
Castor/castoreum (καστόρειον), c; T 649, 675; nutriment and hot medication, T 681
Catmint (καλάμινθος), c; nutriment and medication, T 682
Copper ore. *See* Misu
Cyrenian Juice (κυρηναικός ὀπός), c; T 666

Dill (ἄνηθον), ci; nutriment and medication, T 682

Fat (as suet), T 649
Fennel (νάρθαξ), ci; T 658

Fish, T 634
Fleawort (ψύλλιον), ci; effects of fire on, T 674

Garlic, T 661

Hellebore (ἐλλέβορος), ci; T 684
Hemlock (κωνειόν), ci; T 673, 674; cold medication, T 649; cooling agent, S 776; effects of, S 779; nutriment for fish, medication for humans, T 684
Honey (μέλι), xlvii, ci; T 675; potentially hot, T 669

Kostos (κόστος), ci; as heating medication, T 649

Lettuce (θριδακίνη), cii; T 679; cold medication, T 649; hypnotic, T 585; juice of, T 680–81

Mandragora/mandrake (μανδραγόρας), cii; administered heated, T 673; as a cold

INDEXES

medication, T 649; effect of fire on, T 674; hypnotic, T 585

Mēdian juice (ὀπός Μηδικός), cii; possible harm from, T 666

Milk, xlvii

Misu (μίσυ), cii; as a hot medication, T 649

Mustard (νᾶπυ), cii; hot potentially, T 649; nutriment and medication, T 682

Nitron (νίτρον), cii; hot medication, T 649

Olive oil (ἔλαιον), cii; T 539; as heating, T 660; as hot potentially, T 649, 669

Onions, T 661

Opium (ὄπιον). *See* Poppy

Oregano (ὀρίγανον), ciii; nutriment and medication, T 682

Parthian juice (ὀπόν παρθενίκιον), ciii; as heating, T 666

Pellitory (πυρέθρον), ciii; as a hot medication, T 649

Pennyroyal (γλήχων/βλήχων), ciii; nutriment and medication, T 682

Pickled meat, T 661

Pine resin (ῥητίνη), ciii; as heating, T 659; as hot potentially, T 649, 669

Pitch (πίσσα), ciii; as heating, T 659; as hot potentially, T 649, 669

Plantain (θρυαλίς), ciii; as heating, T 659

Poppy (μήκων), ciii, 12; as heating, T 659; as hot potentially, T 649, 669; as hypnotic, T 585

Pork, T 633, 655

Ptisane/tisane (πτισάνη), civ; acted on more than acting on, T 675; administered heated, T 673

Purslane (ἀνδράχνη), civ; as a cold medication, T 679

Rock alum (χαλκῖτις), civ; hot potentially, T 649

Rocket (εὔζωμον), nutriment and hot medication, T 681

Rosewater (ῥόδινος), civ; T 685

Rue (πήγανον), civ; nutriment and hot medication, T 682

Salamander (σαλαμάνδρα), civ; as cold, T 649

Salt, T 530

Savory (θύμβρα), civ; nutriment and medication, T 682

Soapwort (στρούθειον), cv; as a hot medication, T 649

Spignel (μῆον), cv; as a hot medication, T 649

Spurge (εὐφόρβιον), cv; as a hot medication, T 649

Thyme (θύμον), cv; nutriment and medication, T 682

INDEXES

Vinegar (ὄξος), cv; T 530; as preservative, T 538; whether hot or cold, T 685

Wine (οἶνος), xxviii, xlvii, l, cv; T 585; acted on more than acting on, T 675; for distress and *dysthymia*, S 777; effects on the soul, S 821; as heating, T 538–39; Plato on, S 808–12; Theognis on, S 778

Yellow flag (ἄκορον), cv; as hot when applied, T 649

d
GENERAL

Actuality. *See* Potentiality/actuality
Ambient air, T 529; *krasis* of, S 807–8
Anasarca (ἀνὰ σάρκα), xcv; N 733
Anger (θυμός), T 633; N 747; S 803; A 743
Anthrax (ἄνθραξ), xcv; T 530, 532, 664; N 751
Anxiety (φροντίς), 432; T 633; A 742–43
Apoplexy (ἀποπληξία), xcv; T 582, 661
Arteries, T 578, 580; S 803–4; fluxes in, N 736–37
Atoms/atomist theories, xvi, xxix; S 785

Baldness, T 625–26, 634
Bile, black and yellow, T 603; T yellow, 633; yellow and black, S 776–77
Blood, T 564, 568, 578, 604, 648; Aristotle on, S 791–94; derived from food, T 682–83; evacuation of as cooling, S 776; formation of, T 582; *krasis* of maternal, S 791; maternal blood in genesis, S 795;
in stages of life, T 583–93, 598
Blood vessels, T 568–70, 578, 590, 601; rupture of, A 755. *See also* Atreries; Veins
Bodily states. *See Hexis*; *Schesis*
Bone, T 564, 569, 578, 603; N 735
Brain, T 564, 570, 600–602; S 807; *krasis* of, T 626; seat of rational soul, S 770

Cachexia, A 750
Cancer (καρκίνος), xcv; T 664; N 733, 751
Capacity. *See Dunamis*
Carbuncle. *See* Anthrax
Cataract (ὑπόχυμα), xcv; S 788
Catarrh (κάταρρος), xxxvii, xcv; T 582, 634
Condensation/rarefaction (πυκνός/ἀραιός), T 570; A 746–47
Condition (διάθεσις), 433; A 738
Constitution (κατασκευή), 431–33; A 737–38; the best defined, A 741–42, 749

471

INDEXES

Continuum theories, xv–xvi, xix, xxix, xl, lxiii; T 509–10; S 785

Convulsions. *See* Spasm

Coryza (κόρυζα), xxxvii, xcvi; T 634

Cough (βήξ), xcvi; T 580, 634

Delirium (παραφροσύνη), xcvi; S 777, 787

Dissolution of continuity, N 739, 745

Dropsy (ὕδερος), xcvi; T 522, 667; N 733

Dunamis (δύναμις; capacity), xi; S 769; capacities of medications, T 654; capacities of rational soul, S 770–72; capacities of soul, S 767–69; definition of, lxxviii–lxxxii; in *De temp.*, xxv; as potentiality, T 646–55

Dyskrasia (δυσκρασία), xi–xiv; T 572; N 734; definition of, lxxxv–lxxxviii; in *Ars med.*, xxxvi–xl; in *De caus. morb.*, xxxii; in *De caus. sympt.*, xxxv–xxxvi; in *De meth. med.*, xl–lii; in *De morb. diff.*, xxix–xxxii; in *De san. tuend.*, liii–lxii; in *De sympt. diff.*, xxxv; in *De temp.*, xxii–xxv; healthy/morbid, T 609; non-uniform (anomalous), xxvi–xxvii, xxxiii, 2; in other species, T 535–38; relation to seasons, T 524–31; types of, T 511–18, 556–59

Dysthymia, S 777

Ears, Aristotle on, S 797

Edema (οἴδημα), xcvi; N 733, 751. *See also* Anasarca; Dropsy

Elemental qualities (hot, cold, dry, wet), T 538, 554–55, 560–63, 587, 589–90, 675; S 773; said potentially, T 649, 669; touch in evaluation of, T 598

Elements (earth, air, water, fire), T 648; changes between, T 671

Elephas (ἐλέφας), xcvi; N 733

Energeia (ἐνέργεια; function, action), xi; A 742; as actuality, T 646–55; definition of, lxxviii–lxxxii; in *De temp.*, xxv; paralysis of, A 756

Epilepsy (ἐπιληψία), T 661

Ergon (Action), xi; definition of, lxxviii–lxxxii

Erysipelas (ἐρυσίπελας), xcvi; N 733, 751

Euchymia (εὐχυμία), lxxxiv; T 633; S 814; A 754; definition of, lxxxiv–lxxxv, 431; in regulation of soul, S 821

Euexia (εὐεξία), 432–33; A 740, 748–49, 750, 752–54

Eukrasia (εὐκρασία), xi–xiv; T 519–23, 533, 543, 546, 555, 563, 565, 573–76, 606–7; S 768, 784, 799; A 737–38, 742, 748, 755; of air and water, T 596; definition of, lxxx–lxxxviii; definition of *eukratic* body,

INDEXES

A 745; in *De temp.* xxii–xxv, 2; of humors, A 749; of places and seasons, S 805–9; relation to *eusarkos*, T 567

Eusarkos/polysarkos, lxxxii–lxxxiv; T 541, 567, 607–8, 610

Eyes, Aristotle on, S 796–97

Fever (πυρετός), xcvii; ague (ἠπίαλος), N 733, 749, 751; bilious remittent (καῦσος), N 750; hectic (ἑκτικός), N 733, 743–44, 746–47, 751; malignant intermittent (λιπυρία), N 750; tertian and quartan, N 751

Flesh, T 648; of birds, T 655; hot as cause of inflammation, T 690; non-uniform *dyskrasia* in, N 735

Fluxes, xxvii, xxxiv; T 532; in brain, N 751–52; cause of non-uniform *dyskrasia*, N 736

Function (ἐνέργεια). See *Energeia*

Gangrene (γάγγραινα), xcvii; N 733, 751
Geometry/geometricians, S 813
Grammar/grammarian, T 515
Grief (λύπη), T 633; A 742–43
Gymnastics, S 813; *euexia* associated with, A 752

Hair, T 564, 569, 577; racial characteristics, T 618–20; in relation to humors, T 616–18; in relation to skin state, T 621

Hardness and softness, T 598–99; A 745–46, 748

Heart, T 601–2; S 807; heat in, N 742; in relation to soul, S 772

Heartburn (ὀξυρεγμία), T 634

Heatstroke (ἔγκαυσις), xcvii; N 747–48

Hemiplegia (παραπληξία), T 661

Herpes (ἕρπης), xcvii; N 733

Hexis (a stable state). See *Hexis/Schesis/Euexia*

Hexis/Schesis/Euexia, T 577, 605, 611, 643; N 743; A 738, 750

Homoiomeres/Homoiomerous, xxix–xxi, lvi, 431; T 542; N 735; S 773, 785; A 741–42, 745; Aristotle on, xviii–xix; definition of, lxxxviii–lxxxix

Homonymy, T 542, 554; S 783

Humor (χυμός), xxii; T 603, 616; N 741; black bile, lxvi–lxvii; effects on hair, S 776–77; effects on soul, S 789; and *euchymia*, S 814; flow of, T 630; four types, xxvii, xxxiv, lxv, lxxii–lxxiv; N 751–52; hyaloid, N 749–51; qualities of, T 679

Hypersomnia, T 661

Infancy, T 580
Innate heat (ἔμφυτον θερμόν), T 535–36, 554, 598, 610, 628,

INDEXES

659, 690, 694; increased by nutriments and medications, T 678; innate and acquired, T 628–29

Insomnia (ἀγρυπνία), 431; T 633; A 742–43

Intestines, T 602

Kakochymia (κακοχυμία), 327; T 603; definition of, lxxxiv–lxxxv; effects on soul, S 789; and *euchymia*, S 814; A 743

Kidneys, T 601

Krasis (κρᾶσις), xi–xiv, 327; T 524–31, 556–58, 572, 587; S 779, 795, 821; consequences of, T 604; definition of, lxxxv–lxxxviii; in *De temp.* xxii–xxv; diagnosis of, T 559; hair in, T 611–12, 640–41; of matter, S 773; of medications, T 673; non-uniform, T 622–24; in other species, T 535–38, 547–51; racial variations, T 627–28; in relation to facial features, T 636–37; restored by medications, T 682; of seasons on racial characteristics, S 798–804; skin in, T 613; in stages of life, T 585–86; of stomach, T 678; of whole body, T 638–39

Lethargy (λήθαργος), xxxvii, xcviii; S 777

Ligaments, T 564

Liver, T 601; S 807; discharge of bile from, T 631–32; in digestion, T 655; *krasis* of, T 626; in relation to soul, S 772

Lung, T 600

Mania (μανία), xcviii; S 787–88, 804

Marrow, T 600

Material/matter (ὕλη), A 747–48; and form, S 773, 783; in physical bodies, S 773–77

Medications: acting on and being acted on, T 681; action on body, T 650–55; administration of, T 686–87; deleterious, T 656–57, 670–71; distinction from nutriments, T 656; as nutriments for other species, T 683–84; potency of, T 691; taken internally, T 664–68

Melancholia (μελαγχολία), xcviii; T 522, 641; S 777, 788

Meninges, T 602

Nature (φύσις), T 534, 563–64, 619, 635, 640, 647

Nerves/sinews, T 568–69, 580, 601

Non-uniform *dyskrasia*, 283–84; associated conditions, N 733; differences of, N 746

Numbness (νάρκη), xcviii; T 582

Nutrition/nutriments, T 584, 586, 606; acting on vs. being acted on, T 682; effects on characteristics/soul, S 807–8,